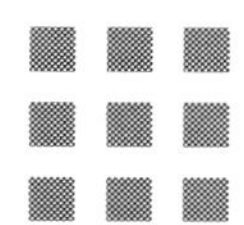

Osmosis Engineering

Osmosis Engineering

Edited by

Nidal Hilal
NYUAD Water Research Center, New York University Abu Dhabi,
Abu Dhabi, United Arab Emirates

Ahmad Fauzi Ismail
Advanced Membrane Technology Research Center (AMTEC),
Universiti Teknologi Malaysia, Skudai, Malaysia

Mohamed Khayet
Department of Structure of Matter, Thermal Physics and Electronics,
Faculty of Physical Sciences, University Complutense of Madrid,
Madrid, Spain

Daniel Johnson
NYUAD Water Research Center, New York University Abu Dhabi,
Abu Dhabi, United Arab Emirates

Elsevier
Radarweg 29, PO Box 211, 1000 AE Amsterdam, Netherlands
The Boulevard, Langford Lane, Kidlington, Oxford OX5 1GB, United Kingdom
50 Hampshire Street, 5th Floor, Cambridge, MA 02139, United States

British Library Cataloguing-in-Publication Data
A catalogue record for this book is available from the British Library

Library of Congress Cataloging-in-Publication Data
A catalog record for this book is available from the Library of Congress

ISBN: 978-0-12-821016-1

For Information on all Elsevier publications
visit our website at https://www.elsevier.com/books-and-journals

Publisher: Joe Hayton
Acquisitions Editor: Kostas Marinakis
Editorial Project Manager: Cole Newman
Production Project Manager: Bharatwaj Varatharajan
Cover Designer: Victoria Pearson

Typeset by MPS Limited, Chennai, India

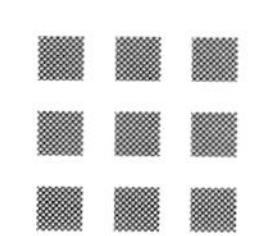

Contents

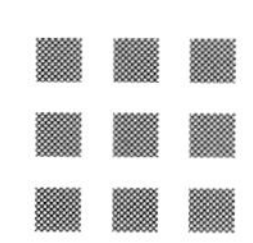

List of contributors

Saber Abdulhamid Alftessi Advanced Membrane Technology Research Center, School of Chemical and Energy Engineering, Faculty of Engineering, Universiti Teknologi Malaysia, Johor, Malaysia

Wei Lun Ang Research Center for Sustainable Process Technology (CESPRO), Faculty of Engineering and Built Environment, Universiti Kebangsaan Malaysia, Bangi, Malaysia; Department of Chemical and Process Engineering, Faculty of Engineering and Built Environment, Universiti Kebangsaan Malaysia, Bangi, Malaysia

M. Essalhi Department of Structure of Matter, Thermal Physics and Electronics, Faculty of Physics, University Complutense of Madrid, Madrid, Spain; Department of Chemistry, Umeå University, Umeå, Sweden

M.C. García-Payo Department of Structure of Matter, Thermal Physics and Electronics, Faculty of Physics, University Complutense of Madrid, Madrid, Spain

Pei Sean Goh Advanced Membrane Technology Research Centre (AMTEC), University of Technology Malaysia, Skudai, Malaysia

Raed Hashaikeh NYUAD Water Research Center, New York University Abu Dhabi, Abu Dhabi, United Arab Emirates

Nidal Hilal NYUAD Water Research Center, New York University Abu Dhabi, Abu Dhabi, United Arab Emirates

Kah Chun Ho Faculty of Engineering, Built Environment, and Information Technology, SEGI UNiversity, Selangor Darul Ehsan, Malaysia

Seungkwan Hong School of Civil, Environmental and Architectural Engineering, Korea University, Seoul, Republic of Korea

Siti Khadijah Hubadillah Advanced Membrane Technology Research Center, School of Chemical and Energy Engineering, Faculty of Engineering, Universiti Teknologi Malaysia, Johor, Malaysia

Ahmad Fauzi Ismail Advanced Membrane Technology Research Center, School of Chemical and Energy Engineering, Faculty of Engineering, Universiti Teknologi Malaysia, Johor, Malaysia

Daniel Johnson NYUAD Water Research Center, New York University Abu Dhabi, Abu Dhabi, United Arab Emirates

V. Karanikola Chemical and Environmental Engineering, University of Arizona, Tucson, AZ, United States

M. Khayet Department of Structure of Matter, Thermal Physics and Electronics, Faculty of Physics, University Complutense of Madrid, Madrid, Spain; Madrid Institute for Advanced Studies of Water (IMDEA Water Institute), Madrid, Spain

N.T. Hassan Kiadeh Department of Chemistry, Umeå University, Umeå, Sweden

Jungbin Kim School of Civil, Environmental and Architectural Engineering, Korea University, Seoul, Republic of Korea

Youngjin Kim Department of Environmental Engineering, College of Science and Technology, Korea University, Sejong-si, Republic of Korea

Jihun Lim School of Civil, Environmental and Architectural Engineering, Korea University, Seoul, Republic of Korea

Twibi Mohamed Advanced Membrane Technology Research Center, School of Chemical and Energy Engineering, Faculty of Engineering, Universiti Teknologi Malaysia, Johor, Malaysia

Abdul Wahab Mohammad Research Center for Sustainable Process Technology (CESPRO), Faculty of Engineering and Built Environment, Universiti Kebangsaan Malaysia, Bangi, Malaysia; Department of Chemical and Process Engineering, Faculty of Engineering and Built Environment, Universiti Kebangsaan Malaysia, Bangi, Malaysia; Chemical Engineering Program, Faculty of Engineering and Built Environment, Universiti Kebangsaan Malaysia, Bangi, Malaysia

Mohd Hafiz Dzarfan Othman Advanced Membrane Technology Research Center, School of Chemical and Energy Engineering, Faculty of Engineering, Universiti Teknologi Malaysia, Johor, Malaysia

Kiho Park School of Engineering, University of Birmingham, Birmingham, United Kingdom

Jeganes Ravi Advanced Membrane Technology Research Center, School of Chemical and Energy Engineering, Faculty of Engineering, Universiti Teknologi Malaysia, Johor, Malaysia

Sarper Sarp Centre for Water Advanced Technologies and Environmental Research (CWATER), College of Engineering, Swansea University, Swansea, United Kingdom

Jing Yao Sum Department of Chemical and Petroleum Engineering, Faculty of Engineering, Technology and Built Environment, UCSI University, Kuala Lumpur, Malaysia

Wafa Suwaileh Research and Development, Qatar Foundation, Doha, Qatar

Nur Diyana Suzaimi Advanced Membrane Technology Research Centre (AMTEC), University of Technology Malaysia, Skudai, Malaysia

Yeit Haan Teow Research Center for Sustainable Process Technology (CESPRO), Faculty of Engineering and Built Environment, Universiti Kebangsaan Malaysia, Bangi, Malaysia; Chemical Engineering Program, Faculty of Engineering and Built Environment, Universiti Kebangsaan Malaysia, Bangi, Malaysia

A.B. Yavuz Department of Structure of Matter, Thermal Physics and Electronics, Faculty of Physics, University Complutense of Madrid, Madrid, Spain; Patnos Sultan Alparslan Natural Sciences and Engineering Faculty, Agri Ibrahim Cecen University, Agri, Turkey

Min Zhan School of Civil, Environmental & Architectural Engineering, Korea University, Seoul, Republic of Korea

Biographies

Professor Nidal Hilal is the Founding Director of NYUAD Water Research Center at the New York University-Abu Dhabi. His research interests lie broadly in the identification of innovative and cost-effective solutions within the fields of nano-water, membrane technology, and water treatment including desalination, colloid engineering, and the nano-engineering applications of atomic force microscopy (AFM). His internationally recognized research has led to the use of AFM in the development of new membranes with optimized properties for difficult separations.

He has published 8 handbooks, 63 invited book chapters, and around 500 articles in refereed scientific literature. He has chaired and delivered lectures at numerous international conferences. In 2005 he was awarded Doctor of Science (DSc) from the University of Wales and the Kuwait Prize for applied science "Water resources development." He is also the Menelaus Medal Winner 2020 which is awarded by the Learned Society of Wales for excellence in engineering and technology.

He is the Editor-in-Chief for the international journal *Desalination*. He sits on the editorial boards of a number of international journals, is an advisory board member of several multinational organizations, and has served on/consulted for industry, government departments, research councils, and universities on an international basis.

Professor Ahmad Fauzi Ismail is the Founding Director of Advanced Membrane Technology Research Center (AMTEC), Universiti Teknologi Malaysia (UTM). He is the author of over 800 papers in refereed journals and over 50 book chapters. He has authored or coauthored 7 books and edited or coedited 9 books, 11 Patents granted and 22 Patents pending. He received more than 130 awards both at national and international level.

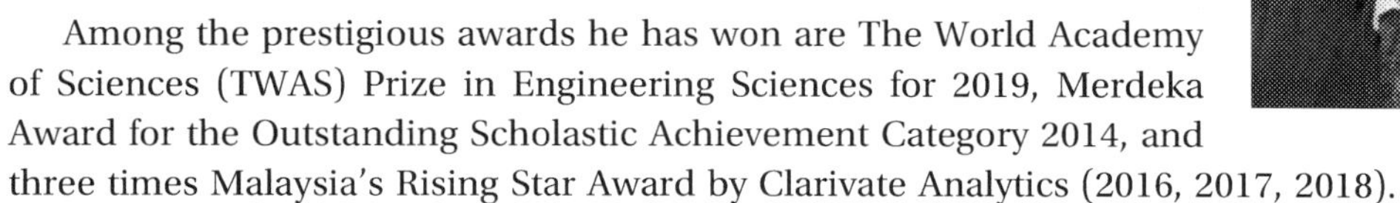

Among the prestigious awards he has won are The World Academy of Sciences (TWAS) Prize in Engineering Sciences for 2019, Merdeka Award for the Outstanding Scholastic Achievement Category 2014, and three times Malaysia's Rising Star Award by Clarivate Analytics (2016, 2017, 2018).

Prof. Fauzi served as the Chief Editor (Water Treatment and Desalination) for *Emergent Materials* journal, Engineering Editor for *The Arabian Journal for Science and Engineering* (AJSE) journal, and as Editor-in-Chief for *Journal of Applied Membrane Science & Technology*, and has served on the advisory and editorial boards for numerous academic journals in the areas of membrane technology and process engineering. Prof. Fauzi's research focuses on the development of polymeric, inorganic, and mixed matrix membranes for water desalination, wastewater treatment, gas separation processes, membrane for fuel cell applications, palm oil refining, hemodialysis membrane, and smart optical fiber for tracking migration of oil flow.

Professor Mohamed Khayet is director of the University Complutense of Madrid (UCM) research group "Membranes and Renewable Energy" affiliated to Campus Moncloa of International Excellence (Madrid, Spain). He is full professor in the Applied Physics area at the Faculty of Physics (UCM) and associated researcher in IMDEA Water (Madrid Institute for Advanced Studies on Water, Spain). He got his PhD in Physical Sciences from the UCM. He realized various research stays in different international institutions (Industrial Membrane Research Institute in Ottawa, Canada; Institute of Nuclear Chemistry and Technology in Warsaw, Poland; Centre for Clean Water Technologies at the University of Nottingham in UK, Singapore Membrane Technology Centre, Nanyang Environment and Water Research Institute, Nanyang Technological University in Singapore, Yale University, University of California Berkeley, etc.). His main research field is membrane science and nanotechnology (design, preparation, modification, and characterization of advanced membranes for water and energy production, hollow fibers and nanofibers, surface science, development of theoretical models, solar thermal and photovoltaic systems, thermal conductivity and diffusivity of phase change materials and nanofluids). He has coordinated various national and international projects funded by different institutions (European Commission, Middle East Desalination Research Center, Spanish Ministries; Companies, etc.). He has published over 200 papers. Among other awards, he received the Prince Sultan Bin Abdulaziz International Prize for Water (PSIPW) in 2012 for his novel and creative works on membrane distillation technology. He is an editor of the journal "*Desalination*" and member of the Editorial Board of various internal journals.

Dr. Daniel Johnson is an assistant professor at New York University Abu Dhabi. He received a BSc in Biochemistry from the University of Sheffield, before pursuing a PhD investigating dissociation of peptide structures using force microscopy at the University of Nottingham. Since then his main research interests have been in the fields of membrane separation research, advanced water treatment, membrane surface characterization, and colloid and interface science, all primarily within the context of improving wastewater treatment technology and processes.

He has made significant contributions to the characterization of process equipment surfaces using AFM, particularly in the development of novel membranes with improved fouling resistance and in the assessment of fouling of polymer filtration membranes and other surfaces of relevance to water purification processes. Recent activity has been in the fields of organic and biological fouling of membranes; development of polymer membranes for removal of heavy metals from contaminated sources; and mitigating scaling in membrane distillation systems for seawater desalination, the development of novel biofouling resistant membranes, and forward osmosis treatment of wastewater. These activities have been aimed at improving water sustainability through enhanced membrane processes.

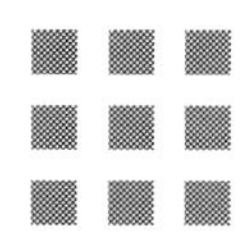

Preface

Osmotic engineering (OE) is a general term used to refer to a family of membrane-based separation processes associated primarily with the production of drinking water from seawater and brackish water, with the treatment of contaminated and wastewater and to less extent with energy production. Osmosis refers to a diffusion-based flow of fluid across a semipermeable membrane, leading to separation of solutes to solvents. OE is a rapidly moving research field with growing commercial application. It includes both isothermal (forward osmosis, FO; pressure retarded osmosis, PRO; reverse osmosis, RO; osmotic distillation, OD) and non-isothermal (osmotic membrane distillation, OMD; thermo-osmosis, TO) membrane separation processes.

Although the nature of osmosis and naturally occurring osmosis-based processes, such as the uptake of water by biological cells, has been known about and studied for centuries at least, the use of osmosis as the heart of engineering processes only really began to take shape in the latter half of the 20th century, particularly with the development of RO for the desalination of seawater. However, in many cases osmotic technology is still not fully mature. As such, much research endeavor is focused on the understanding of osmotic processes and further developing them as commercially viable technologies with a wide range of applications, including desalination of salty water, treatment of wastewater, and concentration of various feedwaters. As such, this book provides a comprehensive overview of the state of the art regarding OE-based research and engineering applications. It covers the underpinning theory, technology development, and commercial applications for each type of osmotic technology, with chapters written by some of the leading researchers in their fields.

We begin with the shared basic principles of osmosis and the determination of osmotic pressures, before describing the individual technologies. For each technology, we start with a description of the basic principles and theoretical treatment underpinning the process, before expanding to discuss the potential applications and recent developments. Much research, and hence recent developments, concern the major bottlenecks to achieving commercial viability for emerging technologies and improvements in process efficiency for more mature technologies. These issues generally concern improving the selectivity of membranes while maintaining high permeability, development of fouling resistant membrane materials or low fouling operating conditions, pretreatment of feedwaters, and posttreatment of product water. These conditions often mean a combination of several processes particularly for complex feedwaters. For these reasons, in this book we devote two chapters to hybrid systems.

With the strain on water supplies in many parts of the globe due to economic growth, population growth, agricultural expansion, and climate change, now is the perfect time to comprehensively cover the family of technologies best able to deal with large-scale water issues. This covers the range from already mature technologies such as RO to emerging processes with the potential for fulfilling niche treatment needs, such as FO, and technologies for the generation of power from salinity gradients, such as PRO.

It is hoped that this book will appeal to a wide range of researchers, technologists, and industrial practitioners working within the myriad water resource fields, as well as bachelor and graduate students and instructors studying and teaching within the environmental, chemical, and civil engineering areas. As far as we are aware, there is no other book to date that comprehensively and specifically addresses the whole family of OE processes.

Nidal Hilal, Ahmad Fauzi Ismail,
Mohammed Khayet, Daniel Johnson

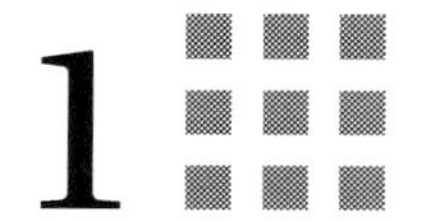

Basic principles of osmosis and osmotic pressure

Daniel Johnson, Raed Hashaikeh, Nidal Hilal

NYUAD WATER RESEARCH CENTER, NEW YORK UNIVERSITY ABU DHABI, ABU DHABI, UNITED ARAB EMIRATES

1.1 Introduction

This book brings together a number of engineering process technologies, which all have the principle of osmotic pressure, or rather differences in osmotic pressure between two solutions, at the heart of their operation. For instance, reverse osmosis requires the application of hydraulic pressure at a magnitude greater than the difference between the feedwater and the permeate water to allow membrane flux to occur against the osmotic pressure gradient. Conversely, in forward osmosis osmotic pressure gradients are harnessed to allow permeate flow from the feedwater to a draw solution of greater osmotic potential. In this chapter, we outline the general principles of osmosis and osmotic pressure, which underpin the technologies discussed in more detail later in this book.

1.2 What is osmotic pressure?

Osmosis, and hence osmotic pressure, is of major importance in a number of natural processes and is responsible for the turgor pressure, which provides solidity to many different types of cell [1,2], is the driving force for water reclamation in the kidneys and transport across many biological barriers [3], and is at least partially responsible for the uptake of water by plant roots [4]. As a phenomenon, it was first recorded by the Jean-Antoine Nollet in 1748 [5], who noted an increase in the volume of an ethanol–water mixture when separated from pure water by a bladder. René Dutrochet later coined the French terms endosmose and exosmose to mean transport into or out of a cell across a membrane, ultimately from the Hellenistic Greek word ἔωσθαι, for thrust or push [6].

Osmosis is often broadly defined as the flow of a fluid across a semipermeable membrane (i.e., one permeable to solvent, but not to solute) from a solution of low solute concentration to one of high concentration (or more precisely from higher to lower chemical potential). Such a process will continue, if left unchecked by external forces, until the chemical potentials of both sides of the membrane become equal. In such a case, the osmotic pressure

Osmosis Engineering. DOI: https://doi.org/10.1016/B978-0-12-821016-1.00011-5

difference between the two solutions can be defined as the hydraulic pressure, which needs to be applied to the system to halt the osmotically generated flow.

In Fig. 1–1, a typical simple setup to demonstrate osmotic pressure is shown. We have an inner chamber containing solute, separated from an outer tank containing pure solvent by a membrane permeable to solvent alone. The solvent will be drawn through the membrane until the solvent in the inner chamber rises to a height at which the hydrostatic pressure of the column of liquid is equal to the osmotic pressure:

$$\Delta p = \Pi = \rho g Z \tag{1.1}$$

where Δp is the hydrostatic pressure difference for a hydrostatic head of height $= z$, ρ is the density of the solution, and g is the acceleration due to gravity. Early osmotic pressure measurements in the 18th and 19th centuries, including those by Dutrochet and Thomas Graham had this principle of operation. A setup of this configuration or one using a U-shaped tube with the two halves separated by the membrane are the two most common configurations used to demonstrate osmotic pressure to students. As is readily apparent, for the solution to reach equilibrium in this case, it will be diluted to some extent. This means that the osmotic pressure that could be assessed by such an apparatus will not be the osmotic pressure of the original solution. Although it is theoretically possible to estimate the initial osmotic pressure before dilution from the concentration change, this is often problematic due to some underlying assumptions, and it is preferable to use a measurement setup where dilution is minimized.

To remove the effect of dilution, we must consider the situation outlined in Fig. 1–2, where a pure solvent (such as water) is separated from a solution with the same solvent by a

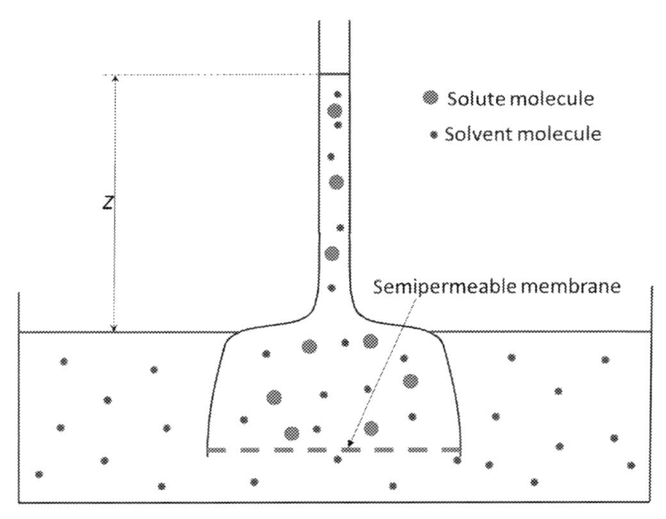

FIGURE 1–1 Two compartments separated by a membrane permeable to solvent but not to solute. Solution will rise within the second solute containing compartment until a height is reached which is proportional to osmotic pressure.

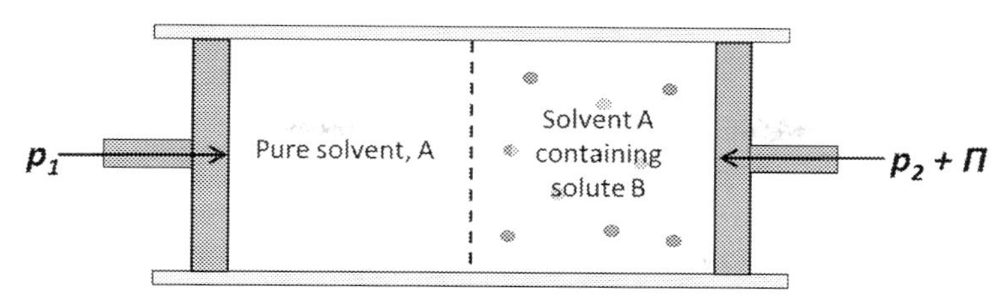

FIGURE 1–2 Two compartments containing pure solvent A and solution of solute molecules in solvent A separated by a rigid membrane permeable to solvent only. At equilibrium, chemical potential of each side is equal with no net flow through the membrane. Pressure applied to pure solvent is equal to pressure applied to solution plus solution osmotic pressure.

rigid barrier, which is only permeable to the solvent molecules. One compartment contains pure solvent A, whereas the other contains a homogenous mixture of solvent A and dissolved substance B. Pistons in either compartment can adjust the hydraulic pressure applied to each compartment. At mechanical equilibrium, that is, when there is no net flow across the membrane, the applied pressure to the first compartment, p_1, will equal to the total pressure in the second solution compartment, which consists of the applied pressure to the second compartment, p_2, plus the osmotic pressure of the solution, Π. If $p_1 > p_2 + \Pi$ then A will flow from compartment 1 to 2. If $p_1 < p_2 + \Pi$ then the flow will be reversed toward compartment 1. As the direction of flow can be changed by an infinitesimally small change in pressure around the equilibrium point, this system must necessarily be a thermodynamically reversible system.

In the 19th century, a physical chemist Jacobus van't Hoff had the insight that the solute molecules in an ideal solution (i.e., one which is highly diluted, with negligible interactions between molecules) should behave in the same manner as those in an ideal gas [7]. This leads to the famous van't Hoff equation, which is analogous to the ideal gas equation. This equation has been expressed in many forms, with some of the most commonly encountered as follows:

$$\Pi = CRTi = \frac{c}{M}RTi = \frac{nk_BT}{V}i = \frac{\chi_S RT}{V_w}i \tag{1.2}$$

where Π is the osmotic pressure, R is the gas constant, T is the absolute temperature, C is the molar concentration, M is the average molecular weight of the solute, k_B is the Boltzmann constant, V is the solution volume, V_w is the partial molar volume of water, χ_s is the mole fraction of the solute molecules, and i is the van't Hoff factor (equal to the number of components the solute separates into upon dissolution). Note that Π relates directly to the number of solute molecules present in a unit volume. This is because the osmotic pressure is a colligative property. This means that, at high dilution at least, it is the number of molecules which is important, not the amount (mass) of material. This connects the osmotic property to other colligative properties, such as vapor pressure, boiling point elevation, and freezing point depression, which allow for some convenient methods for estimating the osmotic pressure of solutions in the laboratory, which will be discussed later.

Note also that it is the number of free atoms, ions, or molecules, which exist *in solution* which is important, not the mass concentration and nor is the exact type of solute molecules. When we have ions dissociating upon dissolution, the number of dissociated ions need also to be considered. This is commonly achieved by adding the van't Hoff factor, i, as a cofactor to the equation. It must also be noted that this equation assumes that the solute molecules are completely rejected by the membrane. In practice, a reflection coefficient may need to be included for cases where the size of the solute is close to or below the molecular weight cutoff of the membrane.

The van't Hoff equation is useful when discussing the osmotic pressure of simple dilute solutions but has a number of drawbacks when discussing either mixtures or concentrated solutions, causing the relationship between concentration and osmotic pressure to deviate from ideality. First, in the case of dissolved ions, for anything over than a very dilute solution (indeed anything not close to infinite dilution) dissociation may not be complete, so Eq. (1.2) needs a further parameter to account for the fraction dissociated. This is further compounded by the changing solution concentration as water passes through the semipermeable membrane. However, the ideal behavior presumed by Eq. (1.2) also assumes that the solute molecules do not interact with each other and that their cumulative volume is not significant (i.e. the partial molar volume is essentially equal to the volume of the solution). In the first case at close to infinite dilution interaction forces between solute are insignificant due to how rarely molecules meet each other, in the second case the partial molar volume is insignificantly small. As solute concentration increases interactions become more common and the volume of the solute becomes a significant fraction of the total volume. Morse and Frantzen suggested a modification for the van't Hoff equation, which used the molal rather than molar concentration of the solution [8]. They found this to better match experimental data for cane sugar. However, as pointed out shortly after by Lewis, this approach only shows a significant difference to the conventional van't Hoff approach for relatively high molecular weight solutes and is still inadequate for most solutions at high concentrations [9].

The deviation of solutions from ideal behavior has often been accounted for empirically in a way that is again analogous to the behavior of gases by using a virial expansion of the form:

$$\Pi = cRT\left(\frac{1}{M} + B_2c + B_3c^2 + B_3c^3 \ldots\right) \tag{1.3}$$

where B_x are the virial coefficients and M is the average molecular weight of the solute. Although there is potentially an infinite number of virial coefficients, due to their declining magnitude at higher iterations, it is only normal the first few which are of any practical significance. Virial coefficients have often been derived from membrane osmometry experiments.

Most membrane osmometry measurements are made using static cells. In general, these consist of two chambers separated by the semipermeable membrane [10–13]. An early such device was the Pfeffer cell (see Fig. 1–3), which was invented in 1877 by Wilhelm Pfeffer for the study of osmotic pressure in plants [14]. Here, the membrane was the cell itself, which was made of porous porcelain onto which had been precipitated a layer of copper ferrocyanide. Pressures inside the cell were measured by observing an attached manometer [15]. A more

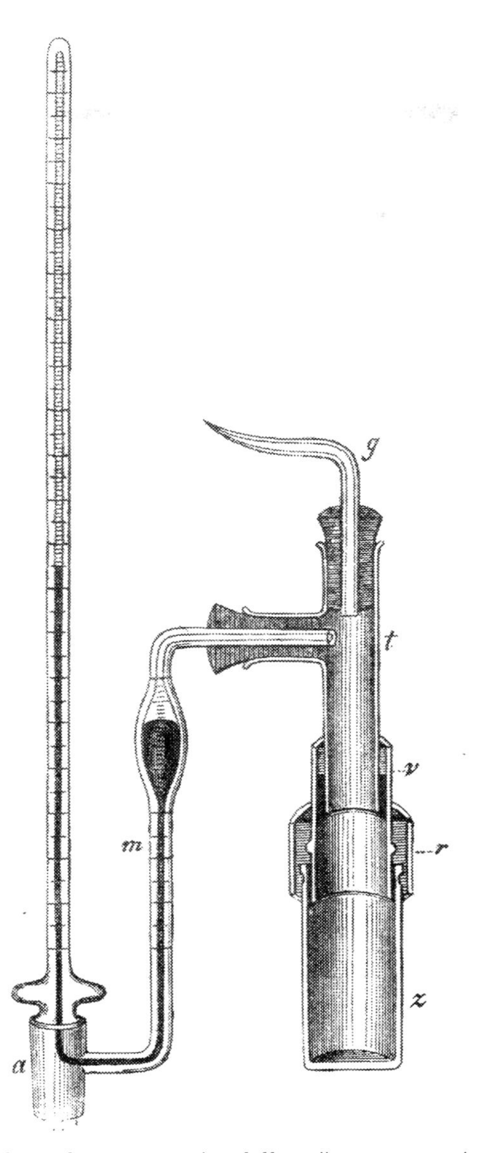

FIGURE 1–3 Diagram of an early form of osmometer, the Pfeffer cell. *Image reproduced from W. Pfeffer, Osmotische untersuchungen: studien zur zellmechanik, W. Engelmann (1877).*

generalized static osmotic pressure cell is shown in Fig. 1–4. One chamber containing the solution to be tested is sealed with a constant volume and a pressure gauge. The other chamber is filled with pure solvent and either immersed in or connected to a water bath, which is open to atmospheric pressure. After allowing some time to pass to allow equilibrium to be reached the osmotic pressure can be read off the pressure gauge. This setup has an advantage over that

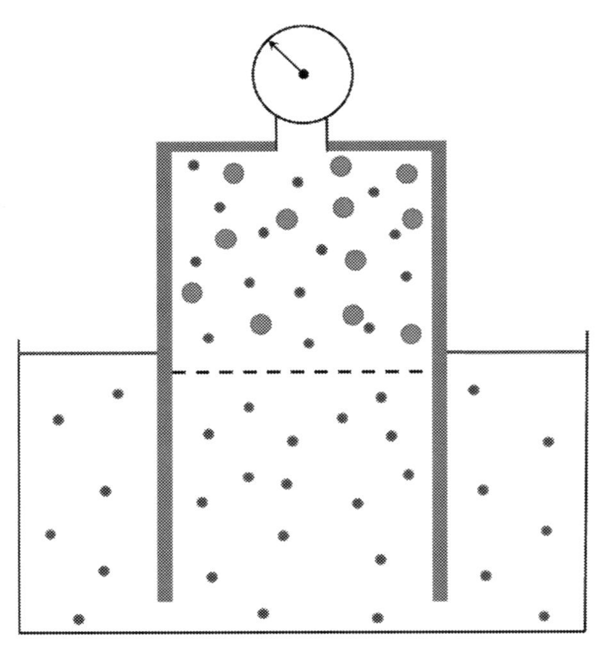

FIGURE 1–4 Simplified schematic of a static membrane osmometer with constant solution volume. Osmotic pressure of the solution is measured directly by a pressure gauge or electronic pressure sensor. The advantage is that minimal flow of solvent allows direct measurement of osmotic pressure of undiluted solution.

presented in Fig. 1–1 in that the volume of solvent flowing across the membrane is kept at a minimum, meaning that the pressure is measured for the undiluted sample. Dynamic measurements can be made by monitoring flow through the membrane and deducing the osmotic pressure from flow rates [16,17] or by adjusting a positive pressure applied to the solution until there is no net flow—at which point the applied pressure will be equal to the osmotic pressure. Alternatively, the osmotic pressure can be determined from dead-end filtration measurements, where the membrane flux rate at several pressure values is used to determine the pressure at which flow is stopped [11,18]. However, for these latter measurements care must be taken to ensure that the permeate volumes measured are much less than the total initial feed volume, or concentration of the feed would be likely to lead to measurement errors.

There are two major problems with directly measuring osmotic pressure through membrane osmometry. The first is merely practical—for static osmometry, the solutions can take a considerable time to reach equilibrium; for filtration measurements, several pressure tests need to be done per sample. As a result, indirect methods using a vapor pressure or freezing point depression measurements may be preferred. The other major problem with membrane osmometry is that the solutes need to be completely retained by the membrane. This means the osmotic pressure of solutions of low molecular weight compounds, particularly monovalent salts, may be difficult to measure due to solute diffusion through the membrane.

A correction may be applied if the reflection coefficient of the solute is known, but this requires further measurement and the reflection coefficient may itself be pressure dependent.

Osmotic pressure has been widely understood in thermodynamic terms, including the original derivation of the van't Hoff equation. There are several ways of phrasing the behavior of osmotic flow, but most often it is done in terms of the competing chemical potentials of the pure solvent and the solution [19–21]. The chemical potential is defined as the change in Gibbs free energy with the change in the number of moles of a particular component for a given temperature, pressure, and composition [22]:

$$\mu_i \equiv \left(\frac{\partial G}{\partial n_i}\right)_{T,p,n_j} \tag{1.4}$$

where μ_i is the chemical potential of substance *i*, G is the Gibbs free energy, and n_i is the number of moles of substance *i*. It is useful to relate the chemical potential to the thermodynamic activity of the solution:

$$\mu_i = \mu_i^0 + RT\ln a_i \tag{1.5}$$

where μ_i^0 is the chemical potential of the pure substance and a_i is the thermodynamic activity. As the activity is essentially an "effective" mole fraction of a substance, it is worth noting that it is related to the actual mole fraction by the activity coefficient, γ_i. It is also of interest that the activity, for an ideal solution, is essentially the ratio of vapor pressures of pure solvent and solute, which allows its easy determination by vapor pressure instrumentation:

$$a_i = \gamma_i \chi_i = \frac{p_i}{p_i^*} \tag{1.6}$$

where p_i and p_i^* are the vapor pressures of the pure substance and as part of a solution, respectively. For an ideal solution $\gamma_i = 1$.

When a change occurs to a chemical system, it then follows that there will be a change in the Gibbs free energy of that system [23]:

$$\Delta G = \Delta H - T\Delta S \tag{1.7}$$

where ΔH is the system enthalpy change and ΔS is the system entropy change. This can be applied to the system as a whole or to specific components of that system, in which case this would be related to the chemical potential of that component. Assuming that no chemical reactions are occurring, ΔH is the enthalpy of mixing. In an ideal solution, this would be zero, but in many cases, non-zero values may occur due to interactions between dissolved components and the solvent. In most cases, the change in a solution's chemical potential when dissolving the solute is due to a change in solvent entropy, due to competition for space between the solvent and solute molecules. This leads to a decrease in ΔG due to the increase in ΔS representing an increased system disorder. It naturally follows from the definition of chemical potential shown in Eq. (1.3) that a decrease in ΔG is the origin of the decrease in chemical potential of the solution compared with pure solvent. As a result, it can be concluded that from a thermodynamic perspective, osmotic pressure is entropic in origin.

At equilibrium, the chemical potentials of the solvent and solution are balanced. In the solution, the presence of the solute molecules reduces the mole fraction of the solvent to χ_w <1 as well as reduces the chemical potential of the solvent. However, this reduction of chemical potential is increased by the addition of osmotic pressure to that applied to the solution. So at equilibrium [20]:

$$\mu_w^0(p) = \mu_w(\chi_w, p + \Pi) \tag{1.8}$$

The chemical potential of the water on the solution side is lower and can be calculated by [20]:

$$\mu_w = \mu_w^0(x_w, p + \Pi) + RT\, ln a_w \tag{1.9}$$

Note that for an ideal solution, the activity is equivalent to the mole fraction, as the activity coefficient in such a case is equal to unity. Change in the Gibbs free energy of a closed system with constant composition can be described by:

$$dG = Vdp - SdT \tag{1.10}$$

where S is the system entropy. Using this to determine how the free energy changes with pressure leads to the following integration:

$$G(p + \Pi) = G(p) + \int_p^{p+\Pi} V dp \tag{1.11}$$

Applying this to the change in chemical potential with osmotic pressure gives:

$$\mu_w(p + \Pi) = \mu_w^0(p) + \int_p^{p+\Pi} V_w dp \tag{1.12}$$

Combining Eqs. (1.7), (1.8), and (1.11) leads to:

$$-RT ln a_w = \int_p^{p+\Pi} V_w dp \tag{1.13}$$

For a practically incompressible liquid, the molar volume is effectively constant and the integral becomes ΠV_w:

$$\Pi = \frac{RT}{V_w} ln\, a_w \tag{1.14}$$

In addition, if the solution is very dilute then this can be approximated to:

$$\Pi = \frac{\chi_s RT}{V_w} \tag{1.15}$$

Note that Eqs. (1.14) and (1.15) are versions of the van’t Hoff equation (Eq. 1.2).

Using the virial coefficients from empirical measurements, an osmotic coefficient, φ, also can be determined, which characterizes the deviation of the solvent from ideal behavior [24]:

$$\varphi = \frac{\mu_w^0 - \mu_w}{RT ln \chi_w} \tag{1.16}$$

The coefficient can then be applied to a modified version of the van’t Hoff equation:

$$\Pi = \varphi CRT \tag{1.17}$$

The osmotic coefficient can be derived empirically for simple electrolytes using Pitzer parameters [25,26].

1.3 Relation of osmotic pressure to other colligative properties

Colligative properties are a range of properties of solutions that are related to their number concentration of solute molecules, at least for dilute or ideal solutions, and are independent of the chemical type. These properties are the osmotic pressure due to solute, depression of freezing point, depression of vapor pressure, and elevation of boiling point. However, it must be noted that in practice, all solutions will show some deviation from ideal behavior, mostly in a concentration-dependent manner, so technically these properties are only really colligative in the strict sense at dilutions where the solution behavior approaches that of an ideal solution.

Colligative properties have a number of features in common. The most important feature is that the presence of solute leads to a reduction in the chemical potential of the solution, compared with the pure solvent. This is expressed previously in Eq. (1.4) and is the common basis of all the colligative properties. For an ideal solution, where $a_w = \chi_w$, then it follows that the solution will have lower chemical potential than pure solvent as $ln\ \chi_w$ will be less for the pure solvent (where $\chi_w = 1$) due to the presence solute. As even ideal solutions, which have zero enthalpy of mixing, are subject to these effects, it follows, according to Eq. (1.6), that it must be of an entropic origin, as has already been mentioned for the specific case of osmotic pressure. The addition of the solute molecules serves to increase the amount of disordering of the solvent molecules, which is reflected in the change in entropy. This serves to both reduce the tendency of molecules of solvent to form a gas, simultaneously lowering the vapor pressure for a given temperature and raising the boiling point, as well as making freezing more difficult and thus lowering the freezing point.

The effect of lowering chemical potential is easy to see when considering a plot of chemical potential versus temperature for a notional substance (see Fig. 1–5). For a pure substance, the chemical potential is the molar Gibbs free energy, which can be expressed in the following form:

$$\left(\frac{\partial \mu}{\partial T}\right) = -S \tag{1.18}$$

As a result, any increase in temperature will lead to a reduction in chemical potential, with the slope equal to the entropy. As the slopes for each phase cross, a phase change occurs to whichever phase has the lowest potential (see dotted line). Assuming that the solute is insoluble in the frozen solvent and nonvolatile, then the addition of solute will not affect the ice and vapor lines. Adding solute to solution lowers the potential of the liquid phase, the crossover points change their position leading to an increase in the boiling point and a decrease in the freezing point.

If the magnitude of one of these colligative properties can be determined and related to concentration or activity, then this can be used to determine the osmotic pressure. Most commonly in the laboratory, the techniques of vapor pressure depression and freezing point depression measurement are used to calculate the osmotic pressure of solutions, due to their relatively high throughput and ease of use compared with membrane-based techniques. In all cases, the colligative property originates in the lowering of the chemical potential of the solution by the addition of the solute.

1.3.1 Vapor pressure depression

Adding solute to a solvent will reduce the vapor pressure, due to the reduced access of solvent molecules to the solvent—vapor interface. Assuming that the solute is nonvolatile, the

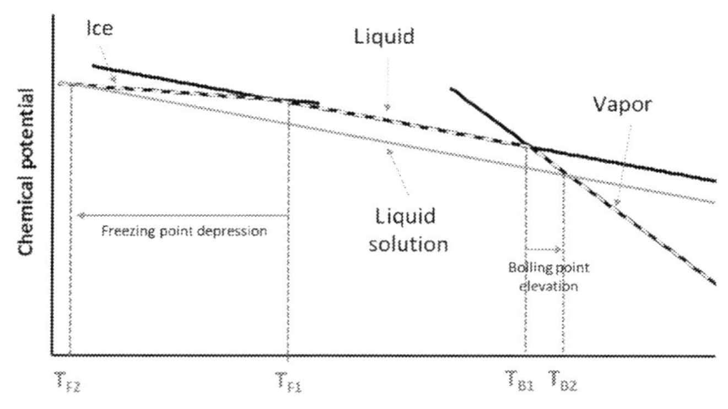

FIGURE 1–5 The presence of solute leads to lowering of chemical potential compared to pure liquid solvent. This effectively lowers the phase change points, simultaneously changing the freezing and boiling temperatures as well as affecting osmotic pressure.

activity of the solution can be related to the ratio of the vapor pressure of the solvent and vapor pressure of the solution:

$$a_i = \frac{p}{p*} \tag{1.19}$$

where p and p^* are the vapor pressures of the pure solvent and solution, respectively, with the ratio p/p^* is always equal to unity or less. Note also that for an ideal solution, such as one whose dilution approaches infinity, Raoult's law will be obeyed, with the ratio of vapor pressures in Eq. (1.19) equal to the mole fraction, χ. This is because the activity is in effect a version of the mole fraction, which accounts for deviation from ideal behavior. The osmotic pressure can then be calculated using Eq. (1.13). This direct relationship between activity and vapor pressure is a major advantage for this technique, which is also insensitive to the presence of undissolved particulates (a problem for freezing point depression osmometry) and does not require a phase change to occur in the sample. However, it is inaccurate when dealing with mixtures containing volatile components [27].

1.3.2 Freezing point depression

Addition of solute molecules to a solvent or solution will necessarily lower the freezing point compared to the pure solvent in proportion to the molar quantity of solute added. The change in freezing point, ΔT, is related to the concentration of a solution by the following relationship:

$$\chi_s \sim \frac{\Delta T \Delta H_{melt}}{RT^{*2}} \tag{1.20}$$

where T^* is the freezing point of the pure solvent, ΔH_{melt} is the latent heat of fusion of the solvent. For measurements using freezing point osmometers, the output value is not osmotic pressure, but osmolality (the molality of solution multiplied by the van't Hoff factor):

$$b \cdot i = \frac{\Delta T}{K_f} \tag{1.21}$$

where b is the osmolality of the solution and K_f is a cryoscopic constant (generally quoted as -1.86 for aqueous solutions, although this value comes with a range of approximations and assumptions behind it [27]). The osmolality can then be used to calculate the osmotic pressure via approximating the molality to the molarity or by calculating the molarity from known solution parameters and then using the van't Hoff equation.

1.3.3 Boiling point elevation

The final colligative property we will look at is the elevation of boiling point by the addition of solute. As was mentioned when discussing reduction in vapor pressure, the presence of a nonvolatile solvent reduces access of solvent molecules to the liquid–vapor interface. As well as reducing vapor pressure this serves to increase the boiling point. The boiling point elevation as a result of solute addition is usually presented similarly to the freezing point depression:

$$b \cdot i = \frac{\Delta T}{K_b} \tag{1.22}$$

where K_b is a boiling point elevation constant. Again, the molality of the solution can be used to estimate the osmotic pressure of the solution.

1.4 Origins of osmotic pressure in solution

Thermodynamic models of osmotic pressure are useful for understanding the macroscale behavior of osmotic systems and for allowing the calculation of relevant parameters from empirical observation, but do nothing to explain the underlying behavior of the solution at the molecular level, which gives rise to osmotic pressure and related phenomena. To understand this, we need to consider kinetic models of solvent and solute molecules, ions, and atoms. Again, insight may be gained by thinking of a dilute solution as analogous to a gas. Pressure arises in a fluid from the kinetic energy of the molecules involved and their collisions with each other and with the walls of their container. In the situation described in Fig. 1–2, both solvent and solute molecules are undergoing thermally driven random motion on both sides of the semipermeable membrane. The number of solvent molecules is effectively the same on both sides of the membrane, especially given the free flow of solvent across the membrane, meaning that the net effect, for dilute solutions at least, is zero. The most generally accepted kinetic model of osmotic pressure is that force which leads to the osmotic pressure in effect derives from the kinetic behavior of the solute molecules and their retention on one side of the membrane.

Consider a solute molecule in solution traveling in a particular direction due to a force acting on it, which is balanced by viscous drag from the surrounding solvent molecules. This viscous drag serves to couple forces experienced by solute molecules to the surrounding fluid and vice versa. For each particle, the force will represent a gradient of potential energy of that particle. For a particular direction, that is, x, with a number concentration of solute molecules, $c(x)$, this leads to a pressure gradient $dp/dx = c(x)$ and a potential gradient $f(x) = -dU\ dx$, where U is the potential energy. If the direction x is selected as the coordinate passing through the membrane channels, then the force $f(x)$ is the force exerted from encounters between the solute molecules and the membrane. Although the force with which the solute molecules experience at the membrane is the same for other walls of any container, the situation with the membrane is different in one fundamental respect—that it is permeable to the solvent. As solute is rejected by the membrane pore, viscous coupling between the solute and solvent molecules will cause a net flow of solvent from the membrane channels into the solute chamber, leading to osmotic flow across the membrane and an arising osmotic pressure difference. This leads to a pressure difference between the membrane pore and the bulk solution of [28]:

$$\Delta p = \frac{n}{V} kT \tag{1.23}$$

where n/V is the number concentration of solute in bulk solution. Note that this is a version of the van't Hoff equation previously expressed.

A major implication of this kinetic model for the basis of osmotic pressure is the central role of the semipermeable membrane in imparting momentum to the solute molecules, which drives the transmembrane flow. As a result, this means that any solution in a simple container does not possess an actual osmotic pressure, it merely has the potential to have an osmotic pressure. For the osmotic pressure to arise, the solution needs to be in a system where it is separated from its solvent or a weaker solution by a membrane permeable to solvent alone.

1.5 Osmotic flow

We have looked at the osmotic pressure of solutions found at equilibrium, that is, the hydraulic pressure that needs to be exerted to stop an osmotically driven flow. But, what about the situation where we have the same pressure applied to either side of the semipermeable membrane (such as atmospheric pressure), giving rise to an osmotically driven flow? We can consider flow of solvent across the semipermeable membrane as being governed by Darcy's law for porous materials:

$$Q = \frac{A_m \, \Delta p}{\mu L R_t} \tag{1.24}$$

where Q is the volumetric flow rate, R_t is the total resistance of the membrane to flow, Δp is the total pressure drop across the membrane, A_m is the cross-sectional effective membrane area, μ is the dynamic viscosity of the fluid (0.89 mPa s for water at 25°C) and L the length of the channels (the membrane thickness if assuming cylindrical pores). Note that this is a driven process, not a diffusion-based process, with the driving force coming from the kinetic energy of the solute molecules. This process has been referred to previously as the rectification of Brownian motion [28]. If the process was diffusion driven, it would be expected to conform to Fick's law, which has been repeatedly shown to not be the case [29].

Given that all other parameters are constant for a given system, the flow rate is determined by the magnitude of the pressure drop across the membrane, which will be the van't Hoff pressure in the absence of any additional hydraulic pressure. This relationship can be reduced to the membrane transport equation commonly used:

$$J_v = Lp(\Delta p - \Delta \Pi) \tag{1.24}$$

where J_v is the volumetric flux, L_p is the hydraulic coefficient, which is dependent on the particular membrane and solvent under examination (often replaced by the pure water permeability coefficient, A, when talking about aqueous systems).

1.6 Reflection coefficient

So far we have assumed that the semipermeable membrane is in all cases completely impermeable to the solute molecules and completely permeable to the solvent. However, in practice this is rarely the case. To account for the imperfect ability of the membrane to hold back solvent, we need to introduce the reflection coefficient (σ). The reflection coefficient is a dimensionless number, which describes the relative interactions between the solute and solvent molecules with the membrane. The value of the reflection coefficient varies from 0 to 1, representing the boundaries of a completely nonselective membrane and a prefect membrane with complete retention of solute respectively. In practice, for membranes with solute rejection of greater than 95%, the rejection coefficient is approximated to 1 and thus ignored.

The reflection coefficient was first described by Staverman who used it as an empirical correction for the deviation of the osmotic pressure of solutions from thermodynamically calculated values when using membranes which were partially permeable to solute [30]. The reflection coefficient can be determined from the ratio of the pressure drop across the membrane to the osmotic pressure difference between the bulk solution and pure solvent or from the volumetric flow across the membrane at zero applied pressure from filtration experiments [31,32]:

$$\sigma = \left(-\frac{\Delta p}{\Delta \Pi}\right)_{J_v=0} = \left(\frac{J_v}{A\,\Delta \Pi}\right)_{\Delta p=0} \tag{1.25}$$

Note that the hydraulic and osmotic pressures here are the bulk values, that is, this approach ignores concentration polarization effects, although this can be to some extent mitigated by the use of stirred cells/cross-flow systems. The addition of the reflection coefficient leads to modification of the membrane transport Eq. (1.24) in the following way for the case of an imperfectly semipermeable membrane:

$$J_v = L_p(\Delta p - \sigma \Delta \Pi) \tag{1.26}$$

Acknowledgement

The authors would like to thank the Royal Society for funding this work through a Royal Society International Collaboration Award (IC160133).

References

[1] J. Philip, The osmotic cell, solute diffusibility, and the plant water economy, Plant. Physiol. 33 (1958) 264.

[2] N.P. Money, On the origin and functions of hyphal walls and turgor pressure, Mycol. Res. 103 (1999) 1360.

[3] F. Kiil, Mechanism of osmosis, Kidney Int. 21 (1982) 303–308.

[4] E. Steudle, Water uptake by plant roots: an integration of views, Plant. Soil. 226 (2000) 45–56.

[5] J.A. Nollet, X. Part of a letter from Abbè Nollet, of the Royal Academy of Science at Paris, and FRS to Martin Folkes Esq; President of the same, concerning electricity, Philos. Trans. R. Soc. Lond. 45 (1748) 187–194.

[6] O.E. Dictionary, "osmosis, n.1", Oxford University Press.

[7] J.H. van't Hoff, Osmotic pressure and chemical equilibrium, Nobel Lecture, 13 (1901).

[8] H.N. Morse, J.C.W. Frazer, The osmotic pressure and freezing-points of solutions of cane-sugar (1905).

[9] G.N. Lewis, The osmotic pressure of concentrated solutions, and the laws of the perfect solution, J. Am. Chem. Soc. 30 (1908) 668–683.

[10] N.O. Chahine, F.H. Chen, C.T. Hung, G.A. Ateshian, Direct measurement of osmotic pressure of glycosaminoglycan solutions by membrane osmometry at room temperature, Biophys. J. 89 (2005) 1543–1550.

[11] H. Nabetani, M. Nakajima, A. Watanabe, S.-i Ikeda, S.-i Nakao, S. Kimura, Development of a new type of membrane osmometer, J. Chem. Eng. Jpn. 25 (1992) 269–274.

[12] A. Grattoni, G. Canavese, F.M. Montevecchi, M. Ferrari, Fast membrane osmometer as alternative to freezing point and vapor pressure osmometry, Anal. Chem. 80 (2008) 2617–2622.

[13] F. Wallner, Membrane osmometer, in: Google Patents (1984).

[14] W. Pfeffer, Osmotische untersuchungen: studien zur zellmechanik, W. Engelmann, (1877).

[15] S. Parker, Pfeffer cell apparatus, in: Embryo Project Encyclopedia, (2017).

[16] R.T.M.R. Berkeley, B.A.E.G.J Hartley, Dynamic" osmotic pressures, Proceedings of the Royal Society of London. Series A, Containing Papers of a Mathematical and Physical Character 82 (1909) 271–275.

[17] R.M. Fuoss, D.J. Mead, Osmotic pressures of polyvinyl chloride solutions by a dynamic method, J. Phys. Chem. 47 (1943) 59–70.

[18] D. Johnson, A.W. Lun, A.W. Mohammed, N. Hilal, Dewatering of POME digestate using lignosulfonate driven forward osmosis, Sep. Purif. Technol. 235 (2020) 116151.

[19] E. Fermi, Thermo-dynamics, Dover Publications, New York, 1936.

[20] P.W. Atkins, J. De Paula, J. Keeler, Atkins' Physical Chemistry, Oxford University Press, 2018.

[21] J.M. Smith, H.C. Van Ness, M.M. Abbott, M.T. Swihart, Introduction to Chemical Engineering Thermodynamics, 8th ed., McGraw Hill, New York, 2018.

[22] W.R. Vieth, Diffusion in and Through Polymers: Principles and Applications, Hanser, New York, 1991.

[23] J.F. Thorpe, M.A. Whiteley, Thorpe's Dictionary of Applied Chemistry, Longmans, Green, 1956.

[24] A. McNaught, A. Wilkinson, IUPAC. Compendium of Chemical Terminology, 2nd ed. (the "Gold Book"). in: Blackwell Scientific Publications, Oxford (1997).

[25] F. Perez-Villasenor, G.A. Iglesias-Silva, K.R. Hall, Osmotic and activity coefficients using a modified Pitzer equation for strong electrolytes 1: 1 and 1: 2 at 298.15 K, Ind. Eng. Chem. Res. 41 (2002) 1031–1037.

[26] K.S. Pitzer, Activity Coefficients in Electrolyte Solutions: 0, CRC press, 2018.

[27] T.E. Sweeney, C.A. Beuchat, Limitations of methods of osmometry: measuring the osmolality of biological fluids, Am. J. Physiol. Regul. Integr. Comp. Physiol. 264 (1993) R469–R480.

[28] P. Nelson, Biological Physics, WH Freeman, New York, 2004.

[29] W. Stein, Transport and Diffusion Across Cell Membranes, Elsevier, 2012.

[30] A. Staverman, The theory of measurement of osmotic pressure, Recl. Trav. Chim. Pays-Bas 70 (1951) 344–352.

[31] J.L. Anderson, D.M. Malone, Mechanism of osmotic flow in porous membranes, Biophys. J. 14 (1974) 957–982.

[32] J. Su, T.-S. Chung, Sublayer structure and reflection coefficient and their effects on concentration polarization and membrane performance in FO processes, J. Membr. Sci. 376 (2011) 214–224.

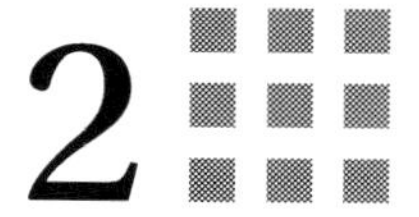

2 Fundamentals and application of reverse osmosis membrane processes

Jungbin Kim[1], Jihun Lim[1], Kiho Park[2], Seungkwan Hong[1]

[1]*SCHOOL OF CIVIL, ENVIRONMENTAL AND ARCHITECTURAL ENGINEERING, KOREA UNIVERSITY, SEOUL, REPUBLIC OF KOREA* [2]*SCHOOL OF ENGINEERING, UNIVERSITY OF BIRMINGHAM, BIRMINGHAM, UNITED KINGDOM*

2.1 Introduction

Traditionally, thermal-based desalination was favored because it is easily adaptable and can produce high-quality water. Because humans have used thermal-based desalination for thousands of years to obtain freshwater by boiling salted feedwater, this traditional method has been expanded to forms such as multistage flash and multieffect distillation [1]. However, with the development of membrane technology [2], membrane-based desalination has attracted increasing attention [3]. Among these membrane-based desalination technologies, reverse osmosis (RO) is one of the most widely utilized technology for water treatment [4–7]. The preference for RO for desalination has increased obviously. The capacities of installed RO plants for seawater desalination are higher than 65% of the overall installed capacities for seawater desalination in 2013 [8]. With the reduction in the overall cost of desalination systems, the capacity for desalination has continuously increased [3]. Under these circumstances, the importance of RO systems has also tremendously increased.

RO systems exhibit multiple advantages such as simple operation, easy scale-up, and low energy consumption [9,10]. In particular, low energy consumption is the most attractive advantage because of the high energy consumption of other water treatment methods. Furthermore, it can be correlated with the energy crisis, which is one of the biggest global issues currently [11,12]. RO systems consume low energy because phase transition is not required to separate feedwater to permeate water (freshwater) and concentrated brine. However, the energy consumption and the required operating pressure in the RO system are directly correlated with the feedwater concentration [4,5]. Therefore RO systems are usually designed and classified depending on the feed concentration.

The feedwater on Earth can be categorized depending on feed concentration. Brackish water, which is usually salted water, is defined based on its range of salinity between

Osmosis Engineering. DOI: https://doi.org/10.1016/B978-0-12-821016-1.00002-4

seawater and freshwater [13–15]. Brackish water, which includes river water, groundwater, and surface water, is among the water resources with the lowest salinity. In other words, the required energy consumption for the treatment of brackish water is the lowest compared to the treatment of other water resources. Thus low energy consumption and low-cost brackish water desalination technologies have been developed to obtain freshwater [4,5,16,17]. However, brackish water resources are not abundant like seawater, and brackish water cannot act as a permanent water resource for the continuously increasing human population [3,18]. Thus choosing seawater desalination for supplying freshwater is inevitable.

Seawater accounts for more than 97% of all water resources on Earth. It can be the ultimate solution for securing sustainable freshwater, especially in regions with water scarcity, such as the Middle East and North Africa (MENA) [19,20]. In addition, the recently installed large-scale seawater reverse osmosis (SWRO) desalination plants can produce freshwater at prices lower than \$0.5/m^3 [21], which is significantly lower than the price of thermal-based desalination and much lower than the price 30 years ago (higher than \$ 2/m^3) [3]. It has attracted many policymakers and stakeholders to invest in SWRO desalination [3,19,20]. However, although the energy efficiency of SWRO is the highest among other mature desalination technologies [22], an important requirement for SWRO plants is still low energy consumption, because with the increase in the overall desalination capacity of the current plants to more than 90 million m^3/day [19], the overall energy consumption in all desalination plants has increased drastically. In addition, the electrical energy required for operating most SWRO plants is supplied by fossil fuels [20]. Because power generation by fossil fuels causes greenhouse gas emissions [23,24], the energy efficiency of SWRO plants needs to be improved for sustainable water supply by desalination systems.

Hypersaline water has a very high osmotic pressure because of its high concentration. Therefore the treatment of hypersaline water by RO requires a very high operating pressure on the feed solution. This required pressure for hypersaline water treatment often exceeds the pressure limit of the RO membrane [25,26]. In addition, because the energy consumption for hypersaline water treatment is very high, a feasible technology cannot be developed because it is not economically viable compared to brackish water and seawater desalination [27–29]. However, the recent increase in the number of SWRO plants has caused environmental problems of brine disposal. Therefore brine treatment systems with zero liquid discharge are required, and the exploration of brine treatment systems by RO has become more attractive [25,28,30]. In addition, shale-gas produced water [31,32], landfill leachate [33], and flue gas desulfurization wastewater [34] contain very high salinity, so effective technologies for the treatment of these types of wastewater should be developed. Recently, a new process called draw solution assisted RO [35], osmotic-enhanced dewatering [31,36], or osmotically assisted RO [37] has been developed. This process can change high-pressure single-stage RO systems to middle-pressure two-stage RO systems. Therefore this process can be effectively utilized for hypersaline water treatment. In membrane development, the Pall disc tube (DT) module design was developed for the treatment of hypersaline water without clogging due to

Table 2–1 Total dissolved solids, applied pressure for RO, and specific energy consumption of RO systems for desalination depending on the type of feed solution [3,19,27].

	TDS range (g/L)	Applied pressure in RO for desalination (bar)	SEC in RO for desalination (kW h/m^3)
Brackish water	0.5–30	20–40	0.5–2.5
Seawater	30–50	50–80	3–5
Hypersaline water	>50	>100	>7

Note: *RO*, Reverse osmosis; *TDS*, total dissolved solid; *SEC*, specific energy consumption.

fouling of highly concentrated brine [27,38]. The maximum pressure range of the DT module is approximately 70–150 bar. Such RO systems can be effectively utilized in the treatment of hypersaline water.

In summary, RO systems can treat various feed solutions effectively, and the feed concentration range, applied pressure, and specific energy consumption (SEC) of the RO process by feed type are summarized in Table 2–1. RO has been utilized as the main technology in osmosis engineering for desalination. RO was developed as an advanced technology for desalination, and research and development to improve the RO system will be continued in the future. The recent development of RO systems focuses on the improvement of energy efficiency for desalination in various feed solutions. To successfully accomplish these goals, fundamental knowledge in RO engineering is essential, especially with respect to osmotic pressure, theoretical minimum energy obtained from osmotic pressure, the basic principles of RO, various RO process configurations, and the basic principles of fouling on the RO membrane.

In this chapter, fundamental knowledge, model equations, design configurations for practical application, and operation index of the RO system are presented. The understanding of the fundamental mechanisms of RO systems is essential for researchers and engineers in the field of desalination to utilize and analyze RO systems more effectively. In addition, this chapter provides comprehensive and valuable information on the basic modeling and operation methodology of RO systems, which will be a useful guide for researchers studying RO systems.

2.2 Principles of RO

2.2.1 Definition of osmotic pressure and RO

Osmotic pressure is defined as the minimum pressure required to prevent the water molecules in less saline water from permeating toward more saline water across a semipermeable membrane [39]. As shown in Fig. 2–1, high-concentrated water and low-concentrated water are placed across a semipermeable membrane initially. Water molecules move freely across the semipermeable membrane, while salt molecules do not. So, there is a net flow of water from the high-concentrated solution to the low-concentrated solution because of the

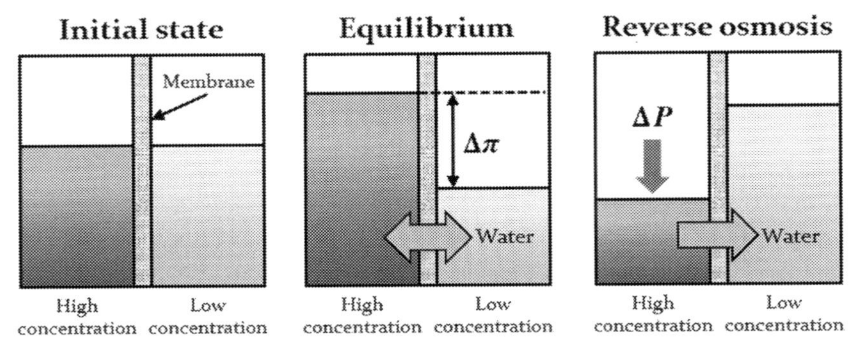

FIGURE 2–1 Water permeation across the semipermeable membrane due to the difference in osmotic pressure from the initial state to the equilibrium state and RO. *RO*, Reverse osmosis.

difference in osmotic pressure between these two solutions. Because of the net water flux, the amount of high-concentrated solution continuously increases. After enough time has passed, the pressure difference caused by the height difference between the two sides becomes the same as the osmotic pressure difference. This is the equilibrium state, as shown in Fig. 2–1. To make water flow from the high-concentrated solution to the low-concentrated solution, a pressure higher than the osmotic pressure should be applied to the high-concentrated solution. Therefore the osmotic pressure is a type of barrier for water molecules to be separated from saline water. In membrane separation, osmotic pressure is very important because it can act as a barometer to estimate the energy required for the separation of water theoretically. Therefore the theoretical minimum energy for separating freshwater from the saline water can be calculated from the osmotic pressure of saline water.

2.2.2 Theoretical minimum energy for separation from osmotic pressure

The Gibbs free energy of mixing can be calculated by the difference in the Gibbs free energy of the initial and final states as follows:

$$\Delta G_{mix} = G_{final} - G_{initial}. \tag{2.1}$$

It is assumed that there are two chemical species in this system. At the initial state, two pure chemical species are separated and are present in different containers. After mixing these two chemical species completely (final state), the Gibbs free energy of the mixture is usually elevated because of the generation of entropy. From the definition of Gibbs free energy at constant temperature and pressure, the Gibbs free energy at the initial and the final states can be obtained as follows:

$$G_{initial} = n_A \mu_A^* + n_B \mu_B^*, \tag{2.2}$$

$$G_{final} = n_A \left\{ \mu_A^* + RT \ln\left(\gamma_A x_A\right) \right\} + n_B \left\{ \mu_B^* + RT \ln\left(\gamma_B x_B\right) \right\}, \tag{2.3}$$

where n is the number of moles of each chemical, μ_i is the chemical potential of i component, x is the mole fraction, and superscript * denotes the property of pure species, R is the gas constant, T is the absolute temperature, and γ is the activity coefficient. Therefore the Gibbs free energy of mixing is obtained as

$$\Delta G_{mix} = n_A RT \ln(\gamma_A x_A) + n_B RT \ln(\gamma_B x_B). \tag{2.4}$$

The theoretical minimum energy for separation has the same value as the Gibbs free energy of mixing but with the opposite sign. If species A and B are water and salt, the infinitesimal change in the theoretical minimum energy for separation can be expressed as

$$d(-\Delta G_{mix}) = d\{-RT[n_w \ln a_w + n_s \ln a_s]\} \tag{2.5}$$

where a is the activity and subscripts $_w$ and $_s$ denote water and salt, respectively. Because salt permeation in membrane separation is usually blocked by a semipermeable membrane, where water molecules can be freely permeated, the infinitesimal change of salt species is almost zero. Therefore Eq. (2.5) can be arranged as follows:

$$d(-\Delta G_{mix}) = -RT\, d(n_w \ln a_w) = -RT(n_w d \ln a_w + \ln a_w dn_w). \tag{2.6}$$

Compared to the change in the number of water molecules, the change in activity of water is very small. Thus the infinitesimal change in activity in water can be negligible, so Eq. (2.6) can be simplified as

$$d(-\Delta G_{mix}) = -RT \ln a_w dn_w. \tag{2.7}$$

From Eq. 2.7, the theoretical minimum energy for separation can be expressed by the osmotic pressure of the feed solution by considering the effect of pressure on the chemical potential at constant temperature as follows:

$$d(-\Delta G_{mix}) = \pi V_m dn_w. \tag{2.8}$$

2.2.3 Permeation mechanism and equations in the RO process

As described in the previous section, the theoretical minimum energy for separation can be directly obtained from the osmotic pressure of the feed solution. However, this minimum energy can only be reached via a reversible process. In other words, the applied pressure should be infinitesimally higher than the osmotic pressure of the feed solution, and the amount of driving force should be maintained until target recovery is reached. In the actual system, however, the applied pressure should be much higher than the osmotic pressure to ensure enough water flux through the RO membrane. The permeated water and salt fluxes across the RO membrane are correlated with membrane selectivity and permeability [40,41].

Therefore the flux equations are usually determined by the driving force and membrane permeability as follows [42,43]:

$$J_w = A(\Delta P - \Delta\pi), \tag{2.9}$$

$$J_s = B(C_m - C_p), \tag{2.10}$$

where J_w is the water flux, A is the water permeability of the RO membrane, ΔP is the applied pressure on the feed solution, $\Delta\pi$ is the osmotic pressure difference between the feed solution and the permeate, J_s is the salt permeability, B is the salt permeability of the RO membrane, C is the concentration, and the subscripts $_m$ and $_p$ denote the membrane surface on the feed solution side and permeate, respectively.

The water and salt permeabilities are estimated from the intrinsic membrane transport properties (P_w for water permeation and P_s for salt permeation) as follows [44]:

$$A = \frac{P_w}{L}\frac{M_w}{RT}, \tag{2.11}$$

$$B = \frac{P_s}{L}, \tag{2.12}$$

where L is the membrane thickness, M_w is the molecular weight. A and B have a trade-off relationship, so the selectivity of the RO membrane can be estimated by the empirical correlation between A and B values as follows [45]:

$$B = \frac{L^\beta}{\lambda}\left(\frac{RT}{M_w}\right)^{\beta+1} A^{\beta+1} \tag{2.13}$$

where β and λ are the correlation parameters that are determined empirically.

2.2.4 Concentration polarization

Although water and salt permeation in the RO system have a linear correlation with the driving force, in actual systems, a nonlinear relationship exists with increasing driving force. The main reason for this nonlinear behavior is the concentration polarization on the surface of the membrane. Concentration polarization is caused by selective permeation across the semipermeable membrane. Therefore the effective concentration at the surface of the RO membrane is always higher than the bulk concentration. Without considering the concentration polarization, the effective osmotic pressure is underestimated, and the water flux and energy consumption in the RO system are also underestimated. The modeling of concentration polarization is derived from the boundary layer film theory [46].

As shown in Fig. 2–2, there are three salt flux terms at the surface of the RO membrane, that is, diffusive flux, convective flux, and salt flux [47]. At the steady state, the flux terms can be arranged to derive the concentration polarization model as follows [48]:

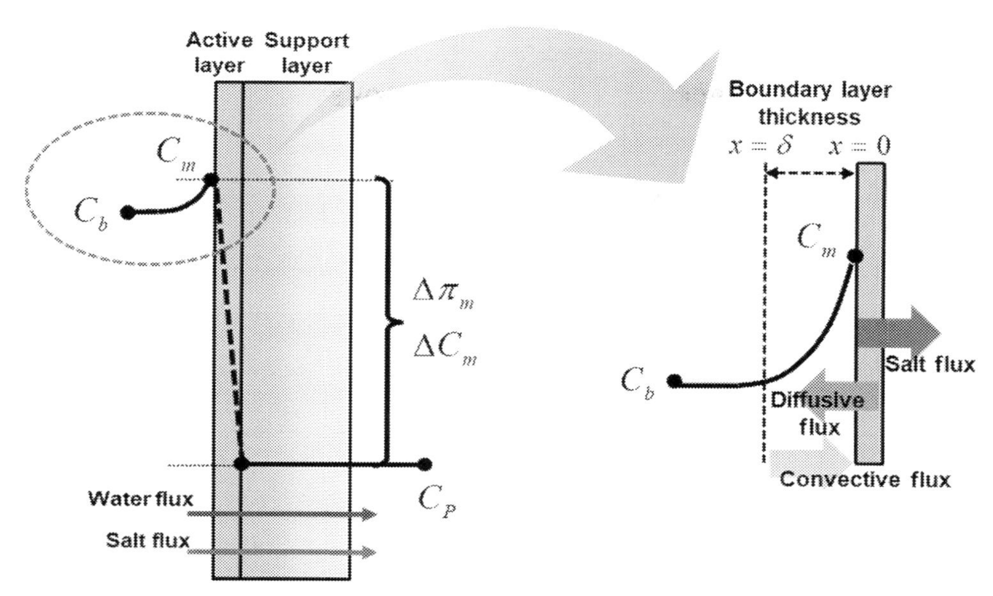

FIGURE 2–2 Schematic diagram of concentration polarization and the detailed salt fluxes on the surface of the RO membrane. *RO*, Reverse osmosis.

$$\underbrace{D\frac{dC(x)}{dx}}_{\text{Diffusive flux}} + \underbrace{J_w C(x)}_{\text{Convectiveflux}} = \underbrace{J_w C_p}_{\text{Saltflux}}, \tag{2.14}$$

where D is the diffusion coefficient. The boundary conditions of this differential equation are $C = C_b$ at $x = \delta$ and $C = C_m$ at $x = 0$. Then, Eq. (2.14) can be solved as

$$\exp\left(\frac{J_w \delta}{D}\right) = \frac{C_m - C_p}{C_F - C_p}. \tag{2.15}$$

From the boundary layer film theory, the ratio D/δ can be replaced by a mass transfer coefficient (k) in each film [49]. Therefore the concentration polarization model can be obtained as a function of water flux and mass transfer coefficient.

$$\exp\left(\frac{J_w}{k}\right) = \frac{C_m - C_p}{C_F - C_p}. \tag{2.16}$$

If the laminar boundary layer is formed and the Schmidt number is unity, Blasius' solution can be utilized to estimate the mass transfer coefficient [50]. However, the practical situation is quite complex, and the flow regime is not limited to the laminar flow region. Therefore an empirical correlation of Sherwood number is usually utilized [48,51]. The Sherwood number is a function of the Reynolds and Schmidt numbers with empirical coefficients as follows:

$$Sh = \frac{kd_h}{D} = aRe^b Sc^c \tag{2.17}$$

where d_h is the hydraulic diameter in the channel. Coefficients a, b, and c are dependent on the flow regime and module structure. A list of coefficients is provided in Table 2–2.

2.2.5 Mass balance and pressure drop equations in the RO process

In the RO membrane module, the concentration of the feed stream increases along the length of the RO module because of water flux, which makes the water flux distribution inside the RO module uneven. To describe this flux distribution, mass balance equations inside the RO module are required. As shown in Fig. 2–3, the overall mass balance equation

Table 2–2 Sherwood number correlation coefficient and the validated conditions [52,53].

Module type	Sherwood number correlation	Validated conditions	References
Flat sheet	$Sh = 1.62\left(Re \times Sc \times \frac{d_h}{L}\right)^{0.33}$	Laminar flow ($Re < 2100$)	[54]
	$Sh = 0.34Re^{0.75}Sc^{0.33}$	$10^4 < Re < 10^5$	[55]
	$Sh = 0.023Re^{0.8}Sc^{0.33}$	$Re > 10^5$	[55]
	$Sh = 0.2Re^{0.57}Sc^{0.4}$	$Re < 50$	[52]
	$Sh = 1.964Re^{0.406}Sc^{0.25}$	$0.7 < Re < 1.7$	[56]
	$Sh = 0.023Re^{0.875}Sc^{0.25}$	$300 < Sc < 700$	[57]
	$Sh = 0.0149Re^{0.88}Sc^{0.33}$	$Sc > 100$	[58]
	$Sh = 0.107Re^{0.9}Sc^{0.5}$	$0.5 < Sc < 10$	[59]
Spiral-wound	$Sh = 0.023Re^{0.875}Sc^{0.25}$	$Re > 2100$	[60]
	$Sh = 0.2Re^{0.57}Sc^{0.40}$	Common commercial spacer	[52]
	$Sh = 0.16Re^{0.605}Sc^{0.42}$	$L/D = 8$, spacer angle = 90 degrees	[52]
Hollow fiber	$Sh = 0.2Re^{0.6}Sc^{0.33}$	$40 < Re < 1000$	[61]
	$Sh = 0.17Re^{0.6}Sc^{0.33}$	$20 < Re < 200$	[62]

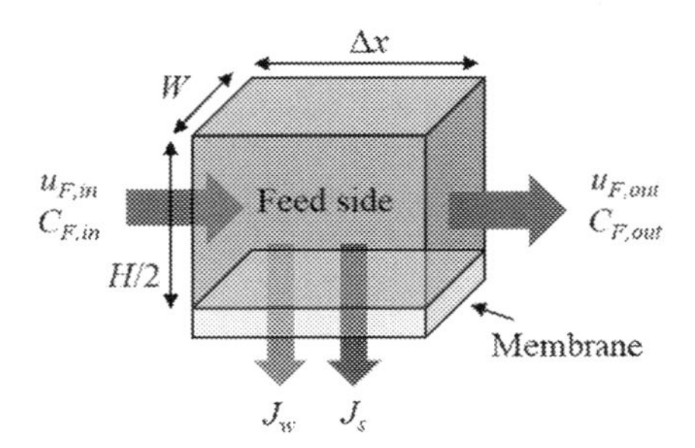

FIGURE 2–3 Schematics of the inlet, outlet flow, and flux terms in the RO module. *RO*, Reverse osmosis.

inside the RO module can be obtained by considering inflow, outflow, and water and salt fluxes as follows [9,35,63]:

$$\underbrace{\frac{dM_F}{dt}}_{\text{rate of accumulation}} = \underbrace{\frac{u_{F,in}WH}{2}\rho_{F,in}}_{\text{inletfeed flow rate}} - \underbrace{\frac{u_{F,out}WH}{2}\rho_{F,out}}_{\text{outlet feed flow rate}} - \underbrace{J_w(W\Delta x)\rho_w}_{\substack{\text{water flow rate}\\ \text{across the membrane}}} - \underbrace{J_s(W\Delta x)}_{\substack{\text{salt flow rate}\\ \text{across the membrane}}} \tag{2.18}$$

where M_F is the overall mass of the feed solution, t is the time domain, u is the linear velocity, ρ is the density, and W, $H/2$, and Δx are the channel width, height, and length, respectively. Subscripts $_{F,in}$, $_{F,out}$, and $_w$ denote the feed inlet, feed outlet, and water, respectively. Under steady state, the rate of accumulation becomes zero. Then, the differential term on the linear velocity can be derived as follows:

$$\frac{d(u_F\rho_F)}{dx} = -\frac{J_w\rho_w}{H/2} - \frac{J_s}{H/2}, \tag{2.19}$$

$$\rho_F\frac{du_F}{dx} + u_F\frac{d\rho_F}{dx} = -\frac{J_w\rho_w}{H/2} - \frac{J_s}{H/2}. \tag{2.20}$$

Because the feed solution density is a function of feed concentration, Eq. (2.20) can be adjusted as follows:

$$\rho_F\frac{du_F}{dx} + u_F\frac{d\rho_F}{dC_F}\frac{dC_F}{dx} = -\frac{J_w\rho_w}{H/2} - \frac{J_s}{H/2}. \tag{2.21}$$

Similarly, the salt mass balance of the feed solution can be derived as [9,35,63]

$$\underbrace{\frac{WH\Delta x}{2}\frac{dC_F}{dt}}_{\text{rate of accumulation}} = \underbrace{\frac{u_{F,in}WH}{2}C_{F,in}}_{\text{inlet salt flow rate}} - \underbrace{\frac{u_{F,out}WH}{2}C_{F,out}}_{\text{outlet salt flow rate}} - \underbrace{J_s(W\Delta x)}_{\substack{\text{salt flow rate}\\ \text{across the membrane}}}. \tag{2.22}$$

The rate of accumulation becomes zero when a steady state is assumed. The differential term on the feed concentration can be obtained as follows:

$$\frac{d(u_FC_F)}{dx} = u_F\frac{dC_F}{dx} + C_F\frac{du_F}{dx} = -\frac{J_s}{H/2}, \tag{2.23}$$

$$u_F\frac{dC_F}{dx} = -\frac{J_s}{H/2} - C_F\frac{du_F}{dx}. \tag{2.24}$$

By combining Eqs. (2.21) and (2.24), Eq. (2.21) can be arranged as follows [9,64]:

$$\rho_F\frac{du_F}{dx} + \frac{d\rho_F}{dC_F}\left(-\frac{J_s}{H/2} - C_F\frac{du_F}{dx}\right) = -\frac{J_w\rho_w}{H/2} - \frac{J_s}{H/2}, \tag{2.25}$$

$$\frac{du_F}{dx} = \frac{-\frac{J_w \rho_w}{H/2} - \frac{J_s}{H/2}\left(1 - \frac{d\rho_F}{dC_F}\right)}{\rho_F - C_F \frac{d\rho_F}{dC_F}}. \tag{2.26}$$

Eqs. (2.24) and (2.26) can be combined to derive the explicit differential term about the C_F.

$$\frac{dC_F}{dx} = -\frac{J_s}{u_F H/2} - \frac{-\frac{C_F J_w \rho_w}{u_F H/2} - \frac{C_F J_s}{u_F H/2}\left(1 - \frac{d\rho_F}{dC_F}\right)}{\rho_F - C_F \frac{d\rho_F}{dC_F}}. \tag{2.27}$$

In the low-concentrated solution, the density of solute can be negligible. Then, the feed solution density can be approximated as $\rho_F = \rho_w + C_F$. If this assumption can be applied to the feed solution, Eqs. (2.26) and (2.27) can be simplified as follows [9,64,65]:

$$\frac{du_F}{dx} = -\frac{J_w}{H/2}, \tag{2.28}$$

$$\frac{dC_F}{dx} = -\frac{J_s}{u_F H/2} + \frac{C_F J_w}{u_F H/2}. \tag{2.29}$$

In addition, the pressure of the feed solution is reduced along the RO module length because of the friction loss in the channel. The pressure drop inside the RO module may be calculated using the pipeline friction correlation as follows [66]:

$$\frac{dP_F}{dx} = -\frac{2\rho f u^2}{d_h}. \tag{2.30}$$

Therefore the feed velocity, concentration, and pressure change along the RO module length can be estimated using differential Eqs. (2.28), (2.29), and (2.30). From the feed velocity change, the amount of water flux in the RO module at the applied pressure can be calculated. The freshwater production rate at a certain amount of applied pressure in the RO module can be estimated, and the recovery can be also calculated using these balance equations.

2.2.6 Energy consumption in the RO process

To operate the RO system, electrical energy should be supplied to the high-pressure pump (HPP) which raises the pressure of the feed stream before entering the RO membrane module. Because the energy consumption in HPP is dependent on the flow rate and target pressure, these variables should be carefully identified to estimate the energy consumption of the RO system. From the suggested model equations in this chapter, the applied pressure and amount of water flux by the applied pressure in the RO system can be determined. Therefore the SEC of the RO process can be calculated from the above model equations.

Before deriving the energy model for the RO system, the importance of an energy recovery device (ERD) should be emphasized. The concentrate stream after the RO module contains a significant amount of pressure energy [67]. Thus the energy needs to be recovered to improve the overall energy efficiency of the RO system. The ERD acts as the main contributor to improve the energy efficiency of RO systems, and the development of ERD has made RO one of the most energy-efficient desalination technologies [4]. Different types of ERDs from Francis turbine to pressure exchangers have been developed [4]. Recently, ERDs of the pressure exchanger type have attained ERD efficiencies of over 95% [68]. Therefore the installation of ERDs in RO systems is indispensable to minimize the SEC. To make up the pressure difference caused by ERD efficiency and pressure drop in the RO module, a booster pump (BP) is placed after the ERD. The schematic flow diagram of the RO process is shown in Fig. 2–4.

The energy consumption model can be obtained from the applied pressure (P_{RO}) and the required feed flow rate (F_{feed}) as follows [9,10,35,65]:

$$E_{RO} = \frac{F_{feed}}{\eta_{pump}}(P_{RO} - P_0) = \frac{F_p}{\eta_{pump} r_{RO}}(P_{RO} - P_0) \quad \text{(without ERD)}, \tag{2.31}$$

where η_{pump} is the pump efficiency, r_{RO} is the RO recovery, F_p is the freshwater production rate, and P_0 is the ambient pressure. In the case of an RO system with ERD, the recovered pressure energy should be considered, which can be obtained by the pressure balance equation [65].

$$\eta_{ERD} F_p \left(\frac{1}{r_{RO}} - 1 \right) (P_{RO} - P_{fric} - P_0) = \frac{F_p}{r_{RO}}(P_{ERD} - P_0), \tag{2.32}$$

where η_{ERD} is the ERD efficiency, P_{fric} is the pressure drop due to friction loss in the RO module, and P_{ERD} is the feed pressure increased by recovering the pressure energy in the concentrate stream. The left side of Eq. (2.32) is the recovered pressure energy, and the right side is the increased amount of pressure energy in the feed stream. Then, the energy consumption in the case of RO systems with ERD can be derived as follows:

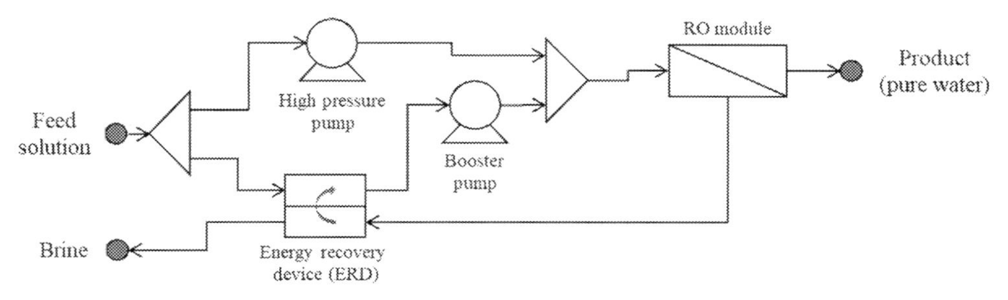

FIGURE 2–4 Flow diagram with pump, ERD, and RO module in the RO process. *ERD*, Energy recovery device; *RO*, reverse osmosis.

$$E_{RO} = \frac{F_{feed}}{\eta_{pump}}(P_{RO} - P_{ERD}) = \frac{F_p}{\eta_{pump} r_{RO}}\left[P_{RO} - \eta_{ERD}(1 - r_{RO})\left(P_{RO} - P_{fric} - P_0\right) - P_0\right] \quad \text{(with ERD).} \tag{2.33}$$

Finally, the SEC of the RO process can be obtained by dividing the energy consumption by the freshwater production rate as follows:

$$SEC_{RO} = \frac{E_{RO}}{F_p} = \frac{1}{\eta_{pump} r_{RO}}(P_{RO} - P_0) \quad \text{(without ERD),} \tag{2.34}$$

$$SEC_{RO} = \frac{E_{RO}}{F_p} = \frac{1}{\eta_{pump} r_{RO}}\left[P_{RO} - \eta_{ERD}(1 - r_{RO})\left(P_{RO} - P_{fric} - P_0\right) - P_0\right] \quad \text{(with ERD).} \tag{2.35}$$

2.3 RO system and design

2.3.1 Single-stage/pass BWRO

Brackish water is water with a total dissolved solid (TDS) of 500–15,000 mg/L. However, brackish water reverse osmosis (BWRO) plants typically treat feeds of 500–10,000 mg/L [69]. Because there are various sources of brackish water, it can be classified with more detailed criteria: water with a TDS of 500–2500 mg/L is classified as low saline and that with a TDS of 2500–10,000 mg/L as high saline. Notably, the osmotic pressure of brackish water is not as high as that of seawater and is still lower than that of seawater even when it is concentrated up to 90% in low-salinity brackish water. Because of the low osmotic pressure, BWRO systems can be operated at a low hydraulic pressure, and system recovery can be increased substantially. When the feed is low-salinity brackish water, a single-stage (or -pass) system can be adopted (Fig. 2–5A and Table 2–3) [69]. However, high recovery cannot be achieved through single-stage/pass BWRO systems even though brackish feed has low osmotic pressure.

2.3.2 Two/multistage BWRO

Two/multistage systems are common in BWRO application for achieving high recovery (Table 2–3). The target recovery is 70%–90% with a water flux of 20–40 L/m^2 h, depending on feed salinity and characteristics [10], and the value is still higher than single-stage (-or pass) operation. The staged system generally follows a 2:1 array for two-stage configuration [where the ratio of pressure vessel (PV) number for the first and second stage is 2:1] and 3:2:1 for three-stage configuration [66].

However, BWRO systems differ depending on the salinity of brackish water. For low-salinity brackish water, two/multistage BWRO systems are commonly used as well as single-stage BWRO systems, and no BPs are installed between the first stage and the second stage (Fig. 2–5B). This is because the hydraulic pressure is high enough to overcome the osmotic pressure of the feed. In contrast, when the feed is high-salinity brackish water, two-stage

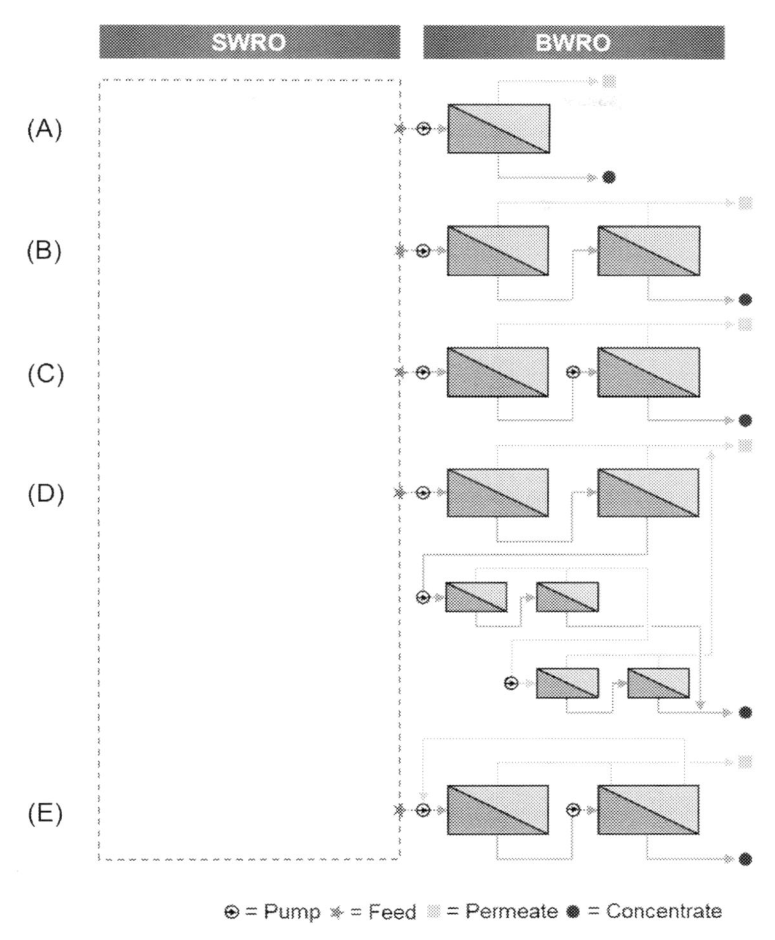

FIGURE 2–5 Scheme of BWRO configurations: (A) single stage/pass, (B) two stage (low-salinity water), (C) two stage (high-salinity water), (D) multistage/cascade (low-salinity water), and (E) PCP. *BWRO*, Brackish water reverse osmosis; *PCP*, permeate circulation process.

BWRO equipped with BPs between the stages is applied to further produce permeate from the second stage (Fig. 2–5C).

Two/multistage BWRO systems can also be implemented in SWRO desalination plants to improve water quality. In two-pass SWRO systems, the first pass is composed of SWRO membranes and the second pass of BWRO membranes [4,10]. As the feed for the second pass is low-salinity brackish, it is usually configured as a two-stage BWRO system without inner BPs. To achieve higher water quantity and quality, several SWRO desalination plants use a cascade BWRO system similar to multistage systems (Fig. 2–5D). The cascade system can be varied, but it is usually a set of BWRO systems where the first-pass permeate is treated by a BWRO multistage system (2–4 stages) to increase water

Table 2–3 Summary of RO configuration.

Feed	Overall RO configuration	SWRO configuration	BWRO configuration
Brackish water	Single stage/pass	N/A	Single stage/pass
	Two stage (low-salinity water)	N/A	Two stage
	Two stage (high-salinity water)	N/A	Two stage
	Multistage/cascade	N/A	Multistage/cascade
Seawater	Single stage/pass	Single stage/pass	N/A
	Two stage	Two stage	N/A
	Full two pass (with two-stage SWRO)	Two stage	Two stage (or cascade)
	Full two pass	Single stage/pass	Two stage (or cascade)
	Partial second pass	Single stage/pass	Two stage (or cascade)
	Split partial second pass	Single stage/pass	Two stage (or cascade)

Notes: For overall RO configuration, it is preferred to name the overall pass configuration first (if the number of pass is multiple) and then specify stage configuration. *RO*, Reverse osmosis; *SWRO*, seawater reverse osmosis; *BWRO*, brackish water reverse osmosis.

recovery. Furthermore, the permeate produced from the rear BWRO stages is further treated by another set of BWRO stages to reduce TDS. In contrast, a permeate circulation process has been applied to BWRO systems (i.e., second pass) of Shuqaiq II SWRO desalination plants where the BWRO rear permeate is circulated back to the feed to improve water quality (Fig. 2–5E) [4,70]. Likewise, BWRO integrated with SWRO is extensively adopted even in SWRO desalination plants.

BWRO systems can be utilized for both, treating brackish water and improving the water quality of SWRO systems. In other words, the role of BWRO will increase in the future desalination market regardless of feed salinity. Because the main benefit of BWRO is a high-recovery operation, BWRO systems should be developed further to maximize their recovery while ensuring permeate quality.

2.3.3 Single-stage/pass SWRO

Typical SWRO desalination plants are operated with 40%–50% recovery with an average water flux of 12–16 L/m^2 h depending on feed characteristics [10]. This target recovery and water flux can be achieved using single stage/pass only, and two/multistage configuration is not required for SWRO. Thus most SWRO configurations are single stage/pass (Fig. 2–6A and Table 2–3; a single pass rather than a single stage would be a more suitable termination for SWRO as pass design is more implemented in SWRO operation). The limited recovery of single-stage/pass SWRO is caused by the high salinity of the feed and high-pressure operation.

High salinity results in both high osmotic pressure and its rapid increase over the process. The SWRO feed becomes concentrated with a high rate after the permeate is produced. As a result, the osmotic pressure of the feed is increased dramatically at the front element, and pressure the same as the hydraulic pressure is reached at the rear elements. When the net driving pressure (NDP) is not positive, a permeate will not be produced through SWRO.

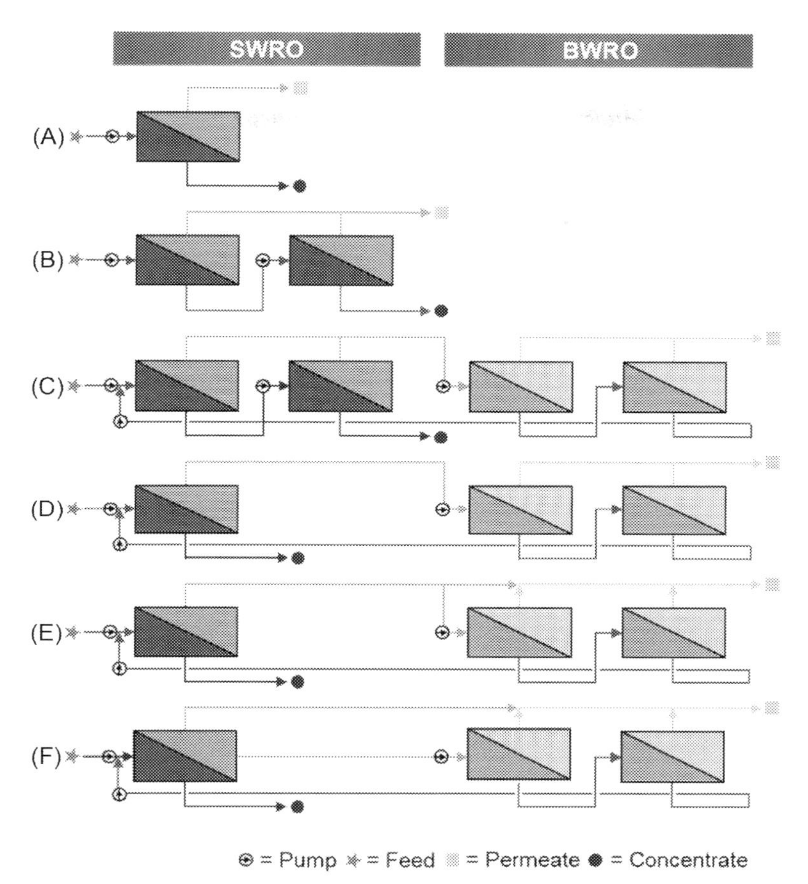

FIGURE 2–6 Scheme of seawater RO configurations: (A) single stage/pass, (B) two stage, (C) full two pass (with two-stage SWRO), (D) full two pass, (E) partial second pass, and (F) SPSP. *RO*, Reverse osmosis; *SWRO*, seawater reverse osmosis; *SPSP*, split partial second pass.

If higher hydraulic pressure is applied, NDP can be increased to become positive, but it also increases the osmotic pressure because of the additional water production. Thus water production is limited with single stage/pass.

Achieving high recovery is also hindered by high-pressure operation because it can damage SWRO membranes and equipment. Most current SWRO membranes are designed to overcome hydraulic pressures up to 80 bar. If the target recovery for SWRO is higher than 50%, the SWRO system should be operated at higher than 80 bar to overcome the inherent osmotic pressure of the feed. If the membranes and other equipment are not strong enough to resist such pressure, the SWRO system will be damaged. To utilize commercialized desalination equipment, it is easier to design single-stage/pass SWRO systems.

Although single-stage/pass systems are a common design for SWRO, the configuration can be varied to compensate for the limitation of single-stage/pass operation and enhance the SWRO performance depending on its use. However, it is notable that single-stage/pass systems are a basic operation even for two-stage or two-pass SWRO; the concentrate of single-stage systems is further treated in two-stage SWROs, while the permeate of single pass is purified in two-pass SWRO.

2.3.4 Two-stage SWRO

For producing a high amount of permeate from the given feed, the SWRO system should be operated with high recovery. However, the achievable recovery is limited when a single stage/pass is only used. The two-stage configuration (Fig. 2–6B) can be a solution to increase water recovery (Table 2–3). In this configuration, additional PVs are installed as the second stage next to the first stage considering a 2:1 array. In the system, the concentrate of the first stage is pressurized and fed to the second stage for additional water production. However, the two-stage configuration is not common for SWROs, unlike BWRO, because the two-stage configuration is disadvantageous for SWRO operation.

In two-stage SWRO, the second stage is operated with extremely high pressure far beyond 80 bar. The second stage receives the concentrate of the first stage, which means that the first-stage concentrate acts as the second-stage feed. However, as the concentration of the second-stage feed is high, the hydraulic pressure should be elevated to overcome the high osmotic pressure. Thus an additional 20–30 bar of hydraulic pressure is applied to the feed, and the second stage is operated under a hydraulic pressure of approximately 100 bars. As standard SWRO membranes and equipment are resistant to 80 bar pressure, the second stage should be equipped with customized products, which eventually increase the capital cost.

In addition, the energy consumption of SWROs is dramatically increased because of high-recovery operation. It has been reported that the optimal recovery for SWRO is approximately 40%–50% that of single-stage SWRO [4]. If the recovery is higher than the value, SEC is elevated. Two-stage SWRO systems include a second stage added to the typical single-stage SWRO (i.e., optimal recovery); thus the increase of water production from the second stage (i.e., additional recovery) is directly associated with the increase in energy consumption. With the benefit of high recovery, two-stage SWRO loses energy efficiency.

Another problem is that water quality can be hampered because of high-recovery operation. As the concentration of the feed for the second stage is high, the concentration of permeate from the second stage can also be high. Although the main benefit of two-stage SWRO systems is the high recovery, the product from two-stage SWRO systems cannot be utilized unless its quality satisfies the criteria. To improve permeate quality, (two-stage) BWRO systems can be integrated with two-stage SWRO systems; this can be classified as two-pass SWRO systems in terms of overall configuration (Fig. 2–6C). An additional process for the two-stage SWRO system would not be useful in terms of energy and maintenance.

The installation of the second stage would increase the cost of water production. In addition to single-stage SWRO, BPs and PVs are installed for second-stage SWRO. Moreover, as the permeate quality of two-stage SWRO systems is not promising, a BWRO system is installed to further improve permeate quality, thus increasing capital expenditures. On the other hand, the SWRO system is operated under a high hydraulic pressure at high water recovery, and the SEC is significantly increased. In addition, an additional BWRO operation is required to improve permeate quality. Therefore the operational expenditure can be elevated for both SWRO and BWRO operation.

Despite achieving high recovery, several disadvantages hinder the application of the two-stage SWRO configuration. Thus most SWRO systems are configured as single-stage systems. However, notably, BWRO in two-pass SWRO is still configured as a two- or multistage configuration.

2.3.5 Two-pass SWRO

Two passes are generally adopted when a pass configuration is implemented in SWRO (Table 2–3). In two-pass SWRO systems, the first pass is configured as a single-stage SWRO, whereas the second pass as a two-stage BWRO. Sometimes, BWRO with a cascade design is employed to increase the recovery of the overall process. Meanwhile, the permeate of the first pass is fed to the second pass to produce a purer permeate, and the concentrate of the second pass is circulated back to the feed for the first pass to dilute the SWRO feed and lower the applied pressure. The main purpose of using two-pass SWRO systems is to improve product quality of the SWRO process, but it also can be used when the feed for the SWRO system is highly concentrated with foulants.

Two-pass SWRO systems can produce permeate with a lower concentration than single-pass SWRO systems by desalting the feed twice. The salt rejection rates for both SWRO and BWRO membranes are higher than 99%, so the rejection rate would be squared when the feed passes two RO processes. As a result, the concentration of a single pass is reported to be 300–500 mg/L, while that of two pass is 15–130 and 115–300 mg/L before and after remineralization, respectively, in a number of SWRO desalination plants (the range of permeate TDS can be varied even among two-pass SWRO depending on its detailed configurations) [4]. With a lowered TDS concentration, the permeate can be utilized for wider purposes.

When the feed has high fouling propensity, two-pass SWRO systems can be employed to prevent membrane fouling and further improve water quality. High water flux is the main cause of organic/colloidal fouling in the front RO elements in a PV [71]. If SWRO systems are operated with a low water flux, the formation of fouling can be mitigated, but the permeate quality is hampered. Thus when the feed contains foulants with high concentration, SWRO systems can be configured as a two-pass system; the first pass is operated with a low water flux to reduce the fouling load for the membranes, and the second pass is equipped to treat the first-pass permeate to improve product water quality.

As the SWRO desalination market is rapidly growing in the MENA region, where salinity and temperature of seawater are both high, two-pass SWRO configurations are preferred to meet the water quality standards. However, operating an additional pass (i.e., second pass) can increase the energy demand of the overall process, and the TDS of the product can also be reduced to much less than the required level. To optimize the system, two-pass configurations can be modified depending on the purposes.

2.3.5.1 *Full two pass*

When a high quality of permeate is required, a full two pass configuration (Fig. 2–6D) can be employed. In such a system, the permeate from the first-pass RO (i.e., SWRO) is wholly supplied to the second-pass RO (i.e., BWRO) without splitting the stream [4,10]. Thus the TDS of the permeate can be efficiently reduced, and the final product may contain a low level of TDS. The full two pass is not a proper configuration for drinking water production as a moderate amount of TDS should be contained in potable water. Instead, this configuration can be effective for industrial water production to meet pure water standards. If a further reduction in TDS of permeate is required, three passes can be implemented even in SWRO. However, the application of three-pass SWRO systems is rare because full two pass systems can still produce a permeate with low TDS [4].

2.3.5.2 *Partial second pass*

The two-pass SWRO configuration, where only a part of the first-pass permeate is desalinated by the second pass, is classified as a partial second pass (Fig. 2–6E) [10]. Streams, which either pass or by-pass the second pass, are sent to the product tank and are blended. A partial second pass is used to control the TDS of the final product; as the amount of permeate sent to the second pass increases, the purity of the final product improves. However, energy consumption is elevated when the treating volume is increased. This configuration is often adopted to SWRO desalination plants for producing drinking water to meet the required water standards, which cannot be achieved by single-pass operation, by adjusting the blending ratio (i.e., the ratio for partial stream). The blending ratio can be increased during the summer as the increased water temperature can deteriorate the permeate quality.

2.3.5.3 *Split partial second pass*

In the front elements of the SWRO process, feed salinity is relatively low and the operating water flux is high. In contrast, the feed is highly saline at the rear elements and permeate is produced with a low water flux. Thus the TDS of the front permeate is low, whereas that of the rear permeate is high [10]. To improve the final water quality, it is more energy efficient to desalinate highly saline permeate only and to mix it with low-salinity permeate from the thermodynamic perspective. Similarly, split partial second pass treats only SWRO rear permeate (i.e., high-salinity permeate) (Fig. 2–6F). The SWRO front permeate is directly delivered to the product tank, whereas the SWRO rear permeate passing the second pass is sent to the product tank.

2.3.6 Internally staged design

Different types of RO membranes in a PV have been implemented in designing RO systems [10,71]. In particular, RO membranes with high-rejection property located in the front, whereas those with a high-flux property are placed in the rear of PV (Fig. 2–7A–D). By doing so, the front elements, which are normally operated at high water fluxes, are operated with lowered water fluxes. Such a design is called an internally staged design (ISD) because

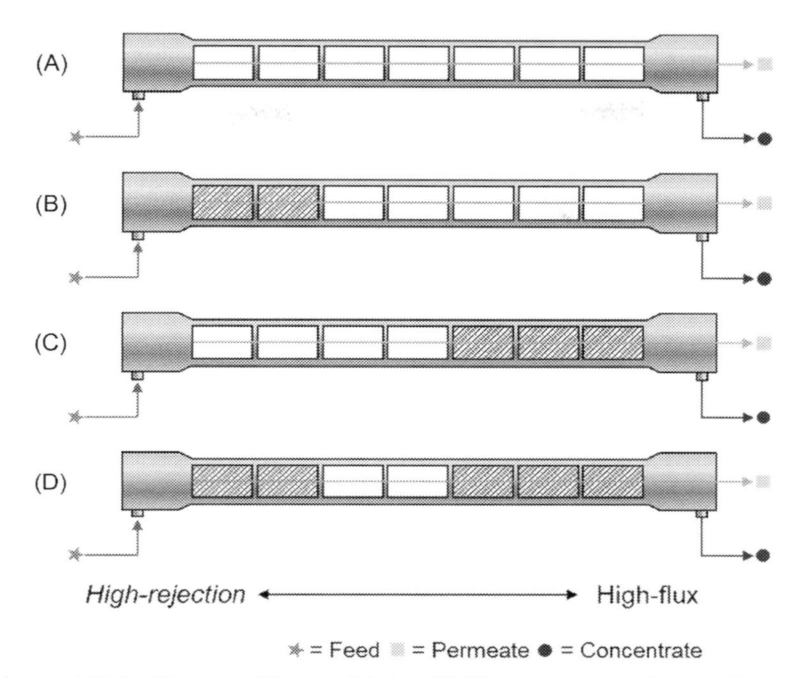

FIGURE 2–7 Scheme of ISD in RO process: (A) normal design, (B) ISD using high-rejection membranes in the front, (C) ISD using high-flux membranes in the rear, and (D) ISD using three different membranes. *ISD*, Internally staged design; *RO*, reverse osmosis.

water flux distribution along the PV (i.e., internally) is similar to that of staged RO design (i.e., externally). RO systems with ISD have several benefits in operation over systems equipped with a single type of membrane.

The fouling propensity of RO membranes can be reduced by lowering the water fluxes of the front elements. As organic/colloidal fouling is strongly associated with high water flux, RO elements in the front of PV are prone to fouling. However, as RO membranes with a low water flux are utilized based on the rule of ISD, water fluxes for the front elements are less likely to increase over the critical flux. Mitigation of fouling using ISD can be beneficial in that the SEC increases over time and water quality deterioration can be lowered.

RO systems can exhibit various performances depending on ISD combinations. When a single type of membrane is used, the performance of the RO system (e.g., SEC and water quality) is determined inherently. To change the performance of the RO system, the operating conditions such as average water flux or recovery should be controlled. Other than changing the operating conditions, membranes with different characteristics can be employed, and this can change the RO performances. By doing so, the RO performance can be optimized based on the performance requirements.

ISD also enhances the RO system to be fully operated by increasing the water permeability of membranes from the middle of the PV. The distribution of water fluxes along a PV is biased in typical RO operation; water fluxes for front elements are high and those for rear elements are low. Thus the load for the front elements is higher than that for other elements, and frequent replacement or membrane rotation is required. ISD can utilize rear elements using membranes with high water permeability so that water production from the rear can be increased without increasing hydraulic pressure. Thus membrane replacement is not required as frequently as for conventional RO membrane placement.

Because of the benefits of ISD, it has been implemented more in SWRO systems than in BWRO systems. This is because the characteristics of SWRO membranes are distinct (e.g., high-flux or high-rejection membranes); thus ISD can be more effective when it is used. El Coloso, Mazarrón, and Las Palmas III SWRO desalination plants are examples of desalination plants that have adopted ISD for SWRO systems [72]. However, ISD still should overcome several operational issues such as replacement difficulty and design optimization [71].

2.3.7 Pressure-center design

The size of HPP is typically determined by the size of the RO train. However, the size of HPPs and BPs can be increased by the pressure-center (or three-center) design. In the pressure-center design (Fig. 2–8), several RO trains (referred to as "banks" for this design) are attached to the main feed line, and the feed is pressurized with larger HPPs rather than smaller and multiple HPPs for typical RO trains. BPs associated with ERDs are also integrated to the main feed line with a larger size. The increase in the pump size improves pump efficiency, and the energy consumed by the RO system can be lowered [4]. In addition, the

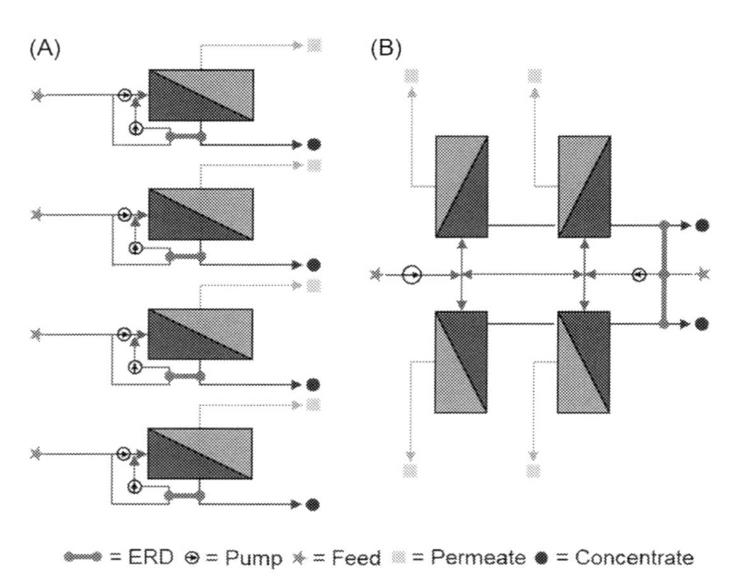

FIGURE 2–8 Scheme of pressure-center design: (A) typical RO trains and (B) RO trains with pressure-center design. *RO*, Reverse osmosis.

pressure-centered RO system is more effective for producing varying amounts of permeate by system integration; thus the operation of the RO system can be flexible compared to that of conventional RO systems. With the benefits of integrating pumps, several SWRO desalination plants in Israel have adopted the pressure-center design in the SWRO process and have successfully operated the system [4,73–75]. However, maintenance can be a problem as RO trains are integrated as an entire system.

2.4 RO fouling

2.4.1 Particulate/colloidal fouling

Particulate or colloidal fouling refers to membrane fouling caused by particles or colloids being deposited on the membrane surface. Generally, these types of foulants are organic and inorganic particles suspended in source water covering a wide variety of matter such as debris, silt, colloidal silica, iron, aluminum, polysaccharides, and NOM. These solids exist in insoluble forms, and although particulate and colloidal foulants are commonly grouped together, they can be categorized according to their sizes. By definition, particulate matter are those with sizes larger than 1 μm, and colloidal matter are those within the size range of 0.001–1 μm (Table 2–4) [76].

Particulate compounds are most easily removed among all foulants because of their relatively large sizes. These solids cannot pass through the RO membrane and are completely retained on the feed side to be concentrated. They form thick cake layers which lead to the degradation of process performance. However, implementation of well-designed conventional pretreatment processes such as coagulation/sedimentation and granular media filtration prior to the RO stage can lead to high removal of these foulants.

Colloidal compounds are foulants that are smaller than their particulate counterparts. They are commonly further classified into organic and inorganic foulants. Organic colloidal foulants include proteins, hydrocarbons, polysaccharides, and NOM, and inorganic colloidal foulants include colloidal silica, iron, aluminum, and manganese [77]. Although colloids do not exist as large flocs like particulate matter, they are still present in the source water in undissolved forms, and when concentrated through separation by the RO membrane, they coalesce together to form precipitates that then lead to the formation of cake layers on the membrane surface [78].

Apart from simple physical parameters like size, the chemical interaction between the foulants and the membrane and the hydrodynamic conditions of operation are important factors for particulate and colloidal fouling. In the pH range of natural seawater, both foulants are known to carry a negative surface charge, and in such case, a RO membrane with a

Table 2–4 Size and composition of particulate and colloidal matter.

Foulant	Approximate size	Composition
Particulate matter	>1 μm	Debris, clay, silt
Colloidal matter	1–0.001 μm	Colloidal silica, iron, polymer, plankton

similar affinity will be less vulnerable to fouling. Likewise, a more hydrophilic and less rough RO membrane will be less susceptible to the same problem [79]. Hydrodynamic conditions, for instance, crossflow velocity and operational pressure, can also significantly alter the fouling propensity. With higher crossflow velocity, foulants will experience difficulty in settling on the membrane surface and will be swept away along with the concentrate stream. When a high-pressure RO system is operated, the applied pressure will not only cause foulants to be pushed toward the membrane, thereby hindering their back diffusion, but will also induce the compaction of preexisting foulant cake layers.

2.4.2 Organic fouling

Organic fouling is caused by organic substances dissolved within the feed solution that may either be microorganisms or naturally occurring or man-made compounds. Some commonly known organic matters include humic acid, bovine serum albumin, sodium alginate, algae, polysaccharides, transparent exopolymer particles (TEPs), antifoaming agents, and polyacrylic polymers [80]. Similar to particulate and colloidal foulants, these organic molecules and microorganisms are relatively large in size, which means that they are easily rejected by RO membranes and even the conventional pretreatment processes. However, operational difficulties regarding the organic foulants are incited by their sticky characteristic. Because of their tendency to stick, organic compounds are frequently adsorbed onto the membrane surface (Fig. 2–9). Unlike the cake layer that is formed by the deposition of particulate or colloidal matter, the sediment of organic foulants will form a gel layer which then leads to the formation of a biofilm [48].

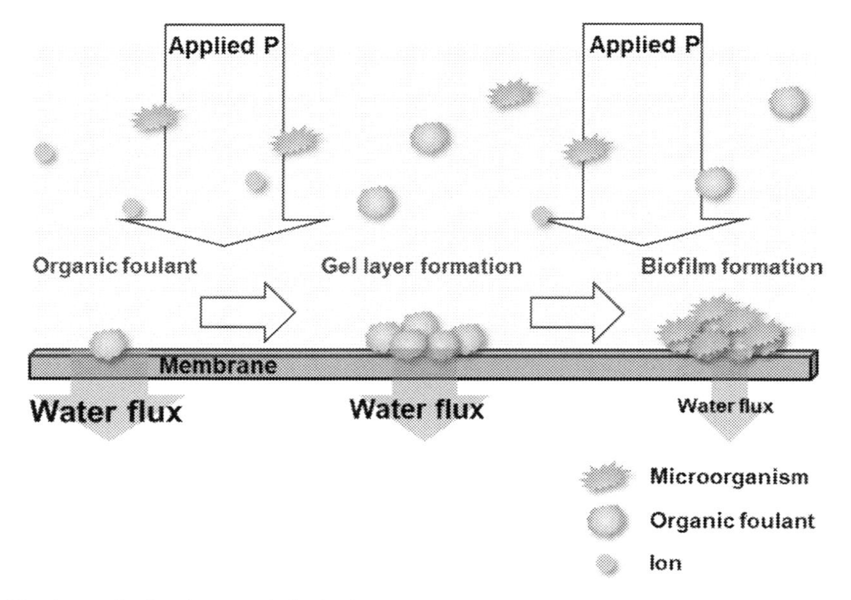

FIGURE 2–9 Membrane fouling by organic foulants.

To better understand the mechanisms of organic fouling, it is important to comprehend the foulant–foulant and foulant–membrane interactions. The initial interaction between the bulk foulants and the membrane surface initiates the deposition of organic foulants. In this context, the hydrophilicity and surface charge of the organic compounds and the membrane are key factors that determine the rate and extent of the formation of the gel layer. Once the coverage of the gel layer on the membrane surface has reached a certain degree, the intermolecular adhesion between the bulk foulants and the foulants on the membrane, in other words, the foulant–foulant interaction, becomes the dominant force. This foulant–foulant interaction is predominantly responsible for deciding the thickness and compactness of the gel layer, which in turn determines the hydraulic resistance that impedes the water permeability of the RO membrane [81].

Organic fouling is mainly attributed to the TEPs. These foulants are organic compounds that are discharged by microorganisms, and they mainly consist of polysaccharides and amino sugars. In open seawater, which is the typical feedwater used in the RO process, biodegradable organic matter, commonly known as algal organic matter (AOM), is excreted by marine algae during algal blooming periods, and TEPs are the high-molecular-weight fractions of those AOM compounds. Because of their extremely adhesive nature, TEPs act as a nutritional platform on which bacteria and other microorganisms can thrive. Furthermore, the same adhesive nature of not only the TEPs but also the organic foulant layer impedes efficient cleaning or backwashing of the fouled membranes. Sufficient operation of pretreatment processes prior to the RO stage is absolutely crucial to minimize the damage.

FIGURE 2–10 Bio-fouled spiral-wound RO membrane. *RO*, Reverse osmosis.

2.4.3 Biofouling

Biofouling, which has become one of the greatest challenges for the RO process, is the subsequent stage to the previous organic fouling (Fig. 2–10). Microorganisms that have accumulated on the membrane surface to form gel layers excrete organic compounds around their cells known as extracellular polymeric substances (EPS) [82]. The surface-active properties of the EPS cause them to remain as a matrix around the microorganisms and act as a type of blanket inside which the microorganisms are entrenched, and this blanket-like layer is commonly known as a biofilm. In addition, the biofilm formed on the membrane surface captures colloidal particles and suspended solids in the feedwater to form a layer several micrometers thick with a high resistance to permeate flow [81].

To reduce the burden required to compensate for the loss of productivity due to the biofilm formation, cleaning or backwashing of the membranes is important, but this is not a simple task when dealing with biofouling. The main components of the EPS, like TEP, are polysaccharides, which means that the EPS also has a highly adhesive nature. Thus in the initial stage of biofouling, each microorganism is surrounded by its own capsule of EPS [83]. However, the EPS capsules stick to each other very quickly to form microcolonies that become irreversibly linked with the membrane surface they have adhered to [84]. At this point, initiation of cleaning or backwashing regimes may have a nominal effect on biofilm removal. The difficulty with controlling biofilm formation is also visible through pretreatment. As the biofilm is a product of microorganisms, pretreatment methods or biocide injection that can remove them to a 3-log reduction level may be utilized to minimize the number of microorganisms reaching the RO membrane. Nonetheless, because no pretreatment process can guarantee 100% removal, the remaining 0.1% of microorganisms can enter the system, and given enough time, they can multiply expansively to form biofilms across the entire membrane surface.

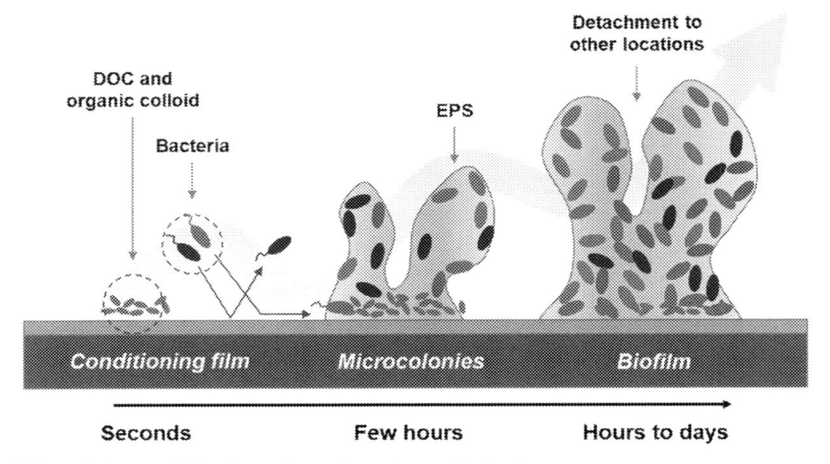

FIGURE 2–11 Different stages of the formation of membrane biofouling.

Another critical characteristic of biofilms is its ability to migrate to different locations. During the final stage of biofilm development, microorganisms detach from the microcolony and disperse into the surrounding environment (Fig. 2–11). Although several different mechanisms such as quorum sensing or environmental signals may take part in the migration, no exact cause has been identified yet [85]. Nevertheless, the detached individual cells or clusters settle at a different location of the RO membrane surface, where they repeat the cycle of biofilm growth.

2.4.4 Scaling

During the process of salt separation by the RO membrane, the concentration of salts in the source water significantly increases. Once the salt concentration exceeds the saturation limit or the solubility product of the individual salt species, they will precipitate to form mineral scales on the membrane surface (Fig. 2–12). As such, scaling is a bigger problem in systems operating with a high recovery rate. This tendency is also verified through the fouling phenomenon experienced by the different RO membrane modules inside a PV. Through autopsy of multiple modules, it has been confirmed that modules near the end of the PV suffer more from problems of scaling because the feedwater that those particular modules come into contact with has a much higher concentration than the feedwater that the modules in the front of the PV encounter.

Mineral ions that form scales in the RO process include various salts such as calcium carbonate, calcium sulfate, barium sulfate, sodium chloride, magnesium, iron oxides, and silicate. Among these, the major participants are calcium carbonate ($CaCO_3$) and magnesium hydroxide [$Mg(OH)_2$], belonging to alkaline hardness scales and calcium sulfate ($CaSO_4$) of nonalkaline hardness scales. Calcium carbonate is perhaps the most prevalent type, which initially exists as calcium and bicarbonate ions in the feedwater. The temperature and pH of the water solution, hardness of calcium, and alkalinity of the bicarbonate all affect the extent

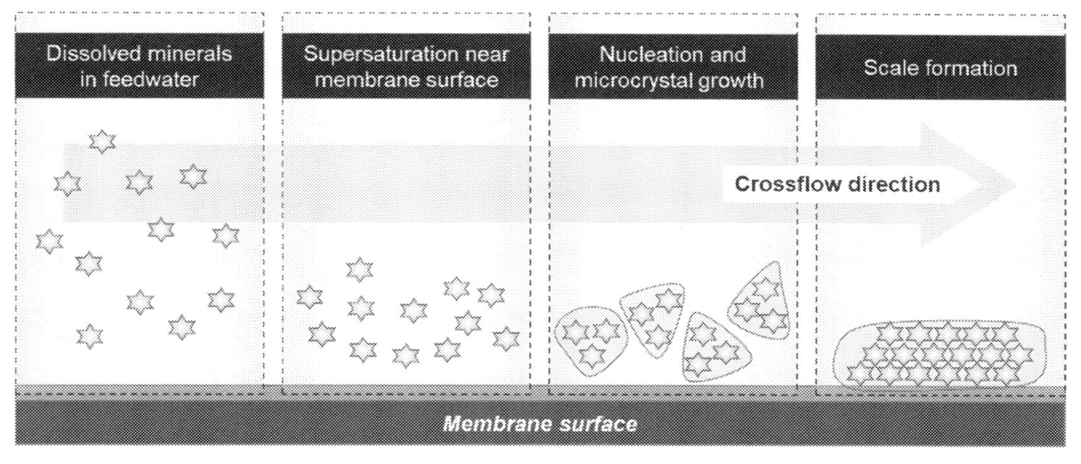

FIGURE 2–12 Different stages of scaling formation.

to which calcium carbonate scaling occurs. However, this form of scaling can be prevented ahead of time by adjusting the feedwater pH to the lower range of 4–6 because under high pH conditions, the bicarbonate in the solution is converted to carbonate, thus elevating the chances of calcium carbonate precipitation [86]. Calcium sulfate, commonly known as gypsum, is another major cause of scaling in the RO process. Calcium sulfate has a low solubility limit even under average temperature conditions of approximately 25°C and becomes even more insoluble under high temperature. Like calcium carbonate formation, the rate of gypsum precipitation increases at a higher pH range because more of the sulfate species are converted to sulfate ions [87].

Discussion on scaling is always closely related to the concentration polarization phenomenon. The RO membrane separation process is commonly divided into two transport regions: the bulk feed solution and the membrane. However, there exists a very thin concentration boundary layer between those two regions, commonly known as the polarized layer or the boundary layer. As the RO process is operated, the concentration of the solution in this immediate polarized layer (C_m) becomes much higher than that of the bulk feed solution (C_b), thereby causing concentration polarization [88].

The degree of concentration polarization depends on several factors, but the most influential of them are crossflow velocity, operating flux, and rejection. Crossflow velocity is a very significant parameter that controls concentration polarization. With low crossflow velocity, foulants will have more time to settle down on the membrane surface, thereby increasing the concentration at the immediate vicinity of the membrane. On the contrary, operating that system at high crossflow velocity will expose the membrane to more foulants, raising the probability of foulant settlement. Therefore an optimal crossflow velocity is critical for stable maintenance of the RO system. The effects of operating flux and rejection on concentration polarization function in a similar manner. With high operating flux, more water will pass through the membrane, which means that the solution near the membrane surface on the feed side will be concentrated at a quicker pace. The trend will intensify the difference in concentrations between the boundary layer and the bulk feed solution. With higher rejection performance of the membrane, more solutes will be retained on the membrane surface, which again leads to the concentration of the boundary layer [89].

When concentration polarization occurs, the system suffers from several consequences. As is widely known, the principal of RO is that a pressure higher than the osmotic pressure of the feed solution is applied to produce clean water. However, with concentration polarization, osmotic pressure starts building up near the membrane surface because of the increase in concentration, which ultimately becomes higher than that of the bulk feed solution. This leads to the loss of effective pressure, which is the driving force for the system, and hence decreases the permeate flux [89]. Concentration polarization also accelerates the formation of the cake layer or gel layer on the membrane surface. As the phenomenon progresses, problems such as solute adsorption, scaling, fouling, and pore blocking, in the case of porous membrane systems, ensue, resulting in the overall degradation of process performance.

2.5 Detection of RO fouling potential

Fouling of the membrane has always been the most prevalent and damage-causing problem in the membrane filtration process. Although methods to eliminate predeposited foulants have been developed, predicting the fouling potential of the feedwater prior to operation is imperative to delay or prevent the fouling of the membrane. As means for prediction, the current water treatment systems implement several fouling indices to assess the quality of the feed and pretreated water, such as the silt density index (SDI) and the modified fouling index (MFI).

2.5.1 Silt density index

The SDI was the earliest of the fouling indices to have been developed and implemented in the water treatment field. The method initially measures the time required to collect 500 mL of feedwater using a filter with a pore size of 0.45 μm at a constant pressure of 207 kPa or approximately 2 bar. Under the same conditions, the feedwater is filtered for 15 min, and afterward, the time required to collect 500 mL of feedwater is measured again [90]. The SDI can then be calculated using the following equation:

$$\mathrm{SDI} = \frac{\left(1 - \frac{t_0}{t_{15}}\right)}{t} \times 100\%, \tag{2.36}$$

where t_0 and t_{15} are the initial time and time required after 15 min to collect 500 mL of feedwater, respectively, and t is the overall duration of the measurement, usually 15 min.

For the RO process, SDI <3 has been set as the recommended threshold for feedwater entering the RO stage. If the feedwater has an SDI value higher than 5, the water has a very high particulate foulant content and is not suitable for direct filtration by the RO membrane [91]. Pretreatment methods including media filtration and cartridge filters or microfiltration (MF)/ultrafiltration (UF) processes are mandatory to reduce the water's fouling propensity. For feedwater with an SDI value lower than 4, the solution is considered to have an intermediate level of fouling potential; such a solution will decrease membrane performance at a slow pace. At an SDI value lower than 3 or 2, the feedwater is considered to have a very low fouling potential, and if the water being measured is pretreated water, the pretreatment scheme can be said to be well-performing. An even lower SDI value is the most ideal situation and will result in close to no degradation of membrane performance [90].

Although the SDI has been widely adopted in both research and in the industry of the water treatment society because of its legal implications, the index has limitations in providing an accurate prediction of the fouling potential of water. First, the standard SDI testing procedure uses 0.45 μm filters, but water typically contains foulants of various sizes, and some particles are smaller than 0.45 μm. In other words, the fouling potential obtained through the SDI measurement can only be said to be partial because it disregards the fouling potential of foulants smaller than 0.45 μm, which are generally biological polymers that cause biofouling. Because smaller particles create more resistance in the fouling layer than bigger particles, the actual degree of fouling on the RO membrane is typically much higher than what the SDI value indicates. Second, the use of 0.45-μm filters indicates that the SDI is based on the assumption of a porous membrane process, which is not the case for RO membranes. As the procedure considers porosity, the fouling measurement considers both the pore-plugging and the cake-formation fouling mechanisms, but the RO membrane is only fouled by the formation of cake layers.

2.5.2 Modified fouling index

Because of the low reliability of the SDI, an MFI was developed for increased accuracy of fouling potential prediction. The initial version of the MFI, $MFI_{0.45}$, utilizes 0.45 μm filters under a constant operating pressure of 2 bar. First, the permeate volume is recorded every 30 s over a 15-min filtration period. Then, a graph is drawn in which t/V is plotted against V (t is the time in seconds to filter a volume of V in L) [92]. Because the MFI is based on the assumption that fouling occurs in the following three steps: (1) pore blocking, (2) formation of cake layer, (3) compression of cake layer, the graph can be represented as shown in Fig. 2–13. As the MFI was developed to be suitable for the RO process, the index can be finally obtained by calculating the slope of the graph in the cake filtration region using the following equations:

$$\frac{t}{V} = \frac{\mu \times R_m}{\Delta P \times A} + \frac{\mu \times \alpha \times C_b}{2 \times \Delta P \times A^2} \times V, \tag{2.37}$$

$$I = \alpha \times C_b, \tag{2.38}$$

$$\mathrm{MFI}_{0.45} = \frac{\mu \times I}{2 \times \Delta P \times A^2}. \tag{2.39}$$

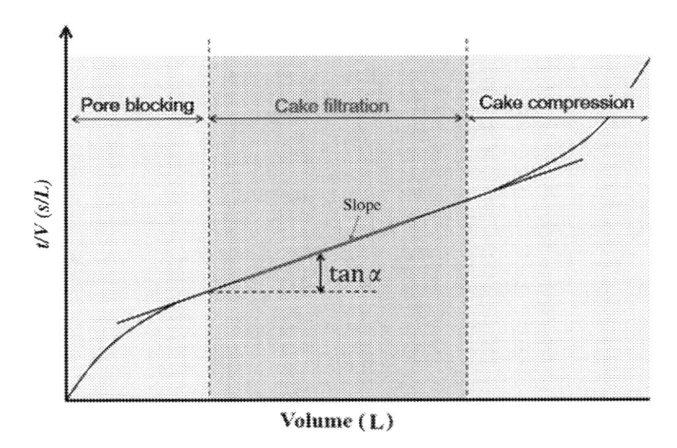

FIGURE 2–13 Schematic diagram of the MFI measurement. *MFI*, Modified fouling index.

However, further research into the MFI revealed that the $MFI_{0.45}$ value measured was often too low to explain the decrease in flux rates of the RO process because the use of 0.45-μm filters led to the neglect of fouling caused by smaller particles, as was the case with the SDI. Hence, a new updated MFI using smaller pore-sized UF membranes to incorporate the fouling potential of small particles was developed called the MFI-UF.

Recently, a concept of multiple MFIs was introduced to analyze the fouling potential of the feedwater in more detail because mechanisms involved in the fouling of a membrane are too complex to be explained by a single MFI measurement. The method, known as the multiple membrane array system (MMAS), utilizes MF, loose UF, and tight UF membranes in series with pore sizes of 0.45 μm, 100 kDa, and 10 kDa, respectively [93]. These three membranes are designed to separate and then measure the fouling potential of particles, colloids, and organic foulants in the feedwater, respectively. The resulting three MFI measurements are particulate-MFI, colloidal-MFI, and organic-MFI, and they have been proven to provide a much more accurate prediction of the fouling potential of water compared to the conventional SDI method or even the single MFI method. In addition, the measurements by the MMAS method precisely reflect the changes in water quality at various pretreatment steps and were well correlated with the trend in flux decline of the RO operation.

Despite the higher reliability, the MFI method has not been adopted as an industrial standard. This is perhaps because the SDI has been applied for a long time before the development of the MFI and because of its ease of measurement. However, the results of the MFI are much more accurate, and the method itself is flexible to be suitable for multiple situations [92]. Thus the wide application of the MFI is suggested to advance the overall operation of the water treatment process.

2.6 Mitigation of RO fouling

2.6.1 Pretreatment processes

Fouling of the membrane decreases both water productivity and solute rejection. To mitigate the impact of fouling on the membrane operation, several techniques are implemented according to the fouling which needs to be reduced (particulate/colloidal, organic, and biofouling and scaling). Among the four types, the most effective way to minimize particulate/colloidal and organic fouling is to remove the foulants that cause the respective fouling beforehand from the feedwater by performing a pretreatment process prior to the membrane separation stage.

The above-mentioned types of fouling occur by the deposition of particulate/colloidal matter or organics on the membrane surface. Because of their relatively large sizes, these foulants can be easily removed by the operation of mechanical and chemical pretreatment techniques [94]. The conventional combination of coagulation and flocculation followed by the sedimentation process is highly capable and shows significant removal capacity of large-sized foulants. The coagulant chemical added in the coagulation stage neutralizes the surface charge of the foulants, which adhere together to initially form microflocs [95]. Currently, the coagulants that are used can be divided into two categories, inorganic (metal) coagulants and polymers. Inorganic coagulants generally require a higher injection dose of 5–20 ppm, whereas polymers can be injected with a much smaller dose of 0.5–1 ppm [96].

With the administration of gentle to vigorous stirring by the flocculation process, more foulants attach to the microflocs and then form large flocs visible to the naked eye. Once agglomeration of the foulants has reached this stage, the water is flown to the sedimentation process where turbidity, suspended solids, and organic compounds are removed via settling [97]. Although the removal mechanism differs, the dissolved air flotation (DAF) process is also often operated in seawater desalination systems with the aim of attaining removal of suspended solids and organics [26]. Subsequently, the media filtration process is used to further polish the feedwater to meet the RO influent standards. For media filtration, particles too small to have been removed in the prior sedimentation or DAF processes are captured by adsorption onto the surface of the media. In the case of gravity media filtration, particles as small as 10–1 μm can be removed, while in the case of pressurized media filtration, particles as small as 0.25 μm can be removed [98].

Finally, as the last polishing stage, the UF membrane process is adopted to remove most particulate/colloidal foulants, thereby significantly lowering the SDI or MFI of the feedwater. Typical UF membranes used in seawater desalination pretreatment have a size of 10–100 kDa, which converts to an approximate pore size of 0.005–0.1 μm. As the size of particulate and colloidal foulants mostly exceeds that of the UF membrane pores, the permeate of the UF process has next to no particulate or colloidal fouling potential [99].

2.6.2 Membrane maintenance

Unlike particulate/colloidal and organic fouling, biofouling and scaling are caused by gradual development. In the case of biofouling, despite a possible 3-log removal of foulants by a well-designed pretreatment process, if only a single organism succeeds in reaching the RO membrane, it can multiply over time to eventually form biofouling on the entire membrane surface. Likewise, as complete removal of the dissolved ions before the RO stage via pretreatment processes is impossible, scales will eventually form on the membrane surface by the ions that have bypassed pretreatment. Conclusively, the optimal strategy to mitigate biofouling and scaling is not preremoval of the foulants but postremoval of fouling once it has occurred through various membrane maintenance and cleaning methods.

Membrane cleaning methods are very diverse but can commonly be divided into physical/chemical cleaning (flushing) and backwashing. However, the spiral-wound membrane, which is the usual membrane type for RO in real scale desalination plants, does not permit reverse flow; thus backwashing cannot be applied [85]. For flushing by physical means, the membranes are cleaned using either the feedwater or the permeate. The water is flown through the system at a higher crossflow velocity than that of the production stage, and rapid water flow causes foulants attached on the membrane surface to detach and be discharged. However, under biofouling circumstances, the organic foulants cover themselves in a mucus-like layer, the biofilm, which not only protects the foulants from the incoming water but also provides extreme adhesion, making detachment difficult. Thus the injection of cleaning chemicals to aid in the cleaning process is crucial.

For the chemical cleaning of the membrane, several agents are adopted and are currently used in multiple desalination plants. These agents are commercially available or directions for their formulas can be obtained from the membrane manufacturers. Generally, the chemicals can be divided into the following five categories: acids, alkalis, chelating agents, oxidants/disinfectants, and surfactants. Among them, alkalis and disinfectants can be used to remove fouling by organic and microbial foulants [100]. Caustic agents, which are alkali cleaning chemicals, hydrolyze organic materials like polysaccharides, the key components of the biofilm. Chlorine is another commonly injected chemical among the disinfectants used to effectively deactivate organics within the RO system. In the past, chlorine disinfection has been used as a pretreatment step because of its impact on the membrane integrity. However, desalination plants that have been operating for a long period have found that the intermittent or the chlorination/dechlorination method of injection has led to the survival of organics now immune to the disinfectant chemical. Therefore in-line shock dosing chlorination using a very high concentration of the chemical is suggested to effectively infiltrate the biofilm and deactivate the susceptible organic foulants within [101].

The removal of scales requires a different chemical cleaning process, which can be divided into acidification and antiscalant injection. Similarly to the effect of alkali chemicals in removing biofouling, the use of acidic chemicals can successfully remove scales because of their solubilization effect. Among the major types of scaling, acidic chemicals can effectively destroy carbonate ions, thereby preventing the formation of calcium carbonate. However, the acidification method causes corrosion of pipes and significantly lowers the pH of the RO permeate. Antiscalant injection is another effective method in preventing scale formation [102]. The agents are mostly synthetic organic compounds containing phosphonates, sulfonates, carboxylic acids, and chelating agents [81]. The antiscalant functions equally for all types of scales, and once it has been administered to the system, it prevents the reaction of carbonate and calcium species in the following three ways.

The first mechanism is "threshold inhibition" in which the antiscalants simply maintain the supersaturation level of salts in the water, so the ions remain in their dissolved forms. The second mechanism is "crystal modification" in which the negative charge of the antiscalant interferes with the electrical charge balance of the forming crystal. The interference causes the crystal scale to form in a less compact and nonadherent manner, and the scale can then be easily removed afterward. The last mechanism is "dispersion" in which certain antiscalants attach onto the surface of scales and impart an overall high negative charge. The coating separates scale crystals from each other because of charge repulsion and even sometimes prevent foulants with a negative surface charge from settling on the membrane surface [103].

Acknowledgment

This work was supported by the Korea Environment Industry & Technology Institute (KEITI) through the Industrial Facilities & Infrastructure Research Program, funded by the Korea Ministry of Environment (MOE) (1485016424).

References

[1] K. Reddy, N. Ghaffour, Overview of the cost of desalinated water and costing methodologies, Desalination 205 (2007) 340–353.

[2] S.S. Shenvi, A.M. Isloor, A. Ismail, A review on RO membrane technology: developments and challenges, Desalination 368 (2015) 10–26.

[3] N. Ghaffour, T.M. Missimer, G.L. Amy, Technical review and evaluation of the economics of water desalination: current and future challenges for better water supply sustainability, Desalination 309 (2013) 197–207.

[4] J. Kim, K. Park, D.R. Yang, S. Hong, A comprehensive review of energy consumption of seawater reverse osmosis desalination plants, Appl. Energy 254 (2019) 113652.

[5] K. Park, J. Kim, D.R. Yang, S. Hong, Towards a low-energy seawater reverse osmosis desalination plant: a review and theoretical analysis for future directions, J. Membr. Sci. 595 (2020) 117607.

[6] L.F. Greenlee, D.F. Lawler, B.D. Freeman, B. Marrot, P. Moulin, Reverse osmosis desalination: water sources, technology, and today's challenges, Water Res. 43 (2009) 2317–2348.

[7] C. Fritzmann, J. Löwenberg, T. Wintgens, T. Melin, State-of-the-art of reverse osmosis desalination, Desalination 216 (2007) 1–76.

[8] S. Miller, H. Shemer, R. Semiat, Energy and environmental issues in desalination, Desalination 366 (2015) 2–8.

[9] K. Park, H. Heo, D.Y. Kim, D.R. Yang, Feasibility study of a forward osmosis/crystallization/reverse osmosis hybrid process with high-temperature operation: modeling, experiments, and energy consumption, J. Membr. Sci. 555 (2018) 206–219.

[10] J. Kim, S. Hong, A novel single-pass reverse osmosis configuration for high-purity water production and low energy consumption in seawater desalination, Desalination 429 (2018) 142–154.

[11] Y.G. Lee, Y.S. Lee, J.J. Jeon, S. Lee, D.R. Yang, I.S. Kim, et al., Artificial neural network model for optimizing operation of a seawater reverse osmosis desalination plant, Desalination 247 (2009) 180–189.

[12] G. Xia, Q. Sun, J. Wang, X. Cao, Y. Yu, L. Wang, Theoretical analysis of a reverse osmosis desalination system driven by solar-powered organic Rankine cycle and wind energy, Desalin. Water Treat. 53 (2015) 876–886.

[13] Q. Ping, Z. Huang, C. Dosoretz, Z. He, Integrated experimental investigation and mathematical modeling of brackish water desalination and wastewater treatment in microbial desalination cells, Water Res. 77 (2015) 13–23.

[14] B. Nesbitt, Handbook of valves and Actuators: Valves Manual International, Elsevier, 2011.

[15] H.T. El-Dessouky, H.M. Ettouney, Fundamentals of Salt Water Desalination, Elsevier, 2002.

[16] M.S. Mohsen, O.R. Al-Jayyousi, Brackish water desalination: an alternative for water supply enhancement in Jordan, Desalination 124 (1999) 163–174.

[17] I.C. Karagiannis, P.G. Soldatos, Water desalination cost literature: review and assessment, Desalination 223 (2008) 448–456.

[18] M. Elimelech, W.A. Phillip, The future of seawater desalination: energy, technology, and the environment, Science 333 (2011) 712–717.

[19] N. Voutchkov, Energy use for membrane seawater desalination—current status and trends, Desalination 431 (2018) 2–14.

[20] Z. Li, R.V. Linares, S. Sarp, et al., Direct and indirect seawater desalination by forward osmosis, Membrane-Based Salinity Gradient Processes for Water Treatment and Power Generation, Elsevier, 2018, pp. 245–272.

[21] F. Kiand, Supply of desalinated water by the private sector: 30 MGD Singapore seawater desalination plant, in: MEDRC International Conference on Desalination Costing, Conference Proceeding, Lemesos, Cyprus, 2004.

[22] S. Lin, M. Elimelech, Staged reverse osmosis operation: configurations, energy efficiency, and application potential, Desalination 366 (2015) 9–14.

[23] M. Martinez-Mate, B. Martin-Gorriz, V. Martínez-Alvarez, M. Soto-García, J. Maestre-Valero, Hydroponic system and desalinated seawater as an alternative farm-productive proposal in water scarcity areas: energy and greenhouse gas emissions analysis of lettuce production in southeast Spain, J. Clean. Prod. 172 (2018) 1298–1310.

[24] T.P. Hendrickson, M. Bruguera, Impacts of groundwater management on energy resources and greenhouse gas emissions in California, Water Res. 141 (2018) 196–207.

[25] X. Chen, N.Y. Yip, Unlocking high-salinity desalination with cascading osmotically mediated reverse osmosis: energy and operating pressure analysis, Environ. Sci. Technol. 52 (2018) 2242–2250.

[26] L. Henthorne, B. Boysen, State-of-the-art of reverse osmosis desalination pretreatment, Desalination 356 (2015) 129–139.

[27] D.M. Davenport, A. Deshmukh, J.R. Werber, M. Elimelech, High-pressure reverse osmosis for energy-efficient hypersaline brine desalination: current status, design considerations, and research needs, Environ. Sci. Tech. Let. 5 (2018) 467–475.

[28] C. Boo, R.K. Winton, K.M. Conway, N.Y. Yip, Membrane-less and non-evaporative desalination of hypersaline brines by temperature swing solvent extraction, Environ. Sci. Technol. Lett. 6 (2019) 359–364.

[29] P.V. Brady, R.J. Kottenstette, T.M. Mayer, M.M. Hightower, Inland desalination: challenges and research needs, J. Contemp. Water Res. Educ. 132 (2005) 46–51.

[30] D.E. López, J.P. Trembly, Desalination of hypersaline brines with joule-heating and chemical pre-treatment: conceptual design and economics, Desalination 415 (2017) 49–57.

[31] J. Kim, J. Kim, J. Kim, S. Hong, Osmotically enhanced dewatering-reverse osmosis (OED-RO) hybrid system: implications for shale gas produced water treatment, J. Membr. Sci. 554 (2018) 282–290.

[32] J. Kim, J. Kim, J. Lim, S. Lee, C. Lee, S. Hong, Cold-cathode X-ray irradiation pre-treatment for fouling control of reverse osmosis (RO) in shale gas produced water (SGPW) treatment, Chem. Eng. J. 374 (2019) 49–58.

[33] S. Renou, J. Givaudan, S. Poulain, F. Dirassouyan, P. Moulin, Landfill leachate treatment: review and opportunity, J. Hazard. Mater. 150 (2008) 468–493.

[34] V. Karanikola, C. Boo, J. Rolf, M. Elimelech, Engineered slippery surface to mitigate gypsum scaling in membrane distillation for treatment of hypersaline industrial wastewaters, Environ. Sci. Technol. 52 (2018) 14362–14370.

[35] K. Park, D.Y. Kim, D.R. Yang, Cost-based feasibility study and sensitivity analysis of a new draw solution assisted reverse osmosis (DSARO) process for seawater desalination, Desalination 422 (2017) 182–193.

[36] J. Kim, D.I. Kim, S. Hong, Analysis of an osmotically-enhanced dewatering process for the treatment of highly saline (waste) waters, J. Membr. Sci. 548 (2018) 685–693.

[37] C.D. Peters, N.P. Hankins, Osmotically assisted reverse osmosis (OARO): five approaches to dewatering saline brines using pressure-driven membrane processes, Desalination 458 (2019) 1–13.

[38] H. Gong, Z. Yan, K. Liang, Z. Jin, K. Wang, Concentrating process of liquid digestate by disk tube-reverse osmosis system, Desalination 326 (2013) 30–36.

[39] F.P. Chinard, The definition of osmotic pressure, J. Chem. Educ. 31 (1954) 66.

[40] W. Zhang, J. Luo, L. Ding, M.Y. Jaffrin, A review on flux decline control strategies in pressure-driven membrane processes, Ind. Eng. Chem. Res. 54 (2015) 2843–2861.

[41] S. Kimura, S. Sourirajan, Analysis of data in reverse osmosis with porous cellulose acetate membranes used, AIChE J. 13 (1967) 497–503.

[42] H.-J. Oh, T.-M. Hwang, S. Lee, A simplified simulation model of RO systems for seawater desalination, Desalination 238 (2009) 128–139.

[43] M. Cheryan, Ultrafiltration and Microfiltration Handbook, CRC Press, 1998.

[44] G.M. Geise, H.B. Park, A.C. Sagle, B.D. Freeman, J.E. McGrath, Water permeability and water/salt selectivity tradeoff in polymers for desalination, J. Membr. Sci. 369 (2011) 130–138.

[45] N.Y. Yip, M. Elimelech, Performance limiting effects in power generation from salinity gradients by pressure retarded osmosis, Environ. Sci. Technol. 45 (2011) 10273–10282.

[46] S. Kim, E.M. Hoek, Modeling concentration polarization in reverse osmosis processes, Desalination 186 (2005) 111–128.

[47] I. Sutzkover, D. Hasson, R. Semiat, Simple technique for measuring the concentration polarization level in a reverse osmosis system, Desalination 131 (2000) 117–127.

[48] M. Qasim, M. Badrelzaman, N.N. Darwish, N.A. Darwish, N. Hilal, Reverse osmosis desalination: a state-of-the-art review, Desalination 459 (2019) 59–104.

[49] P. Luis, Fundamental Modeling of Membrane Systems: Membrane and Process Performance, Elsevier, 2018.

[50] J.R. Welty, C.E. Wicks, G. Rorrer, R.E. Wilson, Fundamentals of Momentum, Heat, and Mass Transfer, John Wiley & Sons, 2009.

[51] S. Sablani, M. Goosen, R. Al-Belushi, M. Wilf, Concentration polarization in ultrafiltration and reverse osmosis: a critical review, Desalination 141 (2001) 269–289.

[52] C.P. Koutsou, S.G. Yiantsios, A.J. Karabelas, A numerical and experimental study of mass transfer in spacer-filled channels: effects of spacer geometrical characteristics and Schmidt number, J. Membr. Sci. 326 (2009) 234–251.

[53] M. Shibuya, M. Yasukawa, S. Goda, H. Sakurai, T. Takahashi, M. Higa, et al., Experimental and theoretical study of a forward osmosis hollow fiber membrane module with a cross-wound configuration, J. Membr. Sci. 504 (2016) 10–19.

[54] E.M. Hoek, A.S. Kim, M. Elimelech, Influence of crossflow membrane filter geometry and shear rate on colloidal fouling in reverse osmosis and nanofiltration separations, Environ. Eng. Sci. 19 (2002) 357–372.

[55] C.O. Bennett, J.E. Myers, Momentum, Heat, and Mass Transfer, McGraw-Hill, New York, 1982.

[56] T. Ishigami, H. Matsuyama, Numerical modeling of concentration polarization in spacer-filled channel with permeation across reverse osmosis membrane, Ind. Eng. Chem. Res. 54 (2015) 1665–1674.

[57] J.P. Hartnett, Recent Advances in Heat and Mass Transfer, McGraw-Hill, 1961, p. 253.

[58] R.H. Notter, C. Sleicher, The eddy diffusivity in the turbulent boundary layer near a wall, Chem. Eng. Sci. 26 (1971) 161–171.

[59] W. Pinczewski, S. Sideman, A model for mass (heat) transfer in turbulent tube flow. Moderate and high Schmidt (Prandtl) numbers, Chem. Eng. Sci. 29 (1974) 1969–1976.

[60] A. Achilli, T.Y. Cath, A.E. Childress, Power generation with pressure retarded osmosis: an experimental and theoretical investigation, J. Membr. Sci. 343 (2009) 42–52.

[61] M. Sekino, Mass transfer characteristics of hollow fiber RO modules, J. Chem. Eng. Jpn. 28 (1995) 843–846.

[62] A. Kumano, H. Matsuyama, Analysis of hollow fiber reverse osmosis membrane module of axial flow type, J. Appl. Polym. Sci. 123 (2012) 463–471.

[63] D.Y. Kim, B. Gu, D.R. Yang, An explicit solution of the mathematical model for osmotic desalination process, Korean J. Chem. Eng. 30 (2013) 1691–1699.

[64] K. Park, D.Y. Kim, Y.H. Jang, M.-g Kim, D.R. Yang, S. Hong, Comprehensive analysis of a hybrid FO/crystallization/RO process for improving its economic feasibility to seawater desalination, Water Res. 171 (2020) 115426.

[65] D.Y. Kim, B. Gu, J.H. Kim, D.R. Yang, Theoretical analysis of a seawater desalination process integrating forward osmosis, crystallization, and reverse osmosis, J. Membr. Sci. 444 (2013) 440–448.

[66] D. Cohen-Tanugi, R.K. McGovern, S.H. Dave, J.H. Lienhard, J.C. Grossman, Quantifying the potential of ultra-permeable membranes for water desalination, Energy Environ. Sci. 7 (2014) 1134–1141.

[67] R.L. Stover, Seawater reverse osmosis with isobaric energy recovery devices, Desalination 203 (2007) 168–175.

[68] S.R. Osipi, A.R. Secchi, C.P. Borges, Cost analysis of forward osmosis and reverse osmosis in a case study, Current Trends and Future Developments on (Bio-) Membranes: Reverse and Forward Osmosis: Principles, Applications, Advances, Elsevier, 2019.

[69] N. Voutchkov, Desalination Engineering: Planning and Design, McGraw Hill Professional, 2012.

[70] N. Nada, T. Attenborough, Y. Ito, Y. Maeda, K. Tokunaga, H. Iwahashi, SWRO drinking water project in Shuqaiq: advanced BWRO, membrane oxidation, and scaling, IDA, Desalin. Water Reuse 3 (2011) 30–39.

[71] J. Kim, S. Hong, Optimizing seawater reverse osmosis with internally staged design to improve product water quality and energy efficiency, J. Membr. Sci. 568 (2018) 76–86.

[72] B. Peñate, L. García-Rodríguez, Reverse osmosis hybrid membrane inter-stage design: a comparative performance assessment, Desalination 281 (2011) 354–363.

[73] E. Spiritos, C. Lipchin, Desalination in Israel, Springer, 2013, pp. 101–123.

[74] B. Sauvet-Goichon, Ashkelon desalination plant—a successful challenge, Desalination 203 (2007) 75–81.

[75] M. Faigon, Y. Egozy, D. Hefer, M. Ilevicky, Y. Pinhas, Hadera desalination plant two years of operation, Desalin. Water Treat. 51 (2013) 132–139.

[76] N. Voutchkov, Pretreatment for Reverse Osmosis Desalination, Elsevier, 2017.

[77] G.P.S. Ibrahim, A.M. Isloor, R. Farnood, Reverse Osmosis Treatment Techniques, Fouling, and Control Strategies, Current Trends and Future Developments on (Bio-) Membranes, Elsevier, 2020, pp. 165–186.

[78] C.Y. Tang, T.H. Chong, A.G. Fane, Colloidal interactions and fouling of NF and RO membranes: a review, Adv. Colloid Interface Sci. 164 (2011) 126–143.

[79] R. Singh, N. Hankins, Emerging Membrane Technology for Sustainable Water Treatment, Elsevier, 2016.

[80] T.A. Saleh, V.K. Gupta, Nanomaterial and Polymer Membranes: Synthesis, Characterization, and Applications, Elsevier, 2016.

[81] A.F. Isamil, K.C. Khulbe, T. Matsuura, RO Membrane Fouling, Reverse Osmosis, Elsevier, 2019, pp. 189–220.

[82] A. Matin, Z. Khan, S.M.J. Zaidi, M.C. Boyce, Biofouling in reverse osmosis membranes for seawater desalination: phenomena and prevention, Desalination 281 (2011) 1–16.

[83] O. Sánchez, Microbial diversity in biofilms from reverse osmosis membranes: a short review, J. Membr. Sci. 545 (2018) 240–249.

[84] C. Xu, W.-C. Chin, P. Lin, H. Chen, M.-H. Chiu, D.C. Waggoner, et al., Comparison of microgels, extracellular polymeric substances (EPS) and transparent exopolymeric particles (TEP) determined in seawater with and without oil, Mar. Chem. 215 (2019) 103667.

[85] J.B. Kaplan, Biofilm dispersal: mechanisms, clinical implications, and potential therapeutic uses, J. Dent. Res. 89 (2010) 205–218.

[86] J. MacAdam, S.A. Parsons, Calcium carbonate scale formation and control, Rev. Environ. Sci. Bio. 3 (2004) 159–169.

[87] M.H. Al-Khaldi, A. AlJuhani, S.H. Al-Mutairi, M.N. Gurmen, New insights into the removal of calcium sulfate scale, in: SPE European Formation Damage Conference, Society of Petroleum Engineers, 2011.

[88] K. Hu, J. Dickson, Membrane Processing for Dairy Ingredient Separation, John Wiley & Sons, 2015.

[89] S. Bhattacharjee, Concentration Polarization: Early Theories, Water Planet, 2017.

[90] S.G.S. Rodriguez, Particulate and Organic Matter Fouling of Seawater Reverse Osmosis Systems: Characterization, Modelling and Applications. (UNESCO-IHE Ph.D. thesis), CRC Press, 2011.

[91] L.N. Sim, T.H. Chong, A.H. Taheri, S.T.V. Sim, L. Lai, W.B. Krantz, et al., A review of fouling indices and monitoring techniques for reverse osmosis, Desalination 434 (2018) 169–188.

[92] Y. Jin, H. Lee, C. Park, S. Hong, ASTM standard modified fouling index for seawater reverse osmosis desalination process: status, limitations, and perspectives, Sep. Purif. Rev. 49 (2020) 55–67.

[93] Y. Yu, S. Lee, K. Hong, S. Hong, Evaluation of membrane fouling potential by multiple membrane array system (MMAS): measurements and applications, J. Membr. Sci. 362 (2010) 279–288.

[94] N. Voutchkov, Desalination Engineering: Operation and Maintenance, McGraw Hill Professional, 2014.

[95] S.F. Anis, R. Hashaikeh, N. Hilal, Reverse osmosis pretreatment technologies and future trends: a comprehensive review, Desalination 452 (2019) 159–195.

[96] M. Badruzzaman, N. Voutchkov, L. Weinrich, J.G. Jacangelo, Selection of pretreatment technologies for seawater reverse osmosis plants: a review, Desalination 449 (2019) 78–91.

[97] S. Jiang, Y. Li, B.P. Ladewig, A review of reverse osmosis membrane fouling and control strategies, Sci. Total. Environ. 595 (2017) 567–583.

[98] N. Voutchkov, Seawater Pretreatment, Water Treatment Academy, 2010.

[99] M. Sillanpää, Natural Organic Matter in Water: Characterization and Treatment Methods, Butterworth-Heinemann, 2014.

[100] R.A. Al-Juboori, T. Yusaf, Biofouling in RO system: mechanisms, monitoring and controlling, Desalination 302 (2012) 1–23.

[101] H. Maddah, A. Chogle, Biofouling in reverse osmosis: phenomena, monitoring, controlling and remediation, Appl. Water Sci. 7 (2017) 2637–2651.

[102] Z. Amjad, K. Demadis, Mineral Scales and Deposits, Elsevier, 2015.

[103] Lenntech, Scaling and Antiscalants 2020.

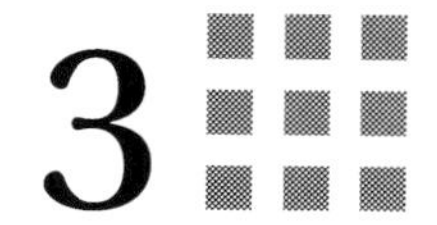

Principles of nanofiltration membrane processes

Yeit Haan Teow[1,2], Jing Yao Sum[3], Kah Chun Ho[4], Abdul Wahab Mohammad[1,2]

[1]RESEARCH CENTER FOR SUSTAINABLE PROCESS TECHNOLOGY (CESPRO), FACULTY OF ENGINEERING AND BUILT ENVIRONMENT, UNIVERSITI KEBANGSAAN MALAYSIA, BANGI, MALAYSIA [2]CHEMICAL ENGINEERING PROGRAM, FACULTY OF ENGINEERING AND BUILT ENVIRONMENT, UNIVERSITI KEBANGSAAN MALAYSIA, BANGI, MALAYSIA [3]DEPARTMENT OF CHEMICAL AND PETROLEUM ENGINEERING, FACULTY OF ENGINEERING, TECHNOLOGY AND BUILT ENVIRONMENT, UCSI UNIVERSITY, KUALA LUMPUR, MALAYSIA [4]FACULTY OF ENGINEERING, BUILT ENVIRONMENT, AND INFORMATION TECHNOLOGY, SEGI UNIVERSITY, SELANGOR DARUL EHSAN, MALAYSIA

3.1 Introduction

Nanofiltration (NF) membrane is a class of pressure-driven membrane, which is an intermediate of ultrafiltration (UF) membrane and reverse osmosis (RO). With the pore size ranging from 0.1 to 1 nm (equivalent to molecular weight cut off of 200–1000 Da), the NF membrane has a high removal rate for multivalent salt ions and organic solutes which have a molecular weight larger than 200 Da, as summarized in Table 3–1 [9]. Hence, the NF membrane is suitable to be used to remove hardness, heavy metals, and large organic matter such as dyes. However, it has a low rejection of monovalent ions and neutral solutes with a molecular weight below 200, which can be removed only through RO membranes [10].

The terminology of NF was introduced in the 1980s. Before this, it was recognized as loose RO or RO/UF hybrid. The first generation of NF was made from cellulose acetate (CA) and its derivatives, which was fabricated through the Loeb–Sourirajan process [11]. Cellulose membranes have been extensively used in seawater desalination. However, the use of cellulose-based RO membranes is extremely energy intensive due to high operating pressure. In the late 1960s, a loose cellulose RO membrane was produced by reducing the temperature of the heat treatment process [12]. The produced membrane demonstrated higher water permeability and removal selectivity toward divalent ions, compared to the membranes cured at a higher temperature. In the 1970s Israel Desalination Engineering demonstrated an RO/UF hybrid process

Osmosis Engineering. DOI: https://doi.org/10.1016/B978-0-12-821016-1.00014-0

Table 3–1 Types of the pressure-driven membrane and their applications.

Types of membrane	Pore size (nm)	Range of operating pressure (bar)	Targeted solutes	References
Microfiltration	50–10,000	0.1–2	Bacteria, colloids/particulate matter, coagulated organic matter	[1–3]
Ultrafiltration	1–50	1–5	Viruses, organic macromolecules	[4,5]
Nanofiltration	0.1–1	5–20	Multivalent ions such as hardness and metal cations dissolved organic carbon, dyes	[6–8]
Reverse Osmosis	Nonporous	10–100	Monovalent ions	–

that had a low rejection of sodium chloride (50%–70%), while having a distinctly high rejection of organic matter ($>$90%) [10]. The loose RO membrane has also been reported to be used for water softening in Florida in 1976 and for removing salts from food-grade dye in 1983 [13,14].

Nevertheless, the cellulose-based membranes have some limitations [15]. Basically, CA and its derivatives tend to be hydrolyzed to form cellulose after a period of time, which causes a loss of salt rejection. On the other hand, the asymmetric membrane is subject to compaction, which results in a loss in permeability. Compaction occurs due to the collapse of the porous structure within the membrane matrix, and it is more significant when high pressure is applied during the filtration process. Since the 1970s, Cadotte and Peterson [16] have developed RO composite membranes via interfacial polymerization (IP) method. They developed a thin-film composite (TFC) membrane based on IP using various types of monomer, including the use of piperazine (PIP) and trimesoyl chloride (TMC), which are the most popular monomers that have been used until today. In comparison with asymmetric cellulose-based membranes, the composite membrane exhibits improved water permeability while retaining high salt rejections. In 1988 the NF name was applied by Peter Eriksson, and since then NF membrane was commercialized by Filmtec Corporation [10].

NF membrane is nowadays usually made from a polymeric material, although some other ceramic-based NF made from alumina and zirconia have been developed for filtration under harsh conditions (high temperature, extreme pH, and presence of organic solvent) since the early 1990s [17,18]. Recently, graphene-based and graphene oxide (GO)-based NF membranes have been developed for desalination, pervaporation, and separation of organics from emulsions [19,20]. Nonetheless, most of the current NF membranes are TFC membranes. As illustrated in Fig. 3–1, typical TFC NF membrane exhibits a multilayered structure which consists of a microporous UF intermediate layer supported on nonwoven fabric, with an ultrathin selective layer mounted on top of it. The selective layer is dense and usually possesses some dissociated functional groups, forming a charged surface that can repel the co-ions in the feed through electrostatic repulsion. Polyamide (PA) has been frequently used as the material of the selective layer due to the excellent characteristics of heat resistance and stability in a wide range of pH. However, it has poor tolerance to chlorine and is susceptible to fouling. Hence, many other types of materials have been developed to improve the

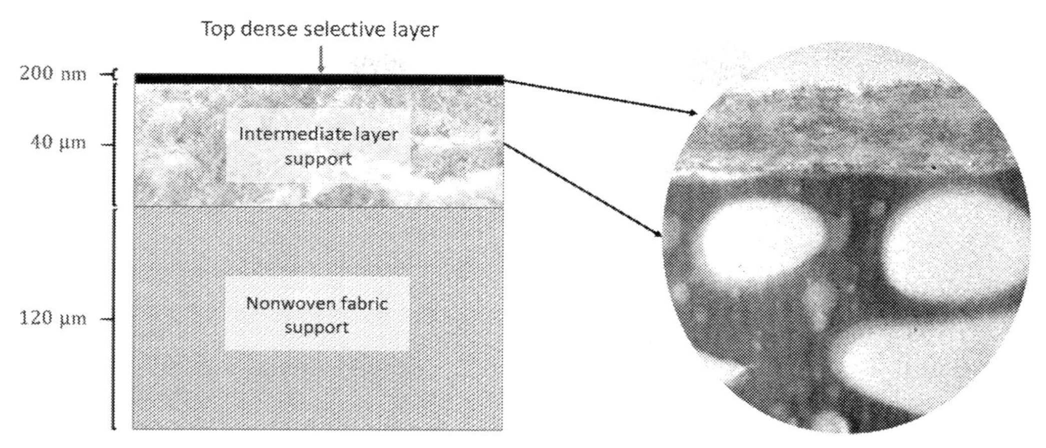

FIGURE 3–1 Schematic cross-section of a TFC NF membrane. *TFC*, Thin-film composite; *NF*, nanofiltration.

performances of the membrane, for instance, polyester, polyetheramide, polyvinyl alcohol, polyurea, and nanocomposite materials [21].

3.2 Basic principle of NF membrane separation process

NF membrane presented as TFC membrane has a multilayered structure, where only the top layer determining the selectivity of the membrane. The selective layer of the NF membrane is dense and consists of dissociated amine and carboxylic acid groups, which makes the membrane surface charged. This makes the separation mechanism of the NF membrane far more complicated than other pressure-driven membranes, as it has a combination of the steric effect, Donnan effect, dielectric effect, transport effect, and adsorption effect between the solutes, fixed charges, and active functional groups on the membrane surface.

3.2.1 Steric effect

Similar to other pressure-driven membranes, steric effect or size exclusion is the predominant rejection mechanism involved in the NF membrane. The transport of uncharged solutes is reasonably well established through numerous studies of UF membranes [22] while the NF studies on the transport of uncharged species have been well elaborated by López et al. [23]. Any solutes with a size larger than the membrane's pore can be effectively removed via size exclusion, whereas smaller solutes (such as water molecules with a diameter of 2.75 Å) may easily pass through the membrane [24]. Fig. 3–2 illustrates the working mechanism of size exclusion in an NF membrane.

The size-exclusion mechanism plays a significant role during the filtration of a neutral solute such as saccharides (glucose, sucrose, and raffinose) [25], polyethylene glycol (PEG) [26], hormones [20], and phenolic compounds [27]. The data obtained throughout the

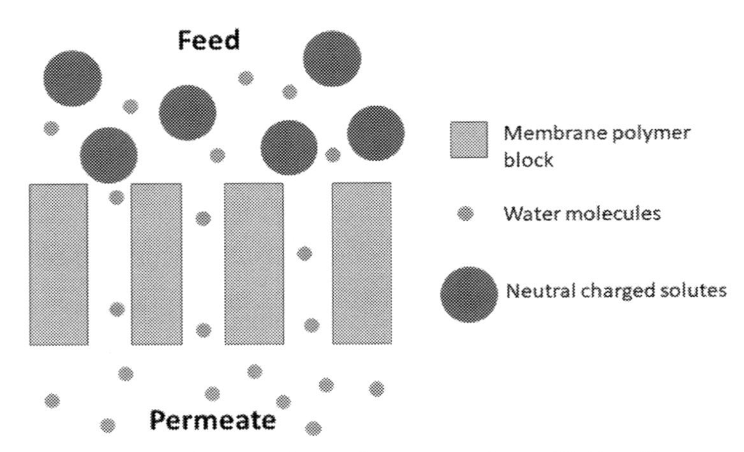

FIGURE 3–2 Schematic of the size-exclusion mechanism in an NF membrane. *NF*, Nanofiltration.

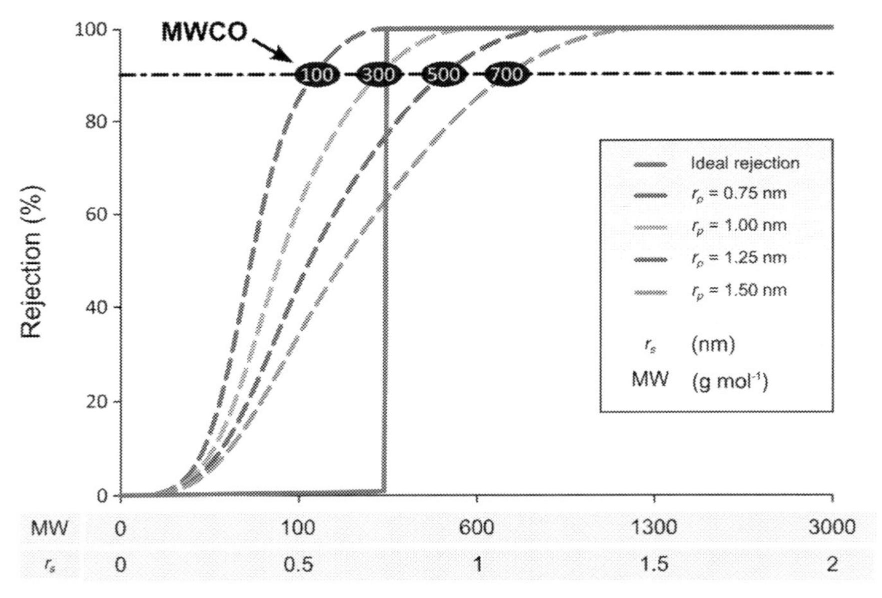

FIGURE 3–3 Rejection profile for uncharged solutes by NF membrane [30]. *NF*, Nanofiltration.

filtration of neutral solute especially PEG and saccharides give important information on determining the molecular weight cut-off and the mean pore size of the membrane [28,29]. In general, NF membranes exhibited sigmoidal curve for rejection as a function of molecular weight as shown in Fig. 3–3 [30]. Fig. 3–3 illustrates the schematic representation of rejection profile for uncharged solutes by NF membrane with uniform pore sizes between 0.75 and 1.5 nm. The dashed lines were obtained by simulation while the solid line represents the desired rejection profile. It was worth to notice that feasible separation of compounds based on molecular weight is only possible with large differences in rejection.

There are several works that reported the influence of solution pH toward the pore opening and the sieving mechanism of the membrane. A slight reduction of rejection and increase of water permeability were observed when the filtration is performed at an elevated pH (in alkaline condition) [31]. This is due to the membrane's pores swelling induced by strong electrostatic repulsion of negative charges residue in the internal pores. Dalwani et al. [32] suggested that high pH leads to the increasing membrane's pore size and effective membrane thickness, with an overall decrease in water flux was observed. However, the finding is contradicted by some other works which they claimed that the pore is larger when pH is at the isoelectric point (IEP) of the membrane [33]. Besides, the temperature of the feed is another factor contributing to the changes in the membrane pore dimension. Sharma et al. [34] reported that increasing temperature could enlarge the membrane pores, which subsequently reduces the steric hindrance of solute. Increasing temperature also reduces the kinematic viscosity and increases the solute diffusivity, hence the rejection of solute will decline overall [35]. On the other hand, the presence of inorganic salts could induce swelling of the membrane pores (only observed in organic membranes) and salting-out effects on the organic solute. In addition, Escoda et al. [36] found that the rejection of PEG is reduced with the increasing concentration of inorganic salts (KCl, LiCl, $MgCl_2$, and K_2SO_4) in the feed which is explained by the increasing mean pore size and size reduction of PEG molecules.

Only a few reports are available to specifically describe the improvement of the NF membrane by strengthening the size-exclusion mechanism. Du et al. [37] modified the membrane pore wall by selectively reducing the size of the large pores through the codeposition of 3-mercapto-1-propanesulfonate. The membrane, which has narrower pore size distribution, exhibits high monovalent ion/divalent ion selectivity. Saenz De Jubera et al. [38] demonstrated a modification of commercial NF membranes using aramide dendrimer. The modified membrane has a tighter pore structure, shows improvement in rejecting low-molecular-weight organic solutes (Rhodamine WT). Nonetheless, a membrane for which the size-exclusion mechanism predominates usually possesses a narrow pore size that sacrifices its water permeability. On the other hand, the permeation flux may decrease further during filtration due to the cake formation on the membrane surface which requires a frequent backwash to restore the membrane performance.

3.2.2 Donnan effect

In addition to size exclusion, the Donnan effect or electrostatic repulsion plays a vital role in determining the selectivity on the ionic solute. Ions usually have a hydrated radius much smaller compared to the membrane's pore; hence, these ions may not be effectively rejected based on size-exclusion mechanisms. Nonetheless, the top dense layer of the NF membrane usually carries charges. The dissociation of a carboxylic acid group forms a negatively charged carboxyl group ($-COO^-$), and the protonation of an amine forms a positively

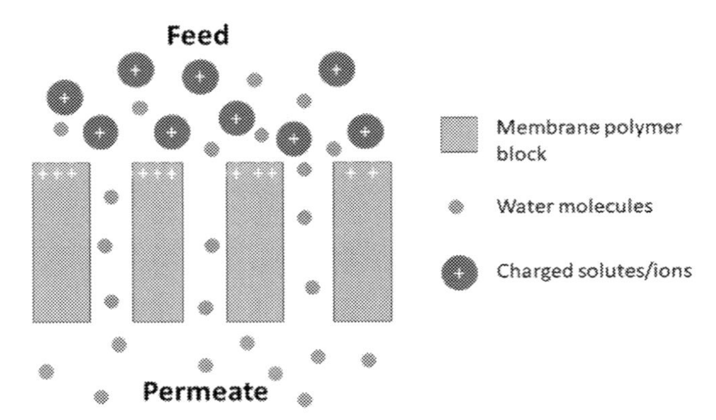

FIGURE 3–4 Schematic of electrostatic repulsion on a positively charged membrane surface.

charged ammonium group (-NH_3^+). Fig. 3–4 illustrates the Donnan potential developed on the membrane surface and how it works in repelling co-ions in the feed. First, an equilibrium exists when the membrane (with the dissociable functional group) is soaked in a salt solution [39]. Since the counter ion is attracted to the fixed charges on the membrane surface, the membrane surface will contain a higher concentration of counter ions than in the bulk phase, while the counter ion has a slightly lower concentration adjacent to the membrane surface. Hence a potential difference is developed at the membrane-bulk interphase, which is called the Donnan potential. Donnan potential plays a significant role in the filtration of ionic species, whereby the counter ion species can be repelled through electrostatic repulsion. The counter ion will be rejected simultaneously to maintain the electroneutrality of the solution. Epsztein et al. [40] examined the role of ionic charge density in the rejection of monovalent anions by negatively charged NF membrane. Their studies suggested that the rejection of fluoride (F^-) is higher than chloride (Cl^-) owing to the higher ionic anion charge density of F^- (−465 kJ/mol) than Cl^- (340 kJ/mol) hence stronger electrostatic repulsion as shown in Fig. 3–5.

The selectivity of the membrane depends on the nature of the membrane and the filtration operating conditions such as feed solution pH and ionic strength. Depending on the types of membrane material and solute in the feed, a change in pH could induce changes in membrane charge properties and the ionic speciation of the solutes. The membrane acquires positive charge at pH values below its IEP and turns negative above its IEP. Commercial NF membrane (NF270) has an impressive rejection of multivalent heavy metal cations below its IEP but has lower metal rejection above the membrane's IEP [41]. Meanwhile, a negatively charged NF membrane was found to be able to reject arsenic effectively at pH values above 7, as divalent anionic $HAsO_4^{2-}$ was the dominant species presented in solution [42]. However, the rejection of salt is minimal at the IEP of the membrane. Only the sieving effect was dominant when the membrane surface is in a neutral state during IEP, hence a lower rejection on salts was observed.

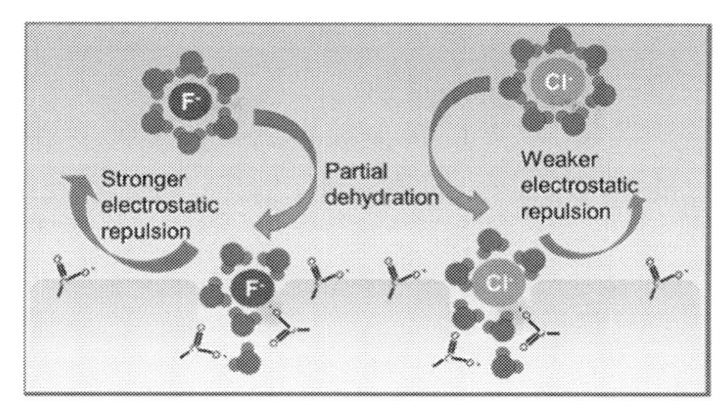

FIGURE 3–5 Proposed mechanism for the role of ionic charge density in Donnan effect by NF membrane [40]. *NF*, Nanofiltration.

On the other hand, ionic strength could impact the selectivity of the membrane. A few works have reported that the rejection of solute declined when there was high ionic strength in the feed. At high ionic strength, the membrane surface is shielded with counter ions that have been attracted onto the membrane surface through electrostatic interaction. This causes the electrostatic repulsion effect to become negligible [43]. The phenomenon is known as the screening effect.

3.2.3 Dielectric effect

The dielectric effect refers to the solvation energy barrier formed when an ion passes from a solvent of one dielectric constant to a solvent of a different dielectric constant. This could happen during the entry of ions at the membrane/solution interface to the confined nanopore suggested by Born theory [44]. Recently, Zhu et al. [45] have investigated the influence of dielectric effect on inhomogeneous fixed charge distribution in negatively charged NF membrane using $CaCl_2$. Their study suggested that electric field behavior and ion concentration distribution inside the nanopore reflect the variation of dielectric effect. Consequently, the dielectric effect affects the repulsive force to co-ions and hence the rejection performance. Typically, transport phenomena of NF process combine the steric effect, Donnan effect, and dielectric effect for a range of separations of mixtures of neutral or charged molecules, as shown in Fig. 3–6 [46].

3.2.4 Transport effect

Transport effect refers to both the convective and diffusive elements in the membrane boundary layer, as shown in Fig. 3–7A. Peclet number, a dimensionless number of ratio of the convective transport J_v and diffusive transport D_i/δ, is commonly used to describe the transport effect in NF membrane [48]. When the value of Peclet number is large ($J_v > D_i/\delta$), the convective flux through the membrane cannot easily be balanced by diffusion in the boundary layer,

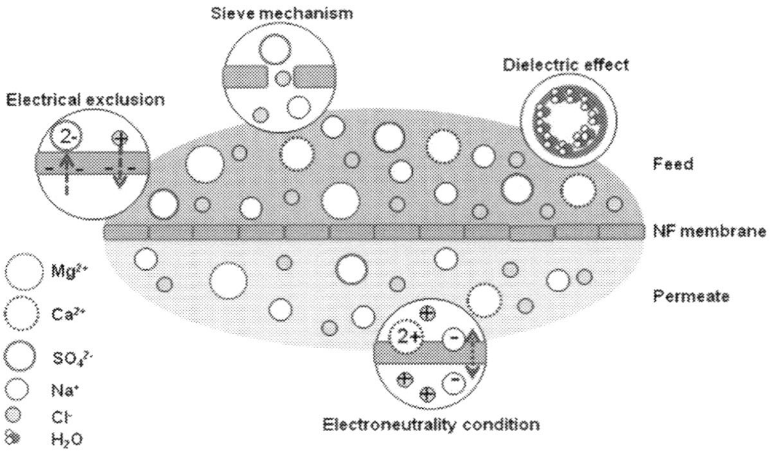

FIGURE 3–6 Schematic illustration of transport phenomena in the NF membrane [46]. *NF*, Nanofiltration.

hence the concentration polarization modulus is high, and vice versa. Fig. 3—7B shows the schematic illustration of external concentration polarization build-up of salt on NF membrane surface. In most cases, concentration polarization is unfavorable to the NF process as it can significantly affect the membrane performance, such as the rapid decline of rejection performance. To mitigate concentration polarization, turbulent mixing of feed stream, modification of membrane module design, and channel spacer geometries can be applied [49].

3.2.5 Adsorption effect

NF membranes can perform separation via adsorption, in addition to size exclusion and electrostatic repulsion. A solute in the feed stream is selectively adsorbed onto the membrane surface via several mechanisms such as surface complexation, electrostatic interaction, or ion exchange. Membranes that are able to perform adsorption should possess a rich amount of active functional groups, such as carboxyl (-COOH), amine (-NH_2), or sulfonate (-SO_3H), which serve as active sites for adsorption. Depending on the strength of the interaction between the solute and the membrane, the captured solutes can be released from the membrane surface/matrix through washing and hence both the solutes and membrane are regenerated for further use.

The membrane which performs adsorptive filtration is more commonly seen in the UF process, where the membrane is produced through the blending of complexing agents into membrane matrix or through surface modification of a membrane substrate. Complexing agent such as poly(ethyleneimine) (PEI) [50], 2-methacryloylamidohistidine, polyacrylic acid [51], chitosan [52], Poly(Amidoamine) dendrimer-like particles [53], Cibacron Blue F3GA [54], and inorganic filler [55] has been reported to produce a composite adsorptive membrane. These membranes, which are enriched with active functional groups, have great affinities toward aqueous metal ions in aqueous solutions. They can be repeatedly used without a loss in adsorption capacity and complexing agents from the polymer matrix. Prior to the

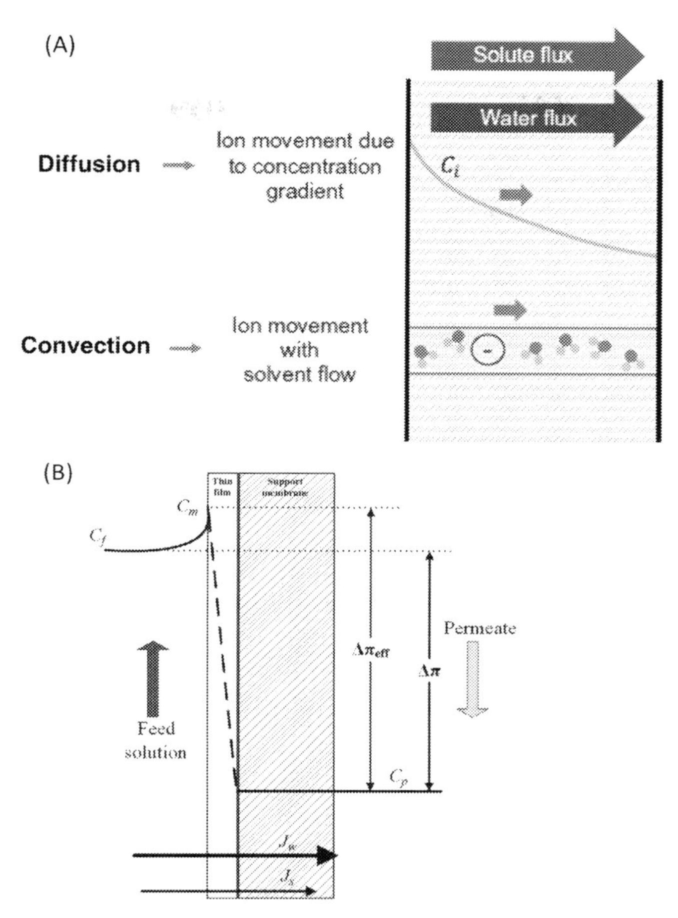

FIGURE 3–7 Schematic illustration of (A) solute transport effect across NF membrane [35], (B) concentration polarization on NF membrane [47]. *NF*, Nanofiltration.

development of the adsorptive composite membrane, the complexing agent is freely dissolved in the feed solution forming a complex with a targeted solute. The formed metal complex is then removed from the solution using a filtration process which is commonly known as micellar-enhanced UF. An adsorptive type membrane simplifies the process by combining both the adsorption and filtration into single steps.

Recently, the concept of adsorptive filtration has been extended to NF. In contrast with UF, the top selective layer of the NF membrane has been tailored and functionalized to make it have a great affinity toward solutes. Sum et al. [56] reported an NF membrane prepared through comonomeric IP using a mixture of PIP and polyamidoamine. The amine-rich and dendritic-like surface membranes are able to selectively adsorb metal cations, such as copper and chromium (IV), from a mixture. The authors found that the membrane showed preferential adsorption of copper over chromium and cadmium, whereby the copper forms a crystal layer on membrane surface which was explained through the Hard Soft acid–base principle. A polyamidoamine dendronized hollow fiber membrane was found to be able to capture heavy metals such as copper, lead, and cadmium, whereby a hydroxide metal crystal was grown on the membrane surface through immersion of membrane into the metal solutions [57].

Adsorption of metal solutes onto the membrane is greatly influenced by the pH of the solution. Low adsorption capacity of metals on the membrane was observed when the filtration was performed in the acidic range [41]. Small and highly diffusive protons (H^+) present at acidic pH are competitively adsorbed onto the membrane surface, which causes less available sites for binding. The membrane surface turns positive, which makes the repulsive forces between the metal cations and the membrane surface increase. However, insoluble metal hydroxide starts to form in a high pH range and is easily deposited on the membrane surface forming a thick scale that creates more hydraulic resistance. Several reports have found that the adsorption capacity of metal reaches a maximum at the IEP of the membrane. In this pH range, the membrane surface is neutral and hence repulsive forces between the solute and membrane are minimal.

It is arguable whether adsorption offers advantages or disadvantages to the membrane filtration process. It is not deniable that adsorption could help to enhance the selectivity. However, the adsorption could bring some adverse effect to the overall efficiency of the filtration process. It is known that the solute is transported across the membrane through the solution-diffusion mechanism. Adsorption of solute onto the membrane surface induces a higher concentration gradient across the membrane, which in fact could promote the partition and diffusion of the solutes across the membrane. In the meantime, the adsorption of solutes onto the membrane potentially creates a concentration polarization layer near the membrane surface or even forms a fouling layer that blocks the water passage across the membrane [58]. The membrane will eventually suffer from flux decline during the filtration process. A washing (through physical and chemical approach) is necessary to remove the fouling layer and restore the membrane performance.

Adsorption of the organic substances onto the membrane progresses through several mechanisms. Hydrophobic interaction is the major contribution to the adsorption of organic solutes onto the membrane surface. Hormone adsorption to the membrane due to hydrophobic interaction is dominant at the beginning of filtration [59]. However, when the membrane is saturated, size exclusion is the predominant separation mechanism. The adsorption drives the organic solutes to partition and diffuses across the membrane matrix which causes a decline in rejection rate [20]. A similar observation was reported when NF 90 and NF 200 were used to treat a feed that contains hydrophobic pharmaceutical residue ibuprofen. A minimal rejection was observed when the value of solution pH was below the pK_a of ibuprofen due to the adsorption and partitioning of ibuprofen through the membrane [60].

3.3 Synthesis and modification of NF membrane

In general, NF membranes can be synthesized using ceramic materials and polymer materials. However, the ceramic-based NF membranes are usually more expensive and fragile than polymer-based NF membranes. On the other hand, polymer-based NF membrane is more flexible and could vary widely in their characteristics according to the preparation conditions such as the concentration of polymer, blending speed, temperature, and nonsolvent selection [61]. This is supported by the number of published studies from 1996 to 2019 shown in Fig. 3–8. It can be seen that the number of published studies on polymer-based NF is almost fourfold higher than those for ceramic-based NF and expected to increase with time. Therefore this section will emphasize the synthesis and modification of NF membrane fabricated from polymeric materials only. In general, the synthesis method can be categorized into phase inversion, IP, and grafting polymerization as discussed in the following sections.

3.3.1 Phase inversion

Phase inversion is a process of fabricating membranes in which an initially homogeneous polymer solution is transformed into a solid structure through a complicated, yet controlled procedure. Fig. 3–9A shows the nonsolvent-induced phase separation (NIPS) which is commonly used for the fabrication of asymmetric NF membranes and the produced NF membrane greatly depends on the membrane fabrication method, as depicted in Fig. 3–9B [63]. During phase inversion, the polymer is first dissolved in solvent with typical concentration of 15–18 wt.% followed by phase inversion in the nonsolvent coagulation bath

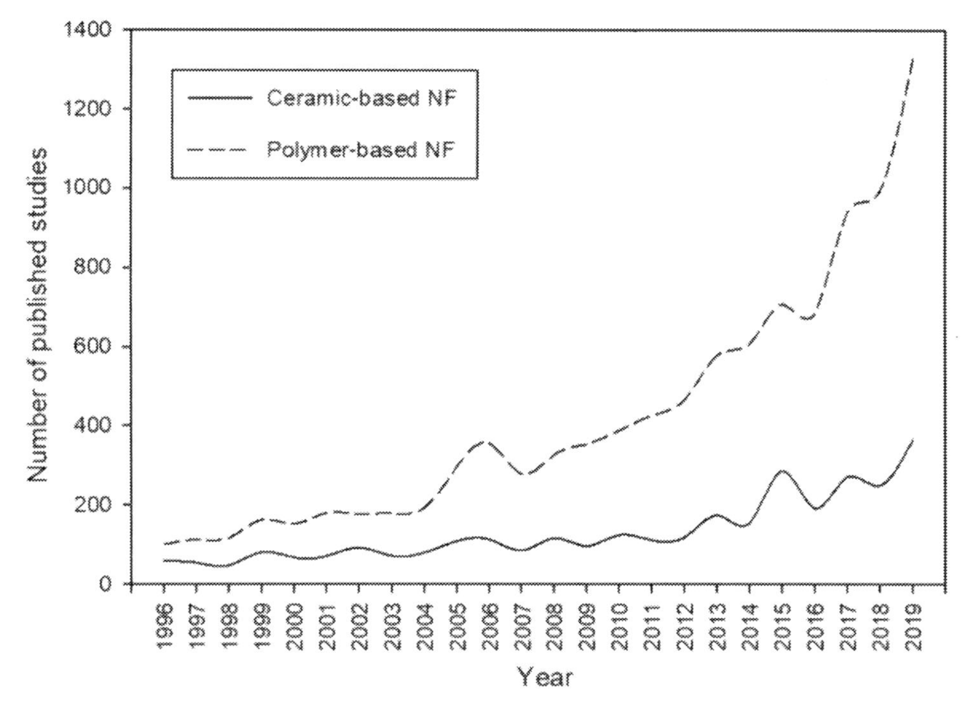

FIGURE 3–8 Number of published research studies linked to the topic of "ceramic nanofiltration" and "polymer nanofiltration" adopted from ScienceDirect database.

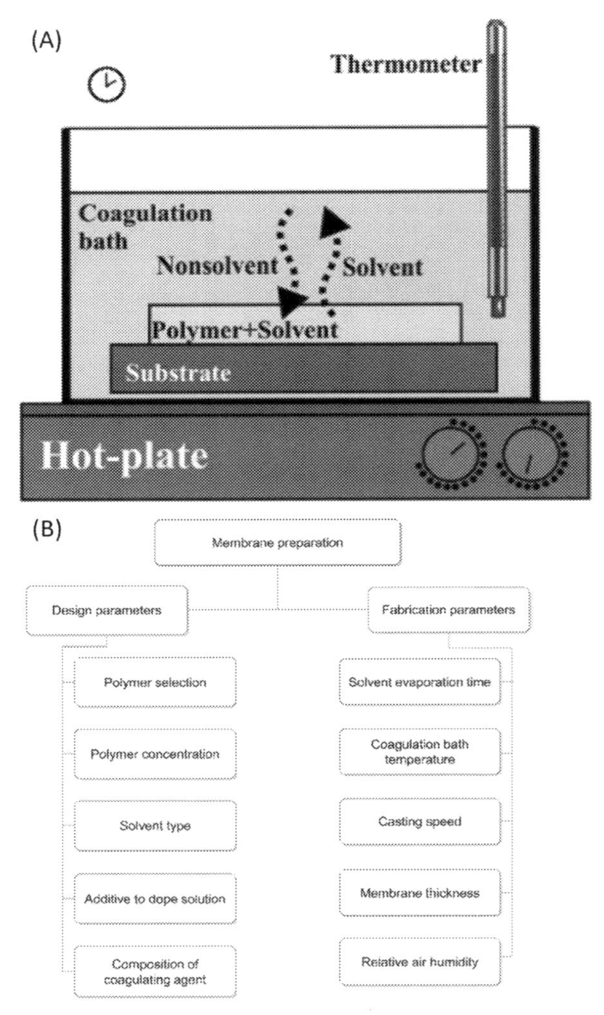

FIGURE 3–9 (A) Schematic illustration of formation of NF membranes via NIPS [62], (B) Classification of prominent design and fabrication parameters involved in the preparation of NF membranes via phase inversion [63]. *NF*, Nanofiltration; *NIPS*, nonsolvent-induced phase separation.

[49,64]. In some cases, a backing layer, usually polyester nonwoven fabric is used to provide mechanical support for the NF membrane.

Typical polymer concentration reported in the literature is 21–30 wt.% depending on the type of polymer and solvent [65]. Shahmirzadi et al. [63] investigated the effects of polymer concentration, solvent type, additives in the dope solution, and composition of coagulating agent to produce poly(ethersulfone) (PES) NF membrane. It was found that polymer concentration of 30 wt.% could offer maximum salt rejection (99.84%) and the water flux could be enhanced by the addition of 1 wt.% citric acid as hydrophilic acids. The research trend shows the incorporation of nanomaterials into the NF membrane during phase inversion. Some examples of the nanomaterials are chitosan-montmorillonite, GO, titanium dioxide (TiO_2), zwitterionic functionalized GO, and metal-organic frameworks (MOFs) [3,65–67]. Gholami and Mahdavi [68] embedded GO and sulfonated GO (SGO) into the PES NF membrane via phase inversion. The study reported that the NF membrane blended with SGO showed higher antifouling ability and rejection of heavy metals toward Cr (97.5%), Cd (87.8%), Cu (76.3%), and Ni (60.3%). This was due to increased surface hydrophilicity represented by lower contact angle (58.4 degrees) and smoother membrane surface demonstrated by lower mean roughness (39.69 nm) compared to the GO-modified membrane and pristine membrane.

3.3.2 Interfacial polymerization

IP is generally performed by polymerizing an extremely thin film layer of polymer at the surface of a microporous support polymer to produce TFC membranes as shown in Fig. 3–10. The microporous support polymer is usually formed by phase inversion as described in Section 3.3.1. During IP, a polycondensation reaction between two monomers is carried out at the boundary of immiscible solutions containing aqueous phase and organic phase. First, the aqueous phase monomer solution, which often comprises PIP or m-phenylenediamine (MPD), is absorbed in the microporous support. Subsequently, the organic phase monomer solution, which usually comprises TMC, is allowed to contact with the saturated porous support. An extremely thin PA barrier layer is formed rapidly at the interface. The schematic diagram of the formation of PA thin film using MPD and TMC monomers is shown in Fig. 3–11A while the schematic representation of the NF membrane synthesized by IP is shown in Fig. 3–11B.

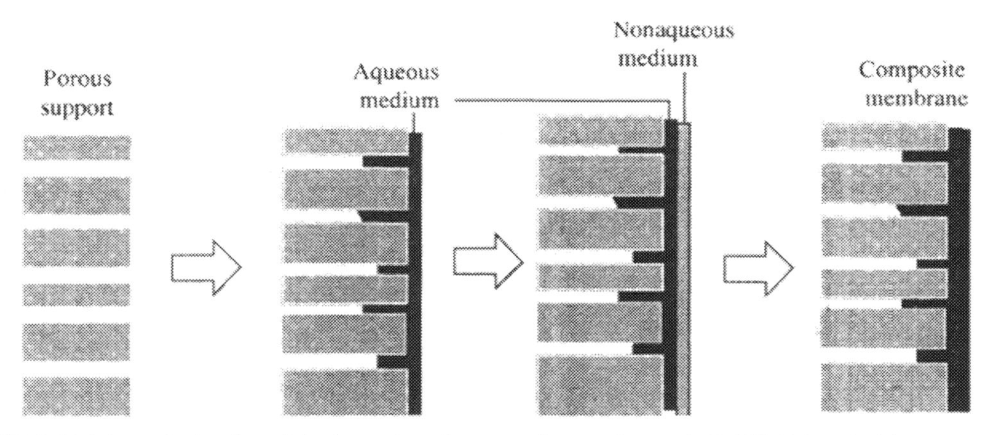

FIGURE 3–10 Schematic drawing of the formation of a composite membrane via IP [69]. *IP*, Interfacial polymerization.

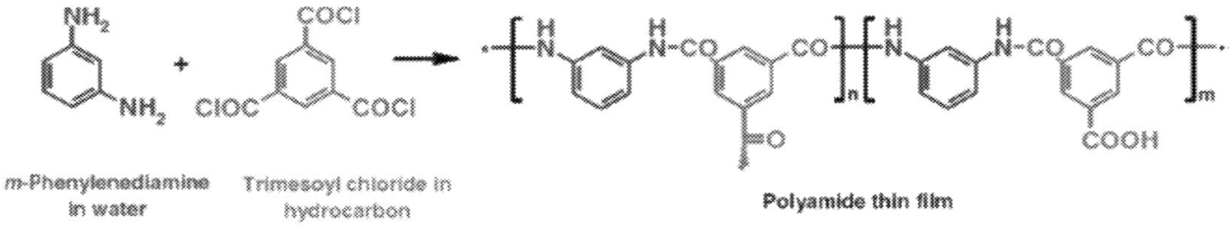

(A)

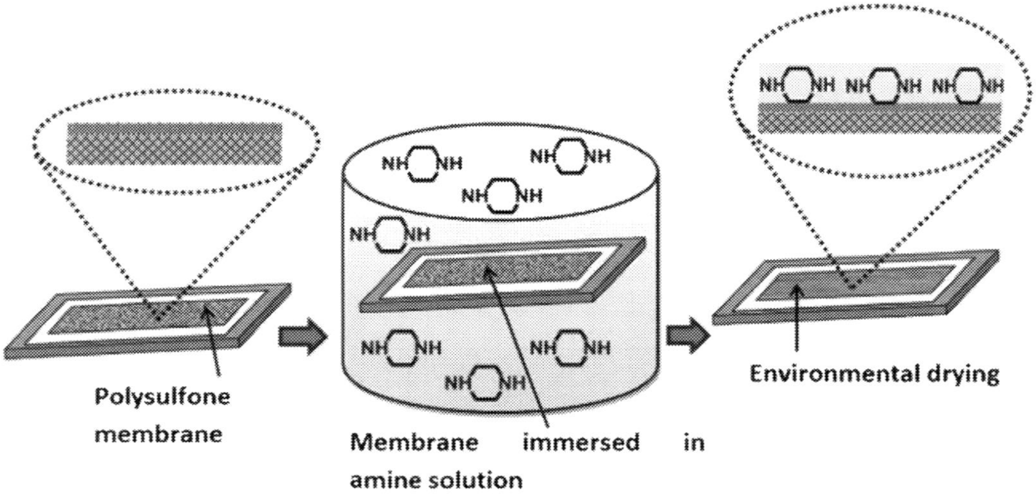

(B)

FIGURE 3–11 (A) Schematic illustration of IP using MPD/TMC monomers [70], (B) schematic representation of NF membrane preparation by IP [71]. *IP*, Interfacial polymerization; *MPD*, m-phenylenediamine; *TMC*, trimesoyl chloride; *NF*, nanofiltration.

3.3.2.1 Monomer

Monomer plays an important role in the formation of NF membrane via IP by controlling membrane pore dimension, chemical resistance, membrane hydrophilicity, active layer thickness, and membrane roughness [72]. Seman et al. [73] studied the effect of different bisphenol A (BPA) concentrations in the aqueous solution. It was observed that the synthesized membrane is more susceptible to irreversible fouling at BPA concentrations of 2% wt./vol. owing to the uniform and dense layer formed by the membrane structure. Besides, other monomers such as zwitterionic amide (N-aminoethyl PIP propane sulfonate, AEPPS), diethylenetriamine (DETA), triethylenetetramine, tetraethylenepentamine, 2,2′-oxybis-ethylamine, pentaerythritol, 4-aminobenzoic acid, 6-aminocaproic acid, 3-aminopropanoic acid, dendrimer trimesoyl amide amine, and 2,2′-bis(1-hydroxyl-1-rifluoromethyl-2,2,2-triflutoethyl)-4,4′-methylenedianiline were also studied to investigate its effects on NF performance [74–81]. Due to the toxicity of amines used in the NF synthesis, Pérez-Manríquez et al. [82] successfully replaced it with the naturally occurring bio-polyphenol morin. The synthesized NF membrane had a permeability of 0.3 L/m^2 h and a rejection of 96% with Brilliant Blue dye. More recently, Rezania et al. [72] synthesized two new sulfonated diamine-diol (SDA) and carboxylated diamine-diol (CDA) monomers to produce the TFC NF membrane. It was reported that the NF membrane produced by SDA/PIP showed the highest salt rejection (97% Na_2SO_4) with flux (50 L/m^2 h) attributed to the hydrophilic sulfonic acid, carboxylic acid, and hydroxyl groups on the NF membrane. Fig. 3–12 shows the two-step synthesis procedure of SDA and CDA diamine monomers.

3.3.2.2 Additives

Other than using different monomers, additives are added to the membrane preparation to enhance the membrane performance in terms of flux, fouling propensity, and separation efficiency. Fig. 3–13 shows the additives used for NF membrane modification: silica (SiO_2),

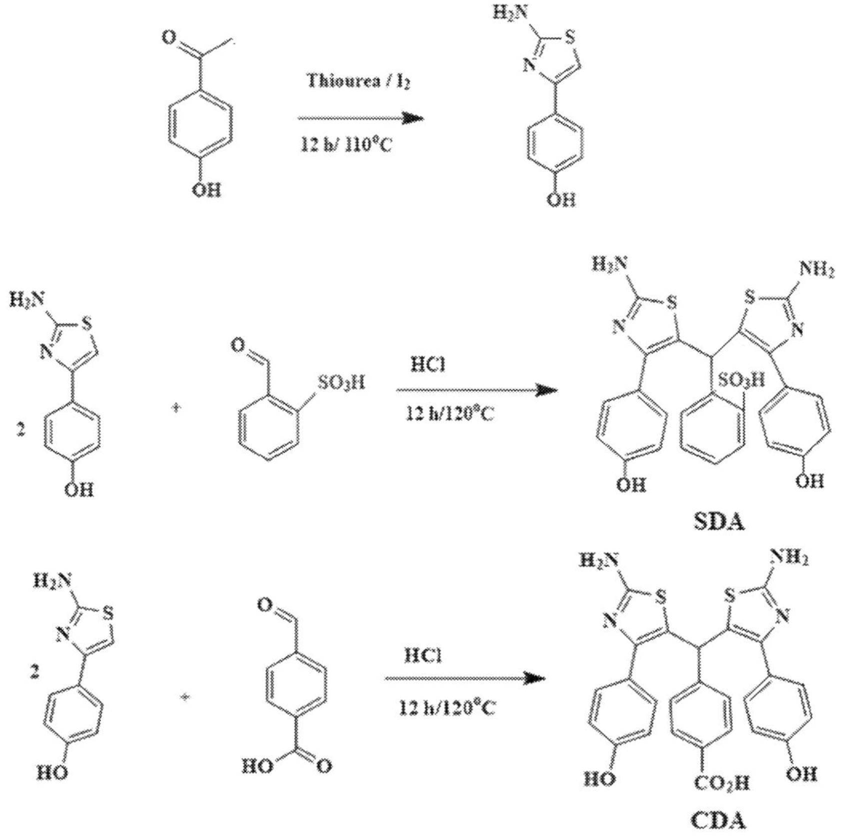

FIGURE 3–12 Two-step synthesis procedure of SDA and CDA diamine monomers [72]. *SDA*, Sulfonated diamine-diol; *CDA*, carboxylated diamine-diol.

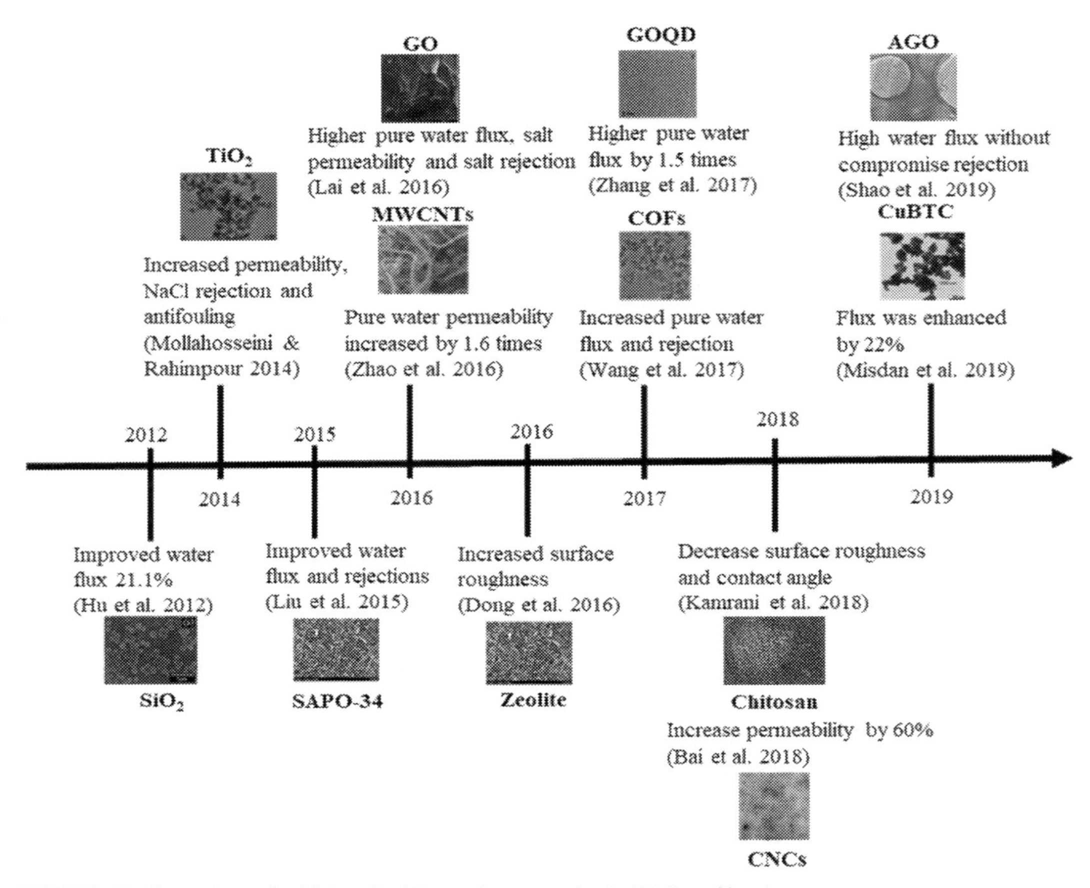

FIGURE 3–13 Chronology of additives for NF membrane synthesis. *NF*, Nanofiltration.

TiO_2, SAPO-34, GO, multiwall carbon nanotubes (MWCNTs), zeolite, GO quantum dot (GOQD), covalent organic frameworks (COFs), chitosan, cellulose nanocrystals, copper benzene-1,3,5-tricarboxylate, and amine-functionalized GO [83–94]. It can be observed that most of the additives are nanomaterials leading to the enhancement of thin-film nanocomposite (TFN) membranes, particularly surface hydrophilicity, charge, porosity, and mechanical stability. Liu et al. [88] have incorporated silicoaluminophosphate (SAPO) zeolite (SAPO-34) into dual-layer (PES/PVDF) hollow fiber substrate. It was shown that the surface hydrophilicity of TFN NF membrane increased characterized by lower contact angle (63 degrees) compared to that of TFC NF membrane (72 degrees).

Carbon-based nanomaterials have been used widely for NF membrane synthesis. Lai et al. [87] fabricated THN NF membrane by incorporating GO into a polysulfone (PSf) microporous substrate. It was reported that 0.3 wt.% GO exhibited the highest water permeability (353.5 L/m^2 h) with high salt rejected (95.2% Na_2SO_4) due to the highly hydrophilic and negative-charged GO nanosheets. Later, Zhang et al. [80] functionalized GO with more carboxyl groups to produce carboxyl-functionalized GO (CFGO) before incorporation into the PA membrane via IP. The CFGO was synthesized by chemical modification (ring opening followed by esterification) to the epoxide ring of GO. They reported that the NF membrane produced using 0.07 wt.% CFGO can produce permeate flux of up to 110.4 L/m^2 h compared to the GO NF membrane (75.5 L/m^2 h) without compromising dye (98.1%) and salt (28.7%) rejection. This is because CFGO has a much better dispersibility due to the deprotonation of carboxyl groups, which significantly increases the hydrophilicity and surface charge density of the NF membranes. GO has been functionalized with different functional groups, such as acyl chlorine, reduced amino, octadecylamine, and amine to improve the functionalities of NF membranes [91,95–97]. Zhang et al. [93] produced a TFN membrane comprising GOQDs dispersed within a tannic acid film through IP. The produced TFN membrane showed a high pure water flux which was 1.5 times more than the pristine TFC membrane with high dye rejection to Congo red (99.8%) and methylene blue (97.6%). Compared to the high aspect ratio of GO, the small size of GOQD created a relatively lower tortuosity for the water transport route, which therefore resulted in a high water permeation rate. Fig. 3–14 shows the schematic illustration of the water transport across TFN membrane incorporated with GOQD.

Recently, more research on MOFs for NF membrane synthesis has been reported [98–100]. The uniform and porous structure of MOFs serves as the molecular sieve or additional pathway for selective permeation. Liao et al. [98] incorporated zeolitic imidazolate framework 8 (ZIF-8) into the PA layer via IP. It was reported that the water permeability and rejection of the TFN membrane increased by 190% and 2%, respectively, compared to the pristine TFC membrane. This is attributed to the surface hydrophilicity and hollow structure of the nanocubes which aids the mass transport in separation mechanism. The synthesis procedure of ZIF-8 with deep eutectic method is shown in Fig. 3–15. Later, Wu et al. [102] introduced polydopamine (PDA)-COF hybrid interlayer on PAN support to produce an ultrathin NF membrane. It was observed that the PA layer has reduced from 79 to 11 nm attributed to the abundant of secondary amine groups of COF that enables facile control over the adsorption/diffusion of PIP molecules during the IP process.

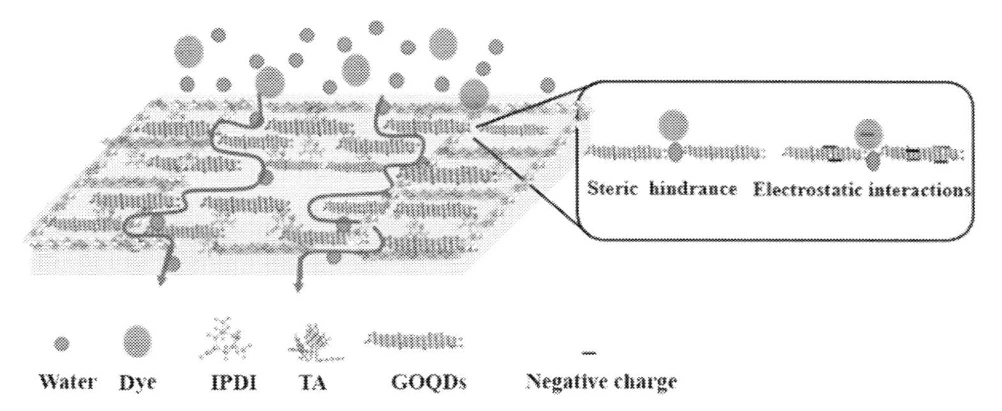

FIGURE 3–14 Synthesis procedure of ZIF-8 with deep eutectic method [93]. *ZIF-8*, Zeolitic imidazolate framework 8.

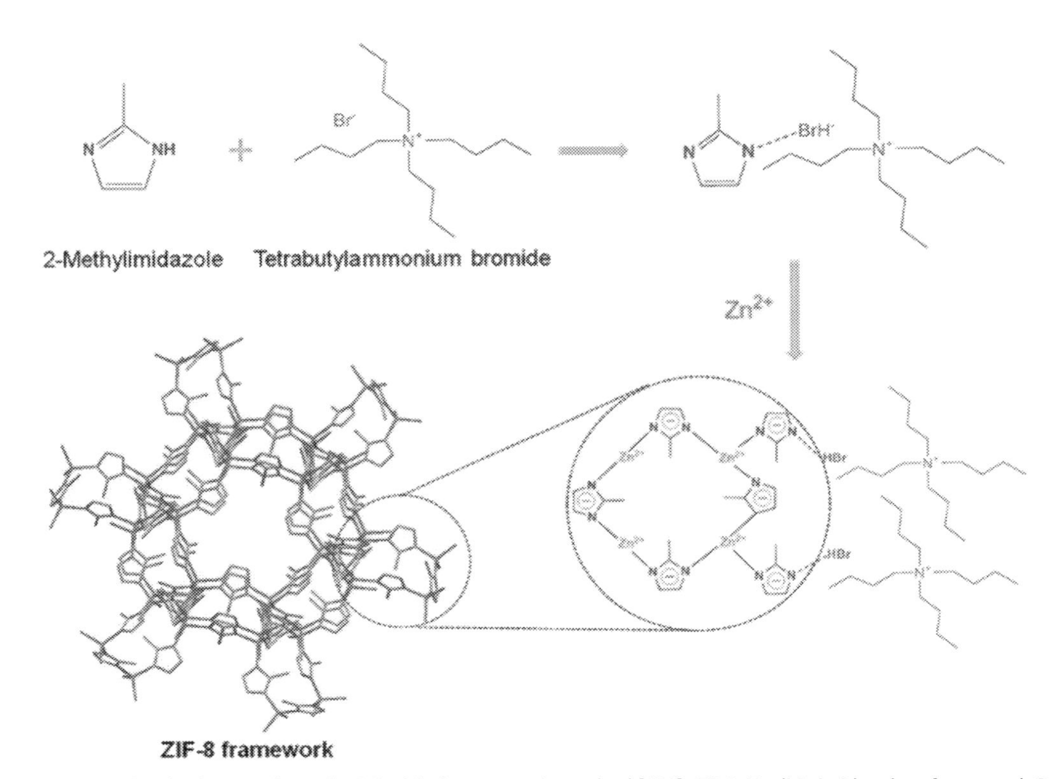

FIGURE 3–15 Synthesis procedure of ZIF-8 with deep eutectic method [101]. *ZIF-8*, Zeolitic imidazolate framework 8.

3.3.2.3 Others

Other NF modifications during IP fabrication including surface coating and codeposition are also employed to produce NF membranes. Surface coating generally refers to the coating of a thin layer across the macroporous support using a filtration-assisted method. Wu et al. [103] coated attapulgite (ATP) nanorods onto the PES membrane through the vacuum filtration process followed by IP using TMC-n-heptane. Fig. 3—16 shows the cross-sectional morphology of pristine PA/PES NF membrane and PA-ATP/PES NF membrane. However, the stability of the membrane is typically compromised due to the scratch-off and agglomeration of additives on the membrane surface [104]. Jiang et al. [105] reported that TiO_2 nanowires that are coated on PEI/commercial P25 membrane can be scratched off from the membrane suggesting that further membrane treatment is needed.

The second technique is the codeposition of PDA. During codeposition, the functional polymer is initially codissolved with dopamine under the weakly alkaline condition for coating onto the substrate. It was followed by the self-oxidization of dopamine, the functional molecules are stably entrapped within the PDA matrix and then firmly codeposited onto the substrate surface to provide the desired functionality. Yang et al. [106] introduced a hydrophilic PDA/PEI interlayer on the UF substrate for the preparation of the TFC NF membrane.

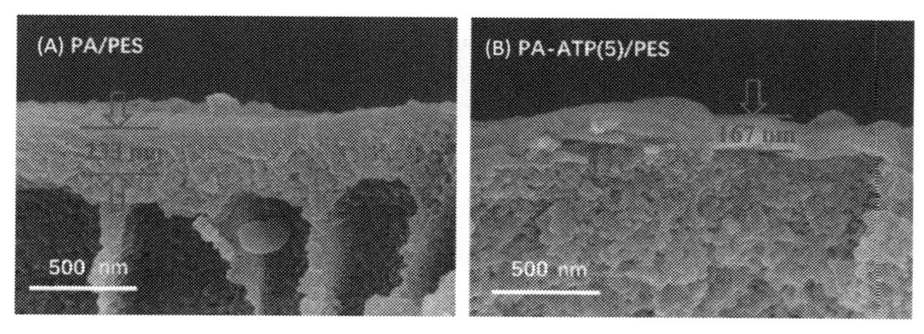

FIGURE 3–16 The cross-sectional morphology of membranes. (A) PA/PES; (B) PA-ATP(5)/PES [103]. *PA*, Polyamide; *PES*, poly(ethersulfone).

The synthesized TFC NF membrane shows high stability as the mussel-inspired interlayer acts as an adhesive between the PA selective layer and the substrate surface. These techniques can be combined to improve the stability/reproducibility of the membrane. For instance, Zhang et al. [107] first codeposited tannic acid and DETA to form a polyphenol interlayer on PSf UF substrate. This was followed by IP of piperazidine and TMC on the polyphenol interlayer to contrast the PA layer. The synthesized NF membrane not only had a defect-free PA layer but also tripled fold of water permeation flux compared to the NF membrane without the interlayer. The schematic diagram for the NF membrane synthesized by codeposition and IP is shown in Fig. 3–17.

3.3.3 Grafting polymerization

In general, grafting polymerization is used to alter the chemical structure of the membrane materials by introducing new functional groups to be covalently attached to a substrate [108]. Typically, grafting polymerization can be performed through UV/photo-grafting, electron beam (EB) irradiation, plasma treatment, and layer-by-layer (LBL) technique.

3.3.3.1 UV/photo-grafting

Among grafting polymerization techniques, UV/photo-grafting has drawn great interest due to its easy operation and low cost. Besides, the fabrication of NF membrane via UV-grafting produces an integral selective layer due to the strong chemical bond to the substrate which provides sufficient mechanical stability under high operating pressure. Zhong et al. [109] fabricated a positively charged NF membrane using a sulfonated polyphenylene sulfone (sPPSU) support via UV-induced grafting using two grafting monomers: diallyldimethylammonium chloride and 2-[(methacryloyloxy)ethyl]trimethyl ammonium chloride (Fig. 3–18). Due to the positive surface charges on NF membrane, a superior rejection of up to 99.98% to Safranin 0 dye was attained. Vatanpour et al. [110] successfully modified commercial PA NF membrane via UV-grafting of acrylic acid and incorporation of carboxylated-MWCNTs. The NF membrane grafted with 50 g/L acrylic acid under 5 min UV exposure showed the high pure water flux of 38.8 L/m^2 h, salt rejection of 93.4%–97.43%, and flux recovery ratio of

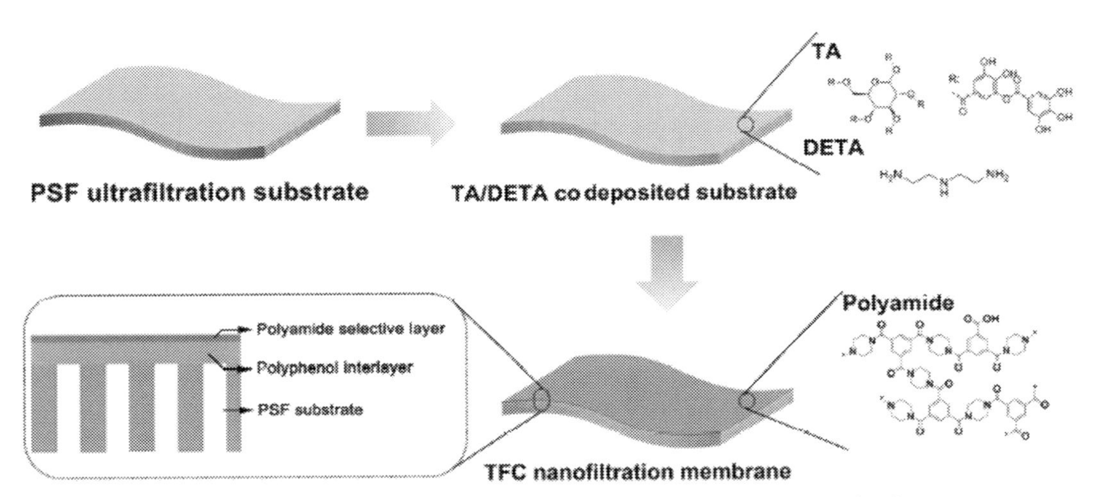

FIGURE 3–17 Schematic diagram for the NF membrane synthesized by codeposition and IP [107]. *NF*, Nanofiltration.

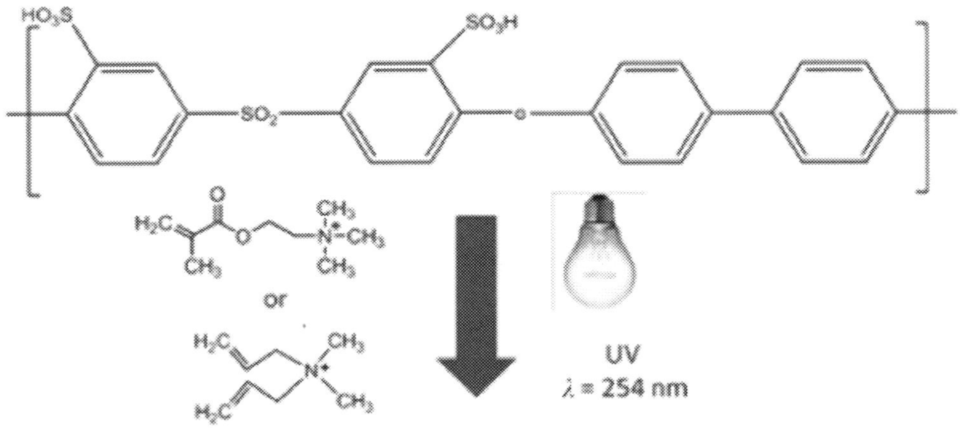

FIGURE 3–18 Schematic diagram of synthesis of NF membrane via UV grafting [109]. *NF*, Nanofiltration.

80.2% during bovine serum albumin filtration. The addition of 0.2 wt.% of carboxylated-MWCNTs to the grafting layer further improved the water flux by 30%. However, UV/photo-grafting technique requires photosensitive polymers including cardo polyetherketone, sPPSU, and PSf which severely limit the applicability of this technique and UV may lead to the damage of membrane substrate because of its high energy at low wavelengths [111,112].

3.3.3.2 EB irradiation

EB irradiation employs high-energy electrons to generate radicals on the membrane surface, followed by polymerization at the radical sites where the monomers are covalently bound to the basic polymer. Some of the monomers that are used for EB irradiation are N-vinyl pyrrolidone, vinyl acetate, N-vinyl pyridine, acrylic acid, and methacrylic acid [69]. Xu et al. [113] studied the effect of EB irradiation dose and monomer concentration on the mean pore size and performance of an NF membrane. They reported that the NF membrane grafted under 80 kGy with 10% of 2-acrylamido-2-methylpropanesulfonic acid solution showed excellent rejection to 0.1 mM Cr(VI) of 95.1% with a permeate flux of 23.8 L/m^2 h. Later, Reinhardt et al. [114] immobilized peptide motifs onto PES membranes via EB irradiation and chemical grafting, as shown in Fig. 3–19. It was found that EB irradiation is more effective to modify membrane in a single-step at the absence of initiators, catalysts, and organic solvents. However, limited research has been carried out using NF membranes employing EB irradiation for grafting polymerization mainly for reasons of cost, availability, and scalability of the technique [115].

3.3.3.3 Plasma treatment

Plasma treatment is an interesting technique to change the surface chemistry of NF membranes leaving their bulk properties primarily unchanged [115]. During plasma treatment, a large variety of macromolecules can be used to react with the formed free radicals on the membrane surface. A simple plasma treatment setup is shown in Fig. 3–20. Typically, inert gas such as helium or argon is introduced at a pressure of 50–100 mTorr and plasma is initiated. Monomer vapor is then introduced to increase the total pressure to 200–300 mTorr. These conditions are maintained for 1–10 min to allow the deposition of a thin polymer film on the membrane sample in the plasma field. The monomer vapor was then polymerized and formed a thin film on the porous film in the reactor chamber. Gao et al. [117] synthesized organic solvent-resistant NF membrane via plasma grafting of high-density PEG on a cross-linked polyimide (cPI) UF substrate, as shown in Fig. 3–21. Their results showed that the NF membrane had a remarkable improvement in the rejection of Rose Bengal from 70.1% to 99.64% compared to the pristine cPI membrane. Nevertheless, plasma treatment

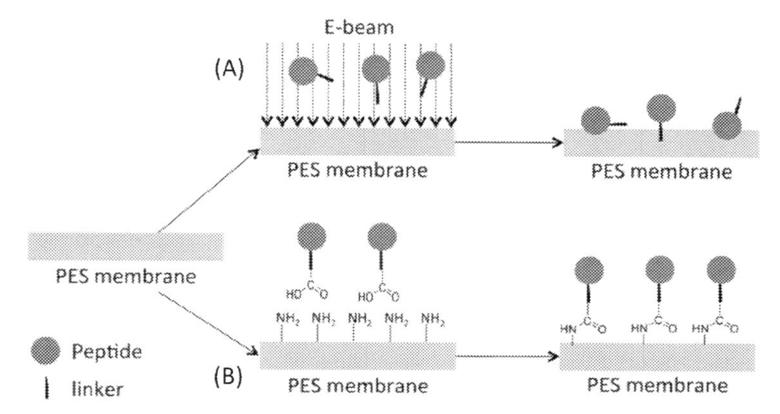

FIGURE 3–19 Schematic illustration of the membrane immobilization using (A) EB irradiation (B) chemical grafting [114]. *EB*, Electron beam.

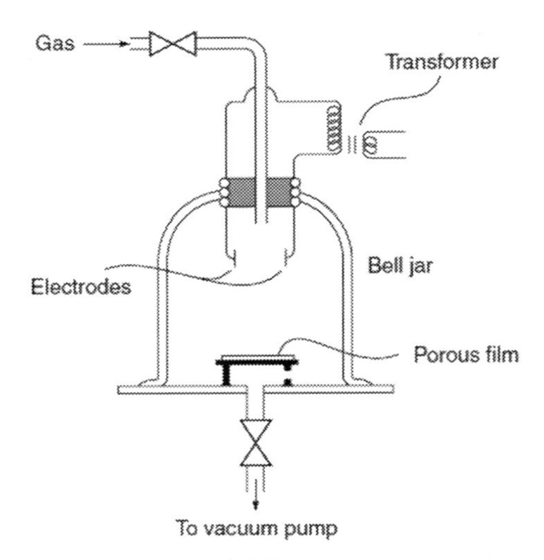

FIGURE 3–20 Simple bell jar plasma treatment setup [116].

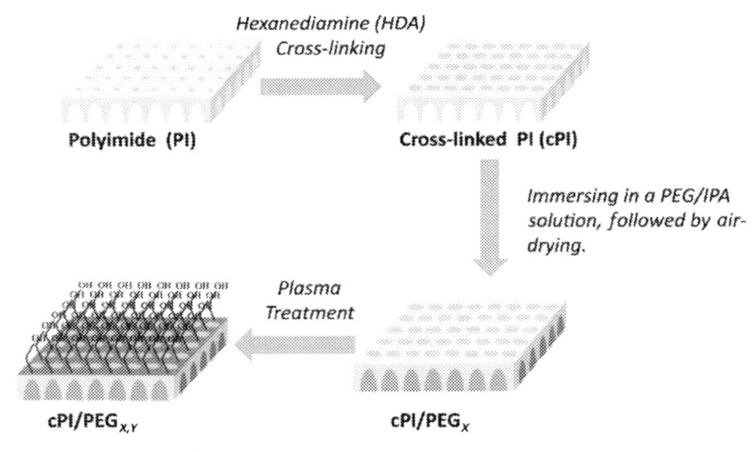

FIGURE 3–21 Schematic illustration of synthesis of NF membrane via plasma treatment [117]. *NF*, Nanofiltration.

usually involves a complex operation process and pore blockage might occur, leading to increased transmembrane resistance and reduced water permeation flux [118].

3.3.3.4 Layer-by-layer

The LBL technique modifies membranes using alternating applications of cationic and anionic polyelectrolytes. The method of NF fabrication and modification using polyelectrolytes has been critically reviewed by Ng et al. [119]. Chen et al. [120] prepared a composite hollow fiber NF membrane with an active layer comprising only one polyelectrolyte bilayer

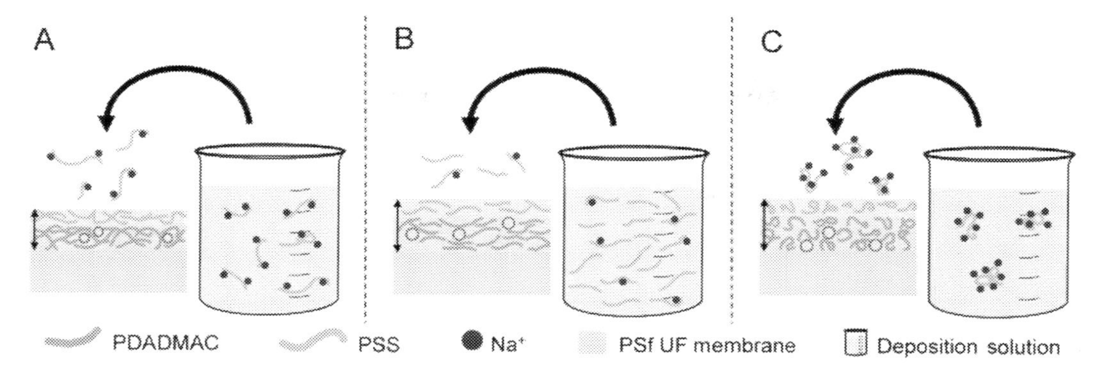

FIGURE 3–22 Schematic illustration of the effects of polyelectrolyte–salt interactions [121].

via LBL deposition of oppositely polyelectrolytes of sodium carboxymethyl cellulose and PEI on a polypropylene hollow fiber substrate. Glutaraldehyde was used as crosslinking agent between each deposition. The optimized NF membrane showed higher water permeability of 14.2 L/m^2 h bar, rejection of 93.2% to $MgCl_2$ at 3 bar compared to several commercial NF membranes. More recently, DuChanois et al. [121] tuned NF membrane pore size and thickness by systematically controlling the polyelectrolytes, poly(diallyldimethylammoniumchloride), and polyanionic poly(sodium 4-styrenesulfonate) to a PSf substrate. Their study investigated the concentration of NaCl salt (0–2.5 M) and polyelectrolyte concentrations (0–20 mM) during the LbL assembly. It was reported that smaller pore sizes of membrane were formed with increasing salt and polyelectrolyte concentrations; however, the pore size increased with further addition of salt and polyelectrolytes. The effect of polyelectrolytes–salt interactions on membrane pore structure is schematically presented in Fig. 3–22.

3.4 Design and operation of NF process

This section discusses the module design (flat-sheet, tubular, spiral-wound, hollow fiber) and operation modes (dead-end, crossflow) of the NF process.

3.4.1 Module design

In general, NF module design can be categorized into flat-sheet, tubular, spiral-wound, and hollow fiber. Comparison and schematic diagram of the membrane module design are shown in Table 3–2 and Fig. 3–23, respectively [123]. The earliest design was the flat-sheet (plate-and-frame) module, soon replaced by other more efficient and less expensive designs, the tubular module, the spiral-wound module, and the hollow fiber module. Currently, the most commonly used NF membrane modules are spiral-wound and tubular. In the spiral-wound module, the flat-sheet membrane is wound around a central collection pipe. The membrane sheets are glued and attached to the permeate channel along the unsealed edge of the leaf. A permeate spacer is positioned between the internal sides of the leaves to

Table 3–2 Comparison of membrane module designs [116].

Parameter	Flat-sheet	Spiral-wound	Tubular	Hollow fiber
Manufacturing cost ($\$/m^2$)	50–200	5–50	50–200	2–10
Concentration polarization fouling control	Good	Moderate	Very good	Poor
Permeate-side pressure drop	Low	Moderate	Low	High
Suitability for high-pressure operation	Marginal	Yes	Marginal	Yes
Limitation to specific types of membrane materials	No	No	No	Yes

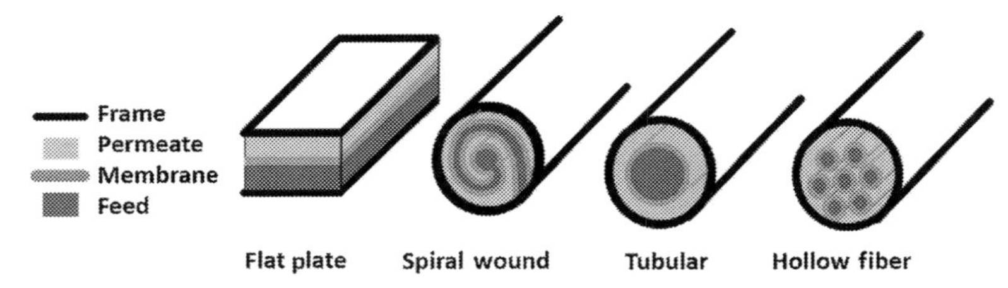

FIGURE 3–23 Schematic diagrams of membrane modules [122].

provide mechanical support and to guide the permeate to the permeation tube, while a feed channel spacer separates the top layers of the membrane. Sairam et al. [124] produced spiral-wound modules of solvent stable integrally skinned asymmetric NF membrane from polyaniline. It was reported that the spiral-wound NF membrane gave high rejections of oligostyrene markers (100%) from acetone, tetrahydrofuran, and N,N-dimethyl formamide solvent that implied a potential application for solvent exchange. However, the spiral-wound NF membrane is highly susceptible to fouling and suffers energy losses caused by head loss in spacer-filled channels, hence cannot be back-flushed [125].

Tubular modules do not require extensive pretreatment and allow for backflushing. Tubular modules are commonly synthesized from inorganic materials, such as ceramics. The feed flows inside the channels, the permeate flows out in a radial direction through the porous support and the active layer, to be finally collected at the outside as shown in Fig. 3–24. Yang et al. [127] produced ZIF-8@GO composite membrane on a tubular ceramic substrate through a vacuum-assisted assembly method. Using the optimum concentration of 0.2 g/L GO and 0.005 WT%ZIF-8@GO, the tubular NF membrane exhibited flux of 6.1 L/m^2 h bar with methanol retention up to 99%. Nevertheless, tubular NF membranes have low packing density and are expensive to manufacture.

Recently, the hollow fiber module has received great attention due to large membrane area per unit membrane volume, ability to self-support, elimination of feed and permeate spacers, less demand for pretreatment, and maintenance in relative to spiral-wound membrane modules [128]. In general, a hollow fiber NF membrane is a viable alternative to both spiral-wound and tubular membranes that combines the advantages of the two modules [129]. It was estimated that an optimized hollow fiber NF module would give a 100% increase in performance

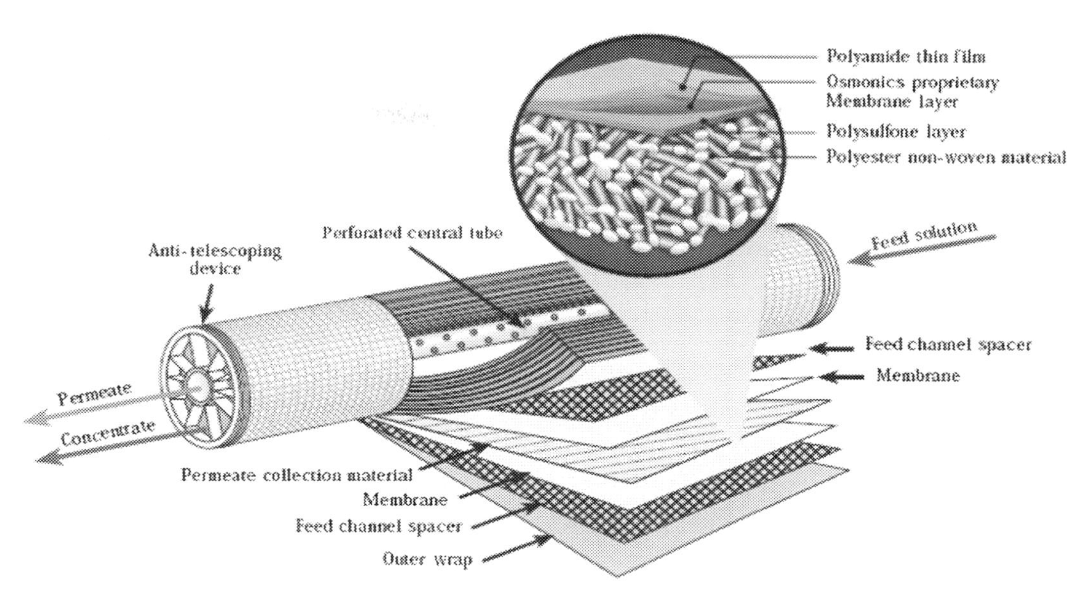

FIGURE 3–24 Cross-section of a spiral-wound module [126].

compared to an optimized spiral-wound module [130]. Yu et al. [128] developed a CA hollow fiber NF membrane by hydrolysis and carboxymethylation of an original cellulose triacetate semipermeable membrane. It was found that the CA hollow fiber NF membrane has a higher pure water permeability (5.2 L/m^2 h bar) and rejection (95.4% Na_2SO_4) compared to the pristine CA semipermeable membrane. This is because hydrolysis had greatly increased the membrane pore size and surface hydrophilicity while carboxylation had increased the membrane surface negative charge. It was further confirmed by Thong et al. [131] using a loose outer-selective hollow fiber NF PES membrane. The NF membrane was synthesized via a single-step spinning process using various doping formulation: total polymer concentration and ratio of polymer. Fig. 3–25 shows the effect of sulfonated PSf (sPSf) on the hollow fiber membrane morphology. As evidenced, when the sPSf concentration increased from 0% to 40%, the length of macrovoids gradually increased until the macrovoids transverse the entire cross-section. Therefore the produced NF membrane showed high water permeability of 13.2 L/m^2 h bar and rejection of more than 94.9% to indigo carmine dye.

3.4.2 Operation

Usually, the NF process can be categorized into two types of operation: (1) dead-end filtration, and (2) crossflow filtration, as shown in Fig. 3–26. Dead-end filtration refers to when the entire feed is directed normal to the membrane area under applied gaseous pressure while the direction of feed flow is parallel to the surface of the membrane for crossflow filtration. In general, crossflow filtration helps to maintain a uniform flow rate of permeate and

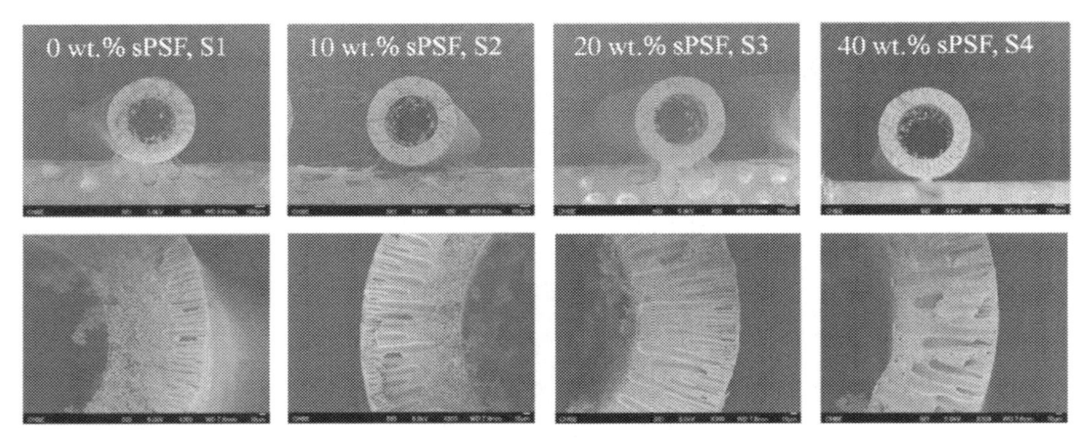

FIGURE 3–25 Effects of sPSf concentration on hollow fiber NF membrane morphology [131]. *sPSF*, Sulfonated PSf.

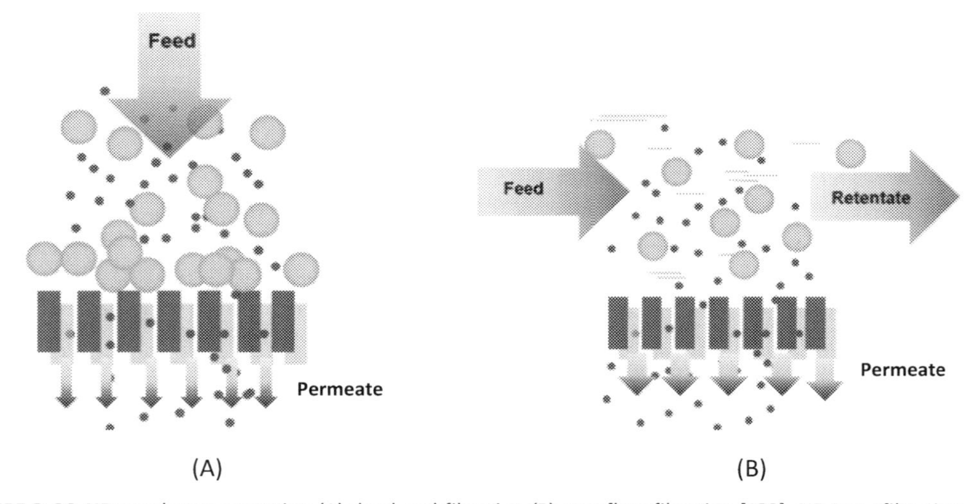

FIGURE 3–26 NF membrane operation (A) dead-end filtration (B) crossflow filtration [132]. *NF*, Nanofiltration.

helps to keep a longer membrane life by reducing membrane fouling. On the other hand, frequent membrane fouling is reported for dead-end filtration due to the accumulation of foulant on the membrane surface. Typically, backwashing of volume 2%–5% of the total feed is required to flush the accumulated foulant to recovery membrane flux [126]. Tsibranska and Tylkowski [132] compared the performance of the NF membrane in concentrating ethanolic extracts from *Sideritis* ssp. L. through dead-end and crossflow filtration. As expected, better conditions for reduced flux decline and tendency to a steady value were obtained by crossflow filtration while rapid flux decline was pronounced for dead-end filtration.

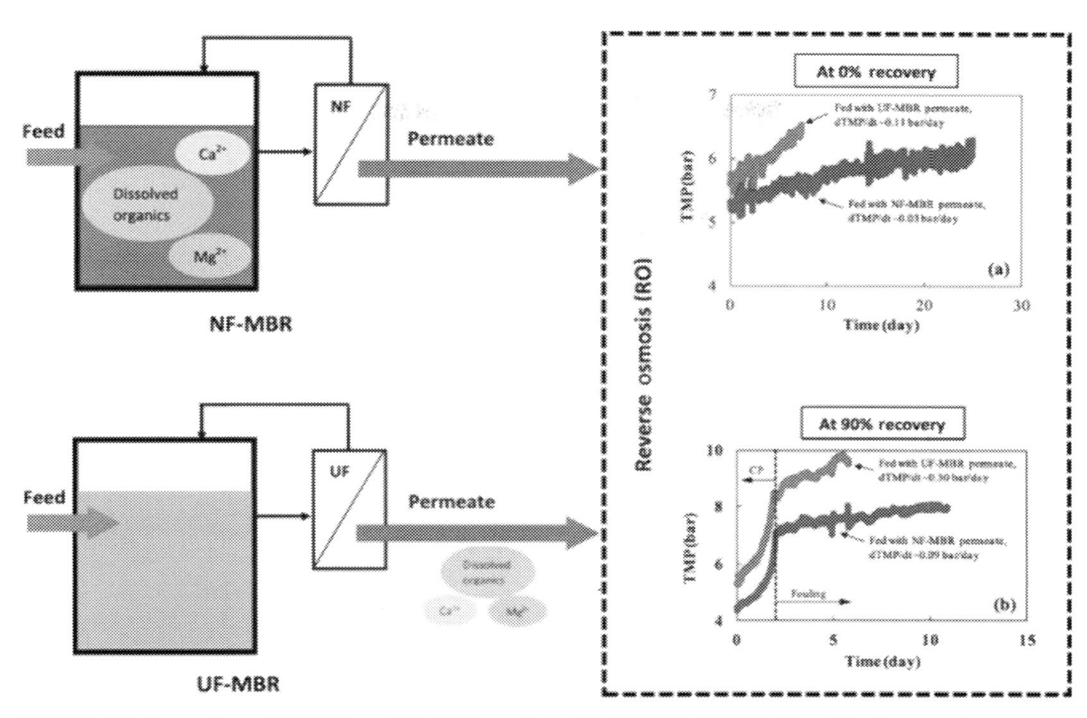

FIGURE 3–27 Comparison of NF-MBR and UF-MBR coupled with RO [133]. *NF-MBR*, Nanofiltration membrane bioreactor; *UF-MBR*, ultrafiltration membrane bioreactor; *RO*, reverse osmosis.

The cumulative permeate volume for crossflow filtration was higher than that of dead-end filtration by 69.04% after 3 h of filtration.

For crossflow filtration, the NF membrane is also employed in membrane bioreactor which is the combination process of the biological reactor with membrane separation. Tay et al. [133] compared the feasibility of UF-MBR and NF-MBR coupled with RO process for water reclamation from municipal wastewater at constant crossflow velocity of 0.2 m/s. The experiment setup and transmembrane pressure (TMP) profile are represented in Fig. 3–27. Their results proved that NF-MBR achieved superior MBR permeate quality operating at a constant flux of 10 L/m^2 h, leading to lower RO fouling rates by 3.3 times as compared to UF-MBR. This is because a lower concentration of dissolved organic carbon was present in the NF-MBR permeate (0.9 mg/L) compared to UF-MBR permeate (8.0 mg/L) that reduced membrane fouling in RO.

3.5 Limitation of the NF membrane applications

This section discusses the limitations of NF membrane and factors contributing to membrane fouling. Several fouling mitigation methods including passive fouling control and active fouling control are also elaborated.

3.5.1 Concentration polarization and membrane fouling

As discussed in Section 3.2.4, the convective transport of solutes toward the NF membrane causes the accumulation of retained solutes on the membrane surface (in a concentrated boundary layer), leading to a drop in permeate flux over time relative to that of pure solvent. This is mainly attributed by concentration polarization and membrane fouling.

As NF membrane selectively removes solutes from feed solution, there will be an accumulation of retained solutes near the membrane surface. This build-up is due to the balance between the convective drag force toward the membrane surface and the back diffusion of solutes from the region near the membrane surface to bulk phase. Thus as more permeate is collected, the retentate concentration increases several folds on the membrane surface and forms a resistive layer hindering the mass-flow across the membrane. This hydrodynamic phenomenon is called concentration polarization which builds up quickly in membrane filtration system. Concentration polarization is inherent to membrane filtration process. However, it is a reversible process where it can be alleviated by operating the membrane filtration system at higher flow rate or continuously agitated the feed solution to maintain the feed solution at uniform concentration profile.

In some instances, even maintaining a fairly uniform feed solution's concentration profile along the membrane filtration process, there is still a significant drop in permeate flux. This is possibly due to membrane fouling. The term membrane fouling includes the totality of phenomena responsible for the decrease of permeate flux over a period of time, except those linked to membrane compaction and mechanical characteristics modification. It is referring to the coupling of reversible/irreversible deposition of retained solutes on or in the membrane through the intermediate step of concentration polarization, which first causes an accumulation or increase in concentration near the membrane surface. Three features commonly found in membrane fouling are (1) gradual decline of membrane permeate flux, (2) gradual decrease of solutes rejection, (3) gradual increment of TMP and pressure difference across the membrane.

Membrane fouling can be divided into reversible fouling and irreversible fouling based on the attachment strength of solute particles onto the membrane surface. Reversible fouling is caused by a gel layer resulted from reversible concentration polarization which can be removed by strong shear force or backwashing. The foulants contribute to reversible fouling include inorganic compounds (iron, silicon, manganese, calcium, and etc.), organic compounds, as well as bacteria, microorganism, and their metabolites. However, formation of strong fouling matrix layer onto membrane surface under continuous filtration process will result in transformation of reversible fouling into an irreversible fouling. Irreversible fouling is a strong adsorption and blockage of solute particles which cannot be removed by simple hydrodynamic cleaning. It is usually much more difficult to reduce fouling in a membrane as loss of permeate flux due to fouling is generally irreversible. Typically, cleaning has to be performed when the permeate flux/salt rejection decreased by 10%–15% or feed pressure/pressure drop increased by 10%–15% [134].

Despite the expansion and successful application of NF membrane system in different industry, a decline in membrane performance over a period of time toward high susceptibility fouling effect is still the most critical limitation of NF membrane in its dignified application as a valuable means of advanced separation and purification technology for large-scale operation. Membrane fouling demands considerable attention as it causes a decay in permeate flux, affect the quality of treated permeate, and increase the operational pressure which lead to higher operational cost due to occasional membrane replacement (membrane replacement involves 20%–30% of the membrane filtration system operational cost) [135].

3.5.2 Factors affecting membrane fouling

As reported by a great variety of publications, membrane separation performance is depended on various factors such as the feed solution chemistry (ionic strength, pH, and water hardness) [136–138], membrane properties (surface hydrophilicity, surface charge, molecular weight cut-off or pore size, and morphology) [118,139,140], foulant characteristics (concentration, surface charge, hydrodynamic particle size, and particle size distribution) [141–143] as well as the membrane filtration system operating conditions and the flow condition close to the membrane surface [144–146]. In practical operation, these factors are often influenced by each other, and therefore present a more complicated effect for membrane fouling.

Identification of foulants is important in membrane fouling control. Many research works have been conducted for a relatively good understanding of foulants properties on membrane fouling phenomenon. Several studies had demonstrated that the natural organic matters (NOMs) is the major foulant controlling the rate and extent of membrane fouling in the application NF membrane system for water industry application [147,148]. Teow [149] found that the flux decline observed during filtration of river water in Malaysia was primarily caused by the deposition of naturally occurring organic macromolecules, particularly humic materials on the membrane surface. Similar results were reported by Shen and Schäfer [148] who noted that a large permeate flux declination across NF membrane for the treatment of natural water sources sampled from Northern Tanzania at constant operational pressure was caused by the deposition of NOMs on the upper surface of membrane. This initial deposition accelerated the subsequent deposition of NOMs.

Some researchers are devoted to studying the effect of membrane surface's hydrophilicity toward fouling phenomenon. It was found that the mechanism accounting for membrane fouling by hydrophobic fraction, hydrophilic part, and transphilic components are concentration polarization, adsorptive fouling, and cake layer formation, respectively [150]. Ho et al. [151] investigated the deterioration of GO/MWCNTs nanocomposite conductive membrane's permeate flux in palm oil mill effluent treatment. It was found that membrane surface hydrophilicity is the desirable surface property of a membrane that can mitigate membrane fouling. This finding was in line with the work of Rosnan et al. [152], who demonstrated that mixed-matrix membrane decorated with zinc oxide-GO nanocomposite material appeared to have a stronger resistance to membrane fouling as compared to pristine PSf membrane, mainly due to its greater surface hydrophilicity in forming a layer of water as the barrier between membrane surface and foulants. Among the methods used to enhance membrane surface hydrophilicity, the addition of hydrophilic additive into membrane polymer solution is considered as the most convenient method to create the impact [153].

Since most of the foulants are negatively charged as well as hydrophobic in nature, in addition to hydrophilic characters, NF membrane possess negative charge on the top selective layer would also reduce the adsorption of foulants on the membrane surface and mitigate membrane fouling. Teow et al. [154] reported that dramatic flux decline and notably higher humic acid deposition was observed on pristine polyvinylidene fluoride (PVDF) membrane compared to negatively charged hydroxyl-functionalized PVDF-TiO_2 mixed-matrix membrane due to the electrostatic interaction between the negatively charged humic acid and polymeric membrane surface, resulting in membrane pore blockage. Additionally, Ho et al. [151] had successfully developed electrically conductive membrane with the incorporation of GO and MWCNTs into membrane matrix via the blending method. The presence of electric field exerted stronger repulsion force in repelling the foulants onto GO/MWCNTs mixed-matrix membrane attributed by negative charge enhancement by GO/MWCNTs nanocomposite. It was reported by Ho et al. [151] where the blockage on membrane surface was greatly reduced without compromising the membrane rejection.

Solution chemistry plays a significant role in determining foulant–foulant and foulant membrane electrostatic double-layer interactions under certain operating conditions, and hence the membrane fouling phenomenon [155]. Some organic substances, which are present naturally in water, are consist of heterogeneous mixture of molecules, held together in supramolecular confirmation by physical bond. Hence, these foulants will exhibit different solubility levels at different pH of the feed solution. Kabsch-Korbutowicz et al. [156] and Yuan and Zydney [157] reported that the significant flux decline at low pH was attributed to the thick humic acid layer deposited on membrane surface as a result of low electrostatic repulsion between humic acid molecules and membrane surface, and between the humic acid solution and the preformed humic acid fouling layer, leading to a strong tendency for humic acid sorption on membrane surface. Similar studies have been carried out by Jones and O'Melia [158] in which they found that humic acid adsorption was decreased as the pH was increased from 4.7 to 10 due to electrostatic interaction. On the other hand, coexisting multivalent cations present in feed solution of NF process for water application could form complex macromolecules with solutes, increase the electrostatic shielding among the charged solutes and thus intensifying membrane fouling [159].

Additionally, the rate and extent of NF membrane fouling are influenced by NF membrane filtration system operating conditions, such as the operational pressure and water flow velocity. Operational pressure of a NF membrane filtration process determines the initial permeate flux and the resulting convective transport of solutes toward the membrane. Ghani et al. [160] performed an experiment for the investigation of performance and fouling propensity of three different types of commercial membranes (NF270, BW30, and XLE) in treating tertiary palm oil mill effluent. This work demonstrated that high permeate flux of NF270 membrane resulted in severe membrane fouling due to higher permeation drag. Although NF270 gives higher initial permeate flux, rapid declination of permeate flux attributed by severe membrane fouling had counteracted its advantages for the long run. On the other hand, high water flow velocity could lessen membrane fouling effect by eliminating the concentration polarization effect near the membrane surface through scouring effect. However, this may be only applicable to the reversible fouling phenomenon.

3.5.3 Fouling mitigation

Even though membrane fouling is an inevitable phenomenon in the NF membrane filtration process, a number of approaches have been studied for the alleviation of membrane fouling and to increase the membrane performance. These methods can be broadly grouped into two categories: passive control (periodic physical or chemical cleaning) and active control (tailoring membrane surface properties).

3.5.3.1 Passive fouling control

Passive fouling control can be divided into physical cleaning and chemical cleaning. In practice, physical cleaning is usually followed by chemical cleaning and this sequence is widely used in membrane fouling control. Physical cleaning is a passive fouling control method employing mechanical forces to dislodge the foulants from membrane surface. This includes change of running mode (forward and reverse flushing), air scouring, and backwashing [161]. The cleaning frequency depends on the fouling severity, feed solution chemistry, membrane configuration, and other factors.

Change of running mode may be practical for reversible membrane fouling control, while backwashing is designed for better operation [162]. During forward flushing, permeate is pumped across the membrane surface at high crossflow velocity in forward direction of membrane feed side. Dense attachment of foulants on the membrane surface is partially removed by the surface shear created at high crossflow velocity. On the other hand, permeate is flushed at reverse direction of membrane feed side during reverse flushing, whereas backwashing is a reversed filtration process in which permeate is flushed back through the membrane to the concentrate side. When high backwashing pressure is applied, the membrane pores are flushed inside out and dislodge the foulants block at membrane pores. Followed by the backwashing, a forward flushing is usually carried out to wash away the detach foulant layer to regain the membrane permeate flux to a certain extent. Air scouring is another passive fouling control method used to provide surface shear or mass transferring to avoid the foulants from compacting on the membrane surface thus minimizing membrane fouling, especially when a high operational pressure is applied. During air scouring cleaning process, bursts of air are injected into forward permeate flush ahead of permeator. This air will disturb the membrane, thus loosening the foulants from the membrane. Air scouring cleaning process is often coupled with rinsing where it can be applied either during the course of filtration to reduce fouling deposition or periodically to remove the already formed deposits [163].

However, membrane fouling is not always totally reversible by physical cleaning. A portion of foulants deposited on the membrane surface and trapped inside the membrane pores are not usually dislodged by physical cleaning, therefore considered to be irreversible fouling, leading to the progressive deterioration of membrane performance. Chemical cleaning with the use of chemical agents or solutions such as acid, alkali, and biocide solution is widely applied as irreversible membrane fouling control to restore permeate flux [164–166]. Nevertheless, frequency of chemical cleaning has to be controlled as regular chemical cleaning might shorten the membrane life. The choice of chemical cleaning agent or solution is critical. The selected chemical cleaning agent or solution has to be able to dissolve the foulants deposited on the membrane surface or into the membrane matrix but not to destroy the membrane surface in order to maintain the membrane surface properties and its performance. Generally, acid solutions (nitric acid, hydrochloric acid, phosphoric acid, citric acid, and sulfuric acid) are used for the control of precipitated salts or scalants, whereas alkali solutions are comparatively effective for organic fouling removal. On the other hand, the biocide solution is responsible for bio-fouling control due to the presence of microorganisms. Table 3–3 summarizes the chemical cleaning agent according to the type of foulants.

3.5.3.2 Active fouling control

As the foulants are directly in contact with the membrane surface and the affinity between solutes (foulants) and the membrane is the main factor in affecting the membrane fouling, modification of membrane material and/or its properties appears as a potential approach for the active control of membrane fouling phenomenon. As reported by Chittrakarn et al. [167], hydrophilicity enhancement of membrane matrix could improve the antifouling and flux enhancement of nascent membrane. This can be done through UV-induced photo-grafting [168], ozone-induced grafting [169], acrylic acid in gas grafting plasma polymerization [170], and CO_2/NH_3 grafting plasma [171]. In the other method, membrane surface modification is done via physical blending of inorganic additives (such as Si, SiO_2, CdS, Al_2O_3, Fe_3O_4, ZrO_2, GO, ZnO, or TiO_2 particles) and hydrophilic polymers (such as CA phthalate, polyvinyl alcohol, or polyvinyl pyrrolidone) into membrane polymer solution. The NF membrane modification methods are briefed in Section 3.3.

Although many attempts have been made to modify membrane surface properties by grafting hydrophilic monomer on the membrane top selective layer, it requires high energy and involved high production cost. Whereas, the impact of physical blending method is too small to obtain satisfactory reduction of membrane fouling as high surface energy of inorganic additives or particles could induce aggregation, resulting in poor distribution in membrane matrix and/or defective pore structure of the membrane [172]. On top of that, once the deposition of foulants has

Table 3–3 Chemical cleaning agent with respective type of foulants [161].

Type of foulant	Chemical agent
Colloidal	NaOH solutions, chelating agents, and surfactants
Organic	NaOH solutions, chelating agents, and surfactants
Metal oxides	Citric acid with low pH or $Na_2S_2O_4$
Silica (SiO_2)	NaOH solutions with high pH
Carbonate scales ($CaCO_3$)	Citric acid or HCl with low pH
Sulfate scales ($CaSO_4$, $BaSO_4$)	HCl solutions or sequestration agents (EDTA)
Biofilms	NaOH solutions, chelating or sequestration agents, surfactants, and disinfectants

taken place, surface modification is no longer effective for preventing membrane fouling. This is understandable because the effect of the solute–membrane interaction is severely reduced after the formation of a layer of deposited foulants. At this stage, membrane surface properties no longer play a role in preventing further deposition of foulants. This implies that there is no membrane that could be totally free from fouling under any circumstances.

Recent studies demonstrated that photocatalytic self-cleaning membrane is an emerging advanced active fouling control strategy for membrane filtration system where TiO_2 photocatalyst is immobilized in membrane polymeric matrix [173–175]. Janssens et al. [173] observed that combination of chemical oxidation and TiO_2-based membrane was managed to mitigate membrane fouling with UV treatment. Direct photolysis was found to be extremely effective to degrade etoposide and paclitaxel from secondary effluent as well as from the highly concentrated retentate produced from NF. The photocatalytic mechanism and process of TiO_2-based membrane are illustrated in Fig. 3–28.

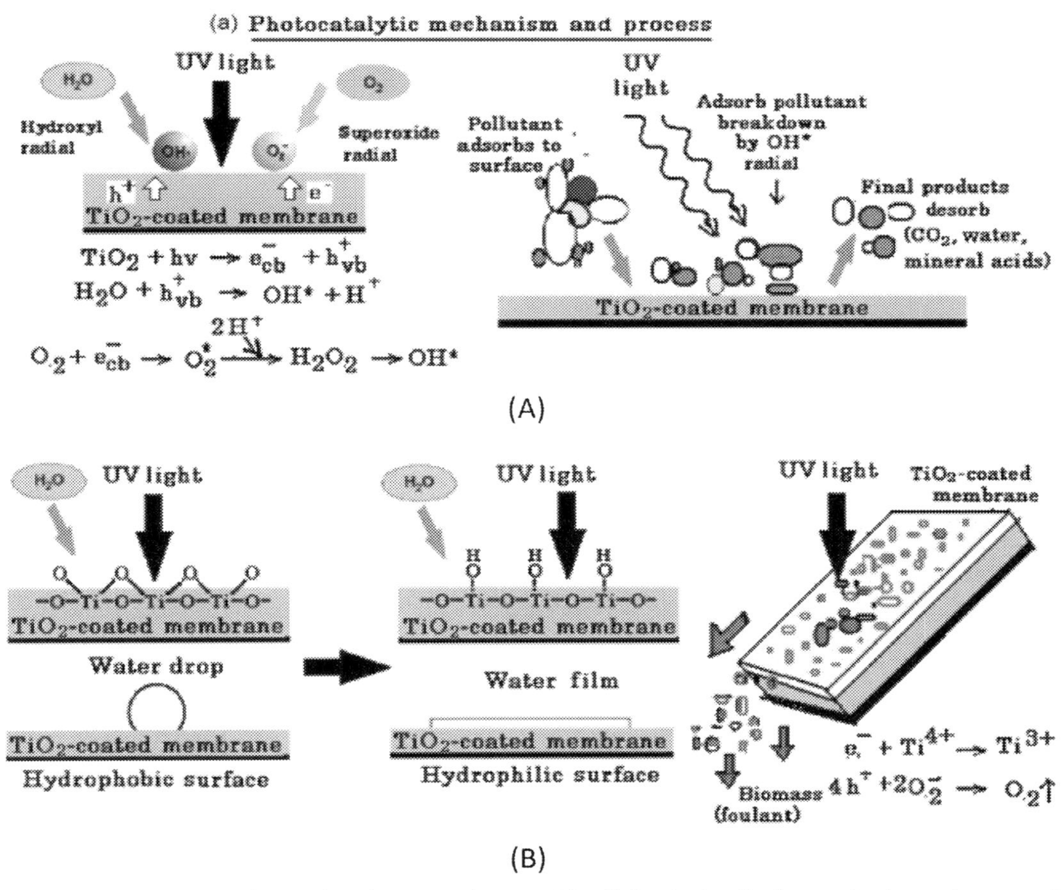

FIGURE 3–28 (A) Photocatalysis and mechanism and process (B) self-cleaning/antifouling mechanism and process of PVDF/TiO_2 membrane [176]. *PVDF*, Polyvinylidene fluoride.

This innovation has been the focus of numerous investigations as it integrates membrane separation, hydrophilic nature of nano-sized TiO_2, and photocatalytic degradation of foulants attached onto the membrane surface in a single system. The development of photocatalytic self-cleaning membrane is a superior active fouling control strategy where it potentially benefits the antifouling performance of a membrane and extends the service life of the membrane [177,178].

3.6 Conclusions

In this chapter, the separation mechanism of NF membrane including the steric effect, Donnan effect, dielectric effect, transport effect, and adsorption effect was summarized. Besides, special attention was devoted to three main approaches that are used to make NF polymeric membranes: phase inversion, IP, and grafting polymerization. Recent advances and developments of the membrane synthesis techniques including the effect of synthesis condition on the performance of NF membrane were also explained. Lastly, the limitation of NF membrane application particularly membrane fouling and fouling mitigation were elaborated. Future study and development of NF membrane should focus on the optimization of the membrane synthesis and operating parameters toward the reduction in membrane fouling and operation cost as well as increases in the membrane lifespan. Life cycle assessment should also be conducted to identify the human health risks, environmental and ecological impacts of the on-going research from material sourcing and handling to membrane process and lastly waste disposal.

References

[1] A.S.C. Chen, J.T. Flynn, R.G. Cook, A.L. Casaday, Removal of oil, grease, and suspended solids from produced water with ceramic crossflow microfiltration, SPE Prod. Eng. 6 (1991) 131–136.

[2] H. Li, A.G. Fane, H.G.L. Coster, S. Vigneswaran, Observation of deposition and removal behaviour of submicron bacteria on the membrane surface during crossflow microfiltration, J. Membr. Sci. 217 (2003) 29–41.

[3] H. Ma, B.S. Hsiao, B. Chu, Functionalized electrospun nanofibrous microfiltration membranes for removal of bacteria and viruses, J. Membr. Sci. 452 (2014) 446–452.

[4] J. Lowe, M.M. Hossain, Application of ultrafiltration membranes for removal of humic acid from drinking water, Desalination 218 (2008) 343–354.

[5] K. Zodrow, L. Brunet, S. Mahendra, D. Li, A. Zhang, Q. Li, et al., Polysulfone ultrafiltration membranes impregnated with silver nanoparticles show improved biofouling resistance and virus removal, Water Res. 43 (2009) 715–723.

[6] B.A.M. Al-Rashdi, D.J. Johnson, N. Hilal, Removal of heavy metal ions by nanofiltration, Desalination 315 (2013) 2–17.

[7] I. Koyuncu, Reactive dye removal in dye/salt mixtures by nanofiltration membranes containing vinylsulphone dyes: effects of feed concentration and cross flow velocity, Desalination 143 (2002) 243–253.

[8] T. Koskela, Removal of Hardness From Groundwater With Nanofiltration, Case Study: Meri-Lapin Vesi Oy, Bachelor of Engineering Thesis, Helsinki Metropolia University of Applied Sciences, 2016.

[9] R.W. Baker, Overview of membrane science and technology, Membrane Technology and Applications, John Wiley & Sons Ltd., 2012, pp. 1–14.

[10] P. Eriksson, Nanofiltration extends the range of membrane filtration, Environ. Prog. 7 (1988) 58–62.

[11] H. Strathmann, K. Kock, P. Amar, R.W. Baker, The formation mechanism of asymmetric membranes, Desalination 16 (1975) 179–203.

[12] D.L. Erickson, J. Glater, J.W. McCutchan, Selective properties of high flux cellulose acetate membranes toward ions found in natural waters, Ind. Eng. Chem. Prod. Res. Dev. 5 (1966) 205–211.

[13] W.J. Conlon, Pilot field test data for prototype ultra low pressure reverse osmosis elements, Desalination 56 (1985) 203–226.

[14] L. Woei Jye, A. Fauzi Ismail, Nanofiltration membranes synthesis, characterization, and applications, J. Membr. Sci. Res. 3 (2017) 1–13.

[15] T. Waite, A. Fane, A. Schäfer, Nanofiltration: Principles and Applications, Journal American Water Works Association, 2005, pp. 121–122.

[16] J.E. Cadotte, R.J. Petersen, Thin-film composite reverse-osmosis membranes: origin, development, and recent advances, ACS Symp. Ser. 153 (1981) 305–326.

[17] A. Larbot, S. Alami-Younssi, M. Persin, J. Sarrazin, L. Cot, Preparation of a γ-alumina nanofiltration membrane, J. Membr. Sci. 97 (1994) 167–173.

[18] T. Tsuru, S.I. Wada, S. Izumi, M. Asaeda, Silica-zirconia membranes for nanofiltration, J. Membr. Sci. 149 (1998) 127–135.

[19] A. Anand, B. Unnikrishnan, J.Y. Mao, H.J. Lin, C.C. Huang, Graphene-based nanofiltration membranes for improving salt rejection, water flux and antifouling—a review, Desalination 429 (2018) 119–133.

[20] L.D. Nghiem, A.I. Schäfer, M. Elimelech, Removal of natural hormones by nanofiltration membranes: measurement, modeling and mechanisms, Environ. Sci. Technol. 38 (2004) 1888–1896.

[21] W.J. Lau, A.F. Ismail, N. Misdan, M.A. Kassim, A recent progress in thin film composite membrane: a review, Desalination 287 (2012) 190–199.

[22] S.M. Miron, P. Dutournié, A. Ponche, Filtration of uncharged solutes: an assessment of steric effect by transport and adsorption modelling, Water (Switz.) 11 (2019) 2173.

[23] J. López, M. Reig, O. Gibert, J.L. Cortina, Increasing sustainability on the metallurgical industry by integration of membrane nanofiltration processes: acid recovery, Sep. Purif. Technol. 226 (2019) 267–277.

[24] A.R.D. Verliefde, E.R. Cornelissen, S.G.J. Heijman, E.M.V. Hoek, G.L. Amy, B. Van Der Bruggen, J.C. van Dijk, Influence of solute-membrane affinity on rejection of uncharged organic solutes by nanofiltration membranes, Environ. Sci. Technol. 43 (2009) 2400–2406.

[25] J.E. Almazán, E.M. Romero-Dondiz, V.B. Rajal, E.F. Castro-Vidaurre, Nanofiltration of glucose: analysis of parameters and membrane characterization, Chem. Eng. Res. Des. 94 (2015) 485–493.

[26] C.T. Cleveland, T.F. Seacord, A.K. Zander, Standardized membrane pore size characterization by polyethylene glycol rejection, J. Environ. Eng. 128 (2002) 399–407.

[27] M.J. López-Muñoz, A. Sotto, J.M. Arsuaga, B. Van der Bruggen, Influence of membrane, solute and solution properties on the retention of phenolic compounds in aqueous solution by nanofiltration membranes, Sep. Purif. Technol. 66 (2009) 194–201.

[28] W.R. Bowen, A.W. Mohammad, N. Hilal, Characterisation of nanofiltration membranes for predictive purposes—use of salts, uncharged solutes and atomic force microscopy, J. Membr. Sci. 126 (1997) 91–105.

[29] K. Sakai, Determination of pore size and pore size distribution, J. Membr. Sci. 96 (1994) 91–130.

[30] J.F. Kim, G. Székely, I.B. Valtcheva, A.G. Livingston, Increasing the sustainability of membrane processes through cascade approach and solvent recovery—pharmaceutical purification case study, Green Chem. 16 (2014) 133–145.

[31] M. Mänttäri, A. Pihlajamäki, M. Nyström, Effect of pH on hydrophilicity and charge and their effect on the filtration efficiency of NF membranes at different pH, J. Membr. Sci. 280 (2006) 311–320.

[32] M. Dalwani, N.E. Benes, G. Bargeman, D. Stamatialis, M. Wessling, Effect of pH on the performance of polyamide/polyacrylonitrile based thin film composite membranes, J. Membr. Sci. 372 (2011) 228–238.

[33] A.E. Childress, M. Elimelech, Relating nanofiltration membrane performance to membrane charge (electrokinetic) characteristics, Environ. Sci. Technol. 34 (2000) 3710–3716.

[34] R.R. Sharma, R. Agrawal, S. Chellam, Temperature effects on sieving characteristics of thin-film composite nanofiltration membranes: pore size distributions and transport parameters, J. Membr. Sci. 223 (2003) 69–87.

[35] Y. Roy, D.M. Warsinger, J.H. Lienhard, Effect of temperature on ion transport in nanofiltration membranes: diffusion, convection and electromigration, Desalination 420 (2017) 241–257.

[36] A. Escoda, P. Fievet, S. Lakard, A. Szymczyk, S. Déon, Influence of salts on the rejection of polyethyleneglycol by an NF organic membrane: pore swelling and salting-out effects, J. Membr. Sci. 347 (2010) 174–182.

[37] Y. Du, Y. Lv, W.Z. Qiu, J. Wu, Z.K. Xu, Nanofiltration membranes with narrowed pore size distribution: via pore wall modification, Chem. Commun. 52 (2016) 8589–8592.

[38] A.M. Saenz De Jubera, Y. Gao, J.S. Moore, D.G. Cahill, B.J. Mariñas, Enhancing the performance of nanofiltration membranes by modifying the active layer with aramide dendrimers, Environ. Sci. Technol. 46 (2012) 9592–9599.

[39] A.E. Childress, M. Elimelech, Effect of solution chemistry on the surface charge of polymeric reverse osmosis and nanofiltration membranes, J. Membr. Sci. 119 (1996) 253–268.

[40] R. Epsztein, E. Shaulsky, N. Dizge, D.M. Warsinger, M. Elimelech, Ionic charge density-dependent donnan exclusion in nanofiltration of monovalent anions, Environ. Sci. Technol. 52 (2018). acs.est.7b06400.

[41] B. Al-Rashdi, C. Somerfield, N. Hilal, Heavy metals removal using adsorption and nanofiltration techniques, Sep. Purif. Rev. 40 (2011) 209–259.

[42] A. Figoli, A. Cassano, A. Criscuoli, M. Salatul Islam Mozumder, M. Tamez Uddin, M. Akhtarul Islam, E. Drioli, Influence of Operating Parameters on the Arsenic Removal by Nanofiltration, Water Res 44 (2015) 97–104.

[43] S.S. Wadekar, R.D. Vidic, Influence of active layer on separation potentials of nanofiltration membranes for inorganic ions, Environ. Sci. Technol. 51 (2017) 5658–5665.

[44] D.L. Oatley, L. Llenas, N.H.M. Aljohani, P.M. Williams, X. Martínez-Lladó, M. Rovira, et al., Investigation of the dielectric properties of nanofiltration membranes, Desalination 315 (2013) 100–106.

[45] Y. Zhu, H. Zhu, G. Li, Z. Mai, Y. Gu, The effect of dielectric exclusion on the rejection performance of inhomogeneously charged polyamide nanofiltration membranes, J. Nanopart. Res. 21 (2019) 217.

[46] J.V. Nicolini, C.P. Borges, H.C. Ferraz, Selective rejection of ions and correlation with surface properties of nanofiltration membranes, Sep. Purif. Technol. 171 (2016) 238–247.

[47] M.C.Y. Wong, K. Martinez, G.Z. Ramon, E.M.V. Hoek, Impacts of operating conditions and solution chemistry on osmotic membrane structure and performance, Desalination 287 (2012) 340–349.

[48] V.N. Burganos, Modeling and simulation of membrane structure and transport properties, Compr. Membr. Sci. Eng. 1 (2010) 29–74.

[49] P. Marchetti, M.F. Jimenez Solomon, G. Szekely, A.G. Livingston, Molecular separation with organic solvent nanofiltration: a critical review, Chem. Rev. 114 (2014) 10735–10806.

[50] H. Bessbousse, T. Rhlalou, J.F. Verchère, L. Lebrun, Removal of heavy metal ions from aqueous solutions by filtration with a novel complexing membrane containing poly(ethyleneimine) in a poly(vinyl alcohol) matrix, J. Membr. Sci. 307 (2008) 249–259.

[51] C. Mbareck, Q.T. Nguyen, O.T. Alaoui, D. Barillier, Elaboration, characterization and application of polysulfone and polyacrylic acid blends as ultrafiltration membranes for removal of some heavy metals from water, J. Hazard. Mater. 171 (2009) 93–101.

[52] E. Salehi, P. Daraei, A. Arabi Shamsabadi, A review on chitosan-based adsorptive membranes, Carbohydr. Polym. 152 (2016) 419–432.

[53] M.R. Kotte, A.T. Kuvarega, M. Cho, B.B. Mamba, M.S. Diallo, Mixed matrix PVDF membranes with in situ synthesized PAMAM dendrimer-like particles: a new class of sorbents for Cu(II) recovery from aqueous solutions by ultrafiltration, Environ. Sci. Technol. 49 (2015) 9431–9442.

[54] A. Denizli, D. Tanyolaç, B. Salih, E. Aydinlar, A. Özdural, E. Pişkin, Adsorption of heavy-metal ions on Cibacron Blue F3GA-immobilized microporous polyvinylbutyral-based affinity membranes, J. Membr. Sci. 137 (1997) 1–8.

[55] P. Tan, Y. Hu, Q. Bi, Competitive adsorption of Cu2 + , Cd2 + and Ni2 + from an aqueous solution on graphene oxide membranes, Colloids Surf. A 509 (2016) 56–64.

[56] J.Y. Sum, A.L. Ahmad, B.S. Ooi, Selective separation of heavy metal ions using amine-rich polyamide TFC membrane, J. Ind. Eng. Chem. 76 (2019) 277–287.

[57] Q. Zhang, N. Wang, L. Zhao, T. Xu, Y. Cheng, Polyamidoamine dendronized hollow fiber membranes in the recovery of heavy metal ions, ACS Appl. Mater. Interfaces 5 (2013) 1907–1912.

[58] B. Van Der Bruggen, L. Braeken, C. Vandecasteele, Flux decline in nanofiltration due to adsorption of organic compounds, Sep. Purif. Technol. 29 (2002) 23–31.

[59] Y. Yoon, P. Westerhoff, S.A. Snyder, E.C. Wert, Nanofiltration and ultrafiltration of endocrine disrupting compounds, pharmaceuticals and personal care products, J. Membr. Sci. 270 (2006) 88–100.

[60] C. Bellona, J.E. Drewes, The role of membrane surface charge and solute physico-chemical properties in the rejection of organic acids by NF membranes, J. Membr. Sci. 249 (2005) 227–234.

[61] A.W. Mohammad, Y.H. Teow, W.C. Chong, K.C. Ho, Hybrid processes: membrane bioreactor, Membrane Separation Principles and Applications, Elsevier, 2019, pp. 401–470.

[62] V.F. Cardoso, G. Botelho, S. Lanceros-Méndez, Nonsolvent induced phase separation preparation of poly(vinylidene fluoride-co-chlorotrifluoroethylene) membranes with tailored morphology, piezoelectric phase content and mechanical properties, Mater. Des. 88 (2015) 390–397.

[63] M.A.A. Shahmirzadi, S.S. Hosseini, G. Ruan, N. Tan, Tailoring PES nanofiltration membranes through systematic investigations of prominent design, fabrication and operational parameters, RSC Adv. 5 (2015) 49080–49097.

[64] J. Zhu, M.T. Tsehaye, J. Wang, A. Uliana, M. Tian, S. Yuan, et al., A rapid deposition of polydopamine coatings induced by iron (III) chloride/hydrogen peroxide for loose nanofiltration, J. Colloid Interface Sci. 523 (2018) 86–97.

[65] M. Safarpour, V. Vatanpour, A. Khataee, Preparation and characterization of graphene oxide/TiO2 blended PES nanofiltration membrane with improved antifouling and separation performance, Desalination 393 (2016) 65–78.

[66] J. Zhu, M. Tian, J. Hou, J. Wang, J. Lin, Y. Zhang, et al., Surface zwitterionic functionalized graphene oxide for a novel loose nanofiltration membrane, J. Mater. Chem. A 4 (2016) 1980–1990.

[67] J. Zhu, M. Tian, Y. Zhang, H. Zhang, J. Liu, Fabrication of a novel "loose" nanofiltration membrane by facile blending with chitosan-montmorillonite nanosheets for dyes purification, Chem. Eng. J. 265 (2015) 184–193.

[68] N. Gholami, H. Mahdavi, Nanofiltration composite membranes of polyethersulfone and graphene oxide and sulfonated graphene oxide, Adv. Polym. Technol. 37 (2018) 3529–3541.

[69] M. Mulder, Basic Principles of Membrane Technology, Springer Netherlands, 1991.

[70] W.J. Lau, A.F. Ismail, P.S. Goh, N. Hilal, B.S. Ooi, Characterization methods of thin film composite nanofiltration membranes, Sep. Purif. Rev. 44 (2015) 135–156.

[71] B.B. Vyas, P. Ray, Preparation of nanofiltration membranes and relating surface chemistry with potential and topography: application in separation and desalting of amino acids, Desalination 362 (2015) 104–116.

[72] H.J. Rezania, V. Vatanpour, A. Shockravi, M. Ehsani, Study of synergetic effect and comparison of novel sulfonated and carboxylated bulky diamine-diol and piperazine in preparation of negative charge NF membrane, Sep. Purif. Technol. (2019) 284–296.

[73] M.N.A. Seman, M. Khayet, N. Hilal, Nanofiltration thin-film composite polyester polyethersulfone-based membranes prepared by interfacial polymerization, J. Membr. Sci. 348 (2010) 109–116.

[74] Q.F. An, W.D. Sun, Q. Zhao, Y.L. Ji, C.J. Gao, Study on a novel nanofiltration membrane prepared by interfacial polymerization with zwitterionic amine monomers, J. Membr. Sci. 431 (2013) 171–179.

[75] W.L. Ang, D. Nordin, A.W. Mohammad, A. Benamor, N. Hilal, Effect of membrane performance including fouling on cost optimization in brackish water desalination process, Chem. Eng. Res. Des. 117 (2017) 401–413.

[76] Y. Li, Y. Su, Y. Dong, X. Zhao, Z. Jiang, R. Zhang, et al., Separation performance of thin-film composite nanofiltration membrane through interfacial polymerization using different amine monomers, Desalination 333 (2014) 59–65.

[77] Y. Li, Y. Su, J. Li, X. Zhao, R. Zhang, X. Fan, et al., Preparation of thin film composite nanofiltration membrane with improved structural stability through the mediation of polydopamine, J. Membr. Sci. 476 (2015) 10–19.

[78] L.F. Liu, X. Huang, X. Zhang, K. Li, Y.L. Ji, C.Y. Yu, et al., Modification of polyamide TFC nanofiltration membrane for improving separation and antifouling properties, RSC Adv. 8 (2018) 15102–15110.

[79] X.D. Weng, Y.L. Ji, R. Ma, F.Y. Zhao, Q.F. An, C.J. Gao, Superhydrophilic and antibacterial zwitterionic polyamide nanofiltration membranes for antibiotics separation, J. Membr. Sci. 510 (2016) 122–130.

[80] H. Zhang, B. Li, J. Pan, Y. Qi, J. Shen, C. Gao, B. Van der Bruggen, Carboxyl-functionalized graphene oxide polyamide nanofiltration membrane for desalination of dye solutions containing monovalent salt, J. Membr. Sci. 539 (2017) 128–137.

[81] B.W. Zhou, H.Z. Zhang, Z.L. Xu, Y.J. Tang, Interfacial polymerization on PES hollow fiber membranes using mixed diamines for nanofiltration removal of salts containing oxyanions and ferric ions, Desalination 394 (2016) 176–184.

[82] L. Pérez-Manríquez, P. Neelakanda, K.V. Peinemann, Morin-based nanofiltration membranes for organic solvent separation processes, J. Membr. Sci. 554 (2018) 1–5.

[83] L. Bai, Y. Liu, N. Bossa, A. Ding, N. Ren, G. Li, et al., Incorporation of cellulose nanocrystals (CNCs) into the polyamide layer of thin-film composite (TFC) nanofiltration membranes for enhanced separation performance and antifouling properties, Environ. Sci. Technol. 52 (2018) 11178–11187.

[84] L.X. Dong, X.C. Huang, Z. Wang, Z. Yang, X.M. Wang, C.Y. Tang, A thin-film nanocomposite nanofiltration membrane prepared on a support with in situ embedded zeolite nanoparticles, Sep. Purif. Technol. 166 (2016) 230–239.

[85] D. Hu, Z.L. Xu, C. Chen, Polypiperazine-amide nanofiltration membrane containing silica nanoparticles prepared by interfacial polymerization, Desalination 301 (2012) 75–81.

[86] M. Kamrani, A. Akbari, A. Yunessnia lehi, Chitosan-modified acrylic nanofiltration membrane for efficient removal of pharmaceutical compounds, J. Environ. Chem. Eng. 6 (2018) 583–587.

[87] G.S. Lai, W.J. Lau, P.S. Goh, A.F. Ismail, N. Yusof, Y.H. Tan, Graphene oxide incorporated thin film nanocomposite nanofiltration membrane for enhanced salt removal performance, Desalination 387 (2016) 14–24.

[88] T.Y. Liu, Z.H. Liu, R.X. Zhang, Y. Wang, B.Vd Bruggen, X.L. Wang, Fabrication of a thin film nanocomposite hollow fiber nanofiltration membrane for wastewater treatment, J. Membr. Sci. 488 (2015) 92–102.

[89] N. Misdan, N. Ramlee, N.H.H. Hairom, S.N.W. Ikhsan, N. Yusof, W.J. Lau, et al., CuBTC metal organic framework incorporation for enhancing separation and antifouling properties of nanofiltration membrane, Chem. Eng. Res. Des. 148 (2019) 227–239.

[90] A. Mollahosseini, A. Rahimpour, Interfacially polymerized thin film nanofiltration membranes on TiO2 coated polysulfone substrate, J. Ind. Eng. Chem. 20 (2014) 1261–1268.

[91] W. Shao, C. Liu, H. Ma, Z. Hong, Q. Xie, Y. Lu, Fabrication of pH-sensitive thin-film nanocomposite nanofiltration membranes with enhanced performance by incorporating amine-functionalized graphene oxide, Appl. Surf. Sci. 487 (2019) 1209–1221.

[92] C. Wang, Z. Li, J. Chen, Z. Li, Y. Yin, L. Cao, et al., Covalent organic framework modified polyamide nanofiltration membrane with enhanced performance for desalination, J. Membr. Sci. 523 (2017) 273–281.

[93] C. Zhang, K. Wei, W. Zhang, Y. Bai, Y. Sun, J. Gu, Graphene oxide quantum dots incorporated into a thin film nanocomposite membrane with high flux and antifouling properties for low-pressure nanofiltration, ACS Appl. Mater. Interfaces 9 (2017) 11082–11094.

[94] F.Y. Zhao, Y.L. Ji, X.D. Weng, Y.F. Mi, C.C. Ye, Q.F. An, et al., High-flux positively charged nanocomposite nanofiltration membranes filled with poly(dopamine) modified multiwall carbon nanotubes, ACS Appl. Mater. Interfaces 8 (2016) 6693–6700.

[95] X. Li, C. Zhao, M. Yang, B. Yang, D. Hou, T. Wang, Reduced graphene oxide-NH2 modified low pressure nanofiltration composite hollow fiber membranes with improved water flux and antifouling capabilities, Appl. Surf. Sci. 419 (2017) 418–428.

[96] P. Wen, Y. Chen, X. Hu, B. Cheng, D. Liu, Y. Zhang, et al., Polyamide thin film composite nanofiltration membrane modified with acyl chlorided graphene oxide, J. Membr. Sci. 535 (2017) 208–220.

[97] S.M. Xue, C.H. Ji, Z.L. Xu, Y.J. Tang, R.H. Li, Chlorine resistant TFN nanofiltration membrane incorporated with octadecylamine-grafted GO and fluorine-containing monomer, J. Membr. Sci. 545 (2018) 185–195.

[98] Z. Liao, X. Fang, J. Xie, Q. Li, D. Wang, X. Sun, et al., Hydrophilic hollow nanocube-functionalized thin film nanocomposite membrane with enhanced nanofiltration performance, ACS Appl. Mater. Interfaces 11 (2019) 5344–5352.

[99] M.R. Mahdavi, M. Delnavaz, V. Vatanpour, J. Farahbakhsh, Effect of blending polypyrrole coated multi-walled carbon nanotube on desalination performance and antifouling property of thin film nanocomposite nanofiltration membranes, Sep. Purif. Technol. 184 (2017) 119–127.

[100] Z. Wang, Z. Wang, S. Lin, H. Jin, S. Gao, Y. Zhu, et al., Nanoparticle-templated nanofiltration membranes for ultrahigh performance desalination, Nat. Commun. 9 (2018).

[101] L. Hu, L. Chen, Y. Fang, A. Wang, C. Chen, Z. Yan, Facile synthesis of zeolitic imidazolate framework-8 (ZIF-8) by forming imidazole-based deep eutectic solvent, Microporous Mesoporous Mater. 268 (2018) 207–215.

[102] M. Wu, J. Yuan, H. Wu, Y. Su, H. Yang, X. You, et al., Ultrathin nanofiltration membrane with polydopamine-covalent organic framework interlayer for enhanced permeability and structural stability, J. Membr. Sci. (2019) 131–141.

[103] M. Wu, T. Ma, Y. Su, H. Wu, X. You, Z. Jiang, et al., Fabrication of composite nanofiltration membrane by incorporating attapulgite nanorods during interfacial polymerization for high water flux and antifouling property, J. Membr. Sci. 544 (2017) 79–87.

[104] R. Hu, R. Zhang, Y. He, G. Zhao, H. Zhu, Graphene oxide-in-polymer nanofiltration membranes with enhanced permeability by interfacial polymerization, J. Membr. Sci. 564 (2018) 813–819.

[105] R. Jiang, W. Wen, J.M. Wu, Titania nanowires coated PEI/P25 membranes for photocatalytic and ultrafiltration applications, New J. Chem. 42 (2018) 3020–3027.

[106] X. Yang, Y. Du, X. Zhang, A. He, Z.K. Xu, Nanofiltration membrane with a mussel-inspired interlayer for improved permeation performance, Langmuir 33 (2017) 2318–2324.

[107] X. Zhang, Y. Lv, H.C. Yang, Y. Du, Z.K. Xu, Polyphenol coating as an interlayer for thin-film composite membranes with enhanced nanofiltration performance, ACS Appl. Mater. Interfaces 8 (2016) 32512–32519.

[108] K.P. Lee, T.C. Arnot, D. Mattia, A review of reverse osmosis membrane materials for desalination-development to date and future potential, J. Membr. Sci. 370 (2011) 1–22.

[109] P.S. Zhong, N. Widjojo, T.S. Chung, M. Weber, C. Maletzko, Positively charged nanofiltration (NF) membranes via UV grafting on sulfonated polyphenylenesulfone (sPPSU) for effective removal of textile dyes from wastewater, J. Membr. Sci. 417-418 (2012) 52–60.

[110] V. Vatanpour, M. Esmaeili, M. Safarpour, A. Ghadimi, J. Adabi, Synergistic effect of carboxylated-MWCNTs on the performance of acrylic acid UV-grafted polyamide nanofiltration membranes, React. Funct. Polym. 134 (2019) 74–84.

[111] M. Paul, S.D. Jons, Chemistry and fabrication of polymeric nanofiltration membranes: a review, Polymer 103 (2016) 417–456.

[112] R. Zhang, Y. Su, X. Zhao, Y. Li, J. Zhao, Z. Jiang, A novel positively charged composite nanofiltration membrane prepared by bio-inspired adhesion of polydopamine and surface grafting of poly(ethylene imine), J. Membr. Sci. 470 (2014) 9–17.

[113] Hm Xu, Jf Wei, Xl Wang, Nanofiltration hollow fiber membranes with high charge density prepared by simultaneous electron beam radiation-induced graft polymerization for removal of Cr(VI), Desalination 346 (2014) 122–130.

[114] A. Reinhardt, I. Thomas, J. Schmauck, R. Giernoth, A. Schulze, I. Neundorf, Electron beam immobilization of novel antimicrobial, short peptide motifs leads to membrane surfaces with promising antibacterial properties, J. Funct. Biomater. 9 (2018) 21.

[115] M. Amirilargani, M. Sadrzadeh, E.J.R. Sudhölter, L.C.P.M. de Smet, Surface modification methods of organic solvent nanofiltration membranes, Chem. Eng. J. 289 (2016) 562–582.

[116] R.W. Baker, Membrane Technology and Applications, 23, John Wiley & Sons Ltd., 2004, p. 588.

[117] Z.F. Gao, G.M. Shi, Y. Cui, T.S. Chung, Organic solvent nanofiltration (OSN) membranes made from plasma grafting of polyethylene glycol on cross-linked polyimide ultrafiltration substrates, J. Membr. Sci. 565 (2018) 169–178.

[118] X. Zhang, C. Liu, J. Yang, C.-Y. Zhu, L. Zhang, Z.-K. Xu, Nanofiltration membranes with hydrophobic microfiltration substrates for robust structure stability and high water permeation flux, J. Membr. Sci. 593 (2020) 117444.

[119] L.Y. Ng, A.W. Mohammad, C.Y. Ng, A review on nanofiltration membrane fabrication and modification using polyelectrolytes: effective ways to develop membrane selective barriers and rejection capability, Adv. Colloid Interface Sci. 197-198 (2013) 85–107.

[120] Q. Chen, P. Yu, W. Huang, S. Yu, M. Liu, C. Gao, High-flux composite hollow fiber nanofiltration membranes fabricated through layer-by-layer deposition of oppositely charged crosslinked polyelectrolytes for dye removal, J. Membr. Sci. 492 (2015) 312–321.

[121] R.M. DuChanois, R. Epsztein, J.A. Trivedi, M. Elimelech, Controlling pore structure of polyelectrolyte multilayer nanofiltration membranes by tuning polyelectrolyte-salt interactions, J. Membr. Sci. (2019) 413–420.

[122] C.M. Salgado, E. Fernández-Fernández, L. Palacio, F.J. Carmona, A. Hernández, P. Prádanos, Application of pervaporation and nanofiltration membrane processes for the elaboration of full flavored low alcohol white wines, Food Bioprod. Process. 101 (2017) 11–21.

[123] D.M. Warsinger, S. Chakraborty, E.W. Tow, M.H. Plumlee, C. Bellona, S. Loutatidou, et al., A review of polymeric membranes and processes for potable water reuse, Prog. Polym. Sci. 81 (2018) 209–237.

[124] M. Sairam, X.X. Loh, Y. Bhole, I. Sereewatthanawut, K. Li, A. Bismarck, et al., Spiral-wound polyaniline membrane modules for organic solvent nanofiltration (OSN), J. Membr. Sci. 349 (2010) 123–129.

[125] J.S. Vrouwenvelder, D.A. Graf von der Schulenburg, J.C. Kruithof, M.L. Johns, M.C.M. van Loosdrecht, Biofouling of spiral-wound nanofiltration and reverse osmosis membranes: a feed spacer problem, Water Res. 43 (2009) 583–594.

[126] M.A. Abdel-Fatah, Nanofiltration systems and applications in wastewater treatment: review article, Ain Shams Eng. J. 9 (2018) 3077–3092.

[127] H. Yang, N. Wang, L. Wang, H.X. Liu, Q.F. An, S. Ji, Vacuum-assisted assembly of ZIF-8@GO composite membranes on ceramic tube with enhanced organic solvent nanofiltration performance, J. Membr. Sci. 545 (2018) 158–166.

[128] S. Yu, Q. Cheng, C. Huang, J. Liu, X. Peng, M. Liu, et al., Cellulose acetate hollow fiber nanofiltration membrane with improved permselectivity prepared through hydrolysis followed by carboxymethylation, J. Membr. Sci. 434 (2013) 44–54.

[129] T. Turken, R. Sengur-Tasdemir, B. Sayinli, G.M. Urper-Bayram, E. Ates-Genceli, V.V. Tarabara, et al., Reinforced thin-film composite nanofiltration membranes: fabrication, characterization, and performance testing, J. Appl. Polym. Sci. 136 (2019) 48001.

[130] J.M. Gohil, P. Ray, A review on semi-aromatic polyamide TFC membranes prepared by interfacial polymerization: potential for water treatment and desalination, Sep. Purif. Technol. 181 (2017) 159–182.

[131] Z. Thong, J. Gao, J.X.Z. Lim, K.Y. Wang, T.S. Chung, Fabrication of loose outer-selective nanofiltration (NF) polyethersulfone (PES) hollow fibers via single-step spinning process for dye removal, Sep. Purif. Technol. 192 (2018) 483–490.

[132] I.H. Tsibranska, B. Tylkowski, Concentration of ethanolic extracts from *Sideritis* ssp. L. by nanofiltration: comparison of dead-end and cross-flow modes, Food Bioprod. Process. 91 (2013) 169–174.

[133] M.F. Tay, C. Liu, E.R. Cornelissen, B. Wu, T.H. Chong, The feasibility of nanofiltration membrane bioreactor (NF-MBR) + reverse osmosis (RO) process for water reclamation: comparison with ultrafiltration membrane bioreactor (UF-MBR) + RO process, Water Res. 129 (2018) 180–189.

[134] T. Srisukphun, C. Chiemchaisri, W. Chiemchaisri, M. Thanuttamavong, Fouling and cleaning of reverse osmosis membrane applied to membrane bioreactor effluent treating textile wastewater, Environ. Eng. Res. 21 (2016) 45–51.

[135] F.H. Butt, F. Rahman, U. Baduruthamal, Characterization of foulants by autopsy of RO desalination membranes, Desalination 114 (1997) 51–64.

[136] M. Larronde-Larretche, X. Jin, Microalgae (*Scenedesmus obliquus*) dewatering using forward osmosis membrane: influence of draw solution chemistry, Algal Res. 15 (2016) 1–8.

[137] J. Lee, S. Jeong, Y. Ye, V. Chen, S. Vigneswaran, T.O. Leiknes, et al., Protein fouling in carbon nanotubes enhanced ultrafiltration membrane: fouling mechanism as a function of pH and ionic strength, Sep. Purif. Technol. 176 (2017) 323–334.

[138] A.W. Zularisam, A. Ahmad, M. Sakinah, A.F. Ismail, T. Matsuura, Role of natural organic matter (NOM), colloidal particles, and solution chemistry on ultrafiltration performance, Sep. Purif. Technol. 78 (2011) 189–200.

[139] T.Y. Haan, W.S. Yee, A.W. Mohammad, Studies on the surface properties and fabrication method of mixed-matrix membrane for textile industry wastewater treatment, Desalin. Water Treat. 135 (2018) 303–313.

[140] A. Soleymani Lashkenrai, M. Najafi, M. Peyravi, M. Jahanshahi, M.T.H. Mosavian, A. Amiri, et al., Direct filtration procedure to attain antibacterial TFC membrane: a facile developing route of membrane surface properties and fouling resistance, Chem. Eng. Res. Des. 149 (2019) 158–168.

[141] S. Mu, S. Wang, S. Liang, K. Xiao, H. Fan, B. Han, et al., Effect of the relative degree of foulant “hydrophobicity” on membrane fouling, J. Membr. Sci. 570-571 (2019) 1–8.

[142] S. Shao, W. Fu, X. Li, D. Shi, Y. Jiang, J. Li, et al., Membrane fouling by the aggregations formed from oppositely charged organic foulants, Water Res. 159 (2019) 95–101.

[143] Y.H. Teow, A.W. Mohammad, S. Ramli, M.S. Sajab, N.I. Mohamad Mazuki, Potential of membrane technology for treatment and reuse of water from Old Mining Lakes, Sains Malays. 47 (2018) 2887–2897.

[144] E. Ferrer-Polonio, K. White, J.A. Mendoza-Roca, A. Bes-Piá, The role of the operating parameters of SBR systems on the SMP production and on membrane fouling reduction, J. Environ. Manage. 228 (2018) 205–212.

[145] Y.H. Teow, Z.H. Wong, M.S. Takriff, A.W. Mohammad, Fouling behaviours of two stages microalgae/membrane filtration system applied to palm oil mill effluent treatment, Membr. Water Treat. 9 (2018) 373–383.

[146] X. Zhao, H. Zhang, J. Wang, Filtering surface water with a polyurethane-based hollow fiber membrane: effects of operating pressure on membrane fouling, Chin. J. Chem. Eng. 22 (2014) 583–589.

[147] A. Gorenflo, D. Velázquez-Padrón, F.H. Frimmel, Nanofiltration of a German groundwater of high hardness and NOM content: performance and costs, Desalination 151 (2003) 253–265.

[148] J. Shen, A.I. Schäfer, Factors affecting fluoride and natural organic matter (NOM) removal from natural waters in Tanzania by nanofiltration/reverse osmosis, Sci. Total Environ. 527-528 (2015) 520–529.

[149] Y.H. Teow, PVDF-TiO2 Mixed-Matrix Membrane With Anti-fouling Properties for Humic Acid Removal, Ph.D dissertation, Universiti Sains Malaysia, Malaysia, 2014.

[150] A.W. Zularisam, A.F. Ismail, M.R. Salim, M. Sakinah, H. Ozaki, The effects of natural organic matter (NOM) fractions on fouling characteristics and flux recovery of ultrafiltration membranes, Desalination 212 (2007) 191–208.

[151] K.C. Ho, Y.H. Teow, A.W. Mohammad, W.L. Ang, P.H. Lee, Development of graphene oxide (GO)/multi-walled carbon nanotubes (MWCNTs) nanocomposite conductive membranes for electrically enhanced fouling mitigation, J. Membr. Sci. 552 (2018) 189–201.

[152] N. Adilah Rosnan, T.Y. Haan, A.W. Mohammad, The effect of ZnO loading for the enhancement of PSF/ZnO-GO mixed matrix membrane performance, Sains Malays. 47 (2018) 2035–2045.

[153] L. Zhao, W.S.W. Ho, Novel reverse osmosis membranes incorporated with a hydrophilic additive for seawater desalination, J. Membr. Sci. 455 (2014) 44–54.

[154] Y.H. Teow, A.A. Latif, J.K. Lim, H.P. Ngang, L.Y. Susan, B.S. Ooi, Hydroxyl functionalized PVDF-TiO2 ultrafiltration membrane and its antifouling properties, J. Appl. Polym. Sci. 132 (2015).

[155] G. Singh, L. Song, Quantifying the effect of ionic strength on colloidal fouling potential in membrane filtration, J. Colloid Interface Sci. 284 (2005) 630–638.

[156] M. Kabsch-Korbutowicz, K. Majewska-Nowak, T. Winnicki, Analysis of membrane fouling in the treatment of water solutions containing humic acids and mineral salts, Desalination 126 (1999) 179–185.

[157] W. Yuan, A.L. Zydney, Humic acid fouling during ultrafiltration, Environ. Sci. Technol. 34 (2000) 5043–5050.

[158] K.L. Jones, C.R. O'Melia, Protein and humic acid adsorption onto hydrophilic membrane surfaces: effects of pH and ionic strength, J. Membr. Sci. 165 (2000) 31–46.

[159] E.E. Chang, Y.C. Chang, C.H. Liang, C.P. Huang, P.C. Chiang, Identifying the rejection mechanism for nanofiltration membranes fouled by humic acid and calcium ions exemplified by acetaminophen, sulfamethoxazole, and triclosan, J. Hazard. Mater. 221-222 (2012) 19–27.

[160] M.S.H. Ghani, T.Y. Haan, A.W. Lun, A.W. Mohammad, R. Ngteni, K.M.M. Yusof, Fouling assessment of tertiary palm oil mill effluent (POME) membrane treatment for water reclamation, J. Water Reuse Desalin. 8 (2017) 412–423. jwrd2017198.

[161] W. Gao, H. Liang, J. Ma, M. Han, Zl Chen, Zs Han, et al., Membrane fouling control in ultrafiltration technology for drinking water production: a review, Desalination 272 (2011) 1–8.

[162] P. Lipp, G. Baidauf, A. Schmitt, B. Theis, Long-term behaviour of UF membranes treating surface water, Water Sci. Technol. Water Supply 3 (2003) 31–37.

[163] Z. Cui, T. Taha, Enhancement of ultrafiltration using gas sparging: a comparison of different membrane modules, J. Chem. Technol. Biotechnol. 78 (2003) 249–253.

[164] A. Aguiar, L. Andrade, L. Grossi, W. Pires, M. Amaral, Acid mine drainage treatment by nanofiltration: a study of membrane fouling, chemical cleaning, and membrane ageing, Sep. Purif. Technol. 192 (2018) 185–195.

[165] M.C. Hacıfazlıoğlu, İ. Parlar, T. Pek, N. Kabay, Evaluation of chemical cleaning to control fouling on nanofiltration and reverse osmosis membranes after desalination of MBR effluent, Desalination 466 (2019) 44–51.

[166] D. Zhao, L. Qiu, J. Song, J. Liu, Z. Wang, Y. Zhu, et al., Efficiencies and mechanisms of chemical cleaning agents for nanofiltration membranes used in produced wastewater desalination, Sci. Total. Environ. 652 (2019) 256–266.

[167] T. Chittrakarn, Y. Tirawanichakul, S. Sirijarukul, C. Yuenyao, Plasma induced graft polymerization of hydrophilic monomers on polysulfone gas separation membrane surfaces, Surf. Coat. Technol. 296 (2016) 157–163.

[168] A.F.H.A. Rahman, M.N.A. Seman, Modification of commercial ultrafiltration and nanofiltration membranes by UV-photografting technique for forward osmosis application, Mater. Today Proc. 17 (2019) 590–598.

[169] K. Pan, P. Fang, B. Cao, Novel composite membranes prepared by interfacial polymerization on polypropylene fiber supports pretreated by ozone-induced polymerization, Desalination 294 (2012) 36–43.

[170] X. Zhao, W. Chen, Y. Su, W. Zhu, J. Peng, Z. Jiang, et al., Hierarchically engineered membrane surfaces with superior antifouling and self-cleaning properties, J. Membr. Sci. 441 (2013) 93–101.

[171] H.Y. Yu, M.X. Hu, Z.K. Xu, J.L. Wang, S.Y. Wang, Surface modification of polypropylene microporous membranes to improve their antifouling property in MBR: NH3 plasma treatment, Sep. Purif. Technol. 45 (2005) 8–15.

[172] H.P. Ngang, A.L. Ahmad, S.C. Low, B.S. Ooi, Preparation of mixed-matrix membranes for micellar enhanced ultrafiltration based on response surface methodology, Desalination 293 (2012) 7–20.

[173] R. Janssens, M.B. Cristovao, M.R. Bronze, J.G. Crespo, V.J. Pereira, P. Luis, Coupling of nanofiltration and UV, UV/TiO2 and UV/H2O2 processes for the removal of anti-cancer drugs from real secondary wastewater effluent, J. Environ. Chem. Eng. 7 (2019) 103351.

[174] H. Zangeneh, A.A. Zinatizadeh, S. Zinadini, M. Feyzi, D.W. Bahnemann, A novel photocatalytic self-cleaning PES nanofiltration membrane incorporating triple metal-nonmetal doped TiO2 (K-B-N-TiO2) for post treatment of biologically treated palm oil mill effluent, React. Funct. Polym. 127 (2018) 139–152.

[175] H. Zangeneh, A.A. Zinatizadeh, S. Zinadini, M. Feyzi, D.W. Bahnemann, Preparation and characterization of a novel photocatalytic self-cleaning PES nanofiltration membrane by embedding a visible-driven photocatalyst boron doped-TiO2–SiO2/CoFe2O4nanoparticles, Sep. Purif. Technol. 209 (2019) 764–775.

[176] R.A. Damodar, S.J. You, H.H. Chou, Study the self cleaning, antibacterial and photocatalytic properties of TiO2 entrapped PVDF membranes, J. Hazard. Mater. 172 (2009) 1321–1328.

[177] Y.H. Teow, B.S. Ooi, A.L. Ahmad, Fouling behaviours of PVDF-TiO2 mixed-matrix membrane applied to humic acid treatment, J. Water Process. Eng. 15 (2017) 89–98.

[178] Y.H. Teow, B.S. Ooi, A.L. Ahmad, J.K. Lim, Mixed-matrix membrane for humic acid removal: influence of different types of TiO_2 on membrane morphology and performance, Int. J. Chem. Eng. Appl. 3 (2012) 374–379.

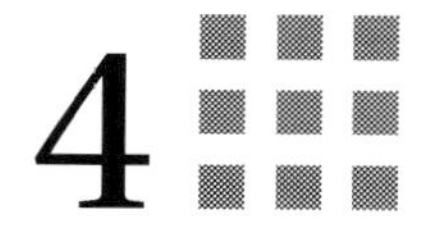

4 Recent development in nanofiltration process applications

Wei Lun Ang[1,2], Abdul Wahab Mohammad[1,2]

[1]RESEARCH CENTER FOR SUSTAINABLE PROCESS TECHNOLOGY (CESPRO), FACULTY OF ENGINEERING AND BUILT ENVIRONMENT, UNIVERSITI KEBANGSAAN MALAYSIA, BANGI, MALAYSIA [2]DEPARTMENT OF CHEMICAL AND PROCESS ENGINEERING, FACULTY OF ENGINEERING AND BUILT ENVIRONMENT, UNIVERSITI KEBANGSAAN MALAYSIA, BANGI, MALAYSIA

4.1 Introduction

In the past few decades, membrane filtration has emerged as one of the most versatile separation technologies that have found widespread applications in various sectors. Among the membrane processes currently present in the market, nanofiltration (NF) membranes are expected to observe increasing interest due to their unique properties and versatility as a separation process [1,2]. NF membranes typically possess a membrane pore size of 0.5−2.0 nm (corresponding to a molecular weight cutoff of 100−5000 Da), which is in between the range of ultrafiltration (UF) and reverse osmosis (RO). NF membranes also exhibit a certain level of surface charge due to the presence of a few particular functional groups on the membrane surface [3]. These characteristics grant the NF membranes the ability to reject impurities based on two main mechanisms: electrostatic repulsion and steric hindrance. Consequently, NF membranes have found niche applications where not only it can remove the impurities that UF membranes fail to but also enjoy the benefits of having higher water flux (lower energy consumption and cost) compared to RO [4,5]. Thus NF has found wide applications across different industrial sectors such as used for water and wastewater treatment, desalination, food industry, biorefinery applications, and organic solvent separation to name a few.

Driven by the versatility and broad acceptance of NF across a range of industrial sectors, NF application market is expected to grow at a fast rate and the global sales are predicted to reach $450 million by 2019 [6]. The development of NF technologies has also received great attention from the researchers and academicians, as indicated by the steady increase of the number of NF research articles published since 2007 [1]. A huge portion ($>55\%$) of the published articles fall into the category of NF application. This shows that the research community is actively investigating the potential and performance of NF membranes in various

Osmosis Engineering. DOI: https://doi.org/10.1016/B978-0-12-821016-1.00010-3

applications as well as to resolve the challenges encountered during the operation process. On the one hand, the capability of NF process can be improved through the fabrication and subsequent modification of NF membranes to enhance its characteristics (mechanical strength, antifouling, and stability) and performance (water flux, impurities rejection, fouling propensity, and cost) [3]. On the other hand, understanding the influence of operating conditions on the NF performance and the design of NF process can also help to minimize the obstacles encountered for a particular industrial application [7]. Hence, this chapter will offer an overview of the application of NF in a wide range of industrial sectors and provide an insight into the latest development of NF technology in those applications.

4.2 Applications of NF membrane process

Owing to its greater permeate flux and being able to function at a lower pressure (lower operating cost), NF membrane process appears to be more attractive and a preferable process to RO in cases where there is no need for complete removal or separation of impurities. NF is typically applied in water-related processes (e.g., wastewater treatment and desalination), either individually or integrated with other technologies for a better treatment efficiency. Recently, NF has also found increasing application potential in the food and biotechnology industries, where the membrane is mainly utilized for purification and concentration purposes. Another branch of NF application is for the separation process dealing with organic solvents. Overall, since NF is an intermediate process of UF and RO, advantages such as efficient removal of dissolved solutes, including multivalent ions and organic compounds of high molar mass, but, with lower pressure requirements and higher flows than RO, have been the consideration factors for its acceptance in various applications [8].

4.2.1 Water and wastewater

The advances and breakthroughs in membrane technology have made it a widely adopted process in water and wastewater treatment plants. Membranes of different variations have found niche applications in water and wastewater treatment industries, depending on the treatment targets. Stringent regulation for wastewater discharge has also driven the adoption of membrane technology to remove the contaminants. Since the membrane technology produces treated water with good quality, the aim of wastewater treatment has also been shifted from meeting discharge regulation to reclaim and reuse the water. Among the membrane variations, NF membrane has received great attention in treating and reclaiming water and wastewater from different sources, which will be discussed below.

Water scarcity has threatened the agriculture sector where irrigation is required to grow the crops. Without sufficient supply of clean water, the agricultural activities will be hindered and food availability for the survival of humanity will also be severely affected. To counter the problem of water shortage and to minimize the competition for clean water resources with other usages (especially drinking water), reclaimed wastewater could be an attractive alternative source of water for agriculture [9]. NF has been integrated into several treatment processes for wastewater reclamation and reuse in agricultural activities. For instance, biologically treated wastewater from sources such as municipal, pig slurry, and baker's yeast manufacturing company has been used as the feedwater to NF process [10–12]. The role played by NF here was for the removal of impurities and pollutants that would impede the reusability of reclaimed water. It was reported that the NF process managed to produce water of good quality with low salinity, pathogenicity, nutrients, and heavy metals, meeting the agricultural irrigation water standard. The filtered water can be reused either as irrigation water or for washing of animals and farmhouses. In the case where the filtered wastewater contains insufficient beneficial ions for irrigation purpose, NF membrane could be used to enrich the divalent ions and nutrients in retentate and further diluted with downstream RO permeate to balance the sodium content, sodium absorption ratio, and electroconductivity (Fig. 4–1) [13].

The dairy industry is another high-water consumption industry, which is closely related to the agriculture sector. The generation of large amounts of liquid waste that contains high concentrations of organic matter and nutrients (mainly carbohydrates, proteins, and fats originated from milk) has been a challenge for conventional treatment systems. Problems related to conventional treatment systems are high production of scum, poor sludge settleability, low resistance to organic shock load, and difficulties in removing nutrients (nitrogen and phosphorus) and degrading fats and oils [14–17]. In accordance with the effort

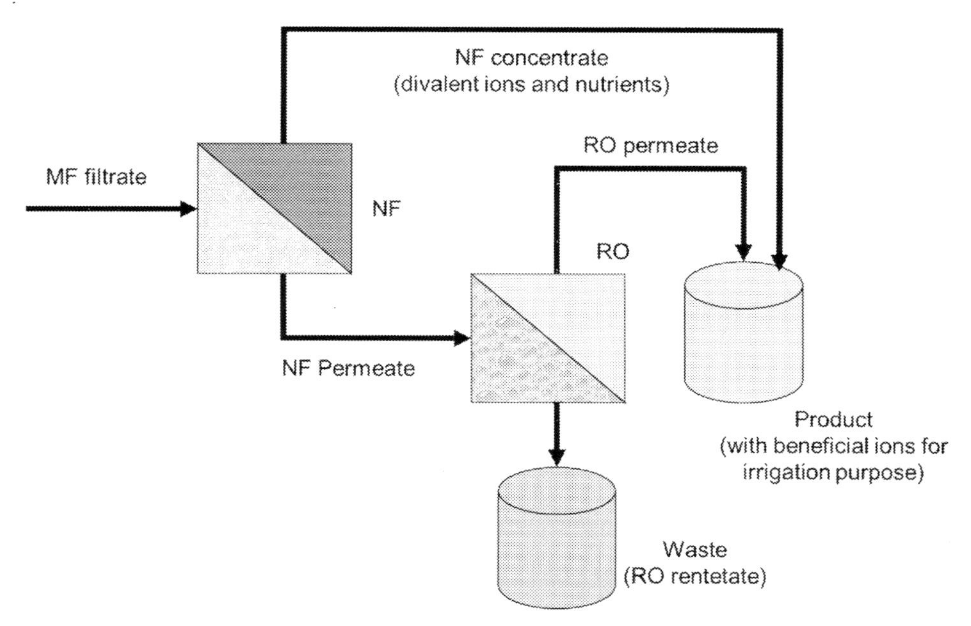

FIGURE 4–1 Integrated MF–NF–RO for water reclamation. *MF*, Microfiltration; *NF*, nanofiltration; *RO*, Reverse osmosis. *Adapted from S.M. Mrayed, P. Sanciolo, L. Zou, G. Leslie, An alternative membrane treatment process to produce low-salt and high-nutrient recycled water suitable for irrigation purposes, Desalination. 274 (2011) 144–149. https://doi.org/10.1016/j.desal.2011.02.003.*

to meet the increasingly stringent regulations for wastewater discharge, NF has also been utilized to treat dairy wastewater and reuse the water in the industry. For instance, Andrade et al. [18] developed an integrated membrane bioreactor (MBR)-NF process to reclaim the water from real dairy industry wastewater. It was reported that the MBR successfully reduced the chemical oxygen demand (COD) and removed the color by 99% and 98.4%, respectively. However, the total solids removal was less effective (47%) due to the larger microfiltration (MF) membrane pore size used in the MBR. Hence, the NF was integrated with MBR to further polish the treated water. With the incorporation of NF as a tertiary treatment process, the overall COD, color, and total solids levels have been successfully removed, with the efficiencies of 99.9%, 99.3%, and 93.1%, respectively. The NF permeate met the standards for water reused in cooling, steam generation, or good manufacturing practices, such as washing floors and trucks and rinsing external areas [19,20]. On the other hand, after passing through both the MBR and NF membrane, the retentate was in good quality for garden irrigation or could be discarded into water bodies since the COD (73 mg/L) was well below the discharge legislation.

The advancement of living standards and the growth of the modern economy have led to the discharge of potential emerging pollutants such as hazardous chemicals, toxic metals, and biowastes to the environment and water sources. These emerging pollutants are originated from various industrial effluents (including pharmaceutical, food, and metal processing) where the industries are responsible to improve the living conditions of the society. These emerging contaminants, such as endocrine-disrupting compounds and pharmaceuticals/personal care products, are unlikely to be removed by conventional wastewater treatment processes, prompting concerns regarding the possible adverse effects on human health and the ecosystem [5,21–24]. NF membrane has been reported to achieve complete or near-complete removal of a wide range of emerging contaminants, with adsorption, size exclusion, and electrostatic repulsion as the dominants removal mechanisms. Liu et al. [25] have shown that NF membrane could fully reject the trace antibiotics in secondary effluent from wastewater treatment plant and the residual antibiotics in NF concentrate could be effectively eliminated by ozone-based advanced oxidation process. Having said that, the retention of emerging contaminants by NF membrane is mainly governed by the physicochemical properties of the contaminants.

The retention of bisphenol A (BPA), ibuprofen (IBP), and salicylic acid by NF membrane exhibited the interaction between the three rejection mechanisms. At pH 7, BPA ($pKa = 9.6–10.2$) exists as uncharged species, whereas IBP ($pKa = 4.9$) and salicylic acid ($pKa = 2.9$) possess a negative charge. Under this condition, the low rejection of BPA (74.1%) was mainly due to the absence of electrostatic repulsion between the membrane surface and BPA, as the only rejection mechanism was the sieving effect (size exclusion) and the hydrophobic BPA had a high affinity to adsorb to the hydrophobic membrane surface [26]. On the other hand, the near-complete rejection (97%–98%) of IBP and salicylic acid could be attributed to both size exclusion and electrostatic repulsion, given the fact that the pollutants possessed the same charge as NF membrane surface. This indicates that by manipulating the charge status of the contaminants (through water pH), the rejection efficiency of a particular contaminant could be enhanced, as reported in the literature [27–30].

Likewise, water quality such as the ionic strength and the presence of divalent cations and organic matter could also affect the contaminants' rejection efficiency by NF process. It was reported that the emerging contaminants could interact with organic macromolecules to form complexes that were easier to be rejected by NF membrane through size exclusion and/or electrostatic repulsion [31–34]. The increase in ionic strength and divalent cation concentrations would change the organic matter confirmation, which subsequently may affect the interaction between the organic matter and emerging contaminants [35–39]. Overall, although NF membrane has been frequently tested for the rejection of pharmaceutically active compounds from water and wastewater, it is not always as effective for complete removal of all types of pharmaceutical compounds, especially when the molecular weight cutoff of contaminants is smaller than the NF membrane [5,23].

Although NF process could remove the pharmaceuticals from the municipal wastewater, the presence of the retained contaminants in the retentate still poses a problem for the water utility [40]. To handle the secondary effluent, NF has been integrated with advanced oxidation processes (such as photo-Fenton, ozonation, and solar photo-Fenton-like Fe(III)-EDDS complex) as tertiary treatment for degrading the pollutants [41,42]. The incorporation of these advanced oxidation processes managed to eliminate more than 90% of the pharmaceuticals in the NF retentate, which helped to minimize the issue of pharmaceutical residues in secondary waste [43]. As shown in Fig. 4–2, the NF membrane helped to concentrate the pharmaceuticals in retentate and subsequently facilitated more efficient degradation by solar photo-Fenton process.

Recently, the detection of hormones in drinking water sources has been a concern for the public due to its potential negative impacts on the flora and fauna as well as human health. NF process is reportedly capable to remove hormones mainly through adsorption of the compounds to the membrane [44–46]. However, the hormones retained in the retentate still possess a problem for the water utility since the hormones still need to be handled properly. To resolve this issue, NF membrane has been integrated with the oxidation process to degrade the hormones in retentate [47,48]. For instance, Lopes et al. [47] integrated NF process with UV irradiation for river surface water treatment in Vale da Pedra, Portuguese. The pilot plant was efficient in delivering high-quality water with recovery rate as high as 91%. The UV photolysis acted as an additional barrier to further remove the resilient compounds such as pesticides and hormones in NF permeate.

The presence of heavy metals in various industrial effluents (e.g., metal plating and mining industries) is a concern for many, as it will cause environmental pollution in addition to being hazardous to living organisms. Various technologies have been developed to treat and manage the wastewaters laden with heavy metal, and this includes membrane process [49,50]. Since most commercial NF membranes are made of synthetic polymers with functional groups that grant the membrane surface charge when comes into contact with water, it has been utilized to reject the heavy metal where to a large extent is present in ionic form. Wang et al. [51] have shown that integrated MF–NF process could remove Cu and Cr from

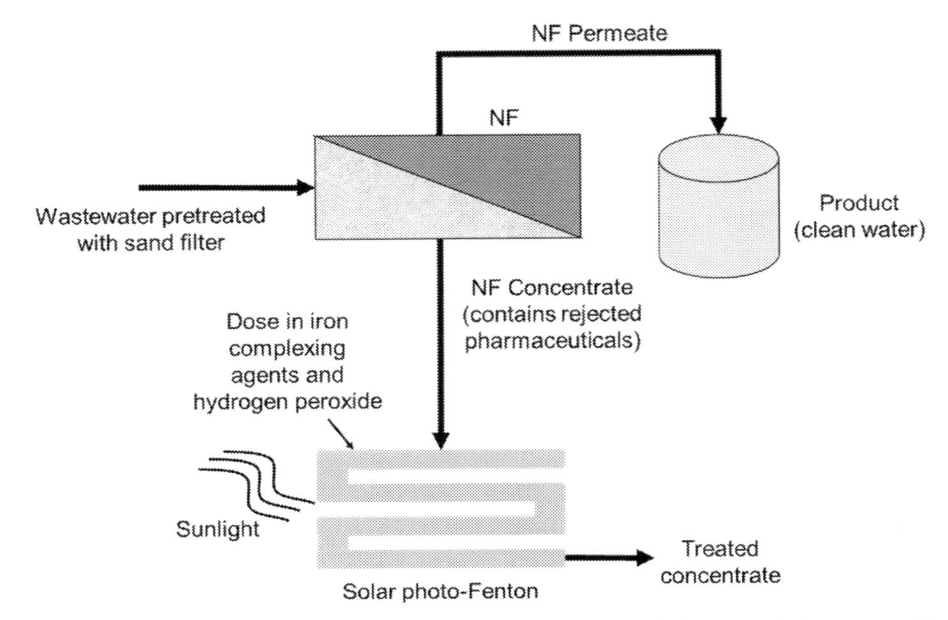

FIGURE 4–2 Integrated NF-solar photo-Fenton process for the degradation of pharmaceuticals. *NF*, Nanofiltration. *Adapted from S. Miralles-Cuevas, I. Oller, J.A.S. Pérez, S. Malato, Removal of pharmaceuticals from MWTP effluent by nanofiltration and solar photo-Fenton using two different iron complexes at neutral pH, Water Res. 64 (2014) 23–31. https://doi.org/10.1016/j.watres.2014.06.032.*

electroplating wastewater with rejection efficiency between 82.8% and 96.6%, depending on the type of NF membranes. The rejection was fully done by NF membrane while MF membrane served as pretreatment to minimize the fouling tendency of subsequent NF process.

The rejection of heavy metals could also be enhanced by modifying the membrane structure and materials. Zhu et al. [52] had fabricated a dual-layer NF hollow fiber membrane for effective removal of heavy metal ions of different charges. By varying the solution pH, the rejection toward $Cr_2O_7^{2-}$ and Pb^{2+} could reach 98% and 93%, respectively. The high removal efficiency could be attributed to the small membrane pore size and low-molecular-weight cut-off, enhanced Donnan exclusion effect associated with the amphoteric membrane property, and low adsorptions of heavy metals on the membrane surface due to membrane hydrophilicity. Gao et al. [53] had synthesized NF membrane with chelating polymers containing negatively charged functional groups for high-efficient heavy metal removal. It was reported that not only did the chelating polymers grant the membrane surface negatively charged but also contribute to the chelating effect of heavy metals. Consequently, the NF membrane exhibited high rejection ($>97\%$) for mixed ions solution (Cu, Ni, Zn, Cr, Cd, and Pb).

Apart from heavy metal pollutants, sulfate and nitrogen compounds are also commonly found in mining wastewater. NF has been employed to remove the sulfate and nitrate since both are charged compounds. Juholin et al. [54] investigated the treatment efficiency of real mining wastewater from different processing stages in rejecting sulfate, manganese, chloride, and nitrate compounds. Overall, the sulfate and manganese rejection was generally high at >91%, but removal of nitrate and chloride ions was unsatisfactory, with the looser membrane (NF 270) that possessed larger molecular-weight cutoff recorded negative rejection for nitrate ions. This indicates that the nitrate compounds permeated through the membrane to the permeate side, a phenomenon that occurred due to the movement of ions to maintain the electroneutrality in the vicinity of a membrane. The divalent sulfate ions are more easily retained compared to monovalent nitrate ions, and subsequently, the latter has higher tendency to pass through the looser NF 270 membrane [55].

Another persistent and challenging contaminant present in wastewater is oil compounds that normally originate from oil and petroleum industries [56]. A full-scale plant consists of biological pretreatment unit, UF, electrodialysis (ED), and NF processes that have been installed in the Daqing oilfield in China to treat the oily produced wastewater. The function of the NF process was to recycle the ED concentrate to improve the water production with a recovery of 75%. However, the facility failed to achieve the target due to the severe membrane fouling (contributed by anionic polyacrylamide and crude oil), reducing the recovery efficiency to 25% [56]. A similar fouling issue has also been reported by Rahimpour et al. [57] in which the NF membrane only managed to reduce the COD and electrical conductivity of real oily wastewater at the efficiencies of 79%–84% and 88%–93%, respectively.

The low-rejection efficiency of NF has also been observed for olive mill wastewater that contains high organic, phenol, and fatty acids. Coskun et al. [58] reported that the NF membrane only reduced 60%–80% of COD compared to 96% achieved by RO membrane, which could be mainly attributed to the larger pore size of NF membrane that resulted in more organic compounds to pass to permeate. On another note, integrated membrane process has been developed to reclaim water and recover phenolic compounds from the olive mill wastewater [59,60]. Pretreatments such as MF and UF have been installed prior to the NF process to remove the suspended solids and impurities that would otherwise foul the NF membrane. It was reported that the NF process capable to produce a retentate enriched with phenolic compounds and permeate (reclaimed water) good enough to be discharged or reused in the mill.

Textile wastewater generally has a complex composition that mainly comprises dyes and inorganic salts (used to enhance dyes uptake of the fabric). Due to this unique combination of contaminants, NF membrane can be used to treat the textile wastewater or it can be employed to fractionate and recover valuable inorganic salts (permeate) and dyes (retentate) for reuse [61–65]. For instance, Cinperi et al. [66] investigated the treatment efficiency of integrated MBR-NF process for woolen textile wastewater for resource recovery. The MBR process reduced 70% of COD, 74% of biochemical oxygen demand (BOD_5), and 86% of total suspended solids, producing permeate with better quality for subsequent NF filtration process. The NF process produced permeates with very low COD and BOD_5 while contained a moderate amount of salts. The reuse of treated water showed no adverse effects on the quality of the product in dying process, suggesting the effectiveness of the integrated MBR-NF in handling textile wastewater. A similar finding has also been reported by Tavangar et al. [67], where electrocoagulation was integrated as a pretreatment prior to the NF process. The electrocoagulation process reduced the membrane fouling propensity and increased the permeate flux of NF membrane by removing dyes with high and medium molecular weight. The NF process then eliminated the leftover color components ($>87\%$) as retentates and produced permeate consists of inorganic salts (rejection less than 4%), where both streams could be potentially reused in the textile industry.

NF membrane process has also been used as an alternative technology to produce drinking water. The first large-scale application of NF in surface water treatment plant is Mery-sur-Oise in France [68]. The water utility included NF at the end of the treatment train (with coagulation, flocculation, settling, and sand filtration as pretreatment) to enhance the removal of organic matter and pesticides. The NF process successfully removed the organic matter and pesticides, with operational costs lower than a traditional plant with refining using ozone and carbon. In another case, García-Vaquero et al. [69] compared the capability of NF (powered by renewable energy) with the conventional treatment system (consists of coagulation, flocculation, sedimentation, filtration, and disinfection) to treat water from the reservoir for the production of drinking water. It was reported that the NF process was more efficient than the conventional treatment system in removing micropollutants, producing treated water with better final water quality.

Aside from surface water (especially river water), groundwater is another alternative source for drinking water production. For instance, more than half of the drinking water produced in the Netherlands was originated from deep groundwater layers. Conventional groundwater treatment (aeration and filtration) could not remove certain impurities, such as hardness and color. Hence, NF has been employed to remove the compounds since it only requires a single-membrane filtration process compared to the combination of softening and carbon filtration for similar treatment purposes [70,71]. Long-term performance of full-scale NF plants operated at 80% recovery for drinking water production from groundwater showed that the performance was stable for very long periods without chemical cleaning [72]. Even with minimal pretreatment, the fouling propensity of the NF membrane was low.

Unlike surface water, groundwater is another important alternative water resource that is less exposed to contamination. However, hazardous contaminants have found its way to the groundwater and rendered it for safe consumption. For instance, the presence of arsenic has been reported worldwide, and it has to be removed since arsenic is carcinogenic for human health. The oxidation pretreatment process has been utilized to convert As(III) to As(V) to increase the rejection of As by NF membrane [73,74]. The increase of the oxidation state of As (from trivalent to pentavalent form) successfully improved the NF rejection efficiency from 50%–63% to 97%–100% [73]. Pal et al. [74] have further integrated coagulation as a stabilization process to deal with the NF retentate (laden with As). The As in the reject stream of NF could be stabilized in the solid matrix through coagulation and coprecipitation with other materials. The study reported that the reaction between As and ferric and calcium ions could remove up to 98% of As in the retentate. Hence, this integrated oxidation—NF—coagulation process could be a solution for ensuring the supply of safe drinking water to the community affected by arsenic-contaminated groundwater areas.

The reclamation of municipal wastewater has been widely recognized as a promising solution to resolve the issues of water shortage. The reclaimed water can be used for landscaping, greening, and industrial uses, where the quality of reclaimed water will be decided by the reuse purpose and subsequently determines the technologies to be employed to treat the wastewater. The conventional biological processes of municipal sewage treatment plants could rarely produce a discharge that meets the stringent standards for water reuse [75]. Hence, membrane processes, such as NF, have been utilized as tertiary treatment technology to further polish the quality of the treated water. Li et al. [76] have demonstrated that NF process was capable to treat the effluent of an oxidation ditch from a municipal wastewater treatment plant and produce permeate with quality meeting the requirements and standards of water reclamation for different uses in Beijing, China. The complete rejection of COD and BOD, as well as near-complete rejection of total organic carbon and salinity by the NF process, enabled the reusability of reclaimed water for urban uses, industrial uses, and farmland irrigation.

Sewage sludge is a promising source for phosphorus recovery as fertilizer because a major percentage of the phosphate from wastewater is transferred into the sludge during the treatment process. Blöcher et al. [77] have developed an integrated phosphorus recovery process consisted of low-pressure wet oxidation, UF, and NF. Organic pollutants in the sewage sludge were oxidized by the low-pressure wet oxidation process while the UF worked as a barrier to remove the suspended solids. Subsequently, this process removed the pollutants and impurities that otherwise would clog the NF membrane and rendered the effectiveness of phosphorus recovery. The UF permeate was then filtered by NF membrane, in which dissolved phosphate ($H_2PO_4^-$), being a monovalent ion, will be collected as permeate while the concentrate is recycled back into the wastewater treatment plant. It was reported that phosphorus recovery of 54% was obtained and the costs of the entire integrated process were in the same range as conventional sewage sludge disposal, with additional benefits such as recovery of valuable phosphorus resource and reduced emission of greenhouse gases due to avoidance of sludge incineration.

Perfluorohexanoic acid is a persistent contaminant produced from the degradation of perfluoroalkyl substances that are widely used in various applications. Membrane technology has been employed to remove the acid from industrial wastewater, although the retention of perfluorohexanoic acid in concentrate stream is another waste that must be treated before disposal. Soriano et al. [78] have proposed an integrated process that consists of NF and electrooxidation to resolve this issue. The results demonstrated that the NF retained up to 99% of perfluorohexanoic acid, and an adequate rejection of divalent ions provided adequate conductivity to the concentrate stream, which facilitated the subsequent electrooxidation process. Electrooxidation with boron-doped diamond electrodes was capable to degrade the pollutants to below the limit of quantification of the analytical technique (98%). The degraded pollutants were then subsequently mineralized as indicated by the reduction of total organic carbon that was higher than 95%. This indicates that the integrated NF-electrooxidation process could be an efficient treatment for perfluoroalkyl substances-impacted waters.

Chon et al. [79] have developed an integrated MBR-NF membrane process for municipal wastewater reclamation, where the focus was given to the removal of trace contaminants. The MBR contributed in terms of the degradation of certain pharmaceuticals and personal care products, such as amlodipine and cefaclor. However, other micropollutants, including acetaminophen and carbamazepine, were not effectively removed by the MBR process. The residual dissolved trace contaminants in MBR permeate were then filtered with NF process, where rejection between 78% and 100% was successfully achieved, although boron was ineffectively removed by the NF since it existed as boric acid at the pH of wastewater.

4.2.2 Desalination

RO desalination process has been generally recognized as one of the best candidates to resolve water shortage, especially for coastal countries. However, the major challenges associated with desalination are membrane fouling and high energy consumption, which have reduced its economic feasibility as the treated water cost is much higher than water treated from surface water (can be more than 25 times higher) [80,81]. Although technological improvements and RO configurations have improved the energy consumption of membrane-based desalination process, the specific energy consumption still remains within high at 2–4 kWh/m^3 [82]. To cope with these challenges, NF has been employed as pretreatment for seawater desalination or as an alternative process for desalination.

The high desalinated water cost (produced from thermal and membrane-based desalination processes) could be attributed to the low water recovery (production), which is limited by the high osmotic pressure of seawater (large amount of dissolved salts). NF has been integrated with both the desalination technologies to assist in the efforts to bring down the water cost. For instance, NF managed to reduce the total dissolved solids or salts in the seawater prior to the RO and multistage flash distiller processes [83–88]. The reduction of salt content enabled the multistage flash plant to operate at higher temperatures and a higher distillate recovery rate. Also, the RO desalination process could achieve higher recovery rate due to the decrease in osmotic pressure. Subsequently, the water production has been increased by 60%, and this resulted in about 30% cost reduction.

NF could also be used to minimize the fouling propensity of desalination process (especially RO). As RO is normally operated at high pressure, the membrane process is vulnerable to fouling caused by the presence of impurities in feedwater [89]. In this case, the superb rejection capability of NF could be employed to ensure the removal of colloidal particles and organic matters before the feedwater is sent to RO modules. The pretreatment process can consist of a single NF process or could be the integration of NF with other processes such as UF to ensure better efficiency of impurities removal. For instance, UF–NF as pretreatment successfully produced a good grade of filtrate with 96.3% removal of total organic carbon in seawater down to the range of 0.06–0.35 mg/L [90]. Apart from the fouling issues arise from the colloidal and organic substances, the ability to reject dissolved ions possessed by NF could help to reduce the scaling potential of RO desalination. By partially rejecting the dissolved ions from entering the RO module, the scaling propensity encountered by RO could also be lowered due to the absence of scales forming ions [91].

The benefits of employing NF as pretreatment in a desalination process have also been supported by energy and exergy analysis [92–94]. Back in the year of 1999, Criscuoli and

Drioli [93] analyzed the energy and exergy of the overall performance of different integration desalination system. They reported that the introduction of NF led to an improvement in the desalination performance while maintaining an almost invariant energy requirement. On the other hand, Liu et al. [92] proposed to reduce the energy consumption and energetic efficiency by utilizing an energy recovery device in the desalination plant. They identified that the greatest exergy destruction of a dual-stage NF seawater desalination was due to the membrane and concentration stream valves. By modifying the system through concentration blending and energy recovery device, the exergy and energy consumptions have been reduced, where the specific energy consumption was cut down to 2.09 kWh/m^3, and the system recover ratio was increased to 43%. Thus with a properly designed simulation model, it could provide useful guidelines in determining the operating conditions of the integrated process.

The exergy and energy simulation results have been supported by a few findings from the real application of NF replacing the RO. The desalination process can be designed with NF replacing the RO modules, considering the advantages of lower operating pressure and higher production of NF process. A dual-stage NF desalination process has been developed by the Long Beach Water Department to replace the RO unit with the primary aim of reducing the energy consumption without sacrificing the water production rate [95]. The prototype desalination plant managed to desalinate the seawater (with salt concentration in between 30,000 and 35,000 ppm) with a capacity of 300,000 gallon-per-day to produce water meeting the drinking water guidelines [96]. More appealing, costing analysis showed that the dual-stage NF process resulted in lower cost of product water compared to RO—NF system, which could be attributed to the lower operating pressure and higher productivity.

NF has also been integrated with electrodialysis with polarity reversal (EDR) process for seawater desalination. In the NF—EDR desalination pilot plant conducted by Liu et al. [82], the seawater was pretreated by sand filtration, oxidation, and reduction as well as UF to remove the impurities that might foul the subsequent NF process. It was reported that the NF process reduced the total dissolved solids of the seawater from 32,749 to 3533 ppm with the energy consumption and recovery of 1.3 kWh/m^3 of water produced and 42%, respectively. The great performance of NF process produced diluted water for subsequent EDR process to further remove the total dissolved solids down to 241 ppm at an energy consumption of 0.52 kWh/m^3 of water produced. This indicated that the total energy consumption for the optimized integrated NF—EDR desalination system was less than 2.15 kWh/m^3 of clean water produced, which was lesser than the existing real seawater RO desalination plants. The proposed integrated NF—EDR process could be a new energy-saving seawater desalination technology in the quest of finding clean water for daily consumption.

NF has also found application in the treatment of brine from seawater desalination process. Disposal of large amount of brine may have adverse impacts on the environment, especially the ocean, and it could be a handling problem for the inland brackish water desalination plant [97]. Since the main composition of brine is NaCl, it could be a potential source to supply NaCl for various industries such as soda and chloralkali production. ED has been a promising technology to purify and recover the NaCl from the brine, although the scaling potential caused by high bivalent ions concentrations remains one challenging obstacle for the process. In this scenario, NF has been employed as a pretreatment prior to ED process to minimize the scaling propensity [98]. Due to its high rejection toward multivalent ions while allowing monovalent ions to pass through, NF can prevent the scaling ions (especially SO_4^{2-}) from entering the ED process. A similar concept has been varied to integrate NF with ED (operated with bipolar membranes) for seawater desalination brine concentration [99]. Instead of producing NaCl, the proposed integrated process further valorized to a strong acid (HCl) and base (NaOH). The NF pretreatment was capable to produce high-purity brine (98% of NaCl) by rejecting the multivalent ions, which would benefit the subsequent use in the ED process.

The desalination capability of NF has also been used to further purify the quality of treated effluent from other plants. For instance, NF was integrated as a tertiary treatment process to further improve the quality of wastewater treated by MBR [100]. The role of NF here was to remove the dissolved ions and pollutants such as pesticides, which failed to be removed by prior treatment processes.

4.2.3 Food industry

Being a selective separation process, NF has been employed in the food industry for concentrating, purifying, and recovering valuable and nutritious compounds or products in the food industry. In this section, the application of NF will be discussed based on the role played by NF in that particular process. In general, NF membrane process has been employed to concentrate, fractionate, recover or purify liquid foods (such as fruit juice and beverage), or nutraceuticals from food processing waste [2,101].

In the dairy and beverage industries, membrane technology has emerged as an alternative method over conventional thermal process for the clarification and concentration of juices or liquid products [102]. The shift to membrane technology could be attributed to its advantages as in less manpower requirement, greater concentration efficiency, shorter processing time, and the better quality preservation as some of the compounds in the juices are susceptible to thermal degradation (subsequently leads to the loss of fragrance, color, and taste) in conventional concentration process [2,102]. For instance, the conventional removal technique (ion-exchange chromatography) for the catalysts (NaCl and boric acid) from the product (lactulose syrup) encountered the challenges as in the need for frequent regeneration of resins, adsorption of lactulose on resins, and low capacity yet expensive material [103]. In this case, NF may be an attractive alternative for the purification of lactulose syrup. NF could be used to purify and concentrate the lactulose syrup since it allowed the catalysts (NaCl and boric acid) to pass through the membrane while retaining the sugar and thereby concentrating the solution.

NF has also been used to concentrate the fruit juices such as watermelon, bergamot, orange, pomegranate, and roselle to increase the amount of useful bioactive compounds (e.g., lycopene, flavonoid, ascorbic acid, anthocyanins, and phenolic contents) and subsequently enabled the consumers to obtain more nutrients [104–108]. Higher contents of bioactive compounds could be correlated with higher antioxidant potential of the samples, which is beneficial to the consumers. The conventional thermal-based concentration methods such as evaporation normally involve high temperature that will destroy the nutritious compounds in the fruit juices and degrade the quality of the products in terms of flavor, color, and taste [2]. Furthermore, the conventional methods consume large amounts of energy, making it unattractive for the fruit juice industry. In this context, NF process appears to be a preferable concentration technique due to its nonthermal operating condition, where the concentration is mainly done by size exclusion. Warczok et al. [109] reported that the total antioxidant activity of juices processed by membrane-based techniques was higher compared to concentration by conventional evaporation processes, supporting the benefits of employing membrane in juice concentration.

Cissé et al. [108] employed an NF membrane to retain and concentrate the anthocyanins in roselle extract. The industrial trial showed that the NF process concentrated the anthocyanins by six times and increased the total soluble solids from 4 to 25 g per 100 g roselle extract. A similar application has also been tested to enrich the polyphenols and anthocyanins in pomegranate juice [106]. The NF process retained much of the polyphenols and anthocyanins in the retentate, up to 84.8% and 90.7%, respectively. The retentate that enriched in phenolic compounds exhibited very high antioxidant activity, which could be potentially used for the formulation of nutraceutical products. On the other hand, glucose and fructose could be recovered from the permeate with efficiency of up to 90% and 93%, respectively.

NF has also been integrated with MF or UF processes for juice clarification. As MF and UF membranes have larger pore sizes, haze precursors, such as proteins, will pass through the membrane and lead to haziness formation in the juices [110]. Therefore the use of NF with smaller pore size could help to remove the haze precursors and at the same time allow the passage of sugars to permeate, maintaining the taste, stability, storage ability, and lightness of the clarified juices. In addition, a pilot-plant scale consisting of integrated UF–NF process has been developed for processing raw sugarcane juice, where the UF acted as a pretreatment to remove the suspended impurities while the NF was used to concentrate the UF permeate and at the same time decrease the contents of reducing sugar, salts, and small pigments in the syrup [111]. The integrated UF–NF membrane process produced a final syrup product with a color value below 800 IU, which could be used as a great source for the production of superior white sugar in the downstream process. Indeed, in the sugar industry, NF was used to recover dextrose from crystallization mother liquors, with the capability of producing produce permeate with dextrose purity of 97% for recovery [112]. This helps the dextrose manufacturing industry to attain a better economic return through the reduction of sugar loss, as the mother liquor is a by-product of the crystallization process that is normally sold as a low-quality syrup.

In alcoholic beverage industries, the control of alcohol and sugar contents in wines by using NF process is getting popular [113]. The growing demands for low-alcohol and alcohol-free drinks (due to health benefits), civil restrictions, and additional taxes in case ethanol content exceeds permissible limits have pushed the industry to reduce the alcohol content in the beverage. Although conventional thermal-based processes can produce wines with very low alcohol content (<0.5 vol.%), the final product will also lose its aroma due to the evaporation of volatile aroma compounds during the ethanol removal [114]. NF was successfully tested on the dealcoholization of wine by reducing the ethanol from 12 vol.% down to 5 vol.% [115]. Although the NF membrane also rejected the aroma compounds, subsequent mixing with pervaporated aroma compounds (obtained via pervaporation before the wine was filtered by NF process) managed to restore the organoleptic properties of the wine. In some cases, the early ripening of grapes would result in the increase of sugar content that subsequently increases the alcohol content of the wines produced. To prevent the excessive alcohol formed in the wines, Salgado et al. [116] employed NF processes to reduce the sugar content in grape must. The results showed that the fermentation of NF-processed grape produced wine with an acceptable reduction of alcohol content (~2 vol.%). Interestingly, the sugar reduction in grape must do not affect much the important compounds (e.g., polyphenols, malic, and tartaric acids) in the wine.

In recent years, the extraction of polyphenols from plants has been actively explored due to its health benefits upon consumption and its economic potential as natural preservatives or bioactive additives in food products and cosmetics [117,118]. NF has been successfully utilized for the fractionation and recovery of phenolic compounds from various sources. Generally, the membrane process would be used to recover the phenolic compounds by allowing the extraction solvent to pass through the membrane while retaining the phenolic substances. This would produce a solution with high concentration of phenolic compounds and at the same time allow the recovery of extraction solvent [119]. For instance, Tylkowski et al. [120] have shown that the NF retentate fraction of ethanolic extracts from *Sideritis* ssp. L. contained 3.4 times higher biologically active compounds, indicating the success of the concentration process. On the other hand, the permeate mainly consists of ethanol that could be reused for the extraction process, helping to reduce the volume of the required solvent for extraction.

In some cases, NF process was integrated with other processes for achieving better overall performance, especially to obtain a higher concentration factor and to reduce membrane fouling propensity. For instance, olive mill wastewater is an environmental issue for the olive oil extraction industry as it contains phenolic compounds that are reported to exhibit antimicrobial and phytotoxic properties [121,122]. The olive mill wastewater can be valorized by recovering the phenols as natural additives in foodstuff and cosmetics [123]. However, direct purification and concentration of phenols via NF process are impractical, as the suspended impurities present in the olive mill wastewater will foul the membrane. To cater this problem, MF and UF membranes have been employed as pretreatment stages to remove the suspended impurities [119,123–125]. The pretreated solution (permeate) would then be channeled to NF process for concentration, where the concentrated phenolic compounds could be used for food, pharmaceutical, and cosmetic applications. A pilot operation conducted by Paraskeva et al. [126] showed that the UF pretreatment process managed to remove 97% of suspended impurities in the olive mill wastewater, producing good quality of permeate for subsequent NF process to concentrate the phenols up to 10 g/L. Fig. 4–3 shows the typical configuration of integrated NF process for phenolic compounds concentration.

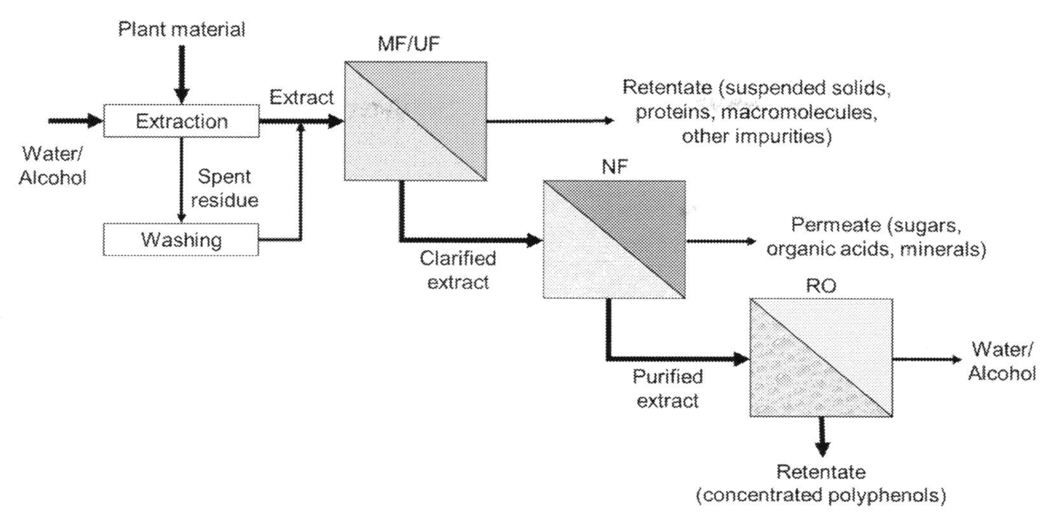

FIGURE 4–3 Integrated membrane process for phenolic compounds recovery from vegetable sources. *Adapted from C. Conidi, E. Drioli, A. Cassano, Membrane-based agro-food production processes for polyphenol separation, purification and concentration, Curr. Opin. Food Sci. 23 (2018) 149–164. https://doi.org/10.1016/j.cofs.2017.10.009.*

Apart from recovering phenols from wastewater, NF has also been used to separate the phenolic compounds from other substances. For instance, Conidi et al. [105,127] employed an integrated process comprised of UF and NF for the purification of phenolic compounds from bergamot juice. The clarified juice was enriched in sugar and organic acids while the phenolic compounds were recovered on the retentate side, as confirmed by the high total antioxidant activity of the NF retentate. By varying the molecular weight cutoff of NF, an alternative separation aim could be achieved. NF membrane showed a high retention (100%) toward polyphenol compounds (chlorogenic acid) and sugars (glucose, fructose, and sucrose) in clarified artichoke wastewater, producing permeate free from phenolic compounds and sugars [128].

The capability of NF to partially reject mineral ions has also been found useful in the demineralization of whey [129]. Whey is the main by-product from cheese production, and it contains valuable protein contents that could be used for hydration, gelation, emulsifying, and foaming process for ice cream manufacture or the production of yogurt [130]. However, the presence of mineral salts such as NaCl has degraded its quality for reuse. NF has been successfully used to remove the salts while concentrating on the valuable components in the whey. This helped to recover the valuable compounds from the industry by-products and at the same time solving the issue of environmental pollution (improper disposal of whey) [131].

4.2.4 Biorefinery applications

Sustainable economic growth is a global development trend where the emphasis has been given to sustainable management of resources for industrial production. As fossil resources are not sustainable, attention has been shifted to biobased resources. The utilization of biobased resources for economic development is part of the biorefinery industry. In general, the biorefinery is defined as the sustainable processing of biomass into a spectrum of marketable products and energy. This concept embraces a wide range of technologies to convert biomass resources into precursor materials which can be converted to value-added biobased products, bioenergy, and biofuels [132,133]. The conversion processes (mainly categorized as thermochemical, biochemical, mechanical, and chemical) will normally generate liquor stream that consists of various precursor materials [134]. This prompts the need to purify or concentrate the targeted precursor materials for downstream processing or utilization. NF membrane, which has a molecular weight cutoff between 200 and 1000 g/mol, plays an important role in this aspect since the compounds normally possess molecular weight of 150–250 g/mol [135].

Lactic acids and amino acids that can be obtained from biorefinery fermentation broth are useful precursors with a wide variety of applications in pharmaceutical, food, and biotechnology products. NF has been utilized to recover these valuable compounds from the fermentation broth, with the aim to produce lactic acids–enriched permeate and amino acids–enriched retentate [136–138]. It was reported that the NF separation efficiency of these compounds using real fermentation broth from the Green Biorefinery pilot plant located in Austria was unsatisfactory [137]. However, the adjustment of broth pH from 3.9 to 2.5 successfully increased the permeation of lactic acids from 33% to 58%. This allowed the separation of sufficient degree for both lactic acids and amino acids for further treatment technologies. In the study conducted by Sikder et al. [138], the lactic acid fermentation broth was produced by fermentation of sugarcane juice by *Lactobacillus plantarum* in a membrane-assisted hybrid reactor system. The fermentation broth was filtered by MF membrane to separate the microbial cell for recycled back to reactor while the permeate was then fed to NF process for concentrating the purity of lactic acid up to 85.6%. It was reported that the integrated membrane process was economically feasible through the recovering and recycling of unconverted sugars (retentate of NF) and microbial cell (retentate of MF).

Xylose is another valuable intermediate product in xylitol production that could be found in hemicellulose hydrolyzate stream. Conventionally, xylose is separated from the hemicellulose hydrolyzate stream by chromatographic method where the separation process is considered tedious and complex. To resolve these issues, NF has been proposed for the recovery of xylose from the hemicellulose hydrolyzate stream. Sjöman et al. [139] reported that NF allowed the passage of xylose into permeate and subsequently led to the enrichment of xylose in permeate with 78%–89% of the total dry solids in permeate was xylose. This can be translated as the increase of xylose purity by 1.4–1.7-fold after NF filtration process, indicating the separation potential of NF for the xylose recovery to the permeate from hemicellulose hydrolyzate.

In the biorefinery of lignocellulosic biomass, several side-products such as furans, phenols, and carboxylic acids have also been produced. Among these, furfural is a strong inhibitor of growth and ethanol fermentation of lignocellulosic biomass [140,141]. On the other hand, furfural itself is also a valuable feedstock for the production of biofuels and biochemical [142]. Hence, the separation of furfural from other compounds could minimize the fermentation inhibition issue while at the same time be recovered as value-added feedstock. Qi et al. [140] have tested the separation efficiency of furfural from the model solution consisting of glucose and xylose. It was reported that the tighter NF membrane (NF 90) achieved 99% rejection of saccharides (xylose and glucose), whereas 20%–40% rejection of furfural depending on the feed pressure (6–20 bar). The looser NF membrane (NF 270) recorded lower rejection of saccharides (60%–90%) but allowed complete permeation of furfural. Thus the authors suggested to employ NF 90 to concentrate the saccharides and followed by the removal of furfural through NF 270, such that this two-stage NF process could ensure a better separation of furfural from the saccharides.

In a typical biorefinery process utilizing cellulosic biomass, cellulase enzyme has been used to break the cellulose to produce glucose for subsequent microbial fermentation. The challenge here is that the cellulase should be separated from the enzymatic hydrolyzate for reuse while the glucose should be concentrated to increase the fermentation efficiency [143]. Qi et al. [144] have developed an integrated UF–NF process to recycle cellulase and concentrate glucose present in lignocellulosic hydrolyzate. The UF process could recycle 73.9% of the cellulase present in the hydrolyzate suspension. This helps to minimize the hydrolysis cost as the recovered cellulase enzymes can be reused in the hydrolysis process. The UF permeate which was rich in glucose would then be further concentrated with NF process. It was reported that the NF process successfully concentrated glucose 3.5 times compared to the level in UF permeate (30.2 g/L increased to 110.2 g/L), which could potentially improve the fermentation efficiency as well as the cost of downstream processing of fermentative product [145].

4.2.5 Organic solvent NF

As discussed previously, NF process is playing a crucial role in both removing hazardous substances and recovering valuable resources from various industries. In the past decade, the application of NF process for organic solvent separation and purification has also acquired increasing attention from membrane scientist. For this emerging application, the NF membrane has been known as organic solvent nanofiltration (OSN), sometimes also referred to as solvent-resistant nanofiltration or organophilic nanofiltration [146,147]. OSN is a pressure-driven separation process that allows solvent molecules to permeate through the membrane while retaining the solute [148,149]. Due to its unique separation performance, OSN has a great potential to be employed for the treatment of organic solvents, such as in fine chemical, pharmaceutical, petrochemical, and food industries [6,146,147]. OSN is becoming a competing technology for the separation involving organic solvent, since it is generally more energy-efficient compared to conventional technologies, such as preparative chromatography, distillation, extraction, or crystallization [148].

In the vegetable oil extraction industry, bio-derived solvents such as terpenes have been proposed as a replacement over conventional solvent (*n*-hexane) which is not environmentally friendly for the industry. However, the high boiling points and heat of vaporization restrict the utilization of terpenes in the industry, since the conventional solvent recovery technology is costly under high temperature and the presence of the antioxidants in the oil is vulnerable to thermal decomposition [150,151]. In this context, Abdellah et al. [152] have shown that by using the OSN membrane, an oil rejection of up to 90% could be achieved for the mixture of canola oil-terpenes solution. Although it could not meet the industry requirement of more than 95% oil rejection and encountered swelling issue (increase of solvent flux), it was suggested further improving the membrane hydrophobicity could potentially resolve this matter.

Spices and aromatic herbs have been known to contain a high content of useful substances that possess potential benefits for human health. The extraction and isolation of active constituents from the plants have been of particular interest since these compounds are useful functional ingredients in the food, cosmetics, and nutraceuticals applications [153]. For instance, rosemary is known to exhibit antitumor, antiviral, antibacterial, antiinflammatory, and antioxidant activities. However, conventional isolation method such as solid–liquid extraction typically converted the extracts into powder form through evaporation, which can cause considerable loss of antioxidant activity of the active constituents (due to thermal degradation) [154,155]. To resolve this issue, OSN membrane has been employed as an alternative extraction method for antioxidant extracts of rosemary [156]. It was reported that the finding was quite promising as reasonable permeate flux (solvent) and almost complete rejection of rosmarinic acid and the other antioxidant constituents of the herb in the retentate were achieved. No significant loss of antioxidant capacity in the retentate was observed. Consequently, the retentate may be applied directly as preservative and functional ingredient in the foods, cosmetics, nutraceuticals, and medicines. Furthermore, the permeate could be reused in the extraction process and is expected to bring additional economic benefits for the extraction process. An integrated OSN process (Fig. 4–4) has also been developed to refine and enrich γ-oryzanol of rice bran oil to enhance its antioxidant capacity [157]. The integrated process successfully enriched the γ-oryzanol content in the rice bran oil by up to 4.1 wt.%, which corresponded to more than a twofold increase in the oil's antioxidant capacity.

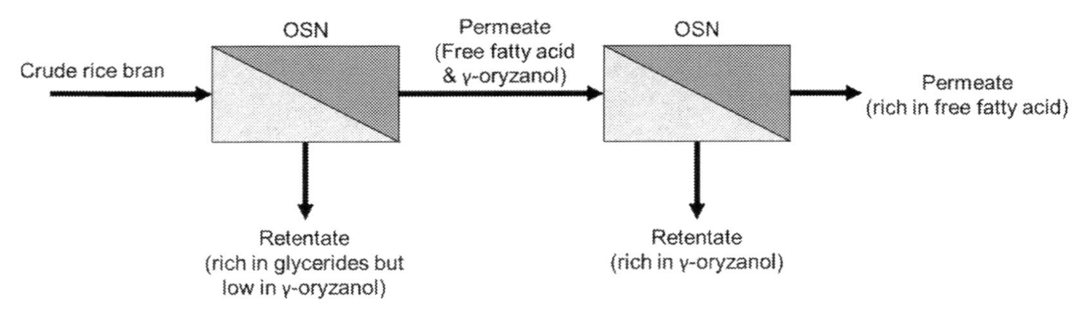

FIGURE 4–4 Integrated OSN process for enrichment of γ-oryzanol of rice bran oil. *OSN*, Organic solvent nanofiltration. *Adapted from I. Sereewatthanawut, I.I.R. Baptista, A.T. Boam, A. Hodgson, A.G. Livingston, Nanofiltration process for the nutritional enrichment and refining of rice bran oil, J. Food Eng. 102 (2011) 16–24. https://doi.org/10.1016/j.jfoodeng.2010.07.020.*

Surfactants have been used as solubilizers to overcome the miscibility gap between the organic compounds (reactants) and aqueous catalyst solution by creating a microemulsion to facilitate the chemical reactions. Although the solubilization helps to speed up the chemical reactions, it possesses a challenge for subsequent phase separation and product purification processes. In this case, the OSN membrane has been used to retain the surfactant (rejection of more than 90%) while achieving solvent flux of 40 LMH [158]. Although the separation efficiency was quite satisfactorily, membrane swelling issue was observed and imparted influence on the membrane performance.

OSN has also been employed to recover rhodium-based complex catalyst from organic solvents of the hydroformylation and hydrogenation reactions. The homogeneous catalysts are used in the reaction to produce high commodities in the surfactant and detergent. However, the commercialization of the reaction involves these catalysts is hampered by the costly distillation method in the recovery of catalysts and the reduced catalytic activity due to destructive conditions during the recovery stage [159]. Peddie et al. [160] demonstrated that OSN could be used to recover the catalyst, where the filtration process was capable of reducing the catalyst concentration in the permeate to near negligible amounts (<5 ppm). Technological evaluation simulation revealed that recovery using the OSN membrane process managed to cut down the energy and costs by 85% and 75%, respectively, compared to the conventional distillation method.

The quality of the pharmaceutical industry end product is highly dependent on the purity of active pharmaceutical ingredients (API) [161]. Existing API purification and isolation technologies suffer from certain challenges such as the difficulty in process control, high consumption of solvents, and loss of API that leads to an increase in the production cost [161,162]. OSN membrane can be integrated with the API production process to concentrate and purify the target molecules at low temperatures, preventing the compounds from the risk of thermal degradation. For instance, Sereewatthanawut et al. [161] employed DuraMem OSN membrane for the purification of API using a dual-membrane diafiltration process. The integrated process achieved two targets at the same time: purified the API and recovered the organic solvent for reuse. The performance was quite encouraging with 99% of the impurities were removed from API and the fresh solvent consumption has been reduced from reuse. In general, higher purity of API can be achieved with more filtration stages. However, increasing the number of filtration stages would lead to higher expenses [163]. Hence, the concept of integrated process as shown in Fig. 4–5 could be adopted to overcome these challenges. The permeate from the first-stage filtration was further treated in the second stage to ensure high purity of polyethylene glycol (PEG) was obtained in the second-stage permeate. Promising results were acquired where the yield of PEG-2000 was increased from 59% (single filtration stage) to 94% (two filtration stages) while maintaining the purity at 98% [164]. Costing analysis of the similar integrated process for solvent recovery and API purification indicated that the increment of product yield from 58% to 95% with significant cost savings up to 92% compared to the conventional single-stage diafiltration [165]. Thus OSN is an attractive process for product purification and successful solvent recovery (lower expense for fresh solvent and lower energy consumption compared to conventional solvent recovery

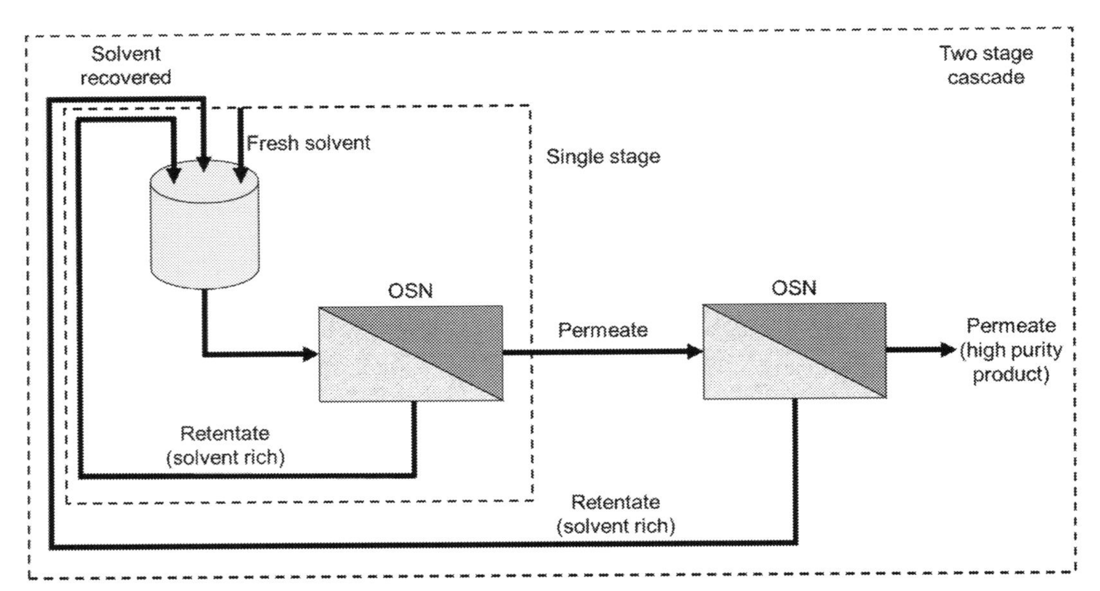

FIGURE 4–5 Single-stage and two-stage cascade diafiltration configuration using OSN membrane for product purification and solvent recovery. *OSN*, Organic solvent nanofiltration. *Adapted from J.F. Kim, A.M. Freitas, I.B. Valtcheva, A.G. Livingston, When the membrane is not enough: a simplified membrane cascade using organic solvent nanofiltration (OSN), Sep. Purif. Technol. 116 (2013) 277–286. https://doi.org/10.1016/j.seppur.2013.05.050.*

process) and subsequently making the whole process to stay economically competitive compared to existing technologies [166,167].

Székely et al. [168] developed an integrated process that consists of OSN and molecularly imprinted polymers for the removal of potentially genotoxic impurity from API. The potentially genotoxic impurity was produced from the hydrolysis of carbodiimides, a substance often used as stabilizing, coupling, and condensing agents in the preparation of peptides and nucleotides (API). Due to the potentially harmful property, the impurity must be removed from API in the postreaction process. It was reported that the OSN process managed to allow 90% of the potentially genotoxic impurity to enter the permeate with a 2.5% loss of the model API. The retentate that contained residual potentially genotoxic impurity was then further treated by molecularly imprinted polymers to enhance the overall removal efficiency of the potentially genotoxic impurity. The proposed concept process could be applied to other cases where ultralow concentrations of contaminants must be reached.

Refining is an energy-intensive process that involves complex separations and the use of large amounts of liquids in the petrochemical industry. One of the processes that OSN plays a role in the petrochemical industry is the recovery of the solvent. ExxonMobil's Beaumont (Texas) refinery process has installed the first large-scale industrial OSN plant in 1998 for the recovery of dewaxing solvents from lube oil filtrates [169–171]. The process was trademarked as MAX-DEWAX and integrated with existing process refinery units. The solvent mixture (toluene and MEK) with high purity (99%) was successfully recovered at refrigeration temperature and could be recycled directly to the chilled feed stream. With just a third of the capital investment, the OSN process managed to reduce the energy consumption per product unit by 20%, increase the average base oil production by 25 vol.%, reduce the volatile organic compounds emissions, and cut down the use of cooling water.

Ionic liquid is a class of nonvolatile organic salts that could help to improve the solubility of many metal complex catalysts and are immiscible with water and many organic solvents [172]. Due to its good capability in dissolving the metal catalysts, chemical reactions could occur in the ionic liquid to produce products that subsequently can be extracted with an organic solvent. However, a small amount of ionic liquid and catalysts would dissolve in the organic solvent that lead to the loss of ionic liquid and catalysts as well as the contamination of products [173]. In this case, the OSN membrane offers the opportunity to recover the ionic liquid and catalysts from the organic solvent for reuse purpose. For instance, Wong et al. [174] employed OSN membrane (STARMEM 122) to recover the palladium catalyst and ionic liquid from organic solvent for reuse in subsequent reaction cycles. Compared to reactions in pure organic solvent, the addition of ionic liquid improved the stability of the palladium catalyst, and this helped to enhance the reaction yields over consecutive recycles. This indicates that with the presence of OSN membrane, not only the production yield can be enhanced (improved reaction rate due to the stability of palladium catalyst) but also cost-saving is possible with the recovery of catalyst and ionic liquid.

Organic (polymeric) and inorganic (ceramic) materials have been widely used for the preparation of OSN membranes, with the former receive more interest due to the advantages like there are a wide variety of available polymers, relatively low price, and ease of fabrication and upscaling potential [147]. However, the main challenge of OSN application is the thermal and chemical stability of OSN membrane under extreme operating conditions, such as harsh and aggressive media, elevated pH, and high temperatures. Since the membrane allows organic solvents to permeate through, the interactions between the solvents and membrane can cause swelling issue (or ultimately even dissolve) that subsequently result in loss of selectivity [175]. Hence, efforts in improving the solvent stability and swelling resilience of OSN membrane through the development of new polymeric materials and surface modifications have been actively explored by membrane scientist. For instance, the incorporation of fillers (metal-organic framework, silica, titania, and graphene) and surface modifications of OSN membrane through various methods (radiation/light-induced modifications, plasma-induced techniques, thin-film formation via polymerization, polyelectrolyte modifications) [146,176,177]. Promising results have been reported where upon modified, the performance of OSN improved in terms of rejection (solute), permeation (solvent flux), and membrane stability and integrity [146,147,176,178].

4.3 Conclusions

The successful use of NF membranes thus far has been due to its uniqueness in having selectivity between the intended separation species. Hence, NF should not be seen as a replacement for RO, instead it should utilize its highly selective characteristics for the application that can benefit from this. The applications of NF in the five main sectors, such as water and wastewater treatment, desalination, food industry, biorefinery applications, and organic solvent separation, have shown the versatility of NF process in treatment, purification, concentration, and recovery. Future studies can be conducted to further enhance the selective property of the NF membrane, which will definitely improve its efficiency in the aforementioned applications.

References

[1] D.L. Oatley-Radcliffe, M. Walters, T.J. Ainscough, P.M. Williams, A.W. Mohammad, N. Hilal, Nanofiltration membranes and processes: a review of research trends over the past decade, J. Water Process. Eng. 19 (2017) 164–171. Available from: https://doi.org/10.1016/j.jwpe.2017.07.026.

[2] K. Nath, H.K. Dave, T.M. Patel, Revisiting the recent applications of nanofiltration in food processing industries: progress and prognosis, Trends Food Sci. Technol. 73 (2018) 12–24. Available from: https://doi.org/10.1016/j.tifs.2018.01.001.

[3] M. Paul, S.D. Jons, Chemistry and fabrication of polymeric nanofiltration membranes: a review, Polymers (Guildf.) 103 (2016) 417–456. Available from: https://doi.org/10.1016/j.polymer.2016.07.085.

[4] M.S. Yorgun, I.A. Balcioglu, O. Saygin, Performance comparison of ultrafiltration, nanofiltration and reverse osmosis on whey treatment, Desalination 229 (2008) 204–216. Available from: https://doi.org/10.1016/j.desal.2007.09.008.

[5] C. Fonseca Couto, L.C. Lange, M.C. Santos Amaral, A critical review on membrane separation processes applied to remove pharmaceutically active compounds from water and wastewater, J. Water Process. Eng. 26 (2018) 156–175. Available from: https://doi.org/10.1016/j.jwpe.2018.10.010.

[6] Y. Ji, W. Qian, Y. Yu, Q. An, L. Liu, Y. Zhou, et al., Recent developments in nanofiltration membranes based on nanomaterials, Chin. J. Chem. Eng. 25 (2017) 1639–1652. Available from: https://doi.org/10.1016/j.cjche.2017.04.014.

[7] A.W. Mohammad, Y.H. Teow, W.L. Ang, Y.T. Chung, D.L. Oatley-Radcliffe, N. Hilal, Nanofiltration membranes review: recent advances and future prospects, Desalination 356 (2015) 226–254. Available from: https://doi.org/10.1016/j.desal.2014.10.043.

[8] C. Suksaroj, M. Héran, C. Allègre, F. Persin, Treatment of textile plant effluent by nanofiltration and/or reverse osmosis for water reuse, Desalination 178 (2005) 333–341. Available from: https://doi.org/10.1016/j.desal.2004.11.043.

[9] C.A. Quist-Jensen, F. Macedonio, E. Drioli, Membrane technology for water production in agriculture: desalination and wastewater reuse, Desalination 364 (2015) 17–32. Available from: https://doi.org/10.1016/j.desal.2015.03.001.

[10] S. Bunani, E. Yörükoğlu, G. Sert, Ü. Yüksel, M. Yüksel, N. Kabay, Application of nanofiltration for reuse of municipal wastewater and quality analysis of product water, Desalination 315 (2013) 33–36. Available from: https://doi.org/10.1016/j.desal.2012.11.015.

[11] G. Balcioğlu, Z.B. Gönder, Recovery of Baker's yeast wastewater with membrane processes for agricultural irrigation purpose: fouling characterization, Chem. Eng. J. 255 (2014) 630–640. Available from: https://doi.org/10.1016/j.cej.2014.06.084.

[12] K. Konieczny, A. Kwiecińska, B. Gworek, The recovery of water from slurry produced in high density livestock farming with the use of membrane processes, Sep. Purif. Technol. 80 (2011) 490–498. Available from: https://doi.org/10.1016/j.seppur.2011.06.002.

[13] S.M. Mrayed, P. Sanciolo, L. Zou, G. Leslie, An alternative membrane treatment process to produce low-salt and high-nutrient recycled water suitable for irrigation purposes, Desalination 274 (2011) 144–149. Available from: https://doi.org/10.1016/j.desal.2011.02.003.

[14] M. Perle, S. Kimchie, G. Shelef, Some biochemical aspects of the anaerobic degradation of dairy wastewater, Water Res. 29 (1995) 1549–1554. Available from: https://doi.org/10.1016/0043-1354(94)00248-6.

[15] F. Carta-Escobar, J. Pereda-Marín, P. Álvarez-Mateos, F. Romero-Guzmán, M.M. Durán-Barrantes, F. Barriga-Mateos, Aerobic purification of dairy wastewater in continuous regime: part I: analysis of the biodegradation process in two reactor configurations, Biochem. Eng. J. 21 (2004) 183–191. Available from: https://doi.org/10.1016/j.bej.2004.06.007.

[16] M.C. Cammarota, D.M.G. Freire, A review on hydrolytic enzymes in the treatment of wastewater with high oil and grease content, Bioresour. Technol. 97 (2006) 2195–2210. Available from: https://doi.org/10.1016/j.biortech.2006.02.030.

[17] M. Vourch, B. Balannec, B. Chaufer, G. Dorange, Nanofiltration and reverse osmosis of model process waters from the dairy industry to produce water for reuse, Desalination 172 (2005) 245–256. Available from: https://doi.org/10.1016/j.desal.2004.07.038.

[18] L.H. Andrade, F.D.S. Mendes, J.C. Espindola, M.C.S. Amaral, Nanofiltration as tertiary treatment for the reuse of dairy wastewater treated by membrane bioreactor, Sep. Purif. Technol. 126 (2014) 21–29. Available from: https://doi.org/10.1016/j.seppur.2014.01.056.

[19] Metcalf & Eddy, Inc., Water Reuse: Issue, Technology and Application, McGraw-Hill, 2007. Available from: http://ssu.ac.ir/cms/fileadmin/user-upload/Daneshkadaha/dbehdasht/markaz-tahghighat-olom-va-fanavarihaye-zist-mohiti/e-book/wastewater-reuse/water-reuse/Untitled14.pdf.

[20] J. Wojdalski, B. Drózdz, J. Piechocki, M. Gaworski, Z. Zander, J. Marjanowski, Determinants of water consumption in the dairy industry, Pol. J. Chem. Technol. 15 (2013) 61–72. Available from: https://doi.org/10.2478/pjct-2013-0025.

[21] S. Kim, K.H. Chu, Y.A.J. Al-Hamadani, C.M. Park, M. Jang, D.H. Kim, et al., Removal of contaminants of emerging concern by membranes in water and wastewater: a review, Chem. Eng. J. 335 (2018) 896–914. Available from: https://doi.org/10.1016/j.cej.2017.11.044.

[22] S.P. Dharupaneedi, S.K. Nataraj, M. Nadagouda, K.R. Reddy, S.S. Shukla, T.M. Aminabhavi, Membrane-based separation of potential emerging pollutants, Sep. Purif. Technol. 210 (2019) 850–866. Available from: https://doi.org/10.1016/j.seppur.2018.09.003.

[23] M. Taheran, S.K. Brar, M. Verma, R.Y. Surampalli, T.C. Zhang, J.R. Valero, Membrane processes for removal of pharmaceutically active compounds (PhACs) from water and wastewaters, Sci. Total Environ. 547 (2016) 60–77. Available from: https://doi.org/10.1016/j.scitotenv.2015.12.139.

[24] X. Jin, J. Hu, S.L. Ong, Removal of natural hormone estrone from secondary effluents using nanofiltration and reverse osmosis, Water Res. 44 (2010) 638–648. Available from: https://doi.org/10.1016/j.watres.2009.09.057.

[25] P. Liu, H. Zhang, Y. Feng, F. Yang, J. Zhang, Removal of trace antibiotics from wastewater: a systematic study of nanofiltration combined with ozone-based advanced oxidation processes, Chem. Eng. J. 240 (2014) 211–220. Available from: https://doi.org/10.1016/j.cej.2013.11.057.

[26] J.H. Kim, P.K. Park, C.H. Lee, H.H. Kwon, Surface modification of nanofiltration membranes to improve the removal of organic micro-pollutants (EDCs and PhACs) in drinking water treatment: graft polymerization and cross-linking followed by functional group substitution, J. Membr. Sci. 321 (2008) 190–198. Available from: https://doi.org/10.1016/j.memsci.2008.04.055.

[27] S.P. Sun, T.A. Hatton, T.S. Chung, Hyperbranched polyethyleneimine induced cross-linking of polyamide-imide nanofiltration hollow fiber membranes for effective removal of ciprofloxacin, Environ. Sci. Technol. 45 (2011) 4003–4009. Available from: https://doi.org/10.1021/es200345q.

[28] A.J.C. Semião, M. Foucher, A.I. Schäfer, Removal of adsorbing estrogenic micropollutants by nanofiltration membranes: part B: model development, J. Membr. Sci. 431 (2013) 257–266. Available from: https://doi.org/10.1016/j.memsci.2012.11.079.

[29] L.D. Nghiem, A.I. Schäfer, M. Elimelech, Role of electrostatic interactions in the retention of pharmaceutically active contaminants by a loose nanofiltration membrane, J. Membr. Sci. 286 (2006) 52–59. Available from: https://doi.org/10.1016/j.memsci.2006.09.011.

[30] B. Van Der Bruggen, J. Schaep, D. Wilms, C. Vandecasteele, Influence of molecular size, polarity and charge on the retention of organic molecules by nanofiltration, J. Membr. Sci. 156 (1999) 29–41. Available from: https://doi.org/10.1016/S0376-7388(98)00326-3.

[31] K.V. Plakas, A.J. Karabelas, T. Wintgens, T. Melin, A study of selected herbicides retention by nanofiltration membranes—the role of organic fouling, J. Membr. Sci. 284 (2006) 291–300. Available from: https://doi.org/10.1016/j.memsci.2006.07.054.

[32] M.A. Zazouli, H. Susanto, S. Nasseri, M. Ulbricht, Influences of solution chemistry and polymeric natural organic matter on the removal of aquatic pharmaceutical residuals by nanofiltration, Water Res. 43 (2009) 3270–3280. Available from: https://doi.org/10.1016/j.watres.2009.04.038.

[33] K. Kimura, T. Iwase, S. Kita, Y. Watanabe, Influence of residual organic macromolecules produced in biological wastewater treatment processes on removal of pharmaceuticals by NF/RO membranes, Water Res. 43 (2009) 3751–3758. Available from: https://doi.org/10.1016/j.watres.2009.05.042.

[34] A. Azaïs, J. Mendret, S. Gassara, E. Petit, A. Deratani, S. Brosillon, Nanofiltration for wastewater reuse: counteractive effects of fouling and matrice on the rejection of pharmaceutical active compounds, Sep. Purif. Technol. 133 (2014) 313–327. Available from: https://doi.org/10.1016/j.seppur.2014.07.007.

[35] C. Zhao, J. Zhang, G. He, T. Wang, D. Hou, Z. Luan, Perfluorooctane sulfonate removal by nanofiltration membrane the role of calcium ions, Chem. Eng. J. 233 (2013) 224–232. Available from: https://doi.org/10.1016/j.cej.2013.08.027.

[36] E.C. Devitt, F. Ducellier, P. Cote, M.R. Wiesner, Effects of natural organic matter and the raw water matrix on the rejection of atrazine by pressure-driven membranes, Water Res. 32 (1998) 2563–2568. Available from: https://doi.org/10.1016/S0043-1354(98)00043-8.

[37] S. Hong, M. Elimelech, Chemical and physical aspects of natural organic matter (NOM) fouling of nanofiltration membranes, J. Membr. Sci. 132 (1997) 159–181. Available from: https://doi.org/10.1016/S0376-7388(97)00060-4.

[38] A.M. Comerton, R.C. Andrews, D.M. Bagley, The influence of natural organic matter and cations on the rejection of endocrine disrupting and pharmaceutically active compounds by nanofiltration, Water Res. 43 (2009) 613–622. Available from: https://doi.org/10.1016/j.watres.2008.11.003.

[39] A.I. Schäfer, L.D. Nghiem, T.D. Waite, Removal of the natural hormone estrone from aqueous solutions using nanofiltration and reverse osmosis, Environ. Sci. Technol. 37 (2003) 182–188. Available from: https://doi.org/10.1021/es0102336.

[40] M. Röhricht, J. Krisam, U. Weise, U.R. Kraus, R.A. Düring, Elimination of pharmaceuticals from wastewater by submerged nanofiltration plate modules, Desalination 250 (2010) 1025–1026. Available from: https://doi.org/10.1016/j.desal.2009.09.098.

[41] N. Klamerth, S. Malato, A. Agüera, A. Fernández-Alba, Photo-Fenton and modified photo-Fenton at neutral pH for the treatment of emerging contaminants in wastewater treatment plant effluents: a comparison, Water Res. 47 (2013) 833–840. Available from: https://doi.org/10.1016/j.watres.2012.11.008.

[42] S. Miralles-Cuevas, F. Audino, I. Oller, R. Sánchez-Moreno, J.A. Sánchez Pérez, S. Malato, Pharmaceuticals removal from natural water by nanofiltration combined with advanced tertiary treatments (solar photo-Fenton, photo-Fenton-like Fe(III)-EDDS complex and ozonation), Sep. Purif. Technol. 122 (2014) 515–522. Available from: https://doi.org/10.1016/j.seppur.2013.12.006.

[43] S. Miralles-Cuevas, I. Oller, J.A.S. Pérez, S. Malato, Removal of pharmaceuticals from MWTP effluent by nanofiltration and solar photo-Fenton using two different iron complexes at neutral pH, Water Res. 64 (2014) 23–31. Available from: https://doi.org/10.1016/j.watres.2014.06.032.

[44] S. Sanches, A. Penetra, A. Rodrigues, E. Ferreira, V.V. Cardoso, M.J. Benoliel, et al., Nanofiltration of hormones and pesticides in different real drinking water sources, Sep. Purif. Technol. 94 (2012) 44–53. Available from: https://doi.org/10.1016/j.seppur.2012.04.003.

[45] A.J.C. Semião, A.I. Schäfer, Estrogenic micropollutant adsorption dynamics onto nanofiltration membranes, J. Membr. Sci. 381 (2011) 132–141. Available from: https://doi.org/10.1016/j.memsci.2011.07.031.

[46] A.I. Schäfer, I. Akanyeti, A.J.C. Semião, Micropollutant sorption to membrane polymers: a review of mechanisms for estrogens, Adv. Colloid Interface Sci. 164 (2011) 100–117. Available from: https://doi.org/10.1016/j.cis.2010.09.006.

[47] M.P. Lopes, C.T. Matos, V.J. Pereira, M.J. Benoliel, M.E. Valério, L.B. Bucha, et al., Production of drinking water using a multi-barrier approach integrating nanofiltration: a pilot scale study, Sep. Purif. Technol. 119 (2013) 112–122. Available from: https://doi.org/10.1016/j.seppur.2013.09.002.

[48] V.J. Pereira, J. Galinha, M.T. Barreto Crespo, C.T. Matos, J.G. Crespo, Integration of nanofiltration, UV photolysis, and advanced oxidation processes for the removal of hormones from surface water sources, Sep. Purif. Technol. 95 (2012) 89–96. Available from: https://doi.org/10.1016/j.seppur.2012.04.013.

[49] S.S. Hosseini, E. Bringas, N.R. Tan, I. Ortiz, M. Ghahramani, M.A. Alaei Shahmirzadi, Recent progress in development of high performance polymeric membranes and materials for metal plating wastewater treatment: a review, J. Water Process. Eng. 9 (2016) 78–110. Available from: https://doi.org/10.1016/j.jwpe.2015.11.005.

[50] J.P. Vareda, A.J.M. Valente, L. Durães, Assessment of heavy metal pollution from anthropogenic activities and remediation strategies: a review, J. Environ. Manage. 246 (2019) 101–118. Available from: https://doi.org/10.1016/j.jenvman.2019.05.126.

[51] Z. Wang, G. Liu, Z. Fan, X. Yang, J. Wang, S. Wang, Experimental study on treatment of electroplating wastewater by nanofiltration, J. Membr. Sci. 305 (2007) 185–195. Available from: https://doi.org/10.1016/j.memsci.2007.08.011.

[52] W.P. Zhu, S.P. Sun, J. Gao, F.J. Fu, T.S. Chung, Dual-layer polybenzimidazole/polyethersulfone (PBI/PES) nanofiltration (NF) hollow fiber membranes for heavy metals removal from wastewater, J. Membr. Sci. 456 (2014) 117–127. Available from: https://doi.org/10.1016/j.memsci.2014.01.001.

[53] J. Gao, S.P. Sun, W.P. Zhu, T.S. Chung, Chelating polymer modified P84 nanofiltration (NF) hollow fiber membranes for high efficient heavy metal removal, Water Res. 63 (2014) 252–261. Available from: https://doi.org/10.1016/j.watres.2014.06.006.

[54] P. Juholin, M.L. Kääriäinen, M. Riihimäki, R. Sliz, J.L. Aguirre, M. Pirilä, et al., Comparison of ALD coated nanofiltration membranes to unmodified commercial membranes in mine wastewater treatment, Sep. Purif. Technol. 192 (2018) 69–77. Available from: https://doi.org/10.1016/j.seppur.2017.09.005.

[55] M. Mänttäri, M. Nyström, Negative retention of organic compounds in nanofiltration, Desalination 199 (2006) 41–42. Available from: https://doi.org/10.1016/j.desal.2006.03.015.

[56] D. Zhao, L. Qiu, J. Song, J. Liu, Z. Wang, Y. Zhu, et al., Efficiencies and mechanisms of chemical cleaning agents for nanofiltration membranes used in produced wastewater desalination, Sci. Total Environ. 652 (2019) 256–266. Available from: https://doi.org/10.1016/j.scitotenv.2018.10.221.

[57] A. Rahimpour, B. Rajaeian, A. Hosienzadeh, S.S. Madaeni, F. Ghoreishi, Treatment of oily wastewater produced by washing of gasoline reserving tanks using self-made and commercial nanofiltration membranes, Desalination 265 (2011) 190–198. Available from: https://doi.org/10.1016/j.desal.2010.07.051.

[58] T. Coskun, E. Debik, N.M. Demir, Treatment of olive mill wastewaters by nanofiltration and reverse osmosis membranes, Desalination 259 (2010) 65–70. Available from: https://doi.org/10.1016/j.desal.2010.04.034.

[59] F. Bazzarelli, E. Piacentini, T. Poerio, R. Mazzei, A. Cassano, L. Giorno, Advances in membrane operations for water purification and biophenols recovery/valorization from OMWWs, J. Membr. Sci. 497 (2016) 402–409. Available from: https://doi.org/10.1016/j.memsci.2015.09.049.

[60] A. Cassano, C. Conidi, L. Giorno, E. Drioli, Fractionation of olive mill wastewaters by membrane separation techniques, J. Hazard. Mater. 248–249 (2013) 185–193. Available from: https://doi.org/10.1016/j.jhazmat.2013.01.006.

[61] J. Huang, K. Zhang, The high flux poly (m-phenylene isophthalamide) nanofiltration membrane for dye purification and desalination, Desalination 282 (2011) 19–26. Available from: https://doi.org/10.1016/j.desal.2011.09.045.

[62] S. Zhao, H. Zhu, Z. Wang, P. Song, M. Ban, X. Song, A loose hybrid nanofiltration membrane fabricated via chelating-assisted in-situ growth of Co/Ni LDHs for dye wastewater treatment, Chem. Eng. J. 353 (2018) 460–471. Available from: https://doi.org/10.1016/j.cej.2018.07.081.

[63] H. Zhang, B. Li, J. Pan, Y. Qi, J. Shen, C. Gao, et al., Carboxyl-functionalized graphene oxide polyamide nanofiltration membrane for desalination of dye solutions containing monovalent salt, J. Membr. Sci. 539 (2017) 128–137. Available from: https://doi.org/10.1016/j.memsci.2017.05.075.

[64] N.H.H. Hairom, A.W. Mohammad, A.A.H. Kadhum, Nanofiltration of hazardous Congo red dye: performance and flux decline analysis, J. Water Process. Eng. 4 (2014) 99–106. Available from: https://doi.org/10.1016/j.jwpe.2014.09.008.

[65] P. Chen, X. Ma, Z. Zhong, F. Zhang, W. Xing, Y. Fan, Performance of ceramic nanofiltration membrane for desalination of dye solutions containing NaCl and Na_2SO_4, Desalination 404 (2017) 102–111. Available from: https://doi.org/10.1016/j.desal.2016.11.014.

[66] N.C. Cinperi, E. Ozturk, N.O. Yigit, M. Kitis, Treatment of woolen textile wastewater using membrane bioreactor, nanofiltration and reverse osmosis for reuse in production processes, J. Clean. Prod. 223 (2019) 837–848. Available from: https://doi.org/10.1016/j.jclepro.2019.03.166.

[67] T. Tavangar, K. Jalali, M.A. Alaei Shahmirzadi, M. Karimi, Toward real textile wastewater treatment: membrane fouling control and effective fractionation of dyes/inorganic salts using a hybrid electrocoagulation—nanofiltration process, Sep. Purif. Technol. 216 (2019) 115–125. Available from: https://doi.org/10.1016/j.seppur.2019.01.070.

[68] B. Cyna, G. Chagneau, G. Bablon, N. Tanghe, Two years of nanofiltration at the Mery-sur-Oise plant, France, Desalination 147 (2002) 69–75.

[69] N. García-Vaquero, E. Lee, R. Jiménez Castañeda, J. Cho, J.A. López-Ramírez, Comparison of drinking water pollutant removal using a nanofiltration pilot plant powered by renewable energy and a conventional treatment facility, Desalination 347 (2014) 94–102. Available from: https://doi.org/10.1016/j.desal.2014.05.036.

[70] W.G.J. Van der Meer, J.C. Van Winkelen, Method for Purifying Water, in Particular Groundwater, Under Anaerobic Conditions, Using a Membrane Filtration Unit, a Device for Purifying Water, as well as Drinking Water Obtained by Such a Method, US Pat. 6395182, 2002.

[71] P. Hiemstra, J. van Paassen, B. Rietman, J. Verdouw, Aerobic versus anaerobic nanofiltration: fouling of membranes, Proceedings of the AWWA Membrane Conference, Long Beach, CA, 1999.

[72] F. Beyer, B.M. Rietman, A. Zwijnenburg, P. van den Brink, J.S. Vrouwenvelder, M. Jarzembowska, et al., Long-term performance and fouling analysis of full-scale direct nanofiltration (NF) installations treating anoxic groundwater, J. Membr. Sci. 468 (2014) 339–348. Available from: https://doi.org/10.1016/j.memsci.2014.06.004.

[73] M. Sen, A. Manna, P. Pal, Removal of arsenic from contaminated groundwater by membrane-integrated hybrid treatment system, J. Membr. Sci. 354 (2010) 108–113. Available from: https://doi.org/10.1016/j.memsci.2010.02.063.

[74] P. Pal, S. Chakrabortty, L. Linnanen, A nanofiltration-coagulation integrated system for separation and stabilization of arsenic from groundwater, Sci. Total Environ. 476–477 (2014) 601–610. Available from: https://doi.org/10.1016/j.scitotenv.2014.01.041.

[75] L. Jin, G. Zhang, H. Tian, Current state of sewage treatment in China, Water Res. 66 (2014) 85–98. Available from: https://doi.org/10.1016/j.watres.2014.08.014.

[76] K. Li, J. Wang, J. Liu, Y. Wei, M. Chen, Advanced treatment of municipal wastewater by nanofiltration: operational optimization and membrane fouling analysis, J. Environ. Sci. (China) 43 (2016) 106–117. Available from: https://doi.org/10.1016/j.jes.2015.09.007.

[77] C. Blöcher, C. Niewersch, T. Melin, Phosphorus recovery from sewage sludge with a hybrid process of low pressure wet oxidation and nanofiltration, Water Res. 46 (2012) 2009–2019. Available from: https://doi.org/10.1016/j.watres.2012.01.022.

[78] Á. Soriano, D. Gorri, A. Urtiaga, Efficient treatment of perfluorohexanoic acid by nanofiltration followed by electrochemical degradation of the NF concentrate, Water Res. 112 (2017) 147–156. Available from: https://doi.org/10.1016/j.watres.2017.01.043.

[79] K. Chon, S. Sarp, S. Lee, J.H. Lee, J.A. Lopez-Ramirez, J. Cho, Evaluation of a membrane bioreactor and nanofiltration for municipal wastewater reclamation: trace contaminant control and fouling mitigation, Desalination 272 (2011) 128–134. Available from: https://doi.org/10.1016/j.desal.2011.01.002.

[80] P. Rao, W.R. Morrow, A. Aghajanzadeh, P. Sheaffer, C. Dollinger, S. Brueske, et al., Energy considerations associated with increased adoption of seawater desalination in the United States, Desalination 445 (2018) 213–224. Available from: https://doi.org/10.1016/j.desal.2018.08.014.

[81] T.N. Bitaw, K. Park, J. Kim, J.W. Chang, D.R. Yang, Low-recovery, -energy-consumption, -emission hybrid systems of seawater desalination: energy optimization and cost analysis, Desalination 468 (2019) 114085. Available from: https://doi.org/10.1016/j.desal.2019.114085.

[82] Y. Liu, J. Wang, L. Wang, An energy-saving “nanofiltration/electrodialysis with polarity reversal (NF/EDR)” integrated membrane process for seawater desalination. Part III. Optimization of the energy consumption in a demonstration operation, Desalination 452 (2019) 230–237. Available from: https://doi.org/10.1016/j.desal.2018.11.015.

[83] M. Abdul-KareemAl-Sofi, Seawater desalination—SWCC experience and vision, Desalination 135 (2001) 121–139. Available from: https://doi.org/10.1016/S0011-9164(01)00145-X.

[84] M.K.M. Al-Sofi, A. Hassan, G. Mustafa, A. Dalvi, Nanofiltration as a means of achieving higher TBT of $\geq$ 120°C in MSF, Desalination 118 (1998) 123–129.

[85] A.M. Hassan, M.A.K. Al-Sofi, A.S. AI-Amoudi, A.T.M. Jamaluddin, A.M. Farooque, A. Rowaili, et al., A new approach to membrane and thermal seawater desalination processes using nanofiltration membranes (Part 1), Desalination 118 (1998) 35–51. Available from: https://doi.org/10.1016/S0011-9164(98)00079-4.

[86] A.M. Hassan, A.M. Farooque, N.M. Kither, A. Rowaili, A demonstration plant based on the new NF-SWRO process, Desalination 131 (2000) 157–171.

[87] M. Pontié, J.S. Derauw, S. Plantier, L. Edouard, L. Bailly, Seawater desalination: nanofiltration—a substitute for reverse osmosis? Desalin. Water Treat. 51 (2013) 485–494. Available from: https://doi.org/10.1080/19443994.2012.714594.

[88] C. Kaya, G. Sert, N. Kabay, M. Arda, M. Yüksel, Ö. Egemen, Pre-treatment with nanofiltration (NF) in seawater desalination—preliminary integrated membrane tests in Urla, Turkey, Desalination 369 (2015) 10–17. Available from: https://doi.org/10.1016/j.desal.2015.04.029.

[89] W.L. Ang, A.W. Mohammad, N. Hilal, C.P. Leo, A review on the applicability of integrated/hybrid membrane processes in water treatment and desalination plants, Desalination 363 (2014) 2–18. Available from: https://doi.org/10.1016/j.desal.2014.03.008.

[90] Y. Song, B. Su, X. Gao, C. Gao, The performance of polyamide nanofiltration membrane for long-term operation in an integrated membrane seawater pre-treatment system, Desalination 296 (2012) 30–36. Available from: https://doi.org/10.1016/j.desal.2012.03.024.

[91] Y.H. Choi, J.H. Kweon, D.I. Kim, S. Lee, Evaluation of various pre-treatment for particle and inorganic fouling control on performance of SWRO, Desalination 247 (2009) 137–147. Available from: https://doi.org/10.1016/j.desal.2008.12.019.

[92] J. Liu, J. Yuan, L. Xie, Z. Ji, Exergy analysis of dual-stage nanofiltration seawater desalination, Energy 62 (2013) 248–254. Available from: https://doi.org/10.1016/j.energy.2013.07.071.

[93] A. Criscuoli, E. Drioli, Energetic and exergetic analysis of an integrated membrane desalination system, Desalination 124 (1999) 243–249. Available from: https://doi.org/10.1016/S0011-9164(99)00109-5.

[94] H. Mehdizadeh, Membrane desalination plants from an energy-exergy viewpoint, Desalination 191 (2006) 200–209. Available from: https://doi.org/10.1016/j.desal.2005.06.037.

[95] C.J. Harrison, Y.A. Le Gouellec, R.C. Cheng, A.E. Childress, Bench-scale testing of nanofiltration for seawater desalination, J. Environ. Eng 133 (2007) 1004–1014. Available from: https://doi.org/10.1061/(ASCE)0733-9372(2007)133:11(1004).

[96] R.C. Cheng, T.J. Tseng, K.L. Wattier, Two-pass nanofiltration seawater desalination prototype testing and evaluation, 2013. https://www.usbr.gov/research/dwpr/reportpdfs/report158.pdf.

[97] A. Hashim, M. Hajjaj, Impact of desalination plants fluid effluents on the integrity of seawater, with the Arabian Gulf in perspective, Desalination 182 (2005) 373–393. Available from: https://doi.org/10.1016/j.desal.2005.04.020.

[98] J. Liu, J. Yuan, Z. Ji, B. Wang, Y. Hao, X. Guo, Concentrating brine from seawater desalination process by nanofiltration-electrodialysis integrated membrane technology, Desalination 390 (2016) 53–61. Available from: https://doi.org/10.1016/j.desal.2016.03.012.

[99] M. Reig, S. Casas, O. Gibert, C. Valderrama, J.L. Cortina, Integration of nanofiltration and bipolar electrodialysis for valorization of seawater desalination brines: production of drinking and waste water treatment chemicals, Desalination 382 (2016) 13–20. Available from: https://doi.org/10.1016/j.desal.2015.12.013.

[100] M.C. Hacıfazlıoğlu, İ. Parlar, T. Pek, N. Kabay, Evaluation of chemical cleaning to control fouling on nanofiltration and reverse osmosis membranes after desalination of MBR effluent, Desalination 466 (2019) 44–51. Available from: https://doi.org/10.1016/j.desal.2019.05.003.

[101] A. Nazir, K. Khan, A. Maan, R. Zia, L. Giorno, K. Schroën, Membrane separation technology for the recovery of nutraceuticals from food industrial streams, Trends Food Sci. Technol. 86 (2019) 426–438. Available from: https://doi.org/10.1016/j.tifs.2019.02.049.

[102] C. Bhattacharjee, V.K. Saxena, S. Dutta, Fruit juice processing using membrane technology: a review, Innov. Food Sci. Emerg. Technol. 43 (2017) 136–153. Available from: https://doi.org/10.1016/j.ifset.2017.08.002.

[103] Z. Zhang, R. Yang, S. Zhang, H. Zhao, X. Hua, Purification of lactulose syrup by using nanofiltration in a diafiltration mode, J. Food Eng. 105 (2011) 112–118. Available from: https://doi.org/10.1016/j.jfoodeng.2011.02.013.

[104] N.A. Arriola, G.D. dos Santos, E.S. Prudêncio, L. Vitali, J.C.C. Petrus, R.D.M. Castanho Amboni, Potential of nanofiltration for the concentration of bioactive compounds from watermelon juice, Int. J. Food Sci. Technol. 49 (2014) 2052–2060. Available from: https://doi.org/10.1111/ijfs.12513.

[105] C. Conidi, A. Cassano, E. Drioli, A membrane-based study for the recovery of polyphenols from bergamot juice, J. Membr. Sci. 375 (2011) 182–190. Available from: https://doi.org/10.1016/j.memsci.2011.03.035.

[106] C. Conidi, A. Cassano, F. Caiazzo, E. Drioli, Separation and purification of phenolic compounds from pomegranate juice by ultrafiltration and nanofiltration membranes, J. Food Eng. 195 (2017) 1–13. Available from: https://doi.org/10.1016/j.jfoodeng.2016.09.017.

[107] N.P. Kelly, A.L. Kelly, J.A. O'Mahony, Strategies for enrichment and purification of polyphenols from fruit-based materials, Trends Food Sci. Technol. 83 (2019) 248–258. Available from: https://doi.org/10.1016/j.tifs.2018.11.010.

[108] M. Cissé, F. Vaillant, D. Pallet, M. Dornier, Selecting ultrafiltration and nanofiltration membranes to concentrate anthocyanins from roselle extract (*Hibiscus sabdariffa* L.), Food Res. Int. 44 (2011) 2607–2614. Available from: https://doi.org/10.1016/j.foodres.2011.04.046.

[109] J. Warczok, M. Ferrando, F. López, C. Güell, Concentration of apple and pear juices by nanofiltration at low pressures, J. Food Eng. 63 (2004) 63–70. Available from: https://doi.org/10.1016/S0260-8774(03)00283-8.

[110] V. Vivekanand, M. Iyer, S. Ajlouni, Clarification and stability enhancement of pear juice using loose nanofiltration, J. Food Process. Technol. 3 (2012) 1000162. Available from: https://doi.org/10.4172/2157-7110.1000162.

[111] J. Luo, X. Hang, W. Zhai, B. Qi, W. Song, X. Chen, et al., Refining sugarcane juice by an integrated membrane process: filtration behavior of polymeric membrane at high temperature, J. Membr. Sci. 509 (2016) 105–115. Available from: https://doi.org/10.1016/j.memsci.2016.02.053.

[112] S. Bandini, L. Nataloni, Nanofiltration for dextrose recovery from crystallization mother liquors: a feasibility study, Sep. Purif. Technol. 139 (2015) 53–62. Available from: https://doi.org/10.1016/j.seppur.2014.10.025.

[113] A. Massot, M. Mietton-Peuchot, C. Peuchot, V. Milisic, Nanofiltration and reverse osmosis in winemaking, Desalination 231 (2008) 283–289. Available from: https://doi.org/10.1016/j.desal.2007.10.032.

[114] E. Gómez-Plaza, J.M. López-Nicolás, J.M. López-Roca, A. Martínez-Cutillas, Dealcoholization of wine. Behaviour of the aroma components during the process, LWT - Food Sci. Technol. 32 (1999) 384–386. Available from: https://doi.org/10.1006/fstl.1999.0565.

[115] M. Catarino, A. Mendes, Dealcoholizing wine by membrane separation processes, Innov. Food Sci. Emerg. Technol. 12 (2011) 330–337. Available from: https://doi.org/10.1016/j.ifset.2011.03.006.

[116] C.M. Salgado, E. Fernández-Fernández, L. Palacio, A. Hernández, P. Prádanos, Alcohol reduction in red and white wines by nanofiltration of musts before fermentation, Food Bioprod. Process. 96 (2015) 285–295. Available from: https://doi.org/10.1016/j.fbp.2015.09.005.

[117] C.M. Galanakis, P. Tsatalas, Z. Charalambous, I.M. Galanakis, Control of microbial growth in bakery products fortified with polyphenols recovered from olive mill wastewater, Environ. Technol. Innov. 10 (2018) 1–15. Available from: https://doi.org/10.1016/j.eti.2018.01.006.

[118] C.M. Galanakis, P. Tsatalas, Z. Charalambous, I.M. Galanakis, Polyphenols recovered from olive mill wastewater as natural preservatives in extra virgin olive oils and refined olive kernel oils, Environ. Technol. Innov. 10 (2018) 62–70. Available from: https://doi.org/10.1016/j.eti.2018.01.012.

[119] C. Conidi, E. Drioli, A. Cassano, Membrane-based agro-food production processes for polyphenol separation, purification and concentration, Curr. Opin. Food Sci. 23 (2018) 149–164. Available from: https://doi.org/10.1016/j.cofs.2017.10.009.

[120] B. Tylkowski, I. Tsibranska, R. Kochanov, G. Peev, M. Giamberini, Concentration of biologically active compounds extracted from *Sideritis* ssp. L. by nanofiltration, Food Bioprod. Process. 89 (2011) 307–314. Available from: https://doi.org/10.1016/j.fbp.2010.11.003.

[121] G. Greco, M.L. Colarieti, G. Toscano, G. Iamarino, M.A. Rao, L. Gianfreda, Mitigation of olive mill wastewater toxicity, J. Agric. Food Chem. 54 (2006) 6776–6782. Available from: https://doi.org/10.1021/jf061084j.

[122] M. Dellagreca, P. Monaco, G. Pinto, A. Pollio, L. Previtera, F. Temussi, Phytotoxicity of low-molecular-weight phenols from olive mill waste waters, Environ. Contam. Toxicol. 67 (2001) 352–359. Available from: https://doi.org/10.1007/s00128-001-0132-9.

[123] C.M. Galanakis, Recovery of high added-value components from food wastes: conventional, emerging technologies and commercialized applications, Trends Food Sci. Technol. 26 (2012) 68–87. Available from: https://doi.org/10.1016/j.tifs.2012.03.003.

[124] C.M. Galanakis, K. Kotsiou, Recovery of bioactive compounds from olive mill waste, Olive Mill Waste: Recent Advances for the Sustainable Management, Elsevier Inc, 2017. Available from: https://doi.org/10.1016/B978-0-12-805314-0.00010-8.

[125] R. Castro-Muñoz, V. Fíla, Membrane-based technologies as an emerging tool for separating high-added-value compounds from natural products, Trends Food Sci. Technol. 82 (2018) 8–20. Available from: https://doi.org/10.1016/j.tifs.2018.09.017.

[126] C.A. Paraskeva, V.G. Papadakis, E. Tsarouchi, D.G. Kanellopoulou, P.G. Koutsoukos, Membrane processing for olive mill wastewater fractionation, Desalination 213 (2007) 218–229. Available from: https://doi.org/10.1016/j.desal.2006.04.087.

[127] C. Conidi, A. Cassano, Recovery of phenolic compounds from bergamot juice by nanofiltration membranes, Desalin. Water Treat. 56 (2015) 3510–3518. Available from: https://doi.org/10.1080/19443994.2014.968219.

[128] C. Conidi, A.D. Rodriguez-Lopez, E.M. Garcia-Castello, A. Cassano, Purification of artichoke polyphenols by using membrane filtration and polymeric resins, Sep. Purif. Technol. 144 (2015) 153–161. Available from: https://doi.org/10.1016/j.seppur.2015.02.025.

[129] K. Pan, Q. Song, L. Wang, B. Cao, A study of demineralization of whey by nanofiltration membrane, Desalination 267 (2011) 217–221. Available from: https://doi.org/10.1016/j.desal.2010.09.029.

[130] E. Tosi, L. Canna, H. Lucero, E. Ré, Foaming properties of sweet whey solutions as modified by thermal treatment, Food Chem. 100 (2007) 794–799. Available from: https://doi.org/10.1016/j.foodchem.2005.11.001.

[131] E. Suárez, A. Lobo, S. Alvarez, F.A. Riera, R. Álvarez, Demineralization of whey and milk ultrafiltration permeate by means of nanofiltration, Desalination 241 (2009) 272–280. Available from: https://doi.org/10.1016/j.desal.2007.11.087.

[132] B. Kamm, P.R. Gruber, M. Kamm, Biorefineries-industrial processes and products, Ullmann's Encyclopedia of Industrial Chemistry, Wiley-VCH Verlag GmbH & Co. KGaA, 2016, pp. 1–37. Available from: https://doi.org/10.1002/14356007.l04-l01.pub2.

[133] F. Cherubini, The biorefinery concept: Using biomass instead of oil for producing energy and chemicals, Energy Convers. Manage. 51 (2010) 1412–1421. Available from: https://doi.org/10.1016/j.enconman.2010.01.015.

[134] C. Abels, F. Carstensen, M. Wessling, Membrane processes in biorefinery applications, J. Membr. Sci. 444 (2013) 285–317. Available from: https://doi.org/10.1016/j.memsci.2013.05.030.

[135] P. Wei, L.H. Cheng, L. Zhang, X.H. Xu, H.L. Chen, C.J. Gao, A review of membrane technology for bioethanol production, Renewable Sustainable Energy Rev. 30 (2014) 388–400. Available from: https://doi.org/10.1016/j.rser.2013.10.017.

[136] M.I. González, S. Alvarez, F.A. Riera, R. Álvarez, Lactic acid recovery from whey ultrafiltrate fermentation broths and artificial solutions by nanofiltration, Desalination 228 (2008) 84–96. Available from: https://doi.org/10.1016/j.desal.2007.08.009.

[137] J. Ecker, T. Raab, M. Harasek, Nanofiltration as key technology for the separation of LA and AA, J. Membr. Sci. 389 (2012) 389–398. Available from: https://doi.org/10.1016/j.memsci.2011.11.004.

[138] J. Sikder, S. Chakraborty, P. Pal, E. Drioli, C. Bhattacharjee, Purification of lactic acid from microfiltrate fermentation broth by cross-flow nanofiltration, Biochem. Eng. J. 69 (2012) 130–137. Available from: https://doi.org/10.1016/j.bej.2012.09.003.

[139] E. Sjöman, M. Mänttäri, M. Nyström, H. Koivikko, H. Heikkilä, Xylose recovery by nanofiltration from different hemicellulose hydrolyzate feeds, J. Membr. Sci. 310 (2008) 268–277. Available from: https://doi.org/10.1016/j.memsci.2007.11.001.

[140] B. Qi, J. Luo, X. Chen, X. Hang, Y. Wan, Separation of furfural from monosaccharides by nanofiltration, Bioresour. Technol. 102 (2011) 7111–7118. Available from: https://doi.org/10.1016/j.biortech.2011.04.041.

[141] R. Pulicharla, L. Lonappan, S.K. Brar, M. Verma, Production of renewable C5 platform chemicals and potential applications, Platform Chemical Biorefinery Future Green Chemistry, Elsevier Inc, 2016. Available from: https://doi.org/10.1016/b978-0-12-802980-0.00011-0.

[142] A.E. Eseyin, P.H. Steele, An overview of the applications of furfural and its derivatives, Int. J. Adv. Chem. 3 (2015) 42. Available from: https://doi.org/10.14419/ijac.v3i2.5048.

[143] B. Qi, X. Chen, Y. Su, Y. Wan, Enzyme adsorption and recycling during hydrolysis of wheat straw lignocellulose, Bioresour. Technol. 102 (2011) 2881–2889. Available from: https://doi.org/10.1016/j.biortech.2010.10.092.

[144] B. Qi, J. Luo, G. Chen, X. Chen, Y. Wan, Application of ultrafiltration and nanofiltration for recycling cellulase and concentrating glucose from enzymatic hydrolyzate of steam exploded wheat straw, Bioresour. Technol. 104 (2012) 466–472. Available from: https://doi.org/10.1016/j.biortech.2011.10.049.

[145] K. Saha, R. Uma Maheswari, J. Sikder, S. Chakraborty, S.S. da Silva, J.C. dos Santos, Membranes as a tool to support biorefineries: applications in enzymatic hydrolysis, fermentation and dehydration for bioethanol production, Renewable Sustainable Energy Rev. 74 (2017) 873–890. Available from: https://doi.org/10.1016/j.rser.2017.03.015.

[146] M. Amirilargani, M. Sadrzadeh, E.J.R. Sudhölter, L.C.P.M. de Smet, Surface modification methods of organic solvent nanofiltration membranes, Chem. Eng. J. 289 (2016) 562–582. Available from: https://doi.org/10.1016/j.cej.2015.12.062.

[147] S. Hermans, H. Mariën, C. Van Goethem, I.F. Vankelecom, Recent developments in thin film (nano) composite membranes for solvent resistant nanofiltration, Curr. Opin. Chem. Eng. 8 (2015) 45–54. Available from: https://doi.org/10.1016/j.coche.2015.01.009.

[148] G. Szekely, M.F. Jimenez-Solomon, P. Marchetti, J.F. Kim, A.G. Livingston, Sustainability assessment of organic solvent nanofiltration: from fabrication to application, Green Chem. 16 (2014) 4440–4473. Available from: https://doi.org/10.1039/c4gc00701h.

[149] P. Vandezande, L.E.M. Gevers, I.F.J. Vankelecom, Solvent resistant nanofiltration: separating on a molecular level, Chem. Soc. Rev 37 (2008) 365–405.

[150] S. Arora, S. Manjula, A.G. Gopala Krishna, R. Subramanian, Membrane processing of crude palm oil, Desalination 191 (2006) 454–466. Available from: https://doi.org/10.1016/j.desal.2005.04.129.

[151] R. Subramanian, M. Nakajima, T. Kawakatsu, Processing vegetable oils using nonporous denser polymeric composite membranes, J. Food Eng. 38 (1998) 41–56. Available from: https://doi.org/10.1007/s11746-004-0901-z.

[152] M.H. Abdellah, L. Liu, C.A. Scholes, B.D. Freeman, S.E. Kentish, Organic solvent nanofiltration of binary vegetable oil/terpene mixtures: experiments and modelling, J. Membr. Sci. 573 (2019) 694–703. Available from: https://doi.org/10.1016/j.memsci.2018.12.026.

[153] E. Hernández-hernández, E. Ponce-alquicira, M.E. Jaramillo-flores, I.G. Legarreta, Antioxidant effect rosemary (*Rosmarinus officinalis* L.) and oregano (*Origanum vulgare* L.) extracts on TBARS and colour of model raw pork batters, Meat Sci. 81 (2009) 410–417. Available from: https://doi.org/10.1016/j.meatsci.2008.09.004.

[154] M. Bonoli, M. Pelillo, G. Lercker, Fast separation and determination of carnosic acid and rosmarinic acid in different rosemary (*Rosmarinus officinalis*) extracts by capillary zone electrophoresis with ultra violet-diode array detection, Chromatographia 57 (2003) 505–512.

[155] S. Basßkan, N. Oztekin, F.B. Erim, Determination of carnosic acid and rosmarinic acid in sage by capillary electrophoresis, Food Chem. 101 (2007) 1748–1752. Available from: https://doi.org/10.1016/j.foodchem.2006.01.033.

[156] D. Peshev, L.G. Peeva, G. Peev, I.I.R. Baptista, A.T. Boam, M. Extraction, et al., Application of organic solvent nanofiltration for concentration of antioxidant extracts of rosemary (*Rosmarinus officiallis* L.), Chem. Eng. Res. Des. 89 (2011) 318–327. Available from: https://doi.org/10.1016/j.cherd.2010.07.002.

[157] I. Sereewatthanawut, I.I.R. Baptista, A.T. Boam, A. Hodgson, A.G. Livingston, Nanofiltration process for the nutritional enrichment and refining of rice bran oil, J. Food Eng. 102 (2011) 16–24. Available from: https://doi.org/10.1016/j.jfoodeng.2010.07.020.

[158] D. Zedel, A. Drews, M. Kraume, Retention of surfactants by organic solvent nanofiltration and influences on organic solvent flux, Sep. Purif. Technol. 158 (2016) 396–408. Available from: https://doi.org/10.1016/j.seppur.2015.12.040.

[159] A. Cano-Odena, P. Vandezande, D. Fournier, W. Van Camp, F.E. Du Prez, I.F.J. Vankelecom, Solvent-resistant nanofiltration for product purification and catalyst recovery in click chemistry reactions, Chem. Eur. J. 16 (2010) 1061–1067. Available from: https://doi.org/10.1002/chem.200901659.

[160] W.L. Peddie, J.N. van Rensburg, H.C.M. Vosloo, P. van der Gryp, Technological evaluation of organic solvent nanofiltration for the recovery of homogeneous hydroformylation catalysts, Chem. Eng. Res. Des. 121 (2017) 219–232. Available from: https://doi.org/10.1016/j.cherd.2017.03.015.

[161] I. Sereewatthanawut, F.W. Lim, Y.S. Bhole, D. Ormerod, A. Horvath, A.T. Boam, et al., Demonstration of molecular purification in polar aprotic solvents by organic solvent nanofiltration, Org. Process Res. Dev. 14 (2010) 600–611.

[162] G. Székely, J. Bandarra, W. Heggie, B. Sellergren, F. Castelo, Organic solvent nanofiltration: a platform for removal of genotoxins from active pharmaceutical ingredients, J. Membr. Sci. 381 (2011) 21–33. Available from: https://doi.org/10.1016/j.memsci.2011.07.007.

[163] W. Eugene, A.G. Livingston, C. Ates, A. Merschaert, Continuous solute fractionation with membrane cascades—a high productivity alternative to diafiltration, Sep. Purif. Technol. 102 (2013) 1–14. Available from: https://doi.org/10.1016/j.seppur.2012.09.017.

[164] J.F. Kim, A.M. Freitas, I.B. Valtcheva, A.G. Livingston, When the membrane is not enough: a simplified membrane cascade using organic solvent nanofiltration (OSN), Sep. Purif. Technol. 116 (2013) 277–286. Available from: https://doi.org/10.1016/j.seppur.2013.05.050.

[165] J.F. Kim, G. Székely, I.B. Valtcheva, A.G. Livingston, Increasing the sustainability of membrane processes through cascade approach and solvent recovery—pharmaceutical purification case study, Green Chem. 16 (2014) 133–145. Available from: https://doi.org/10.1039/c3gc41402g.

[166] E.M. Rundquista, C.J. Pink, A.G. Livingston, Organic solvent nanofiltration: a potential alternative to distillation for solvent recovery from crystallisation mother liquors, Green Chem. 14 (2012) 2197–2205. Available from: https://doi.org/10.1039/C2GC35216H.

[167] G. Székely, M. Gil, B. Sellergren, W. Heggie, F.C. Ferreira, Environmental and economic analysis for selection and engineering sustainable API degenotoxification processes, Green Chem. 15 (2013) 210–225. Available from: https://doi.org/10.1039/C2GC36239B.

[168] G. Székely, J. Bandarra, W. Heggie, B. Sellergren, F.C. Ferreira, A hybrid approach to reach stringent low genotoxic impurity contents in active pharmaceutical ingredients: combining molecularly imprinted polymers and organic solvent nanofiltration for removal of 1,3-diisopropylurea, Sep. Purif. Technol. 86 (2012) 79–87. Available from: https://doi.org/10.1016/j.seppur.2011.10.023.

[169] L.S. White, A.R. Nitsch, Solvent recovery from lube oil filtrates with a polyimide membrane, J. Membr. Sci. 179 (2000) 267–274.

[170] L.S. White, Development of large-scale applications in organic solvent nanofiltration and pervaporation for chemical and refining processes, J. Membr. Sci. 286 (2006) 26–35. Available from: https://doi.org/10.1016/j.memsci.2006.09.006.

[171] R.M. Gould, L.S. White, C.R. Wildemuth, Membrane separation in solvent lube dewaxing, Environ. Prog. 20 (2001) 1–5.

[172] J.F. Brennecke, E.J. Maginn, Ionic liquids: innovative fluids for chemical processing, AIChE J. 47 (2001) 2384–2389.

[173] S. Han, H.-T. Wong, A.G. Livingston, Application of organic solvent nanofiltration to separation of ionic liquids and products from ionic liquid mediated reactions, Chem. Eng. Res. Des. 83 (2005) 309–316. Available from: https://doi.org/10.1205/cherd.04247.

[174] H. Wong, C.J. Pink, F.C. Ferreira, A.G. Livingston, Recovery and reuse of ionic liquids and palladium catalyst for Suzuki reactions using organic solvent nanofiltration, Green Chem. 8 (2006) 373–379. Available from: https://doi.org/10.1039/b516778g.

[175] G.M. Shi, M.H. Davood Abadi Farahani, J.Y. Liu, T.S. Chung, Separation of vegetable oil compounds and solvent recovery using commercial organic solvent nanofiltration membranes, J. Membr. Sci. 588 (2019) 117202. Available from: https://doi.org/10.1016/j.memsci.2019.117202.

[176] P. Chuntanalerg, S. Bureekaew, C. Klaysom, W.-J. Lau, K. Faungnawakij, Nanomaterial-incorporated nanofiltration membranes for organic solvent recovery, Advanced Nanomaterials for Membrane Synthesis and Its Applications, Elsevier Inc, 2019. 10.1016/b978-0-12-814503-6.00007-0.

[177] T. Gao, L. Huang, C. Li, G. Xu, G. Shi, Graphene membranes with tuneable nanochannels by intercalating self-assembled porphyrin molecules for organic solvent nanofiltration, Carbon 124 (2017) 263–270. Available from: https://doi.org/10.1016/j.carbon.2017.08.042.

[178] J. Chau, P. Basak, J. Kaur, Y. Hu, K.K. Sirkar, Performance of a composite membrane of a perfluorodioxole copolymer in organic solvent nanofiltration, Sep. Purif. Technol. 199 (2018) 233–241. Available from: https://doi.org/10.1016/j.seppur.2018.01.054.

5 Principles of forward osmosis

Wafa Suwaileh[1], Daniel Johnson[2], Nidal Hilal[2]

[1]RESEARCH AND DEVELOPMENT, QATAR FOUNDATION, DOHA, QATAR [2]NYUAD WATER RESEARCH CENTER, NEW YORK UNIVERSITY ABU DHABI, ABU DHABI, UNITED ARAB EMIRATES

5.1 Introduction

In recent years, forward osmosis (FO) has gained much attention in both academia and industrial research and development [1–3]. FO is a process driven by differences in osmotic pressure between a feed and draw solution, as opposed to many other membrane processes, such as reverse osmosis (RO), where hydraulic pressure is applied. The FO process needs a semipermeable membrane that segregates the feed solution, with relatively low osmotic pressure and a more concentrated draw solution [4]. As the main operating cost for systems such as RO is the energy needed to pump solutions at high pressure, it is generally expected that FO will have much lower energy costs. A general scheme for FO operation principles is shown in Fig. 5–1 [5]. Due to water permeation across the membrane, the concentration of the feed solution will increase over time, whereas the draw solution will become diluted.

5.2 Water flux in FO

Volumetric flux across a semipermeable membrane across an osmotic pressure gradient was discussed in Chapter 1, Basic Principles of Osmosis and Osmotic Pressure, with Eqs. (1.24) and (1.26) presenting the effects of opposing osmotic and hydraulic pressure gradients and reflection coefficient. The pure water permeability of the membrane, A, is important for quantifying the separation properties of a particular membrane. Thus an increase in the water permeability indicates high pure water flux. As far as the osmotic pressure gradient is the main driving force for the water transport, the water flux equation can be rewritten as [6]:

$$J_w = A(\pi_{D,b} - \pi_{F,b}) \tag{5.1}$$

where J_w is the volumetric water flux and $\pi_{D,b}$ and $\pi_{F,b}$ are the bulk osmotic pressure of the draw feed solutions, respectively. This equation presumes that the draw solute is completely rejected by the FO membrane as well as it does not consider the deposition of solute close

Osmosis Engineering. DOI: https://doi.org/10.1016/B978-0-12-821016-1.00008-5

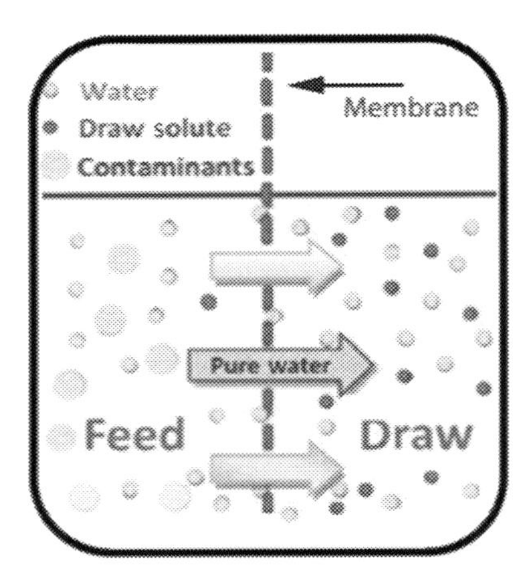

FIGURE 5–1 A description of solution transport mechanism through an asymmetric membrane in FO system. *FO*, forward osmosis.

to the membrane surface [concentration polarization (CP)]. The water flux is affected by the resistance of the solute diffusion through the membrane (K), and hence the water flux equation can be rewritten as [7]:

$$J_w = \frac{1}{K} \ln\left(\frac{\pi_{D,b}}{\pi_{F,b}}\right) \tag{5.2}$$

where the parameter K is inversely proportional to the solute diffusion coefficient (D_s) and can determine the ease of movement of the solute in the support layer. Because the osmotic pressure here correlates linearly to the draw solution concentration, Eq. (5.2) can be rewritten to relate the solute resistivity (K) that represents the dilutive internal concentration polarization (ICP) as follows [6,8]:

$$K = \left(\frac{1}{J_w}\right) \ln \frac{B + A\pi_{D,b}}{B + J_w + A\pi_{F,m}} \tag{5.3}$$

where A and B are the water permeability and solute permeability coefficients, respectively, whereas B can be calculated using:

$$B = \left(\frac{(1 - R)J_w}{R}\right) \tag{5.4}$$

where R is the salt rejection which can be measured from the concentration difference between the initial feed solution and the permeate salt concentrations [2]. The water flux depends on the osmotic driving force on both sides of the membrane and the mass transfer resistance of the solute in the membrane. Thus the solute resistivity correlates to the water flux

[8]. Additionally, the membrane permeability coefficient depends on both the osmotic pressure of the draw solution and the membrane structure configuration. Moreover, the diffusion of a solute within a tortuous support layer is referred to as the effective diffusion coefficient (D_{eff}). In this respect, the resistance of the solute diffusion through the membrane (K) and the solute diffusion coefficient are inversely correlated [8,9]. This parameter is lower than the theoretical solute diffusion coefficient (D_s) and can be obtained from Fick's Law [10,11]. The D_{eff} factor is also related to the membrane thickness (t), tortuosity (τ), and porosity (ε), which can be written as follows:

$$D_{eff} = \frac{t\tau}{\varepsilon K} \tag{5.5}$$

D_{eff} is also related to the constrictivity parameter (δ). Moreover, the effective diffusion coefficient for a mixture of salt solution varies from that for an individual salt solution due to the mutual diffusion, charge of solute, ion size, and characteristics of the porous support layer. The resulting equation of D_{eff} is given by:

$$D_{eff} = \frac{K_d\, d_h}{1.85\left(R_e \times S_c \times \frac{d_h}{L}\right)^{0.33}} \tag{5.6}$$

where d_h, K_d, R_e, S_c, and L describe the hydraulic diameter of the feed channel, the mass transfer coefficient, the Reynolds number of the flow, the Schmidt number, and the length of the channel, respectively, while the term $\left(R_e \times S_c \times \frac{d_h}{L}\right)^{0.33}$ is defined as the Sherwood number (Sh).

The mass transfer coefficients for the membrane support layer (k_m), feed solution side (k_f), and draw solution side (k_d) can be used to compute the maximum water flux. The correlation between the water flux and the natural logarithm of the ratio of the osmotic pressures π_D and π_F is written as

$$J_w = \cong \frac{k_m k_f k_d}{k_m k_f + k_m k_d + k_f k_d} \ln\left(\frac{\pi_D}{\pi_F}\right) \tag{5.7}$$

Jeon et al. [12] modeled the water flux of a spiral wound module including the effective osmotic pressure ($\Delta\pi_{eff}$), transmembrane pressure (Δp), and the water permeability coefficient (A) by the following formula:

$$J_w = A(\Delta\pi_{eff} + \Delta p) = A(\pi_3 - \pi_2 + \Delta p) \tag{5.8}$$

Gruber et al. [13] applied a computational fluid dynamic (CFD) model to determine the water flux of an asymmetric membrane in an FO system. The relationship between the solute concentration and the osmotic pressure was assumed to be linear, and the water flux equation can be written as:

$$J_w = \frac{1}{K} \ln \frac{B + A\pi_{D,m}}{B + J_w + A\pi_{F,m}} n_d \tag{5.9}$$

where $\pi_{D,m}$ and $\pi_{F,m}$ denote the osmotic pressure at the membrane surface on the draw and feed solutions sides, respectively, and n_d is the unit normal vector on the porous boundary. This model is commonly used to determine the influence of the support layer on the water permeation. It also describes the velocity boundary condition in the regular direction of the membrane. Assuming $B \ll A\pi_{D,m}$ and $B \ll |J_w|$, Eq. (5.9) can be reduced to:

$$J_w = A(\pi_{D,m^{e^{-|J_w|K}}} - \pi_{F,m})n_d \tag{5.10}$$

Deshmukh et al. [14] derived a water flux formula and assessed the influence of membrane intrinsic transport characteristics on the performance of an FO membrane module. By considering the water permeability coefficient, feed concentration, and draw concentration at the membrane–solution interface, $C_{F,m}$, and $C_{D,m}$, the local water flux formula across the membrane can be written as:

$$J_w = A\left(\pi_{D,m} - \pi_{F,m}\right) = \upsilon AR_g T\left(C_{D,m} - C_{F,m}\right) \tag{5.11}$$

Taherian et al. [15] developed an agent-based modeling method to predict the water flux that was influenced by both external concentration polarization (ECP) and ICP phenomena across both sides of the membrane in the FO system. The water flux was quantified by:

$$J_w = A\left[\pi_{D,b} \exp\left(-J_{water} k_d\right) - \pi_{F,b} \exp\left(\frac{J_{water}}{k_f}\right)\right] \tag{5.12}$$

The results from the simulation model were a good fit to the empirical data. This simulation protocol implied that the most influencing parameters on the water flux were the length of the membrane module and the structural parameter. When the membrane module is very long and the structural parameter is large, the water permeation reduces remarkably. It is worth noting that this modeling protocol provided an easy method for estimating the theoretical water flux under different operating conditions and allowed the evaluation of various draw solutions in the FO process.

You et al. [16] revised the above model to consider the transmembrane temperature difference. It may be noted that the temperature gradient through the membrane surface varies from that in the bulk solution as a result of a heat transfer mechanism. With the van't Hoff theory to quantify the osmotic pressure, the temperature is in relation to the solution concentration. Eq. (5.12) is modified as:

$$J_w = A\left(\pi_i - \pi_{F,m}\right) = A\beta R\left(C_i T_i - C_{F,m} T_{F,m}\right) \tag{5.13}$$

where β, C_i, T_i, $C_{F,m}$, and $T_{F,m}$ are the van't Hoff coefficient, the concentration at the interface of the support–selective layer, the temperature at the interface of the support–selective layer, the concentration at the membrane surface, and the temperature at the membrane surface, respectively.

5.3 Practical challenges in FO process

5.3.1 Concentration polarization

CP in FO membranes refers to the change in concentration of solute ions close to the membrane surface or within the membrane structure compared with the bulk solution concentrations. In general, this results in an osmotic pressure differential that occurs on the selective layer which is greatly lower than the osmotic pressure difference of the bulk solution [11,17], leading to significant deviations in the flux for FO filtration processes from that predicted from bulk solution properties. In addition to solute concentration, solute characteristics, membrane properties, and operating conditions are the crucial factors that influence the CP [18].

CP can be generally divided into two categories, ECP and ICP depending on the orientation of the asymmetric membrane. The membrane can be placed with the selective layer oriented toward the feed side or the draw solution side. Although most experiments are carried out with the selective layer facing the feed (commonly referred to as active layer-feed side (AL-FS) or FO mode), many studies also consider the effect of membrane orientation, with the support layer facing the feed water (referred to in the literature as active layer-draw side (AL-DS) or pressure retarded osmosis (PRO) mode). According to Zhao [3,19], both the ECP and ICP could be produced in the FO configuration as shown in Fig. 5–2. ICP occurs within the porous support layer of an asymmetric membrane, whereas ECP occurs external to the membrane on either side. ICP may be either dilutive or concentrative depending upon which configuration is used (i.e., for the Al-FS configuration, draw solution within the support layer will become diluted, and for Al-DS, feed solution within the support layer may become concentrated). In either case, CP will serve to reduce the osmotic pressure difference across the membrane from that of the bulk solutions. On the other hand, the concentrative ECP can be accelerated by the concentrated reverse draw solute transported across the selective layer leading to decline in the osmotic pressure across the selective layer [19]. The reverse solute flux will affect the osmotic pressure difference across the selective layer, reducing the osmotic driving force and hence permeation flux [20,21].

The ratio of the osmotic pressure value across the selective layer to that measured in the bulk solution $(-J_w/K_d)$ can be used to describe the ECP modulus [22]. The ratio of the osmotic pressure at the support–selective layer boundary to that calculated across the selective layer is regarded as the ICP modulus $(-J_wS/K_d)$. The dilutive ECP/ICP is usually between 0 and 1, whereas the concentrative ECP/ICP is higher than 1. If the value of any of these ratios is more than 1, severe polarization occurs. CP phenomena have been studied using different theoretical approaches to predict the water flux of a particular FO process or configuration. These modeling approaches can be classified into numerical simulation [23,24], finite elements method [25,26], and CFD [27]. The following subsections elucidate the theoretical modeling of both the ICP and ECP in the FO process.

5.3.1.1 External concentration polarization

ECP is defined as the local difference of the effective solute concentration gradient across the membrane surface [28]. It exhibits a severe effect on the water permeability and the efficiency of

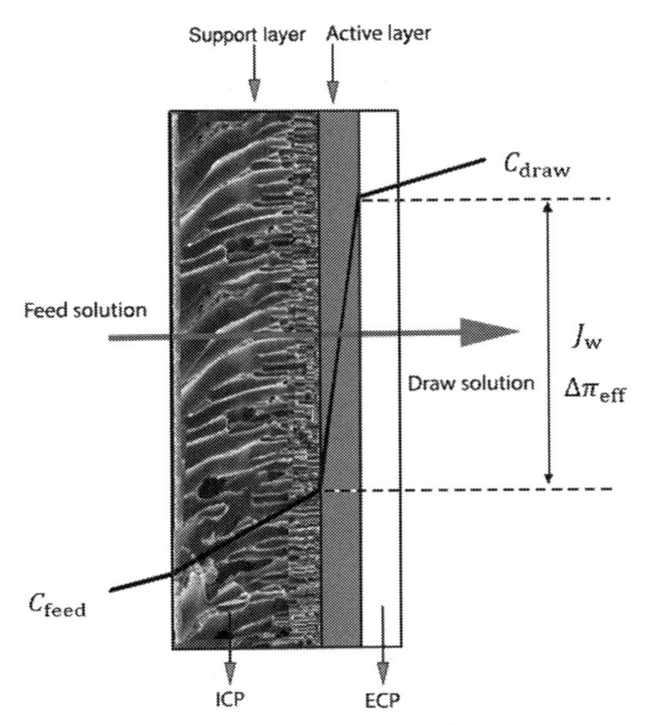

FIGURE 5–2 Diagram showing the internal concentration polarization (ICP) and external concentration polarization (ECP) occurred within an FO membrane. C_F and C_D describe the feed and draw solutions concentration, respectively. $\Delta\pi_{eff}$ is the effective driving force, whereas J_w is the water permeation. *FO*, forward osmosis.

the separation properties of the water–solute. According to the membrane orientation, it can be categorized into concentrative and dilutive ECP [28]. Concentrative ECP occurs on the feed side as the solute is being rejected while the water flows through the membrane, and hence the concentration of the solute increases on the membrane surface. Conversely, the dilutive ECP occurs due to the permeation of water flux through the membrane to the draw stream and dilutes the draw solution at the membrane surface. Both cases have an adverse effect on the osmotic driving force [3,17,29]. The ECP on the selective layer can be mitigated by increasing the cross-flow velocity of the feed side or by introducing a suitably designed spacer, both of which will increase turbulence close to the membrane [30,31]. The ECP can be a serious problem if the salt content in the feed solution exceeds its saturation limit in the solvent leading to scaling on the active layer, causing a massive deterioration in the water permeation.

Zhao et al. [3] reported that, when assuming a low concentration of feed solute, the osmotic pressure could be replaced by the concentration. Hence, the concentrative ECP modulus ($C\rho$) could be expressed as:

$$\frac{\pi m\_feed}{\pi b\_feed} = \exp\left(\frac{J_w}{k_{feed}}\right) \tag{5.14}$$

where k_{feed} refers to a mass transfer coefficient which describes the relationship of the osmotic pressure of the solution on the membrane surface to the osmotic pressure in the bulk solution (πm_feed, πb_feed, respectively). The dilutive ECP modulus can be expressed as:

$$\frac{\pi m\_\text{draw}}{\pi b\_\text{draw}} = \exp\left(-\frac{J_w}{k_{draw}}\right) \tag{5.15}$$

where k_{draw} refers to the mass transfer coefficient of the draw solute. Both terms (πm_draw and πb_draw) relate the osmotic pressures of the draw solute on the surface and in the bulk solution. Both the dilutive ICP and concentrative ECP occur in the Al-FS configuration normally employed for FO processes. Considering the increase in the water permeation which further impacts the ECP during the operation, and if the salt flux is assumed to be insignificant ($B = 0$), the fundamental water flux equation can be rewritten to include these parameters and both CP effects as follows [9]:

$$J_w = A\left[\pi_{D,b} \exp\left(-\frac{J_w}{k}\right) - \pi_{F,b} \exp\left(\frac{J_w}{k}\right)\right] \tag{5.16}$$

This equation indicates that high bulk osmotic force potential causes lower water flux, while being self-limiting [9]. The experimentally determined flux values are commonly lower than the theoretical flux value obtained as a result of the reduction in the osmotic driving power. It is of practical importance to mitigate the ECP in order to optimize the FO system.

It is generally accepted that Eq. (5.29) is only valid for DI water feed solution in the FO system and is inaccurate to quantify the water flux in actual application due to fouling effects. Other researchers have considered the influence of fouling on the membrane performance. Eq. (5.43) describes the water flux incorporated cake-enhanced concentration polarization (CECP) [32].

$$J_w = A\left[\left(\pi_D + \frac{B}{A}\right)e^{-(J_w/K_m)} - \left(\pi_F + \frac{B}{A}\right)e^{-(J_w/K_{CECP})}\right] \tag{5.17}$$

The water and salt permeability coefficients depend on the coefficients of a membrane (subscript *me*) and fouling layer (subscript *la*) as written in the following terms:

$$\frac{1}{A} = \frac{1}{A_{me}} + \frac{1}{A_{la}} \tag{5.18}$$

$$\frac{1}{B} = \frac{1}{B_{me}} + \frac{1}{B_{la}} \tag{5.19}$$

When running the FO experiments over a long time, the K_{CECP} coefficient influences the water permeability as low K_{CECP} exacerbates concentrative ECP impacts on the water permeability of the membrane and vice versa. Other studies in the literature have quantitatively modeled the mass transfer coefficient following various theoretical approaches. Tan et al. [33] developed a modified computational model that used an iterative calculation software considering the suction effect on the mass transfer coefficient for the dilutive ECP. It was noted that the suction impact is increased leading to an enhanced ECP effect. An equation for the solute-mass balance for a rectangular draw solution channel was presented as follows:

$$u\frac{\partial C}{\partial x} + J_w\frac{\partial C}{\partial y} = \overline{D}\frac{\partial^2 C}{\partial y^2} \tag{5.20}$$

where u is the axial flow velocity, $\overline{D}$ is the mean diffusivity of the solute through the interface between the membrane and bulk solution, C is the molar solute concentration, y is the distance perpendicular to the membrane, and x is the distance along the membrane. The axial velocity profile in the draw channel was given as:

$$u = \frac{3}{2}u_0\left(1 - \frac{(y-h)^2}{h^2}\right) = \frac{3u_{0y}}{h} \tag{5.21}$$

where u_0 is the average bulk velocity, and h is the channel half-height. Taking into account the boundary conditions and a dimensionless concentration, the equation can be written as:

$$\frac{\partial^2 C*}{\partial \eta^2} = -\left[\frac{\eta^2}{3} - J_w\left(\frac{3u_0}{D}hx\right)^{\frac{1}{3}}\right]\frac{dC*}{d\eta} \tag{5.22}$$

where C^* is a dimensionless solute concentration defined as ratio of molar concentration to bulk molar concentration, and η is a dimensionless parameter defined by:

$$\eta = y\sqrt[3]{\frac{u_0}{hxD}} \tag{5.23}$$

Setting the boundary conditions ($C^* = 1,\ \eta = \infty$), and using solute-mass balance on the membrane surface of longitudinal direction, Eq. (5.22) can be rearranged to give the average water flux along the membrane:

$$J_w = \frac{1}{L}\int_0^L M\left(\frac{3u_0 D^2}{hx}\right)^{\frac{1}{3}} dx = \frac{3}{2}M\left(\frac{3u_0 D^2}{hL}\right)^{\frac{1}{3}} \tag{5.24}$$

where M is given by

$$M = 0.2912Q \tag{5.25}$$

and Q is given by:

$$Q = \frac{Pe_w}{(R_e\ S_c d_h)^{\frac{1}{3}}} \tag{5.26}$$

By introducing the solute-mass balance parameter across the selective layer on the draw solution stream to obtain the parameter K, the equation can be redescribed as:

$$K\left(C_{D,b}-C_{D,w}\right)=D\left(\frac{\partial C}{\partial y}\right)_{y=0} \tag{5.27}$$

where K is given by:

$$K=\frac{1}{p}\left(\frac{u_0 D^2}{hx}\right)^{\frac{1}{3}} \tag{5.28}$$

The average Sherwood number as a function of the length, L, can be expressed as:

$$\acute{S}h=3.434\frac{\left({}^{Re}Shd_h/L\right)^{\frac{1}{3}}}{p} \tag{5.29}$$

The value of the lumped parameter, Q, ranged from 0 to 10. Eq. (5.29) suggests that the behavior of $1/p$ and the Sherwood number can be affected by variation in the Q value (Tan and Ng [33]). This shows the impact of the dilution factor on the ECP layer. Therefore a polynomial correlation is useful to avoid the complexity in the equation, which can be expressed as in Eq. (5.30) and hence quantifying the water flux in respect to the dilutive ECP in FO system can be expressed as:

$$\acute{S}h=1.849\left({}^{Re}Shd_h/L\right)^{\frac{1}{3}}\left(1.002-0.0319Q+0.000334Q^2-0.001Q^3\right) \tag{5.30}$$

$$\frac{C_{D,w}}{C_{D,b}}=\frac{K}{J_w}+K \tag{5.31}$$

Moreover, to quantify the water flux in respect to the concentrative ECP in an FO system, the following equations can be used:

$$\acute{S}h=1.849\left({}^{Re}Shd_h/L\right)^{\frac{1}{3}}\left(0.997+0.0315Q+0.022Q^2-0.008Q^3\right) \tag{5.32}$$

$$\frac{C_{F,w}}{C_{F,b}}=\frac{K}{J_w}-K \tag{5.33}$$

As the membrane has a typically asymmetric structure, the solute molecules are also transported within the porous support layer resulting in more significant CP in this porous structure, namely ICP.

5.3.1.2 Internal concentration polarization

ICP is governed by the rate of diffusion of the draw solute within the support layer, which affects the concentration of solute at the active layer/support layer interface [5,34]. There are two different categories depending on the orientation of the membrane, concentrated and dilutive ICP. The concentrative ICP occurs when the selective layer is facing the draw solution. It occurs due to the transportation of solute from the feed solution (FS) within the support layer due to convective water flux. Conversely, the water flux in dilutive ICP occurs when the selective layer is adjacent to the FS and is due to the dilution of draw solution within the support layer due to water permeation from the feed side, leading to a reduction in the osmotic pressure gradient resulting in a lowered water permeation rate [35]. ICP is difficult to minimize even if the flow velocity or turbulence is increased [8]. It has been reported that when a high mass transfer resistance is produced in the support layer, the solute particles precipitate in the support layer [36] resulting in this irreversible component. Previous studies reported that this ICP effect causes a decrease in the water flux by more than 80% [18]. Tan et al. [33] determined that this phenomenon was correlated with the support layer structure. The support layer tends to be relatively thick and tortuous to give mechanical strength to the membrane. However, the high thickness and tortuosity hinders the diffusion of the solute. The structural parameter may also influence the resistance to mass transfer. The dilutive ICP forms the most severe impact in the FO system.

The diffusion rate of solute molecules and ions is also an important factor in ICP, with a higher solute diffusion rate more effective at countering dilutive ICP due to the increased backflow of solute molecules into the support layer from bulk solution. Therefore the diffusion rate should be taken into account when selecting a draw agent in case dilutive ICP needs to be taken into account. Similarly, the modulus of the dilutive ICP, including the osmotic pressure across the selective layer ($\pi_{D,i}$) and the osmotic pressure of the bulk draw solution ($\pi_{D,b}$), was previously expressed by McCutcheon et al. [9] as follows:

$$\frac{\pi_{D,i}}{\pi_{D,b}} = \exp\left(-\frac{J_w}{K}\right) \tag{5.34}$$

The passage of the fluid flux occurs through open pores which depend on the porosity, the void structure (ε) in the support layer, and tortuosity (τ). When the structural parameter (S) is high, it leads to an increase in the mass resistance and ICP by decreasing the water flux as well as the viscosity and effective diffusivity (D) of the solute [28,37,38]. Accordingly, the ICP derivation can be modified to include an effective diffusivity factor (D) that represents the interaction between the support layer matrix and the draw solute as follows:

$$K_s = \frac{t\tau}{D\varepsilon} \tag{5.35}$$

Another important parameter is the solute flow resistivity in the support layer, which is modeled for the dilutive and concentrative ICP by the following relations [33]:

$$K = \frac{1}{J_w}\ln\frac{B + A\pi_{D,b}}{B + J_w + A\pi_{F,w}} \tag{5.36}$$

$$K = \frac{1}{J_w}\ln\frac{B + A\pi_{D,w} - J_w}{B + A\pi_{F,b}} \tag{5.37}$$

It must be noted that in addition to these factors, the magnitude of ICP is also influenced by the interaction of the solute with the polymer matrix of the support layer [33].

The most commonly used water flux model containing the dilutive and concentrative factors in the literature was reported by Tiraferri et al. [39]. It takes into account the bulk solute concentration of the draw solution, C_D, feed solution concentration, C_F, and the corresponding osmotic pressures, π_D and π_F, diffusion coefficient, mass transfer coefficient k_f, and relating all to the water and solute permeability coefficients as follows [22]:

$$J_w = A\left\{\frac{\pi_D \exp\left(-\frac{J_w S}{D}\right) - \pi_F \exp\left(\frac{J_w}{K}\right)}{1 + \frac{B}{J_w}\left[\exp\left(\frac{J_w}{K}\right) - \exp\left(-\frac{J_w S}{D}\right)\right]}\right\} \tag{5.38}$$

Alternatively, another model has been developed taking into account all the mass transfer parameters in the FO system [40]. This model considers the external boundaries on both sides of the membrane, the skin, and the sponge layers employing a diffusive–convective mass transport equation for every single sublayer. Hence, the concentration profile and mass transport rate for the draw solution side is expressed as.

$$-J_s = -\left(\frac{DdC}{\delta dy} - vC\right) \tag{5.39}$$

where v is the convective water velocity, and y is the salt mass transfer rate as it flows through the boundary layer to the membrane layer. The negative sign of the solute flux indicates the opposite flow of the salt flux to that of the water flux.

To obtain the water flux and the solute transfer in FO mode, a modified version of Eq. (5.39) was derived to represent the solute mass transport within the boundary layer at the membrane layer for the water flux and is given by:

$$-J_s = J_w F = J_w \frac{C_b - C_{m}^{e^{(J_w/k_d)}}}{1 - e^{(J_w/k_d)}} \equiv \beta_d\left(C_b - m^{e^{(J_w/k_d)}}\right) \tag{5.40}$$

where

$$\beta_s = \frac{J_w}{1 - e^{(J_w/k_d)}} \tag{5.41}$$

It is important to include the mass transfer resistance of the permeate side boundary layer. Subsequently, the diffusive and convective flows for the permeate side can be combined as follows:

$$-J_s = \beta_f\left(C_{sp} - C_f^{e^{(J_w/k_d)}}\right) \tag{5.42}$$

with

$$\beta_f = \frac{J_w}{1 - e^{(J_w/k_d)}} \tag{5.43}$$

The mass transfer of the solute flux can be expressed as follows:

$$J_s = B(C_i - C_m) \tag{5.44}$$

Likewise, the differential mass balance equation was used to describe the solute flux within the sponge layer as follows:

$$-J_s = \beta_f\left(C_{sk} - C_{sp}^{(J_w K)}\right) \equiv \beta_f\left(C_{sk} - C_{sp}^{(J_w S/D)}\right) \tag{5.45}$$

where

$$\beta_f = \frac{J_w}{1 - e^{(J_w S/D)}} \tag{5.46}$$

By solving for the internal concentration parameters (C_{sk}), substituting $C_m = C_b$, and $C_{sp} = C_f$, the final water flux equation is represented as:

$$J_w = A\left[\pi_{D,b}\left(\frac{C_{D,i}}{C_{D,b}} - \frac{C_{F,m}}{C_{D,b}}\right) - \Delta p\right] \tag{5.47}$$

where $C_{D,i}$ and $C_{D,b}$ are defined as the draw solute concentration across the interface between the selective layer, the support layer, and the draw solution concentration at bulk draw solution stream, respectively. $C_{F,m}$ denotes the draw solution concentration at the feed stream against the selective layer surface. Δp is the transmembrane pressure.

Attarde et al. [41] has proposed a solution-diffusion model to describe the membrane mass transfer mechanism, whereas the Spiegler–Kedem model is used to determine the ICP taking into account the local mass transfer across the selective layer of a spiral wound membrane as follows:

$$\frac{C_{F,m}}{C_{D,m}} = \frac{[1 - \exp(F)]\exp(J_w K)\frac{C_{F,b}}{C_{D,m}} + (1 - \sigma)[1 - \exp(J_w K)]}{[1 - \exp(F) + (1 - \sigma)\exp(F)[1 - \exp(J_w K)]]} \tag{5.48}$$

where K, and σ are the solute resistivity and the reflection coefficient, and $C_{F,m}$ and $C_{D,m}$ denote the feed and draw solution concentrations, respectively. Eq. (5.49) can be applied for both models if the reflection coefficient equals one.

$$\frac{C_{F,m}}{C_{D,m}} = \frac{B[\exp(J_w K) - 1] + \frac{C_{F,b}}{C_{D,m}} J_w \exp(J_w K)}{B[\exp(J_w K) - 1] + J_v} \tag{5.49}$$

It can also be used for nonideal membranes in which the ICP impacts might be avoided by reducing the thickness, porosity, and tortuosity of the support layer.

5.3.2 Reverse solute flux

The reverse solute flux refers to the transport of solute from the draw solution to the feed solution. In principle, forward diffusion refers to the flow of solute solution from feed stream to the draw solution, whereas reverse diffusion is the backflow of a solute from draw solution to the feed stream (solute leakage) [34]. When the draw solute contains low-molecular weight of salt, reverse solute flux appears unavoidable in FO processes because of the concentration difference between the feed and draw solutions. The transport of draw solute within the membrane is conventionally modeled by Fick's Law of diffusion as follows [10,11]:

$$J_s = B\Delta C \tag{5.50}$$

where B and ΔC describes the solute permeability coefficient and the solute concentration gradient across the membrane, respectively. The solute flux across the membrane–solution interface transfers in two different directions, a diffusive and a convective flux transfer (see Fig. 5–3) [20,42]. Both the diffusive and the convective flux are commonly expressed by the following equation:

$$J_s = -J_w C + D^s \frac{dC}{dx} \tag{5.51}$$

where D^s denotes the solute diffusion coefficient, and x is the distance from interface between the active layer and the support layer, whereas C denotes the solute concentration at the same point. Another crucial variable is the solute concentration on the support layer surface ($C_{D,m}$) which sets the boundary conditions, and by integrating, the equation can be expressed as:

$$J_s{}^S = \frac{J_w\left[\frac{J_w t_s \tau}{D\varepsilon}\right] C_{D,i}{}^s - C_{D,b}}{\exp\left(\frac{J_w t_s \tau}{D\varepsilon}\right) - 1} \tag{5.52}$$

where $J_s{}^s$ denotes the solute flux in the support layer, C_D defines the draw solute concentration at the support layer side of the active layer interface, $C_{D,b}$ is the bulk draw solute concentration, and t_s is the support layer thickness. The reverse solute flux through the selective layer can be expressed as follows:

$$J_s{}^A = -\frac{D^A}{t_A}\left(C_{D,i}{}^A - 0\right) \tag{5.53}$$

where J_s^A denotes the solute flux in the active layer, D^A describes the draw solute diffusion coefficient, $C_{D,i}{}^A$ defines the draw solute concentration on the active layer side of the active layer–support layer interface, and t_A defines the active layer thickness. A simplified combined new expression is reported in Ref. [43] as follows:

$$J_s = \frac{J_w C_{D,b}}{1 - \left(1 + \frac{J_w t_A}{D^A H}\right)\exp\left(\frac{J_w t_s \tau}{D\varepsilon}\right)} = \frac{J_w C_{D,b}}{1 - \left(1 + \frac{J_w}{B^A}\right)\exp\left(\frac{J_w S}{D}\right)} \tag{5.54}$$

It is important to integrate the $C_{F,m}$ and $C_{D,m}$ as their concentrations vary across the length of the membrane module [14]. t_A, D^A, and H are the active layer thickness, the draw

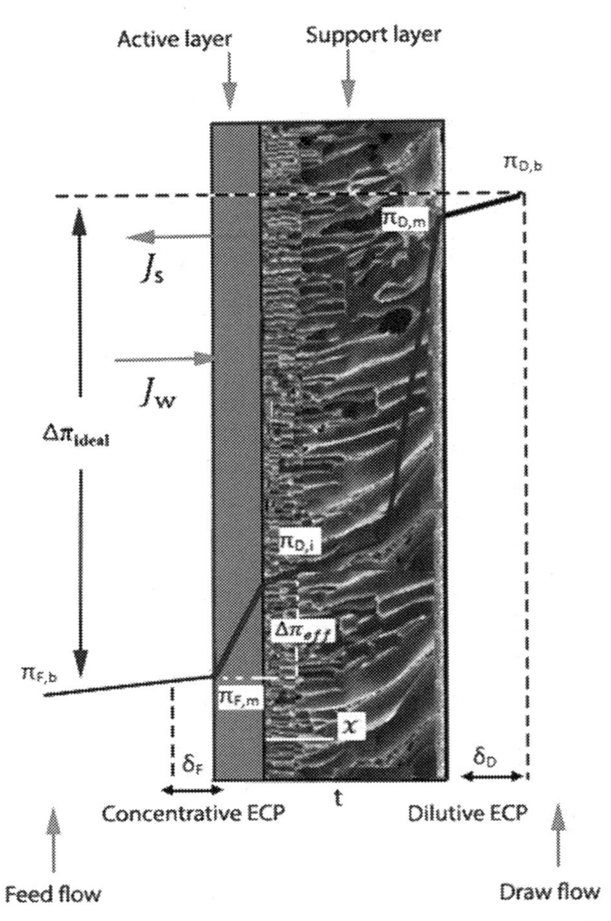

FIGURE 5–3 Showing concentrative and dilutive concentration external concentration polarization and the concentration boundary thickness created on both sides of an asymmetric membrane in the FO process. *FO*, forward osmosis.

solute diffusion coefficient in the selective layer, and the partition coefficient indicating the relative concentration in each phase, respectively.

By considering the solute permeability coefficient, the local net solute flux along the membrane module can be modeled as:

$$J_s = B(C_{D,m} - C_{F,m}) \tag{5.55}$$

In the case of a spiral wound module, the reverse solute flux is a function of hydraulic pressure. Eq. (5.55) can be modified to include the van't Hoff coefficient, the universal gas constant, and the absolute temperature as follows:

$$J_s = B(C_3 - C_2) = \frac{B}{\beta R_g T}(J_w/A - \Delta p) \tag{5.56}$$

where C_2 and C_3 are the salt concentration at the feed solution–support layer interface and at the interface among the selective layer and support layer. The solute flux can also be calculated using the experimental data of the water permeability coefficient, the solute permeability coefficient, the proportionality factor ($\varnothing$), and the water flux measured from Eq. (5.34) [27]. It should be noted that the solute concentration is correlated to the osmotic pressure which is written as:

$$J_s = -\frac{B}{\varnothing A} J_w \tag{5.57}$$

Additionally, another reliable model for calculating the reverse solute flux for a draw solution containing multiple salt species was reported in Ref. [20], which considers the adverse effect of CP on both sides of a membrane, and is given by:

$$J_s = B\left(\frac{C_{D,b} + \frac{J_s}{J_w}}{\exp(J_w K)\exp\left(\frac{J_w}{k_d}\right)} - \left(C_{F,b} + \frac{J_s}{J_w}\right)\exp\left(\frac{J_w}{k_f}\right)\right) \tag{5.58}$$

The classical reverse solute flux model in the literature that considers all important parameters such as membrane intrinsic parameters, the hydrodynamic factors for the cell channel, the concentrative ECP, the dilutive ICP, the bulk solute concentration of the draw solution, C_D, feed solution concentration, C_F, and the draw solution (DS), FS osmotic pressures, π_D and π_F , were reported in Ref. [39]. This equation describes the draw solution concentration at the support–active layer boundary considering the dilutive ECP and ICP impact that are described by the terms $-J_w/K$ and $-J_w S/D$. This model is expressed as:

$$J_w = A\left\{\frac{C_D \exp\left(-\frac{J_w S}{D}\right) - C_F \exp\left(\frac{J_w}{K}\right)}{1 + \frac{B}{J_w}\left[\exp\left(\frac{J_w}{k_f}\right) - \exp\left(-\frac{J_w S}{D}\right)\right]}\right\} \tag{5.59}$$

Because the draw solution concentration is assumed to be linearly proportional to the osmotic pressure, the van't Hoff relation can also be applied to calculate the osmotic water flux [22]. Eq. (5.59) can be written as follows:

$$J_w = A\left\{\frac{\pi_{D,b}\exp\left[-J_w \frac{1}{k_d} + \frac{S}{D_D}\right] - \pi_{F,b}\exp\left(\frac{J_w}{k_f}\right)}{1 + \frac{B}{J_w}\left\{\exp\left(\frac{J_w}{k_f}\right) - \exp\left[-J_w\left(\frac{1}{k_d} + \frac{S}{D_D}\right)\right]\right\}}\right\} \tag{5.60}$$

Experimentally, the reverse solute flux can be determined from the slope of the plot of $1/C_p$ against $1/t$, which can be described as $V_{Fo}/(J_s A_m)$ as explained in Ref. [43]. The derived equation is expressed as:

$$\frac{1}{C_F} = \frac{V_{Fo}}{J_s A_m}\left(\frac{1}{t}\right) - \frac{J_w}{J_s} \tag{5.61}$$

This simple model has been proven to be effective because it is possible to select a suitable membrane and the best draw solution that generates excellent water flux and minimum reverse salt flux. Another factor that should be considered is the specific reverse solute flux that is represented by the ratio of the reverse solution flux (mg/m^2/h) to the forward water flux (L/m^2/h) as reported in Ref. [18]. This ratio is named $J_{specific}$ and is used to evaluate the performance of FO where a high ratio indicates low selectivity and membrane efficiency.

Even though the nature of solute mass transfer seems to be complex, this relationship is more accurate and preferable to determine the loss of draw solute in an FO system. In a process when a high rejection rate is required, multivalent ions with low diffusion coefficients are favorable. Understanding the special criteria of the draw solution is of great importance for selecting a suitable draw solution that allows sustainable FO applications.

Acknowledgments

The authors would like to thank the Royal Society for funding this work through a Royal Society International Collaboration Award (IC160133).

References

[1] W.A. Suwaileh, D.J. Johnson, S. Sarp, N. Hilal, Advances in forward osmosis membranes: altering the sub-layer structure via recent fabrication and chemical modification approaches, Desalination 436 (2018) 176–201.

[2] B. Kim, G. Gwak, S. Hong, Review on methodology for determining forward osmosis (FO) membrane characteristics: water permeability (A), solute permeability (B), and structural parameter (S), Desalination 422 (2017) 5–16.

[3] S. Zhao, L. Zou, C.Y. Tang, D. Mulcahy, Recent developments in forward osmosis: opportunities and challenges, J. Membr. Sci. 396 (2012) 1–21.

[4] S. Phuntsho, S. Vigneswaran, J. Kandasamy, S. Hong, S. Lee, H.K. Shon, Influence of temperature and temperature difference in the performance of forward osmosis desalination process, J. Membr. Sci. 415–416 (2012) 734–744.

[5] N. Akther, A. Sodiq, A. Giwa, S. Daer, H.A. Arafat, S.W. Hasan, Recent advancements in forward osmosis desalination: a review, Chem. Eng. J. 281 (2015) 502–522.

[6] A. Achilli, T.Y. Cath, A.E. Childress, Selection of inorganic-based draw solutions for forward osmosis applications, J. Membr. Sci. 364 (2010) 233–241.

[7] S. Loeb, L. Titelman, E. Korngold, J. Freiman, Effect of porous support fabric on osmosis through a Loeb-Sourirajan type asymmetric membrane, J. Membr. Sci. 129 (1997) 243–249.

[8] G.T. Gray, J.R. McCutcheon, M. Elimelech, Internal concentration polarization in forward osmosis: role of membrane orientation, Desalination 197 (2006) 1–8.

[9] J.R. McCutcheon, M. Elimelech, Influence of concentrative and dilutive internal concentration polarization on flux behavior in forward osmosis, J. Membr. Sci. 284 (2006) 237–247.

[10] P.H. Nelson, Biophysics and Physiological Modeling, first ed., Cambridge University Press, UK, 2015.

[11] N. Hancock, T. Cath, Solute coupled diffusion in osmotically driven membrane processes, Environ. Sci. Technol. 43 (2009) 6769–6775.

[12] J. Jeon, J. Jung, S. Lee, J.Y. Choi, S. Kim, A simple modeling approach for a forward osmosis system with a spiral wound module, Desalination 433 (2018) 120–131.

[13] M.F. Gruber, U. Aslak, C. Hélix-Nielsen, et al., CFD model for optimization of forward osmosis and reverse osmosis membrane modules, Sep. Purif. Technol. 158 (2016) 183–192.

[14] A. Deshmukh, N.Y. Yip, S. Lin, M. Elimelech, Desalination by forward osmosis: identifying performance limiting parameters through module-scale modeling, J. Membr. Sci. 491 (2015) 159–167.

[15] M. Taherian, S.M. Mousavi, Modeling and simulation of forward osmosis process using agent-based model system, Comput. Chem. Eng. 100 (2017) 104–118.

[16] S. You, X. Wang, M. Zhong, Y. Zhong, C. Yu, N. Ren, Temperature as a factor affecting transmembrane water flux in forward osmosis: steady-state modeling and experimental validation, Chem. Eng. J. 198–199 (2012) 52–60.

[17] T. Cath, A. Childress, M. Elimelech, Forward osmosis: principles, applications, and recent developments, J. Membr. Sci. 281 (1–2) (2006) 70–87.

[18] S. Kim, E.M.V. Hoek, Modeling concentration polarization in reverse osmosis processes, Desalination 186 (2005) 111–128.

[19] Y. Jiao, C. Zhao, Y. Kang, Ch Yang, Microfluidics-based fundamental characterization of external concentration polarization in forward osmosis, Microfluid. Nanofluid. 23 (2019) 36.

[20] C. Suh, S. Lee, Modeling reverse draw solute flux in forward osmosis with external concentration polarization in both sides of the draw and feed solution, J. Membr. Sci. 427 (2013) 365–374.

[21] D.J. Johnson, W.A. Suwaileh, A.W. Mohammed, N. Hilal, Osmotic's potential: an overview of draw solutes for forward osmosis, Desalination 434 (2018) 100–120.

[22] N.N. Bui, J.T. Arena, J.R. McCutcheon, Proper accounting of mass transfer resistances in forward osmosis: improving the accuracy of model predictions of structural parameter, J. Membr. Sci. 492 (2015) 289–302.

[23] D.H. Jung, J. Lee, Y.G. Lee, M. Park, S. Lee, D.R. Yang, et al., Simulation of forward osmosis membrane process: effect of membrane orientation and flow direction of feed and draw solutions, Desalination 277 (2011) 83–91.

[24] S.-M. Shim, W.-S. Kim, A numerical study on the performance prediction of forward osmosis process, J. Mech. Sci. Technol. 27 (2013) 1179–1189.

[25] W. Li, Y. Gao, C.Y. Tang, Network modeling for studying the effect of support structure on internal concentration polarization during forward osmosis: model development and theoretical analysis with FEM, J. Membr. Sci. 379 (2011) 307–321.

[26] A. Sagiv, R. Semiat, Finite element analysis of forward osmosis process using NaCl solutions, J. Membr. Sci. 379 (2011) 86–96.

[27] M.F. Gruber, C.J. Johnson, C.Y. Tang, M.H. Jensen, L. Yde, C. Hélix-Nielsen, Computational fluid dynamics simulations of flow and concentration polarization in forward osmosis membrane systems, J. Membr. Sci. 379 (2011) 488–495.

[28] Q. Ge, M. Ling, T. Chung, Draw solutions for forward osmosis processes: developments, challenges, and prospects for the future, J. Membr. Sci. 442 (2013) 225–237.

[29] K.Y. Wang, R.C. Ong, T. Chung, Double-skinned forward osmosis membranes for reducing internal concentration polarization within the porous sublayer, Ind. Eng. Chem. Res. 49 (2010) 4824–4831.

[30] S.S. Sablani, M.F.A. Goosen, R. Al-Belushi, M. Wilf, Concentration polarization in ultrafiltration and reverse osmosis: a critical review, Desalination 141 (2001) 269–289.

[31] M.F.A. Goosen, S.S. Sablani, S.S. Al-Maskari, R.H. Al-Belushi, M. Wilp, Effect of feed temperature on permeate flux and mass transfer coefficient in spiral-wound reverse osmosis systems, Desalination 144 (2002) 367–372.

[32] Z. Wang, J. Zheng, J. Tang, X. Wang, Z. Wu, A pilot-scale forward osmosis membrane system for concentrating low-strength municipal wastewater: performance and implications, Sci. Rep. 6 (2016) 21653. Available from: https://doi.org/10.1038/srep21653.

[33] C.H. Tan, H.Y. Ng, Revised external and internal concentration polarization models to improve flux prediction in forward osmosis process, Desalination 309 (2013) 125–140.

[34] K. Lutchmiah, A.R.D. Verliefde, K. Roest, L.C. Rietveld, E.R. Cornelissen, Forward osmosis for application in wastewater treatment: a review, Water Res. 58 (2014) 179–197.

[35] H. Du, Osmotically Driven Membrane Processes—Approach, Development and Current Status, First ed., Intech Open Limited, London, UK, 2018. Available from: http://doi.org/10.5772/intechopen.68607. ISBN: 978-953-51-3922-5.

[36] J. Wang, D.S. Dlamini, A.K. Mishra, M.T.M. Pendergast, M.C.Y. Wong, B.B. Mamba, et al., A critical review of transport through osmotic membranes, J. Membr. Sci. 454 (2014) 516–537.

[37] N.Y. Yip, A. Tiraferri, W.A. Phillip, J.D. Schiffman, M. Elimelech, High performance thin-film composite forward osmosis membrane, Environ. Sci. Technol. 44 (2010) 3812–3818.

[38] A. Tiraferri, N.Y. Yip, W.A. Phillip, J.D. Schiffman, M. Elimelech, Relating performance of thin-film composite forward osmosis membranes to support layer formation and structure, J. Membr. Sci. 367 (2011) 340–352.

[39] A. Tiraferri, N.Y. Yip, A.P. Straub, S.R. Castrillon, M. Elimelech, A method for the simultaneous determination of transport and structural parameters of forward osmosis membranes, J. Membr. Sci. 444 (2013) 523–538.

[40] E. Nagy, A general resistance-in-series salt and water flux models for forward osmosis and pressure-retarded osmosis for energy generation, J. Membr. Sci. 460 (2014) 71–81.

[41] D. Attarde, M. Jain, Sh. K. Gupta, Modeling of a forward osmosis and a pressure-retarded osmosis spiral wound module using the Spiegler-Kedem model and experimental validation, Sep. Purif. Technol. 164 (2016) 182–197.

[42] B. Kim, S. Lee, S. Hong, A novel analysis of reverse draw and feed solute fluxes in forward osmosis membrane process, Desalination 352 (2014) 128–135.

[43] W.A. Phillip, J.S. Yong, M. Elimelech, Reverse draw solute permeation in forward osmosis: modeling and experiments, Environ. Sci. Technol. 44 (2010) 5170–5176.

6 Recent developments in forward osmosis and its implication in expanding applications

Min Zhan[1], Youngjin Kim[2], Seungkwan Hong[1]

[1]SCHOOL OF CIVIL, ENVIRONMENTAL AND ARCHITECTURAL ENGINEERING, KOREA UNIVERSITY, SEOUL, REPUBLIC OF KOREA [2]DEPARTMENT OF ENVIRONMENTAL ENGINEERING, COLLEGE OF SCIENCE AND TECHNOLOGY, KOREA UNIVERSITY, SEJONG-SI, REPUBLIC OF KOREA

6.1 Introduction

Forward osmosis (FO) is an emerging innovation in the field of membrane-based separation technology and has attracted significant attention over the last decade [1,2]. Distinct from reverse osmosis (RO), which is a pressure-driven membrane process, FO is driven by the natural osmotic gradient for permeation of water through a semipermeable membrane [3]. A concentrated solution having a high osmotic potential (known as draw solution, DS) transfers water from a saline solution with a low osmotic pressure (the feed solution, FS); thus the FS is concentrated and the DS is diluted [4].

Because of its low hydraulic-pressure requirement, FO possesses several distinct advantages, namely, (1) it consumes less energy, (2) involves lower fouling, and (3) can achieve high rejection of various solutes and pollutants [5]. These apparent advantages have encouraged many researchers to focus on the development of FO, and promising results have been reported in various applications including direct desalination (seawater and brackish water desalination), indirect desalination (wastewater reclamation), wastewater treatment (municipal wastewater and industrial wastewater treatment), brine concentration [zero liquid discharge (ZLD)], and food processing [6].

Extensive research has been conducted to resolve the challenges encountered by the FO process. Meanwhile, advances in technology have resulted in an improved performance of the FO process with regard to FO membranes, DS, and their application [7]. Numerous studies have summarized the state-of-the-art FO technologies. Fig. 6–1 summarizes the literature in terms of publication numbers, nature of research articles, and citations on FO in the last decade. Fig. 6–1A shows that the manuscript types of the publications can be split into five major categories: articles (1765; 81%), reviews (151; 7%), conference papers (149; 7%), book chapters (60; 3%), and others (48; 2%). Research content classification and citation data are

Osmosis Engineering. DOI: https://doi.org/10.1016/B978-0-12-821016-1.00009-7

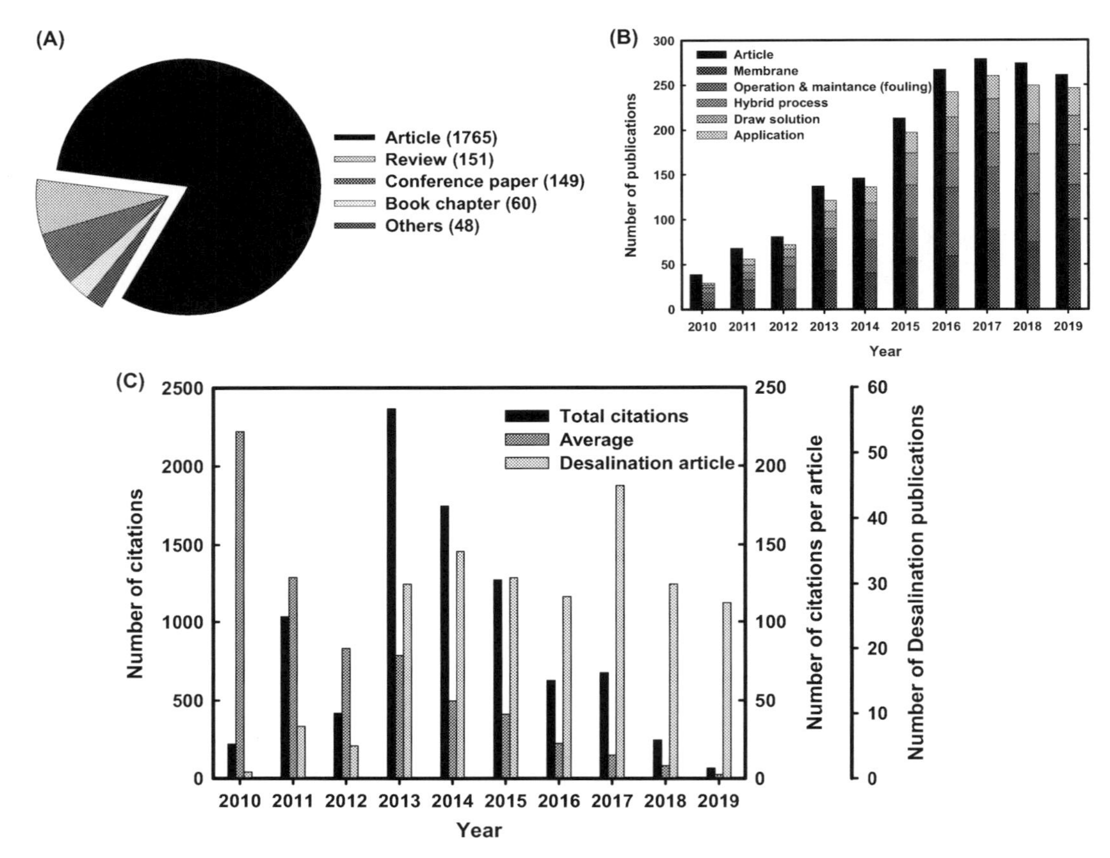

FIGURE 6–1 (A) Academic literature database from 2010 to 2019 was extracted from Scopus with keywords "Forward Osmosis" and "Direct Osmosis"; (B) nature of research article publications with five categories "Membrane," "Operation & Maintenance (fouling)," "Hybrid Process," "DS," and "Application"; (C) total and normalized citations for Desalination.

provided in Fig. 6–1B and C. Among the five different categories, majority of the highly cited papers are on the application of the FO process in various industries; this indicates that the FO technology has drawn the attention of researchers working in various industries. Moreover, a detailed study on Desalination publication in FO is summarized in Table 6–1, which is based on data in Fig. 6–1C. The three major focus areas of the papers are the FO membrane, hybrid processing, and application of the FO process. Overall, the published papers reflect the trends of FO research and the areas that are most concerning in the research community. Unsurprisingly, the areas of interest resonate with the key aspects and challenges that are associated with the FO process.

Over the last few years, efforts to promote FO have been narrowed to focus on establishing more commercial credibility. All companies have reported increased activity, some of which have experienced an increase in sales. The market is dominated by FTSH2O, Porifera, Oasys, and Modern Water (Fig. 6–2) [8]. However, the FO market is small. None of the

Table 6–1 Citation details of Desalination publications in forward osmosis research.

Article	Publication number	Publications on Desalination	Citation	Normalized citation
Membrane	515	75	2499	33
Operation and maintenance (fouling)	400	68	2020	30
Hybrid process	261	45	2023	45
DS	237	33	875	27
Application	195	19	1266	67
Total	1608	240	8683	36

DS, Draw solution.

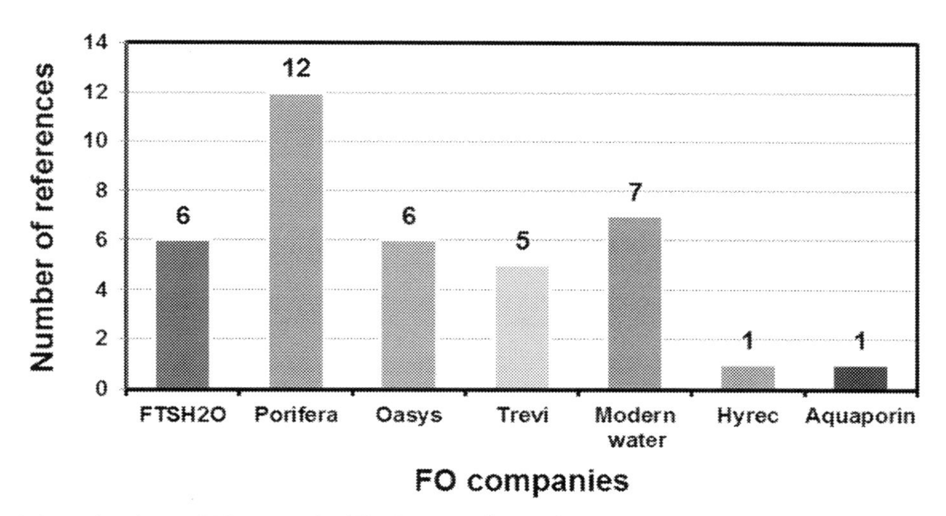

FIGURE 6–2 Market share of FO companies [8]. *FO*, Forward osmosis.

companies have more than a few commercial installations; therefore a few breakthroughs and contracts from other companies have been covered in this report, which would make a considerable difference with regard to the market share.

FO as a technology is gaining market traction with regard to ZLD/brine concentration, production of valuable products by dewatering, minimization of waste hauling costs, concentrations of difficult-to-treat wastes, landfill leachate, and specialized desalination opportunities [8]. Within the next 5 years, FO will most probably be picked up by larger firms as a product offering applications such as brine management and water extraction solutions within a suite of water treatment. The evaporator market will be more challenged by FO; however, the addressable market for FO will be more realistic than the current optimistic projections.

However, it cannot be denied that FO is still an immature technology because some remaining limitations have hampered its real-world implementation. A key challenge to the further development of FO is based on the draw solute, which is directly linked with the two performance-affecting components [9]. The ideal draw solute should meet several important criteria, including (1) high osmotic pressure, (2) low reverse solute flux, and (3) easy/cost-effective recovery from the diluted DS. Various materials including conventional inorganic salts, organic solutes, and newly developed chemicals have been studied and evaluated in the search for an optimized draw solute; however, it was revealed that each material exhibits serious drawbacks. The best draw solute that can be adopted for practical utilization with regard to industrial application has not been successfully identified yet [10].

Another crucial barrier to successful FO is the recovery process of the DS at a low-energy cost. DS regeneration depends on the used DS. Because additional treatment technologies require energy, the energy demand of the hybrid FO processes has to be considered [2]. A review of promising approaches for DS recovery such as pressure-driven membrane processes, thermal-driven separation, precipitation, stimulus–response method, and direct use has been recently published by our research group. Nevertheless, the development of better draw agents is still in progress, and the practical application of FO with the development of an optimized draw solute is expected to be realized in the near future.

Finally, although full-scale implementation of FO in seawater desalination shows that FO is an applicable treatment technology, more pilot-scale investigations are necessary to prove the technical practicability of FO [4]. With regard to the industrial applications, only basic proof-of-principle studies have been conducted at the lab scale. The upscaling of pilot- or full-scale studies will be the next step to optimize the operation and implement FO in applications pertaining to the treatment of industrial water and wastewater.

In this chapter a review of various applications has been performed for the production and treatment of water using the FO process. A comprehensive and critical review on the existing hybrid FO systems and their performances in different applications is presented. To find an appropriate niche for FO the review organized past researches on the FO process and suggested optimized applications and prospects for future research areas on hybrid FO systems. This review also includes a brief overview of the different types of fouling mechanisms that can be found in FO processes, the possible fouling mitigation and cleaning methods to enhance FO efficiency, and the use of advanced membrane fouling characterization methods.

6.2 Forward osmosis

6.2.1 Theoretical background

Osmosis is the natural phenomenon in which water molecules are transported from the FS of higher water chemical potential (i.e., lower solute concentration) to a DS of lower water chemical potential (i.e., higher solute concentration) through a semipermeable membrane (Fig. 6–3). The membrane allows water transport but rejects most solute molecules or ions. Therefore FO utilizes this osmotic pressure differential as the driving force to extract water molecules across the membrane rather than hydraulic pressure differential as used in RO. Pressure-retarded osmosis (PRO) is similar to RO because it applies pressure, but the net water flux is still in the same direction to FO because the driving force in PRO is an osmotic pressure gradient through the membrane (i.e., $\Delta\pi > \Delta P$).

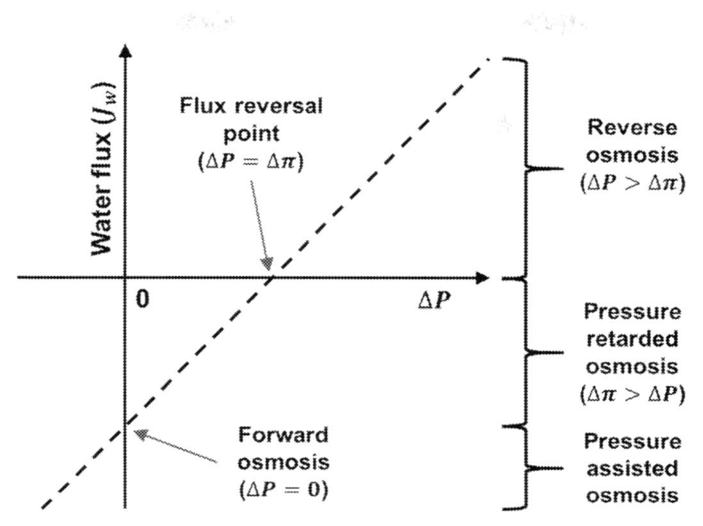

FIGURE 6–3 Direction and magnitude of water flux as a function of applied pressure in FO, PRO, PAO, and RO. *FO*, Forward osmosis; *PAO*, pressure-assisted osmosis; *PRO*, pressure-retarded osmosis; *RO*, reverse osmosis.

In FO, water molecules pass through the membrane, whereas salts are concentrated on the membrane surface, which is called external concentration polarization (ECP). In the case of a symmetric membrane (Fig. 6–4A), salt concentration increases in the feed side (i.e., concentrative ECP) and decreases in the draw side (i.e., dilutive ECP) because water permeates from the FS to DS. This ECP can reduce the driving force by decreasing the effective concentration gradient across the active layer of the FO membrane. By improving the hydrodynamic properties such as crossflow velocity the effective concentration gradient can increase and thus water flux can be enhanced. The concentrative and dilutive ECP can be written as shown in Eqs. (6.1) and (6.2). Therefore water flux in FO can be obtained using Eq. (6.3) by considering both ECPs.

$$\frac{\pi_{F,m}}{\pi_{F,b}} = \exp\left(\frac{J_w}{k}\right) \tag{6.1}$$

$$\frac{\pi_{D,m}}{\pi_{D,b}} = \exp\left(-\frac{J_w}{k}\right) \tag{6.2}$$

$$J_w = \mathrm{A}\left[\pi_{D,m} - \pi_{F,m}\right] = \mathrm{A}\left[\pi_{D,b}\exp\left(-\frac{J_w}{k}\right) - \pi_{F,b}\exp\left(\frac{J_w}{k}\right)\right] \tag{6.3}$$

where A is the water-permeable coefficient; $\pi_{F,b}$ and $\pi_{D,b}$ are the bulk concentrations of the FS and DS, respectively; $\pi_{F,m}$ and $\pi_{D,m}$ are the concentrations of the FS and DS, respectively, at the active layer; J_w is the water flux; and k is the mass transfer coefficient in the channel. In the case of an asymmetric membrane (i.e., consisting of an active layer and a support layer) with the active layer facing the FS (also called FO mode or active layer (AL)-FS mode), only concentrative ECP occurs, as shown in Fig. 6–4B. The support layer has a highly porous

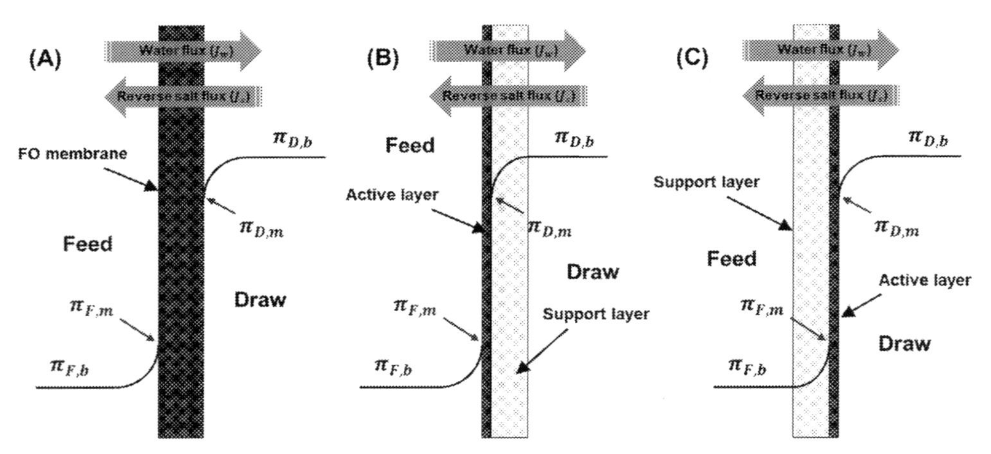

FIGURE 6–4 Schematic diagram of concentration profiles in FO: (A) a symmetric dense membrane, (B) an asymmetric membrane with a dense active layer facing the feed solution (i.e., FO mode), and (C) an asymmetric membrane with the dense active layer facing the draw solution (i.e., PRO mode). *FO*, Forward osmosis; *PRO*, pressure-retarded osmosis.

structure and disturbs the diffusion of salts, which accelerates the dilution of the DS inside the support layer. This is called dilutive internal concentration polarization (ICP). In the asymmetric membrane, ICP, rather than ECP, is a major factor for the lower-than-expected water flux in FO [11]. Hence, its structure (i.e., porosity, tortuosity, and thickness) should be optimized. Lee et al. [12] and Loeb et al. [13] derived an expression for this phenomenon (Eq. 6.4). The solute resistivity for diffusion within the support layer can be written as Eq. (6.5) according to its definition [11]. Furthermore, the dilutive ICP modulus can be written as Eq. (6.6), and hence, water flux in FO can be obtained using Eq. (6.7).

$$K = \left(\frac{1}{J_w}\right)\ln\left(\frac{B + A\pi_{D,b}}{B + J_w + A\pi_{F,m}}\right) \tag{6.4}$$

$$K = \frac{t\tau}{D\varepsilon} \tag{6.5}$$

$$\frac{\pi_{D,m}}{\pi_{D,b}} = \exp(-J_w K) \tag{6.6}$$

$$J_w = A\left[\pi_{D,b}\exp(-J_w K) - \pi_{F,b}\exp\left(\frac{J_w}{k}\right)\right] \tag{6.7}$$

where B is the salt permeable coefficient, K is the salt-resistant coefficient within the support layer, D is the diffusion coefficient, t is the thickness of the support layer, and τ is tortuosity. In contrast, in the case of an asymmetric membrane with an active layer facing the DS (i.e., PRO mode or AL-DS mode), dilutive ECP and concentrative ICP occur, as shown in Fig. 6–4C, because of the opposite direction of water flux to that in Fig. 6–4B. The solute resistivity for diffusion can be obtained using Eq. (6.8) [13]. The concentrative ICP modulus can be written as Eq. (6.9), and hence, water flux in FO can be written as Eq. (6.10) [11].

$$K = \left(\frac{1}{J_w}\right)\ln\left(\frac{B + A\pi_{D,m} - J_w}{B + A\pi_{F,b}}\right) \tag{6.8}$$

$$\frac{\pi_{F,m}}{\pi_{F,b}} = \exp\,(J_w K) \tag{6.9}$$

$$J_w = A\left[\pi_{D,b}\exp\left(-\frac{J_w}{k}\right) - \pi_{F,b}\exp\,(J_w K)\right] \tag{6.10}$$

Yip et al. derived the theoretical model to accurately predict water flux in PRO [14], and Tiraferri et al. modified this equation into the following equation for FO [15]:

$$J_w = A\left\{\frac{\pi_{D,b}\exp\left(-\left(J_w S/D\right)\right) - \pi_{F,b}\exp\left(J_w/k\right)}{1 + \left(B/J_w\right)\left[\exp\left(J_w/k\right) - \exp\left(-\left(J_w S/D\right)\right)\right]}\right\} \tag{6.11}$$

where S is the structural parameter ($t\tau/\varepsilon$). If FO is operated under the membrane orientation shown in Fig. 6–4C, the porous support layer can be blocked by foulants, possibly leading to significant flux decline [16]. Therefore double-skinned FO membranes were designed to prevent foulants from entering the porous support layer sandwiched between the two skins. Water flux can be obtained for double-skinned FO membrane using the following equation [17]:

$$J_w = K\left(\frac{A\pi_{D,b} - \left(B/B_{D,b}\right)J_w + B}{A\pi_{F,b} + \left(B/B_{F,b}\right)J_w + B}\right) \tag{6.12}$$

FO utilizes highly concentrated DS to induce an osmotic pressure gradient across the active layer, as presented in Fig. 6–4. Thus the draw solutes are readily diffused to FS owing to the concentration gradient between the FS and DS in FO, which is called reverse salt flux (RSF). Phillip et al. developed a theoretical model, shown in the following equation, that describes the reverse diffusion of DS across an asymmetric membrane (FO mode) in FO [18]:

$$J_s = \frac{J_w C_{D,b}}{1 - \left(1 + \left(J_w t_A/D^A H\right)\right)\exp\left(J_w t_s \tau/D\varepsilon\right)} = \frac{J_w C_{D,b}}{1 - \left(1 + \left(J_w/B\right)\right)\exp\left(J_w S/D\right)} \tag{6.13}$$

where D^A is the diffusion coefficient of draw solutes in the active layer, and t_A and t_s are the respective thicknesses of the active layer and the support layer, respectively. Tiraferri et al. proposed a model for the prediction of RSF as shown in the following equation [15]:

$$J_s = B\left\{\frac{C_{D,b}\,\exp\left(-\left(J_w S/D\right)\right) - C_{F,b}\,\exp\left(J_w/k\right)}{1 + \left(B/J_w\right)\left[\exp\left(J_w/k\right) - \exp\left(-\left(J_w S/D\right)\right)\right]}\right\} \tag{6.14}$$

These models considered concentrative ECP and dilutive ICP for theoretical RSF, as presented in Fig. 6–4B, even though dilutive ECP can occur on the support layer. This is because ICP influences the water flux more seriously than ECP [11]. Nevertheless, ECP on the support layer can be still a potential obstacle to precisely predict RSF. Therefore Suh and Lee developed a more accurate FO model, given by Eq. (6.15), which can simulate RSF and ECP simultaneously [19].

$$J_s = B\left(\frac{C_{D,b} + J_s/J_w}{\exp(J_w K)\exp(J_w/k_D)} - (C_{F,b} + J_s/J_w)\exp\left(\frac{J_w}{k_F}\right)\right) \tag{6.15}$$

where k_D is the mass transfer coefficient of the DS on the support layer. Because this reverse diffusion of draw solutes is inevitable in FO, it can seriously affect FO performance through accelerated cake-enhanced osmotic pressure (A-CEOP) [20], membrane scaling [21], and biofouling [22].

Solution temperature is an important operation parameter in FO because FO is often coupled with thermal desalination technologies for DS regeneration and clean water production [23]. Therefore many researchers have proposed theoretical models to simulate water flux when changing temperatures of FS and DS [24,25]. If the draw and feed temperatures are different, heat transfer occurs in the FO system. First, heat transfer from bulk feed to the adjacent feed boundary layer occurs, as shown Eq. (6.16). Then, heat transfer across the membrane occurs, which can be expressed as Eq. (6.17). From the nonporous active layer and the porous support layer, thermal conductivity can be calculated using Eq. (6.18). Lastly, heat transfer from the adjacent draw boundary layer to the feed bulk occurs, given by Eq. (6.19). At the steady-state condition the overall heat transfer in FO should be identical to that in Eq. (6.20).

$$Q_F = h_F(T_{F,b} - T_{F,m}) \tag{6.16}$$

$$Q_m = (J_w c_p + k_{overall})(T_{F,m} - T_{D,m}) \tag{6.17}$$

$$k_{overall} = \left(\frac{t_A}{k_m} + \frac{t_S}{\varepsilon k_w + (1-\varepsilon)k_m}\right)^{-1} \tag{6.18}$$

$$Q_D = h_D(T_{D,b} - T_{D,m}) \tag{6.19}$$

$$Q_F = Q_m = Q_D \tag{6.20}$$

where c_p is the heat capacity, h_F and h_D are the respective convection heat transfer coefficient of the FS and DS, respectively; $k_{overall}$, k_m, and k_w are the overall thermal conductivity and those of the membrane and water, respectively; $T_{F,b}$ and $T_{D,b}$ are the bulk temperatures of the FS and DS, respectively; $T_{F,m}$ and $T_{F,m}$ are the temperatures of the FS and DS at the membrane surface, respectively; and Q_F, Q_m, and Q_D are the respective heat transfers through the FS, the membrane, and DS. By solving Eqs. (6.11), (6.14), and (6.16)–(6.20) simultaneously, the theoretical water flux and RSF in FO can be obtained when the feed and draw temperatures are different.

In addition to the simulation of water flux and RSF in FO the transport of target contaminants can be predicted. Jin et al. suggested a theoretical model for boric acid permeation in FO using the following equation [26]:

$$\frac{J_B}{c_{f,B}} = \frac{B_B \exp\left(J_w/K_{m,B}\right)}{1 + \left(\left(B_B \exp\left(J_w/K_{m,B}\right)\right)/J_w\right)} = \frac{1}{\left(1/\left(B_B \exp\left(J_w/K_{m,B}\right)\right)\right) + \left(1/J_w\right)} \tag{6.21}$$

where B_B is the permeable coefficient of boric acid, J_B is the flux of boric acid, and $K_{m,B}$ is the mass resistant coefficient of boric acid within the support layer. The pore hindrance model was employed to predict the rejection of contaminants by size exclusion with an assumption that the FO membrane consisted of several cylindrical capillary tubes with the same radius, where the spherical solute particles can penetrate through these FO membrane pores [27]. Real rejection was determined using the following equation [27,28]:

$$R_r = 1 - \frac{C_p}{C_m} = 1 - \frac{\varphi K_c}{1 - \exp\left(-Pe\left(1 - \varphi K_c\right)\right)} \tag{6.22}$$

where R_r refers to the real rejection of the FO membrane, C_p is the permeate concentration, K_c is the hydrodynamic hindrance coefficient for convection, φ is the distribution coefficient, and Pe is the membrane Peclet number. The distribution coefficient (Eq. 6.23) is related to the ratio of the contaminant radius to the membrane pore radius (Eq. 6.24):

$$\varphi = (1 - \lambda)^2 \tag{6.23}$$

$$\lambda = \frac{r_s}{r_p} \tag{6.24}$$

The Peclet number is defined as the ratio of the convective transport rate to the diffusive transport rate and can be obtained from the following equation:

$$Pe = \frac{K_c J_w t^A}{K_d D \varepsilon^A} \tag{6.25}$$

where K_d refers to the hydrodynamic hindrance coefficient for diffusion and ε^A refers to the active layer effective porosity. The hydrodynamic hindrance coefficients for convection and diffusion can be determined via Eqs. (6.26) and (6.27), respectively, which were proposed by Bungay and Brenner [29]. Diffusion may be more dominant than convection in determining the solute transports when λ is close to 1.

$$K_c = \frac{(2 - \varphi)K_s}{2K_t} \tag{6.26}$$

$$K_d = \frac{6\pi}{K_t} \tag{6.27}$$

6.2.2 Process description

FO is an osmotically driven separation process, which uses a semipermeable membrane. It utilizes a high-concentration DS and a low-concentration FS to create an osmotic pressure gradient as the driving force. FO can provide high rejection of contaminants, low-fouling propensity, high-fouling reversibility, and low-energy requirement [30,31]. The absence of high hydraulic pressure leads to high-fouling reversibility of the FO membrane. Lee et al. observed that the high crossflow velocity completely recovered the water flux during organic fouling in FO, whereas no water flux recovery was observed in RO [20]. Mi and Elimelech demonstrated that organic fouling in FO can be readily removed by hydraulic flushing [32]. They also found that FO has better fouling reversibility, particularly when cellulose triacetate (CTA) FO membranes are used [33]. However, the combined fouling of organic matter and colloids showed much lower reversibility than organic fouling and colloidal fouling, respectively, particularly with calcium ions in the FS [34]. However, irreversible membrane fouling has been also reported by many researchers. Thin-film composite (TFC) polyamide (PA) FO membranes exhibited not only severer organic fouling but also lower fouling reversibility than CTA FO membranes because of the different physicochemical surface properties [35]. Both FO membranes showed similar flux decline, but the flux recovery of the CTA FO membrane was much higher [33]. Furthermore, when real wastewater was used, the CTA FO membrane showed irreversible fouling. When treating real secondary effluents, the water flux was significantly restored by hydraulic flushing, but biopolymer-like substances remained on the FO membrane surface even after consecutive hydraulic flushing [36]. In addition, membrane orientation has a significant impact on fouling reversibility because of the highly porous structure of the support layer [16,37].

A high-concentration gradient as the driving force generates reverse diffusion of draw solutes toward the FS. As RSF moves in the opposite direction of the solute flux, it hinders the flux of organic micropollutants (OMPs) as illustrated in Fig. 6–5. Thus FO has a higher

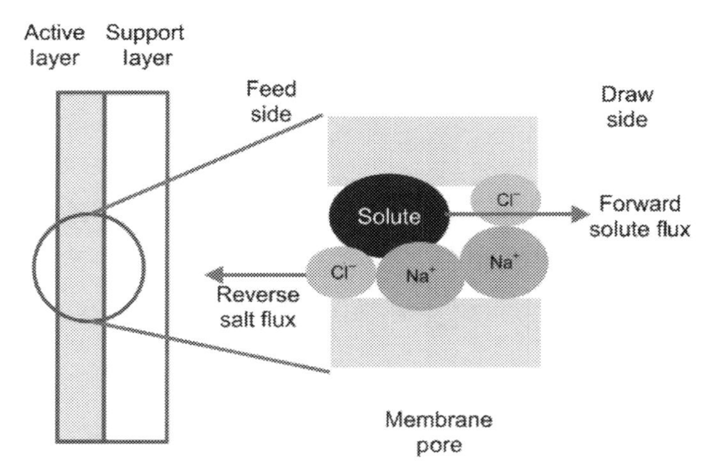

FIGURE 6–5 Schematic diagram representing the retarded forward diffusion of feed solutes in the FO process by the reverse draw solutes [38]. *FO*, Forward osmosis.

OMP removal efficiency than RO [38,39]. Operational parameters (i.e., water flux, solution pH, membrane orientation, and working temperature) influence the rejection propensity [40–42]. OMP removal of the FO process at the PRO mode can be much less efficient than that of the FO mode [41,43]. In the FO mode, ICP not only increases the OMP concentration inside the membrane support layer but also severely restricts the backward diffusion (mass transfer) of the OMPs into the FS. Thus OMP flux is enhanced and the efficiency of its removal is reduced [41,44]. High feed temperature reduces OMP rejection because of the enhanced diffusivity in OMPs, whereas OMP rejection increases at high DS temperatures because of the dilution effect caused by enhanced water flux and the hindrance effect induced by the slightly increased RSF [40]. Solution pH significantly influences the ionic OMP rejection [43]. The TFC FO membrane based on PA exhibits much better OMP rejection because of pore hydration than FO membranes based on CTA, even though TFC membranes have larger pore sizes [27]. Membrane fouling also significantly influences OMP rejection [41,45]. Interfacial interactions between foulants and the membrane are forces that affect membrane fouling [46,47], potentially influencing OMP rejection. The surface characteristics of the fouling layer in a fouled FO membrane can influence OMP rejection [48]. The characteristics of the fouling/cake layer may be an important factor affecting OMP rejection because of their different resistance depending on their composition [49,50].

However, because concentrated DS can be simply converted into diluted DS, additional desalting processes are required to produce pure water, while the diluted DS is reconcentrated and regenerated for sustainable operation [51,52]. Therefore FO can be coupled with other desalting processes such as distillation [53], membrane distillation (MD) [54], nanofiltration (NF) [55], or RO [56], depending on the DS type to regenerate the concentrated DS and produce freshwater (Fig. 6–6). McGinnis and Elimelech proposed ammonia–carbon

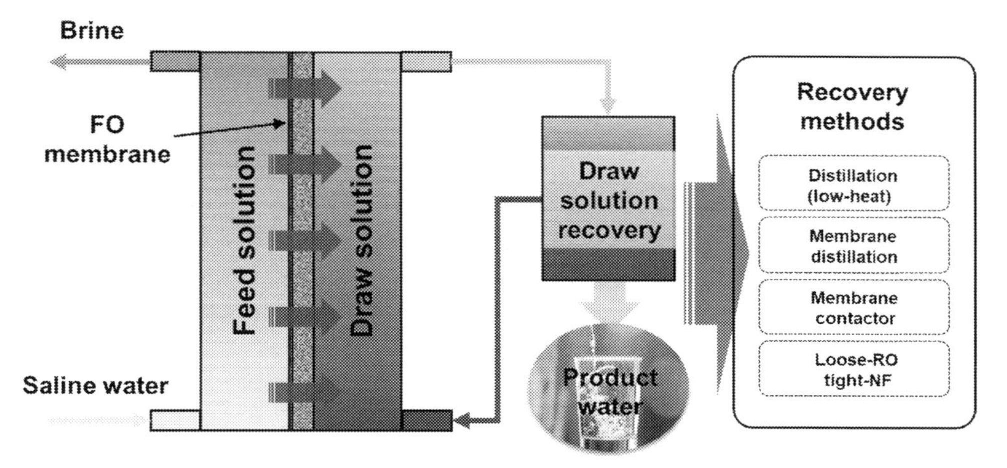

FIGURE 6–6 Schematic diagram of the FO process combined with a DS recovery process for water and wastewater treatments. For DS recovery, various desalting processes such as distillation, MD, MC, NF, and RO can be used depending on DS type. *DS*, Draw solution; *FO*, forward osmosis; *MC*, membrane contactor; *MD*, membrane distillation; *NF*, nanofiltration; *RO*, reverse osmosis.

dioxide as DS and used distillation to separate the DS from the product water for solute recycling within the FO system [53]. In FO the DS should have thermodynamic properties such as high osmotic pressure, low viscosity, high water solubility, high diffusivity, and a small molecular weight [57,58], which makes inorganic salts the most appropriate draw solute [59]. Therefore NaCl is the most commonly used DS, and MD is employed for DS regeneration [60–62]. The details of the FO process are discussed in the following sections.

6.3 Technological factors

6.3.1 Forward osmosis membrane

Various membranes have been used for the FO process in both flat sheet and hollow fiber configurations, including cellulose acetate, CTA, TFC, and biomimetic membrane [63]. The first-generation of FO membranes such as nonwoven and embedded support FO membranes are asymmetric CTA membranes commercialized by HTI (Hydration Technologies Innovations). Among the FO membranes, cellulosic membranes have been extensively employed for the FO process; however, such an FO process suffers from low selectivity and is prone to biological attacks and chemical hydrolysis. The second generation of FO membranes is TFCs that were first commercialized by HTI and Oasys Water [64]. More importantly, the configuration of TFC allows high flexibility in the structural design because the properties of the selective layer and substrate layer can be tuned separately to cater to specific needs.

The FO membrane with specific design modifications of the conventional RO membrane module in the traditional spiral wound and plate-and-frame configurations has been tested at the pilot scale. Fig. 6–7 demonstrates the development history of the FO membrane to make them feasible, including the selection of the material and the module type. As one of the main service providers of FO, Oasys Water has ventured into fully integrated FO systems completed

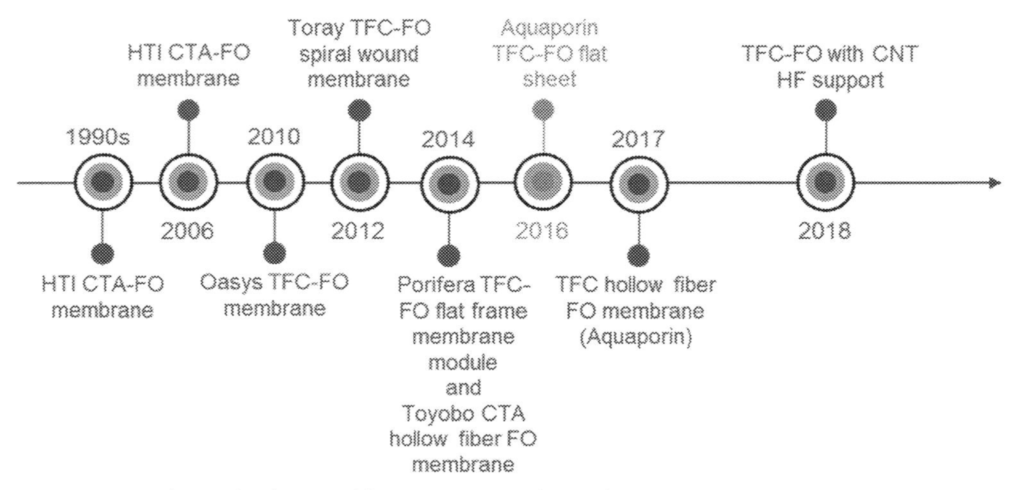

FIGURE 6–7 FO membrane development history. *FO*, Forward osmosis.

with thermally regenerated ammonium carbonate DS. More recently, Aquaporin A/S deployed a new configuration of hollow fiber membranes that feature a high packing density compared with the common spiral wound configuration. Because the first commercial FO membrane was introduced to the market more than a decade ago, currently more global FO membrane suppliers have entered the market to provide more competitive solutions [64].

Notwithstanding the commercialization of bespoke FO products and the significant market potential (purportedly up to $6b), thus far, the majority of reported large-scale practical FO studies have been based on either commercial RO membranes, not necessarily the best suited to FO applications, or on bespoke hand-cast FO membranes prepared on a small scale [65]. There are currently only six global FO membrane suppliers (Fluid Technology Solutions, Modern water, Oasys water, Porifera, Toyobo, and Trevi Systems), with a comparable number of products in the latter stages of development or commercialization (Table 6–2). Thus, despite the increasing number of research publications (Fig. 6–1), examples of full-scale applications remain scarce and incompletely reported.

6.3.2 Draw solution

Because FO utilizes chemical potential difference as the driving force for water permeation, the selection of an optimal DS and its recovery method are the key components for successful application of the FO technology. Finding an appropriate draw solute is a crucial issue for the future of osmosis-based membrane systems. Using such a solute, not only a significant improvement in process efficiency but also a radical cost reduction in such systems would be achieved, and thereby practical commercialization could be realized [9]. The requirements of draw solutes (DSs) for FO are (1) high water flux, (2) low RSF, and (3) easy and low-cost recovery from the diluted DS [66].

Table 6–2 Summary of current commercial forward osmosis (FO) membranes [64,65].

Supplier/manufacturer	Membrane	Configuration	Location
Modern Water	Undefined	SWo[a]	United Kingdom
BLUE–tech	Undefined	SWo, FS[b]	The Netherlands
Aquaporin A/S	Aquaporin	SWo, HF[c]	Denmark
Aquaporin Asia	Aquaporin	SWo	Singapore
Toray	FO membrane	SWo	Korea
Toyobo	N.A.	HF	Japan
Fluid Technology Solutions	CTA	SWo	United States
Oasys	TFC	SWo	United States
Trevi Systems	N.A.	SWo	United States
Porifera	TFC	FS	United States
Hydration Technology Innovations[d]	CTA, TFC	FS	United States

CTA, Cellulose triacetate; *FS*, feed solution; *HF*, hollow fiber; *SWo*, spiral wound; *TFC*, thin-film composite.
[a]Spiral wound.
[b]Flat sheet.
[c]Hollow fiber.
[d]Out of business.

The first criterion is that the DS should have high enough osmotic pressure to produce high water flux. In general, the osmotic pressure of a solution can be calculated by the Van't Hoff equation (Eq. 6.28):

$$\pi = iCR_gT \tag{6.28}$$

where i is the Van't Hoff coefficient, C is the molar concentration, R is the ideal gas constant, and T is the absolute temperature. The molar concentration is usually inversely proportional to the molecular weight of the solute. Therefore low molecular weight solutes can easily generate osmotic pressures at viscosities close to that of pure water (Fig. 6–8) and are less susceptible to the ICP effect because of their relatively high diffusivities [67].

Until now, various new DSs have been developed and evaluated to maximize the efficiency of the FO process, but none of them have been successfully commercialized. A summary of FO DSs is provided in Fig. 6–9. There are more than 500 inorganic compounds that can be potentially used as DSs; 14 were chosen and investigated in a previous study by Achilli et al. [59]. Other investigations have studied the applicability of dissolved gases or even nanoparticles as suitable DSs in tailored FO applications. Organic solutes such as ethanol, glucose, and fructose can also be used as osmotic DSs and be tailored to obtain the desired physiochemical properties; however, simple organic DSs typically yield low osmotic pressure and water flux. Organic DSs might be advantageous because they may be biodegradable and are well rejected by reconcentration technologies such as RO; however, these DSs are still experimental and require further research and development.

The second criterion in the selection of a draw solute is the minimal RSF during the FO process. Because of the difference in the concentration across the membrane, the DS diffuses back into the FS. This RSF will decrease the effective osmotic pressure. Because low molecular weight solutes more easily pass through the membrane, there is a tradeoff between

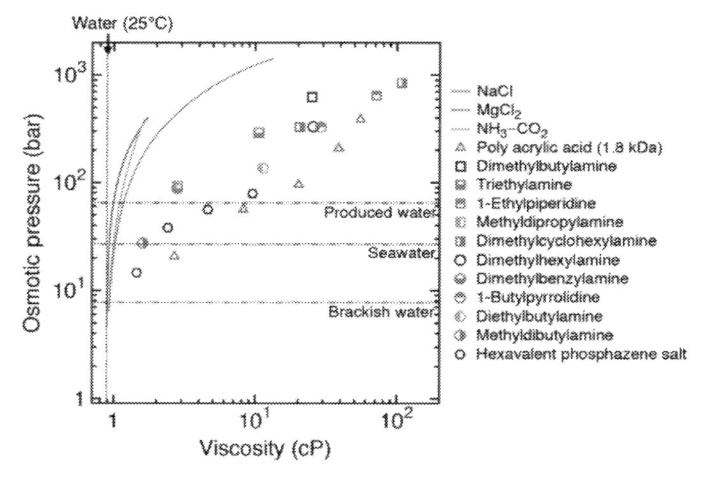

FIGURE 6–8 Relationship between osmotic pressure and viscosity for different DSs [67]. *DS*, Draw solution.

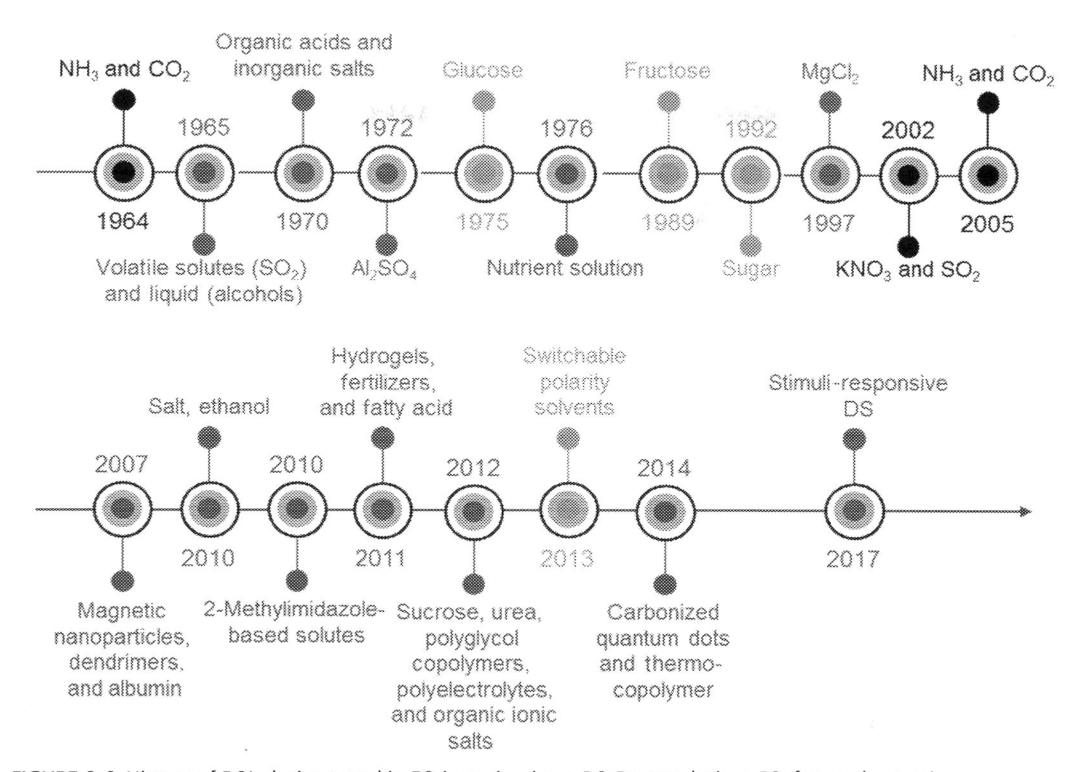

FIGURE 6–9 History of DS/solutions used in FO investigations. *DS*, Draw solution; *FO*, forward osmosis.

osmotic pressure and RSF. For an osmotic membrane bioreactor (OMBR) system, the accumulation of the draw solute in the bioreactor may have a toxic influence on the microbial community [9]. The trace amount of the draw solute in the product water is also important in FO for drinking water production and has to be lower than the maximum allowable level. Therefore minimization of the RSF is crucial for general applications and should thus be considered the first step when selecting a draw solute.

The last criterion is that the DS can be easily separated from the product water and is economically feasible to regenerate. Until now, various DSs and their recovery methods have been developed and investigated, as described in Table 6–3. Usually, saline gradient-based membrane processes are coupled with another process to form a closed-loop system. For instance, FO-based desalination includes at least two steps: (1) water extraction from the saline feedwater by a DS and (2) production of clean water by separating the diluted DS with the aid of pressure-driven membrane processes such as RO and NF or thermal-driven membrane technologies including MD. Saline gradient–driven membrane processes are obviously energy efficient, but the posttreatments required for separation of pure water and DSs, as investigated so far, are generally energy intensive, requiring either hydraulic pressure or heating.

Table 6–3 Overview of draw solutions (DSs) and recovery methods in forward osmosis.

Recovery methods	Typical DSs	Remarks	References
Heating	NH_3 and CO_2, SPS, etc.	Energy intensive	[68,69]
Membrane distillation	2-Methylimidazole-based solutes, organic ionic salts, etc.	High energy cost unless using waste heat	[70–72]
Pressure-driven membrane processes (UF, NF, RO, etc.)	Organic acids and inorganic salts, polyglycol copolymers, polyelectrolytes, etc.	Low water recovery and poor solute rejection	[59,67,73–76]
Precipitation	Al_2SO_4	Not economical and feasible	[77,78]
Stimulus–response Method	Magnetic nanoparticles, hydrogels, etc.	Complex and ineffective	[79–82]
Direct use	Fertilizers	Economic viability and flexible, but need to be commercialized	[83–86]

NF, Nanofiltration; *RO*, reverse osmosis; *SPS*, switchable polarity solvents; *UF*, ultrafiltration.

6.4 Understanding of fouling in forward osmosis

6.4.1 Operation without hydraulic pressure

In RO, as shown in Fig. 6–10, a hydraulic pressure drives water through the semipermeable membrane against the osmotic pressure gradient between the more saline feedwater (high osmotic pressure) and the less saline permeate water (low osmotic pressure) streams. The output of RO is purified water (the permeate) and a concentrated impaired water source (RO brine) [20,30]. In contrast, FO utilizes this osmotic pressure differential as the driving force to extract water molecules across the membrane rather than hydraulic pressure differential as used in RO. Therefore flux recovery in the FO mode is much higher than that in the RO mode under similar cleaning conditions, although the rate of membrane flux decline is similar in the two modes [87].

The structure of the fouling layer in FO is mainly affected by the permeation drag of the convective permeate flow induced by the osmotic pressure driving force, whereas in RO, in addition to the permeation drag, the applied hydraulic pressure compresses the deformable organic molecules and the organic-colloidal clusters that accumulate on the membrane surface. Consequently, as shown in Fig. 6–10, the fouling layer structure in FO is likely to be looser, sparser, and thicker than that in RO. The fouling reversibility of FO was attributed to the less compact organic fouling layer formed in the FO mode due to the lack of hydraulic pressure [20]. In addition, FO can achieve a higher recovery rate than RO because the driving force in FO is the osmotic pressure difference, whereas the applied hydraulic pressure in RO should overcome the osmotic pressure of the concentrate [88].

6.4.2 Bidirectional diffusion

FO utilizes a highly concentrated DS to induce an osmotic pressure gradient across the active layer, as presented in Fig. 6–11. Thus DSs are readily diffused to the FS due to the

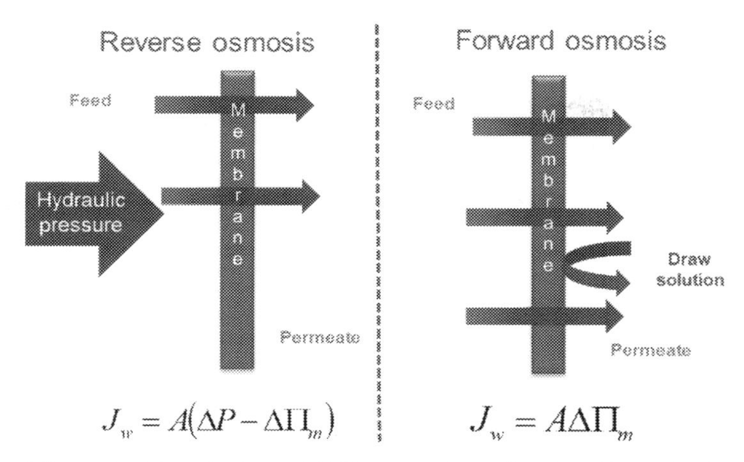

FIGURE 6–10 Solvent (i.e., water) flows in FO and RO. The general equation describing water transport in FO, and PRO is $J_w = A(\Delta P - \Delta\pi_m)$. In FO, ΔP is approximately zero and water diffuses to the more saline side of the membrane. In RO, water diffuses to the less saline side because of hydraulic pressure ($\Delta P > \Delta\pi_m$). *FO*, Forward osmosis; *PRO*, pressure-retarded osmosis; *RO*, reverse osmosis.

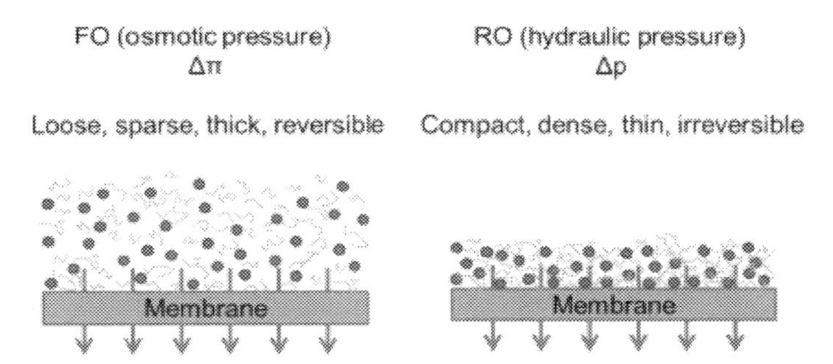

FIGURE 6–11 Proposed combined fouling layer structure in FO and RO mode [34]. *FO*, Forward osmosis; *RO*, reverse osmosis.

high-concentration gradient between FS and DS in FO unlike RO [18]. Because this reverse diffusion of DSs, called RSF, is inevitable in FO, it can seriously affect FO performance. It is well known that RSF contributes to flux decline through A-CEOP, as shown in Fig. 6–12 [20]. In FO, the DS is highly concentrated, and thus salt moves from the draw to the feed. The salt is captured by the fouling layer that accelerates CEOP. As a result, the surface concentration of the feed side significantly increases, and the permeate flux decreases.

In RO, the presence of multivalent cations such as Ca^{2+} or Mg^{2+} in FS can induce severe membrane fouling by forming organic-divalent complexes [89]. Similarly, in FO, the reverse diffusion of the draw solute cations can also promote foulant–membrane and foulant–foulant interactions, thereby enhancing membrane fouling. When examining the

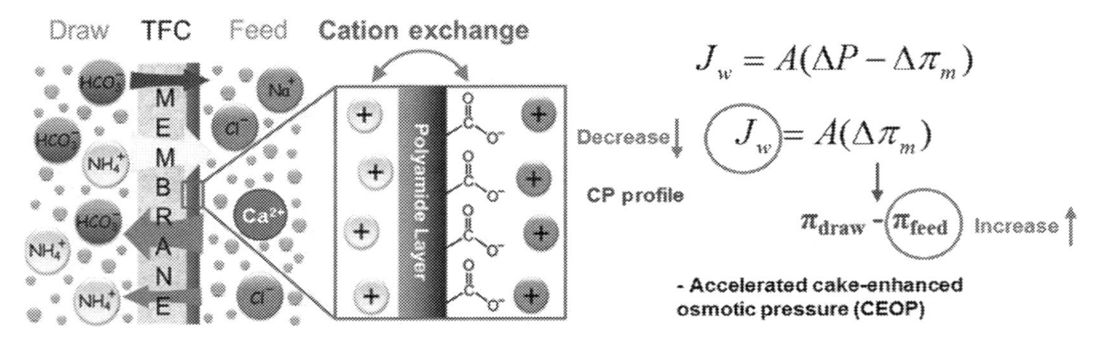

FIGURE 6–12 Schematic diagram of A-CEOP [2,20]. *A-CEOP*, Accelerated cake-enhanced osmotic pressure.

various inorganic DSs (i.e., NaCl, $MgCl_2$, $CaCl_2$, and $Ca(NO_3)_2$), the reverse solute diffusion of draw solute (especially divalent cation) changes the FS chemistry, thus enhancing membrane fouling by alginate, the extent of which is related to the rate of the reverse draw solute diffusion and its ability to interact with the foulant [90]. Reverse divalent cation diffusion also severely affects FO biofouling. Reverse Ca^{2+} diffusion leads to a significantly more serious water flux decline in comparison with reverse Mg^{2+} permeation. Unlike magnesium, reverse Ca^{2+} permeation dramatically alters biofilm architecture and composition, where extracellular polymeric substances form a thicker, denser, and more stable biofilm. For efficient recovery of DS in FO, ammonia–carbon dioxide (NH_3–CO_2) was suggested as a novel DS [91]. HCO_3^- in the NH_3–CO_2 DS was reversely transported to FS and chemically reacted with Ca^{2+} when NH_3–CO_2 was used as the DS to treat the Ca^{2+}-containing FS. Thus $CaCO_3$ scales were formed on the active layer (Fig. 6–12). On the other hand, there are a few studies to utilize an inevitable phenomenon (i.e., reverse diffusion of draw solutes) of the FO process to treat high scaling potential wastewater using an antiscalant as DS. Poly(aspartic acid sodium salt) (PAspNa) as a novel DS was proposed because PAspNa exhibited superiority in high water flux and low RSF in FO. They suggested that the reversely flowed PAspNa in FO could act as an antiscalant for the scaling control [2,62].

6.4.3 Fouling control and cleaning in forward osmosis

Several efforts have been undertaken to address the issue of fouling, especially with respect to cleaning. However, effective fouling control techniques for FO are still lacking. Several studies have reported that fouled FO membranes can be easily cleaned by a simple change in the hydrodynamic conditions, without using any chemical agents. Most of these studies have used model foulants, where fouling exhibited reversibility by changing the hydrodynamic conditions, such as high crossflow velocity flushing with DI water. However, when treating complex feeds such as real wastewater, the potential irreversible membrane fouling cannot be mitigated by merely changing the hydrodynamic conditions, thus requiring chemical cleaning. The fouling control strategies in different FO studies are summarized in Table 6–4.

Table 6–4 Fouling control strategies in different forward osmosis studies.

Mitigation method	Fouling type	Model foulants/feedwater	Membrane	Fouling reversibility	Reference
High crossflow velocity	Organic	Soluble algal product	CTA, TFC	Irreversible for CTA but reversible for TFC	[92]
	Organic and colloidal	Sodium alginate, BSA, and Suwannee River humic acid. Silica (20 and 300 nm)	CTA	Reversible	[20]
	Inorganic	$CaSO_4$	CTA	Reversible	[93]
	Colloidal	Silica (10–20 nm)	CTA	Partially reversible (75%)	[93]
	Organic–inorganic–colloidal–biofouling	Oily wastewater	CTA	Irreversible	[94]
Osmotic backwashing	Organic	Humic acid and alginate	CTA, TFC, Porifera	Reversible	[95]
	Organic–inorganic–colloidal–biofouling	Oily wastewater	CTA	95% recovery	[94]
	Organic–inorganic–colloidal–biofouling	Drilling wastewater from shale gas	CTA	Reversible	[96]
Air scouring	Biofouling and organic, inorganic	Municipal secondary wastewater Synthetic municipal wastewater	CTA	Not effective	[87]
	Organic	Sodium alginate, 50 mM NaCl, 0.5 mM $CaCl_2$	CTA, TFC	Reversible and fastest reversibility with bubbled DI water	[32]
Feed spacer	Biofouling	*Chlorella sorokiniana* with NaCl and/or $MgCl_2$	CTA	Less reversible in the presence of Mg^{2+} ions in feed or draw	[97]
	Organic–inorganic	Sodium alginate, BSA, and Suwannee River natural organic matter with synthetic wastewater	CTA	Reversible	[98]
Chemical cleaning: HCl	Organic–inorganic–colloidal–biofouling	Oily wastewater	CTA	90% recovery but narrow down the pores	[94]
EDTA				90% recovery but potential membrane damage	
NaClO				85% recovery but potential membrane damage	
Surfactant				100% recovery but adhere to the membrane surface	

BSA, bovine serum albumin; *CTA*, Cellulose triacetate; *DI*, deionized; *TFC*, thin-film composite.

The easiest method to clean fouled FO membranes involves flushing it with deionized (DI) water (or 50 mM NaCl) using high crossflow velocity. However, fouling in FO can also become irreversible when intermolecular adhesion forces among the contaminants overcome the hydrodynamic interactions between the membrane surface and the contaminants [32,99–101]. Another popular cleaning method for osmotically driven membrane processes, known as osmotic backwashing, in which the direction of water flow across the semipermeable membrane is reversed, has been used by several researchers in OMBRs. However, under severe fouling conditions (oil and gas wastewater), a direct observation over the microscope of the osmotically backwashed membrane revealed that loosely bound foulants were effectively removed, and those absorbed to the membrane surface were not entirely removed by osmotic backwashing.

Air scouring is another effective and widely used technique for fouling mitigation, even for natural organic matter fouling (90% recovery of flux) [32]. For biofouling control, air scouring can mitigate biofilm growth; however, the biofilm may grow back rapidly under favorable conditions. Air scouring is also an expensive cleaning protocol and can be a serious drawback to the economic sustainability of the FO process [87]. Feed spacers can also minimize fouling propensity. Spacers can enhance the initial flux performance and reduce the fouling deposition of microalgae on the FO membrane. Spacer thickness also plays a role in minimizing biofilm formation [97]. Thicker spacers are reported to exhibit better performance than thinner spacers. However, thicker spacers in the presence of lower crossflow velocities are reported to promote organic and colloidal fouling and reduce permeate flux [98]. Sparging of CO_2 saturated solution was proposed as a novel cleaning agent for polymerized silica fouling. CO_2 saturated solution can not only release CO_2 bubbles but also cause a pH drop, which helps in controlling membrane fouling [100].

Several other researchers have proposed chemical cleaning protocols for wastewater-fouled FO membranes. Chemical cleaning with surfactants is the most effective way to restore flux completely in this chapter. Chemical cleaning, however, is not ideal because it entails extra energy consumption; thus alternatives should be investigated. Moreover, the effectiveness of chemical cleaning is potentially constrained by the compatibility of the membrane material with the chemical agent [94]. For instance, TFC membranes in general cannot tolerate oxidizing agents such as chlorine or Alconox. Chemical cleaning can also shorten membrane life, has environmental constraints because of the waste chemical disposal, and can increase operational costs. Apart from these disadvantages, some researchers have claimed that chemical cleaning can only remove or dissolve the cake- or gel-type fouling layers and cannot remove foulants inside the membrane pores [102].

6.5 Exploiting advantages of forward osmosis in its applications

FO has a range of potential benefits, mainly because of the low hydraulic pressure required by this osmotically driven process. The potential benefits of FO as used in the two main processes (i.e., feed concentration and draw dilution process) are illustrated in Fig. 6–13. Recent studies have demonstrated that membrane fouling in FO is relatively low, more reversible, and can be minimized by optimizing the hydrodynamics. Various contaminants can be effectively rejected via the FO process. Thus FO also has the potential to help achieve high water flux and water recovery because of the high osmotic pressure gradient across the membrane when considering feed concentration [9,103–105]. High water recovery could help reduce the volume of various feed influents to produce high-quality products in the food and beverage industry and pharmaceutical processing, could ensure available resource recovery of industrial wastewater, and could minimize the environmental impact of contaminants of emerging concern (CEC) [4,106–108].

In addition, FO holds the promise of helping achieve low-energy consumption during the draw dilution process, thereby lowering the costs, if suitable DSs and their regeneration methods can be economically and technically developed. This could be one of the most attractive aspects of FO, especially under the stress of energy crises. Even though the recovery of draw solute remains challenging, FO has a special edge over other membrane processes, such as RO, where the fate of the diluted draw solute is irrelevant [64,109]. FO has been studied for various applications in which the separation of draw solute from water is not required, for example, emergency water supply (hydration bags), fertilization, and osmotic pumps. Furthermore, early FO studies have focused on finding efficient draw solute recovery methods, and therefore the development of hybrid FO systems has begun (i.e., FO coupled with another physical or chemical separation process) [31]. Hybrid FO systems could potentially consume less total energy compared to a stand-alone separation process especially for applications

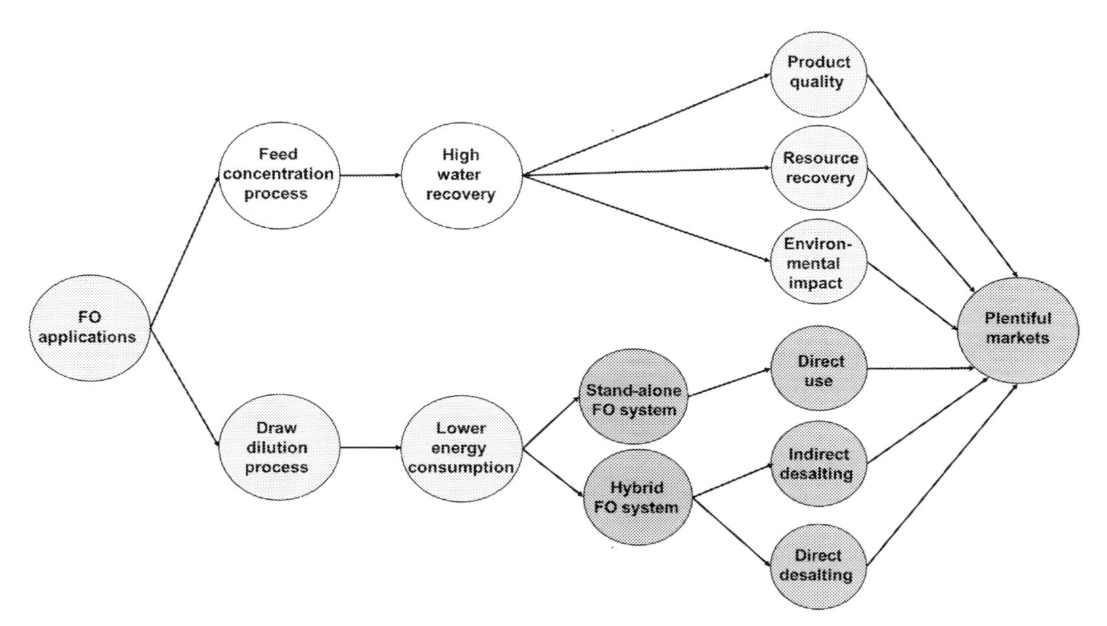

FIGURE 6–13 Potential benefits of FO used in feed concentration and draw dilution process. *FO*, Forward osmosis.

where the source of water for treatment has extreme characteristics. The FO process can be effectively used as an advanced pretreatment process in hybrid desalination systems to reduce inorganic scaling and organic fouling in the combined membrane or thermal process. The practical application of osmotic dilution, which simultaneously treats impaired water while indirectly diluting seawater before the RO process, is one good example and is receiving increasing attention for its energy reduction potential [98,110,111]. During the last few decades, as an emerging technology leading to a low-energy desalination process, a series of novel DSs have been proposed together with various remarkable recovery technologies depending on the physicochemical and thermodynamic properties of DSs [58,112].

6.5.1 Feed concentration process with high water recovery

6.5.1.1 High-quality product

Compared with the conventional evaporative concentration techniques, FO can provide advantages in maintaining the physical properties (e.g., color, taste, aroma, and nutrition) of the liquid food without deteriorating its quality. Therefore the diversified applications of FO process have been studied. Moreover, FO results in increased rejection and less membrane fouling in comparison to pressure-driven membrane processes such as RO. FO has been used for the concentration and recovery of many products, such as the food and beverage industry (juice, dairy industry, organic acids, sugar, drinks, etc.) and pharmaceutical and chemical processing (protein, dye, ethanol, etc.). The results obtained by several FO researchers are summarized in Table 6–5. Because most pharmaceutical products are heat sensitive and have larger molecular sizes than water and are mono and multivalent ions, FO can be very suitable for their

Table 6–5 High-quality products of forward osmosis in the food and beverage and pharmaceutical industries.

Applications	Product	Remarks	References
Food and beverage industry	Juice (food, fruit, etc.)	Retention of fresh fruit flavor and high sugar content concentration	[113–116]
	Dairy industry	Wastewater recycling or reuse and economic	[117–120]
	Organic acids (butyric acid, tannic acid, etc.)	Volume reduction	[121–124]
	Sugar (sucrose, xylose, etc.)	High water recovery and rejection	[125–127]
	Drinks (tea, coffee, etc.)	Minimum color degradation and low cooked taste	[128,129]
Pharmaceutical or chemical industry	Protein (BSA, lysozyme, etc.)	Retention of nutritional components	[130,131]
	Dye (acid orange 8)	High water recovery	[132]
	Ethanol	Lower reverse salt flux and higher separation factor	[133–137]
	Anthocyanin	High bioactive quality	[138]

BSA, bovine serum albumin.

athermal dehydration. Compared with traditionally chemical (extraction) or thermal (evaporation, distillation, or crystallization) dewater treatment, the FO membrane separation process is a simpler, more environment friendly, more efficacious process.

6.5.1.2 Effective resource recovery

Resource recovery is the extraction of worthy materials among FSs for recycle and upcycle. Recovery of resources can bring maximum benefits from the disposal of waste and postpone the depletion of rare resources. For example, the function of wastewater plants has transformed from the treatment of wastewater into the recovery of energy and resources in recent years. Nitrogen (N), phosphorus (P), and potassium (K) are critical nutrients for intensive agricultural production, but their long-term availability and cost of extraction (P and K) are significant concerns for the future [139]. Hence, FO dewatering is the appropriate method to separate certain materials from the FS and to concentrate them with high purity.

In municipal and industrial wastewater treatment, OMBR is an emerging technology integrating a FO process into an MBR [109,140]. Although the treatment performances of the three major OMBR system configurations differ, as shown in Table 6–6, the viability of OMBR has been demonstrated in some special applications such as the recovery of nutrients from wastewater and the removal of the trace organic matter and nanoparticles from wastewater [6,151].

In some other industries, wastewater streams are formed that contain valuable substances, including heavy metals from the manufacturing industry [152–156], algal biomass for biofuel production [97,157–159], and precious metal from the electronic industry [5,160]. These substances are often toxic or harmful and have to be removed from the wastewater. Their resource recovery processes offer the chance to regain and recycle these substances back into the production process. However, research aiming at full-scale implementation of those industry branches is still scarce. Besides, the technical applicability and energetic and economic benefits of FO also need to be critically evaluated.

6.5.1.3 Minimal environmental impact

As climate change, population growth, and overconsumption continue to strain overburdened freshwater resources, water authorities are considering and implementing water

Table 6–6 Three major system configurations of forward osmosis (FO) technology in wastewater treatment.

FO system configurations	Name	Remarks	References
Aerobic osmotic membrane bioreactor	Ae-OMBR	High organic matter, TN, TP removal efficiency but can cause blockages	[6,141–144]
Anaerobic osmotic membrane bioreactor	An-OMBR		[145–147]
Preconcentration		Low removal efficiency but lower fouling propensity	[148–150]

TN, total nitrogen; *TP*, total phosphorus.

recycling schemes. However, ubiquitous CECs, such as endocrine-disrupting compounds (EDCs), pharmaceutically active compounds, personal care products (PPCPs), disinfection byproducts, and industrial chemicals in water resources are significantly concerning because of the human and environmental health threats due to physiologically active minerals at trace concentrations [108]. The potential representative sources and routes of CECs in the natural environment are described in Fig. 6–14 [161]. Several membrane processes, particularly pressure-driven membrane processes including RO, NF, UF, microfiltration (MF), and more recently, FO, have already been applied for CEC removal or destruction [44,162–167]. Removal efficiencies of more than 90% can be achieved when many of these technologies are applied, but the efficiency largely depends on CEC hydrophobicity, charge, and biological or chemical degradability [108,168,169].

FO has been extensively investigated in recent years for water and wastewater treatment. The main advantages of FO are the production of high-quality permeate due to the high removal of various CECs and the ability to operate under an osmotic driving force without requiring a hydraulic pressure difference. The bench-scale FO retention of 23 nonionic and ionic EDCs and PPCPs was 40%–98%, which depended primarily on size and charge (80%–98% for positively and negatively charged compounds and 40%–90% for nonionic compounds) [170]. However, numerous studies have been limited to synthetic solutions or limited solution pH/conductivity ranges and operating conditions [108,171].

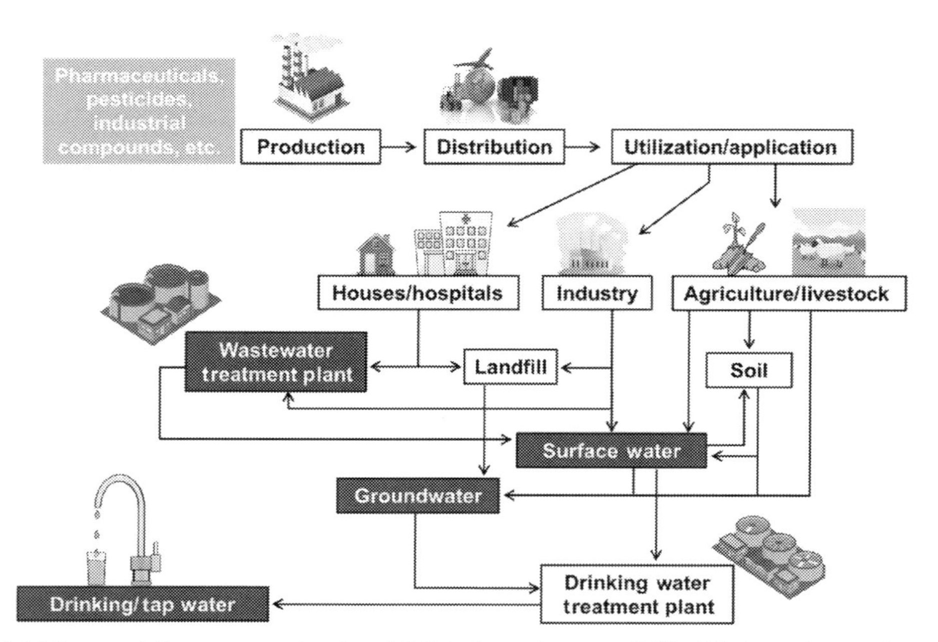

FIGURE 6–14 Representative sources and routes of CECs in the environment [161]. *CEC*, Contaminants of emerging concern.

6.5.2 Draw dilution process with lower energy consumption

6.5.2.1 Stand-alone forward osmosis system: direct use

FO has been studied for various applications in which the separation of draw solute from water is not required, for example, nutrient water, fertilization, and polymer flooding in oil fields [9]. FO membranes were first commercialized by HTI as hydration bags for portable water use in emergencies [30]. The combination of DS and FS is a sugar solution and water resources (e.g., river or lake water), respectively. Because the FO membrane has high solute rejection capability (exceeding 95% for NaCl), the diluted sugar solution is drinkable [172]. This process can be also used in space and ultralight (microlight) aviation where drinkable water must be obtained from human raw sewage.

In addition to the applications described earlier, the ability of FO processes to spontaneously and selectively permeate water without applying hydraulic pressure has been applied extensively. One such innovative system is the fertilizer drawn FO (FDFO) which is distinguishable in terms of DS use [83,84,86,173]. The design concept of the fertilizer driven FO desalination process for fertigation is shown in Fig. 6–15. The novelty of the FDFO approach is that the final diluted fertilizer DS can be directly used for irrigation or fertigation because it contains nutrients for plant growth [84,174]. Various types of fertilizers have been developed to find the acceptable final nutrient concentration. There is an obvious silver lining for the practical commercialization of FO in those applications, but the corresponding technologies are not yet ready for the real-world industry.

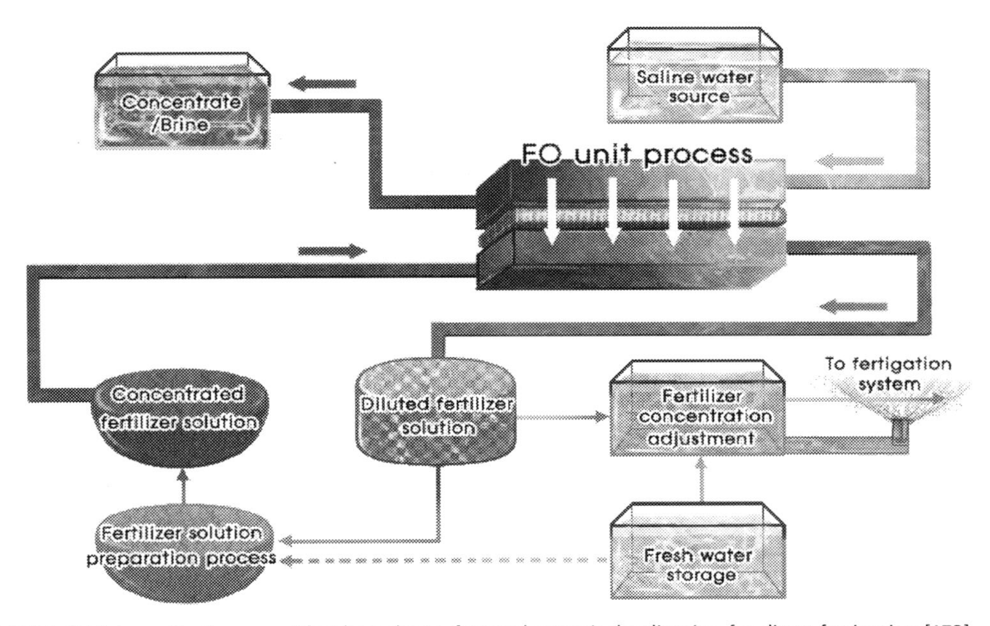

FIGURE 6–15 Schematic diagram of fertilizer drawn forward osmosis desalination for direct fertigation [173].

6.5.2.2 Hybrid forward osmosis systems

Stand-alone FO, as an energy-efficient option for desalination, is now considered overly idealistic and is impractical in the continued absence of a competitive process to easily regenerate the diluted DS and recover the product water. For the use of independent FO to generate portable water, the recent concept of coupling FO with another process to realize hybrid FO-based systems for desalination has been proposed to capitalize FO in a more advantageous manner. Generally, when seawater or the water to be desalted is used as feedwater for the FO process, we seemingly refer to it as direct desalination. Conversely, indirect desalination refers to a system where seawater or water to be desalted is used as the DS instead.

6.5.2.2.1 Indirect desalting process along with wastewater reclamation

A schematic diagram of the indirect desalination system is shown in Fig. 6–16. Wastewater is first treated with FO-utilizing seawater as the DS with much less energy use. Then, osmotically diluted seawater is processed by subsequent RO that is operated at a much lower pressure [110]. This process is expected to considerably enhance the sustainability of both wastewater reclamation and seawater desalination by reducing energy consumption as well as environmental impact. Therefore the theoretical thermodynamic energy required for indirect desalination to produce pure water must be lower than that for direct RO desalination. In addition, there are still technical barriers to be solved: (1) low water flux in FO. Because wastewater includes some chemicals and salts, the feed layer of FO must prevent fouling within the support layer. Therefore FO flux becomes lower [175,176]. (2) Cross leakage. On account of the absence of a perfect FO membrane, cross leakage from both the DS and FS sides to the other side should be considered [172].

6.5.2.2.2 Direct desalting process for draw solute recovery

After FO-based direct desalination was first proposed using recoverable DS in 2005, the direct desalination systems have received attention as an alternative to current desalination technologies including RO membranes and evaporation methods. As mentioned earlier, the major part of energy consumption in an FO process is used for the DS recovery thus making

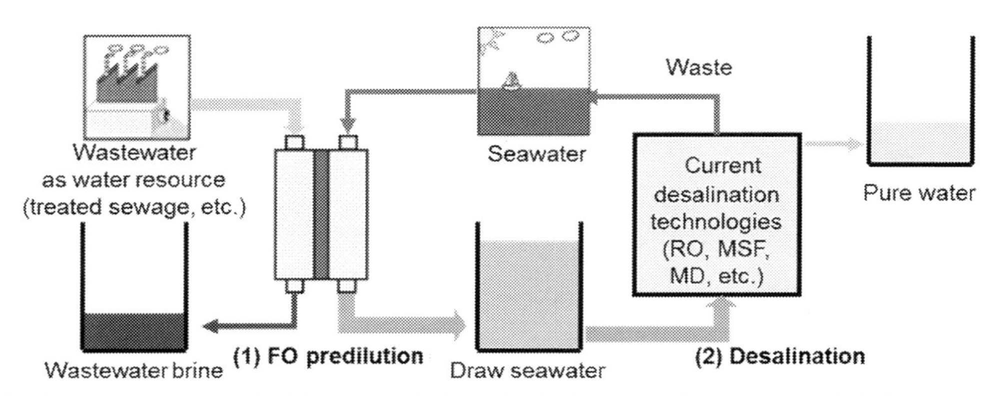

FIGURE 6–16 Flow diagram of the FO system for indirect desalination and water reclamation [172]. *FO*, Forward osmosis.

it important. In recent years, various novel DSs have been proposed together with various remarkable recovery technologies besides traditional DSs such as sodium chloride (NaCl) to advance the FO technology for various applications (Table 6–3) [59]. This section discusses all reported processes for the DS recovery unit, including pressure-driven membrane systems, thermal recovery, and the stimulus–response method.

Pressure-driven membrane processes such as RO, NF, and ultrafiltration (UF) have been widely investigated as representative technologies for the reconcentration of diluted ionic to macromolecular-based DS (Fig. 6–17) [9]. RO was first proposed as a posttreatment option for draw solute recovery among the different membrane systems because of its high water recovery rate as well as its high rejection of various salts. FO–RO hybrid processes also have demonstrated the potential of energy cost saving for high salinity stream treatment compared to stand-alone RO [112]. However, the RO recovery process requires high hydraulic pressure, typically 35–100 bar, and thus the operating costs are still high [177–179]. NF and UF were therefore suggested for the recovery of DSs because they require less energy input compared to RO and other thermal-driven processes. Some organic and/or inorganic salts, polyelectrolytes, and synthetic materials with relatively large molecular size were separated by NF or UF for their reconstitution, and the corresponding studies demonstrated encouraging results; a successful operation could be achieved with a sufficiently high rejection rate in the recovery process [75,180–182]. The recovery of DSs by low pressure-driven membrane systems, for example, NF and UF, thus, has a certain potential as a posttreatment for FO, especially in terms of energy consumption. However, the practical feasibility of real-world commercialization should be systematically examined through further research and development [9,183].

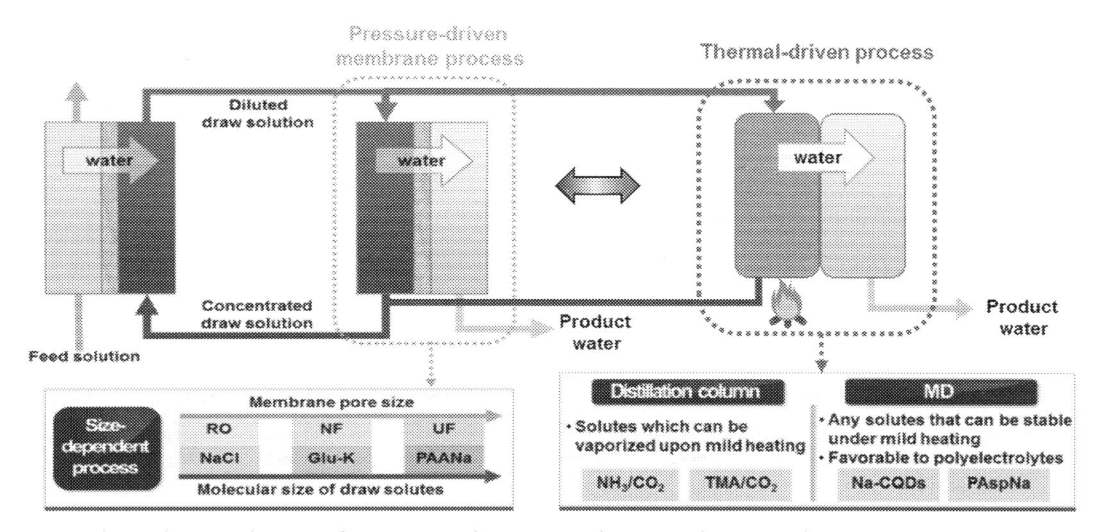

FIGURE 6–17 Schematic diagram of a system combining FO and pressure-driven membrane processes (including RO, NF, UF, and MF) and thermal-driven process (including a distillation column and MD) [9]. *FO*, Forward osmosis; *MD*, membrane distillation; *MF*, microfiltration; *NF*, nanofiltration; *RO*, reverse osmosis; *UF*, ultrafiltration.

Thermal separation is mostly used for volatile compounds and dissolved gas-based DSs. Other types of thermal recovery such as MD and thermosensitive materials are categorized as membrane processes and stimuli-responsive separation, respectively [58,71,184]. First, thermal separation has been considered the most practical recovery method, and it remains the only commercially available process. Most of the thermolytic compounds, including NH_4HCO_3, can be recovered by low-temperature distillation or membrane-gas separation because they can be vaporized upon mild heating (around 60°C) and thus can be easily separated from water [69,185]. Another thermal-based process for draw solute recovery is MD, which uses a hydrophobic porous membrane through which only steam can be diffused from the hot FS to cold water with only a temperature gradient difference of 10°C–20°C. In MD, the separation of draw solute from water is theoretically achievable and could consequently lead to low-energy consumption from the utilization of low-grade heat such as industrial waste heat [130,186,187]. However, the technology is immature and accordingly several obstacles remain, such as (1) low water flux because of nonoptimized membranes, (2) performance deterioration caused by inherent phenomena including temperature polarization, (3) membrane wetting and scaling problems, and (4) a high loss of heat by conduction.

To overcome the limitations of classical DSs, synthetic materials, such as stimulus-responsive polymers, nanogels, polyelectrolytes, hydrogels, and magnetic nanoparticles, have been developed [112]. In general, the recovery of synthetic materials is performed via stimulus response by applying a magnetic field, an electric field, heat, light, and other stimuli depending on the stimulus response of materials. However, these methods have not been thoroughly researched, and to date, only simple demonstrations of these new concepts have been conducted. In addition, they may be complicated and not effective during practical applications in a continuous FO process.

6.6 Conclusion and perspectives

We conclude that despite the surge in the number of publications on FO, extremely few studies have elucidated the fouling process with real wastewater. We believe that further research is necessitated to investigate fouling in more detail; this can be achieved by using real wastewater or seawater feeds. In this chapter a review of applications for the FO process was carried out in terms of production and treatment. The need for a reduction in the total energy cost of the FO process has been identified; however, the recovery of the DS used as the driving force is a major obstacle for achieving a practical utilization of the FO process. To overcome this problem, several attempts have been evaluated and applied for various purposes.

In the food industry, concentration of liquids such as juice and extracted components is important to increase the mobility of liquid products. With regard to this, the FO process has shown possibilities through promising concentration results. Thus the technology can be adopted as a potential method for concentrating liquor, especially juice and extracted sugar cane solutions. Moreover, the emphasis was on achieving a high rejection and recovery. These properties were applied for concentrating high-value products; this facilitated the use of FO to be applied for producing concentrated organic compounds such as pigments, enzymes, and pharmaceuticals, which need high purity and concentration. The FO process showed satisfactory performance in terms of concentration rate and energy consumption; however, the development of novel FO membrane with low RSF is required for high purity production.

Despite the promising potential for desalination, there are some challenges associated with FO. One of the main obstacles for industrial application is that FO always exhibits a lower energy efficient than that of RO. Although water molecules from an FS are naturally transported into the DS in the first FO step, further reconcentration of the diluted DS is required in the subsequent process. Because minimum energy consumption to achieve a certain water recovery rate is equal to the osmotic pressure of the concentrate (or brine), a higher concentration of the DS requires more energy than the stand-alone RO process in the subsequent reconcentration process from a thermodynamic point of view.

Finally, the FO process has arisen as a distinguishable technology for water treatment. However, since optimized applications have not been developed yet, the FO system needs to be improved in terms of higher productivity and a low amount of salt passage. The improvements in the performance of the FO membrane will significantly expand the implications of the FO process.

Acknowledgment

This work was supported by the Korea Environment Industry & Technology Institute (KEITI) through the Industrial Facilities & Infrastructure Research Program, funded by the Korea Ministry of Environment (MOE) (1485016282).

References

[1] B. Kim, G. Gwak, S. Hong, Review on methodology for determining forward osmosis (FO) membrane characteristics: water permeability (A), solute permeability (B), and structural parameter (S), Desalination 422 (2017) 5–16.

[2] G. Gwak, S. Hong, New approach for scaling control in forward osmosis (FO) by using an antiscalant-blended draw solution, J. Membr. Sci. 530 (2017) 95–103.

[3] F. Wicaksana, A.G. Fane, C. Tang, R. Wang, Nature meets technology: forward osmosis membrane technology, in: C. Hélix-Nielsen (Ed.), Biomimetic Membranes for Sensor and Separation Applications, Springer, Dordrecht, 2012, pp. 21–42.

[4] A. Haupt, A. Lerch, Forward osmosis application in manufacturing industries: a short review, Membranes (Basel) 8 (2018) 47.

[5] N.C. Nguyen, S.-S. Chen, Y.-T. Weng, H. Thi Nguyen, S.S. Ray, C.-W. Li, et al., Iodide recovery from thin film transistor liquid crystal display plants by using potassium hydroxide-driven forward osmosis, J. Membr. Sci. 520 (2016) 214–220.

[6] X. Wang, V.W.C. Chang, C.Y. Tang, Osmotic membrane bioreactor (OMBR) technology for wastewater treatment and reclamation: advances, challenges, and prospects for the future, J. Membr. Sci. 504 (2016) 113–132.

[7] W.L. Ang, A. Wahab Mohammad, D. Johnson, N. Hilal, Forward osmosis research trends in desalination and wastewater treatment: a review of research trends over the past decade, J. Water Process. Eng. 31 (2019) 100886.

[8] Desalination, Forward osmosis: is it beginning to live up to the hype?. <https://www.waterworld.com/international/desalination/article/16201130/forward-osmosis-is-it-beginning-to-live-up-to-the-hype>.

[9] G. Gwak, S. Hong, Chapter 3—Draw solute selection, in: S. Sarp, N. Hilal (Eds.), Membrane-Based Salinity Gradient Processes for Water Treatment and Power Generation, Elsevier, 2018, pp. 87–122.

[10] G. Gwak, B. Jung, S. Han, S. Hong, Evaluation of poly(aspartic acid sodium salt) as a draw solute for forward osmosis, Water Res. 80 (2015) 294–305.

[11] J.R. McCutcheon, M. Elimelech, Influence of concentrative and dilutive internal concentration polarization on flux behavior in forward osmosis, J. Membr. Sci. 284 (2006) 237–247.

[12] K.L. Lee, R.W. Baker, H.K. Lonsdale, Membranes for power generation by pressure-retarded osmosis, J. Membr. Sci. 8 (1981) 141–171.

[13] S. Loeb, L. Titelman, E. Korngold, J. Freiman, Effect of porous support fabric on osmosis through a Loeb-Sourirajan type asymmetric membrane, J. Membr. Sci. 129 (1997) 243–249.

[14] N.Y. Yip, A. Tiraferri, W.A. Phillip, J.D. Schiffman, L.A. Hoover, Y.C. Kim, et al., Thin-film composite pressure retarded osmosis membranes for sustainable power generation from salinity gradients, Environ. Sci. Technol. 45 (2011) 4360–4369.

[15] A. Tiraferri, N.Y. Yip, A.P. Straub, S. Romero-Vargas Castrillon, M. Elimelech, A method for the simultaneous determination of transport and structural parameters of forward osmosis membranes, J. Membr. Sci. 444 (2013) 523–538.

[16] Y. Kim, S. Li, L. Chekli, Y.C. Woo, C.-H. Wei, S. Phuntsho, et al., Assessing the removal of organic micro-pollutants from anaerobic membrane bioreactor effluent by fertilizer-drawn forward osmosis, J. Membr. Sci. 533 (2017) 84–95.

[17] C.Y. Tang, Q. She, W.C.L. Lay, R. Wang, R. Field, A.G. Fane, Modeling double-skinned FO membranes, Desalination 283 (2011) 178–186.

[18] W.A. Phillip, J.S. Yong, M. Elimelech, Reverse draw solute permeation in forward osmosis: modeling and experiments, Environ. Sci. Technol. 44 (2010) 5170–5176.

[19] C. Suh, S. Lee, Modeling reverse draw solute flux in forward osmosis with external concentration polarization in both sides of the draw and feed solution, J. Membr. Sci. 427 (2013) 365–374.

[20] S. Lee, C. Boo, M. Elimelech, S. Hong, Comparison of fouling behavior in forward osmosis (FO) and reverse osmosis (RO), J. Membr. Sci. 365 (2010) 34–39.

[21] Y. Kim, Y.C. Woo, S. Phuntsho, L.D. Nghiem, H.K. Shon, S. Hong, Evaluation of fertilizer-drawn forward osmosis for coal seam gas reverse osmosis brine treatment and sustainable agricultural reuse, J. Membr. Sci. 537 (2017) 22–31.

[22] S. Li, Y. Kim, L. Chekli, S. Phuntsho, H.K. Shon, T. Leiknes, et al., Impact of reverse nutrient diffusion on membrane biofouling in fertilizer-drawn forward osmosis, J. Membr. Sci. 539 (2017) 108–115.

[23] Y. Kim, S. Lee, H.K. Shon, S. Hong, Organic fouling mechanisms in forward osmosis membrane process under elevated feed and draw solution temperatures, Desalination 355 (2015) 169–177.

[24] S.-J. You, X.-H. Wang, M. Zhong, Y.-J. Zhong, C. Yu, N.-Q. Ren, Temperature as a factor affecting transmembrane water flux in forward osmosis: steady-state modeling and experimental validation, Chem. Eng. J. 198–199 (2012) 52–60.

[25] J. Lee, N. Ghaffour, Predicting the performance of large-scale forward osmosis module using spatial variation model: effect of operating parameters including temperature, Desalination 469 (2019) 114095.

[26] X. Jin, C.Y. Tang, Y. Gu, Q. She, S. Qi, Boric acid permeation in forward osmosis membrane processes: modeling, experiments, and implications, Environ. Sci. Technol. 45 (2011) 2323–2330.

[27] M. Xie, L.D. Nghiem, W.E. Price, M. Elimelech, Relating rejection of trace organic contaminants to membrane properties in forward osmosis: measurements, modelling and implications, Water Res. 49 (2014) 265–274.

[28] L.D. Nghiem, A.I. Schäfer, M. Elimelech, Removal of natural hormones by nanofiltration membranes: measurement, modeling, and mechanisms, Environ. Sci. Technol. 38 (2004) 1888–1896.

[29] P.M. Bungay, H. Brenner, The motion of a closely-fitting sphere in a fluid-filled tube, Int. J. Multiphase Flow 1 (1973) 25–56.

[30] T.Y. Cath, A.E. Childress, M. Elimelech, Forward osmosis: principles, applications, and recent developments, J. Membr. Sci. 281 (2006) 70–87.

[31] L. Chekli, S. Phuntsho, J.E. Kim, J. Kim, J.Y. Choi, J.-S. Choi, et al., A comprehensive review of hybrid forward osmosis systems: performance, applications and future prospects, J. Membr. Sci. 497 (2016) 430–449.

[32] B. Mi, M. Elimelech, Organic fouling of forward osmosis membranes: fouling reversibility and cleaning without chemical reagents, J. Membr. Sci. 348 (2010) 337–345.

[33] B. Mi, M. Elimelech, Silica scaling and scaling reversibility in forward osmosis, Desalination 312 (2013) 75–81.

[34] Y. Kim, M. Elimelech, H.K. Shon, S. Hong, Combined organic and colloidal fouling in forward osmosis: fouling reversibility and the role of applied pressure, J. Membr. Sci. 460 (2014) 206–212.

[35] N.M. Mazlan, P. Marchetti, H.A. Maples, B. Gu, S. Karan, A. Bismarck, et al., Organic fouling behaviour of structurally and chemically different forward osmosis membranes – a study of cellulose triacetate and thin film composite membranes, J. Membr. Sci. 520 (2016) 247–261.

[36] B.G. Choi, D.I. Kim, S. Hong, Fouling evaluation and mechanisms in a FO-RO hybrid process for direct potable reuse, J. Membr. Sci. 520 (2016) 89–98.

[37] R. Honda, W. Rukapan, H. Komura, Y. Teraoka, M. Noguchi, E.M.V. Hoek, Effects of membrane orientation on fouling characteristics of forward osmosis membrane in concentration of microalgae culture, Bioresour. Technol. 197 (2015) 429–433.

[38] M. Xie, L.D. Nghiem, W.E. Price, M. Elimelech, Comparison of the removal of hydrophobic trace organic contaminants by forward osmosis and reverse osmosis, Water Res. 46 (2012) 2683–2692.

[39] C. Kim, S. Lee, H.K. Shon, M. Elimelech, S. Hong, Boron transport in forward osmosis: measurements, mechanisms, and comparison with reverse osmosis, J. Membr. Sci. 419–420 (2012) 42–48.

[40] M. Xie, W.E. Price, L.D. Nghiem, M. Elimelech, Effects of feed and draw solution temperature and transmembrane temperature difference on the rejection of trace organic contaminants by forward osmosis, J. Membr. Sci. 438 (2013) 57–64.

[41] X. Jin, Q. She, X. Ang, C.Y. Tang, Removal of boron and arsenic by forward osmosis membrane: influence of membrane orientation and organic fouling, J. Membr. Sci. 389 (2012) 182–187.

[42] S. Jamil, S. Jeong, S. Vigneswaran, Application of pressure assisted forward osmosis for water purification and reuse of reverse osmosis concentrate from a water reclamation plant, Sep. Purif. Technol. 171 (2016) 182–190.

[43] M. Xie, W.E. Price, L.D. Nghiem, Rejection of pharmaceutically active compounds by forward osmosis: role of solution pH and membrane orientation, Sep. Purif. Technol. 93 (2012) 107–114.

[44] A.A. Alturki, J.A. McDonald, S.J. Khan, W.E. Price, L.D. Nghiem, M. Elimelech, Removal of trace organic contaminants by the forward osmosis process, Sep. Purif. Technol. 103 (2013) 258–266.

[45] M. Xie, L.D. Nghiem, W.E. Price, M. Elimelech, Impact of organic and colloidal fouling on trace organic contaminant rejection by forward osmosis: role of initial permeate flux, Desalination 336 (2014) 146–152.

[46] J. Teng, M. Zhang, K.-T. Leung, J. Chen, H. Hong, H. Lin, et al., A unified thermodynamic mechanism underlying fouling behaviors of soluble microbial products (SMPs) in a membrane bioreactor, Water Res. 149 (2019) 477–487.

[47] X. Qu, X. Cai, M. Zhang, H. Lin, Z. Leihong, B.-Q. Liao, A facile method for simulating randomly rough membrane surface associated with interface behaviors, Appl. Surf. Sci. 427 (2018) 915–921.

[48] R. Valladares Linares, V. Yangali-Quintanilla, Z. Li, G. Amy, Rejection of micropollutants by clean and fouled forward osmosis membrane, Water Res. 45 (2011) 6737–6744.

[49] M. Zhang, H. Hong, H. Lin, L. Shen, H. Yu, G. Ma, et al., Mechanistic insights into alginate fouling caused by calcium ions based on terahertz time-domain spectra analyses and DFT calculations, Water Res. 129 (2018) 337–346.

[50] J. Teng, L. Shen, G. Yu, F. Wang, F. Li, X. Zhou, et al., Mechanism analyses of high specific filtration resistance of gel and roles of gel elasticity related with membrane fouling in a membrane bioreactor, Bioresour. Technol. 257 (2018) 39–46.

[51] M. Xie, L.D. Nghiem, W.E. Price, M. Elimelech, A forward osmosis–membrane distillation hybrid process for direct sewer mining: system performance and limitations, Environ. Sci. Technol. 47 (2013) 13486–13493.

[52] T.N. Bitaw, K. Park, D.R. Yang, Optimization on a new hybrid forward osmosis-electrodialysis-reverse osmosis seawater desalination process, Desalination 398 (2016) 265–281.

[53] R.L. McGinnis, M. Elimelech, Energy requirements of ammonia–carbon dioxide forward osmosis desalination, Desalination 207 (2007) 370–382.

[54] Q. Liu, C. Liu, L. Zhao, W. Ma, H. Liu, J. Ma, Integrated forward osmosis-membrane distillation process for human urine treatment, Water Res. 91 (2016) 45–54.

[55] S. Phuntsho, J.E. Kim, S. Hong, N. Ghaffour, T. Leiknes, J.Y. Choi, et al., A closed-loop forward osmosis-nanofiltration hybrid system: understanding process implications through full-scale simulation, Desalination 421 (2017) 169–178.

[56] R.A. Maltos, J. Regnery, N. Almaraz, S. Fox, M. Schutter, T.J. Cath, et al., Produced water impact on membrane integrity during extended pilot testing of forward osmosis–reverse osmosis treatment, Desalination 440 (2018) 99–110.

[57] Q. Ge, M. Ling, T.-S. Chung, Draw solutions for forward osmosis processes: developments, challenges, and prospects for the future, J. Membr. Sci. 442 (2013) 225–237.

[58] H. Luo, Q. Wang, T.C. Zhang, T. Tao, A. Zhou, L. Chen, et al., A review on the recovery methods of draw solutes in forward osmosis, J. Water Process. Eng. 4 (2014) 212–223.

[59] A. Achilli, T.Y. Cath, A.E. Childress, Selection of inorganic-based draw solutions for forward osmosis applications, J. Membr. Sci. 364 (2010) 233–241.

[60] S. Lee, Y. Kim, A.S. Kim, S. Hong, Evaluation of membrane-based desalting processes for RO brine treatment, Desalin. Water Treat. 57 (2016) 7432–7439.

[61] S. Zhang, P. Wang, X. Fu, T.-S. Chung, Sustainable water recovery from oily wastewater via forward osmosis-membrane distillation (FO-MD), Water Res. 52 (2014) 112–121.

[62] S. Lee, Y. Kim, S. Hong, Treatment of industrial wastewater produced by desulfurization process in a coal-fired power plant via FO-MD hybrid process, Chemosphere 210 (2018) 44–51.

[63] W.A. Suwaileh, D.J. Johnson, S. Sarp, N. Hilal, Advances in forward osmosis membranes: altering the sub-layer structure via recent fabrication and chemical modification approaches, Desalination 436 (2018) 176–201.

[64] P.S. Goh, A.F. Ismail, B.C. Ng, M.S. Abdullah, Recent progresses of forward osmosis membranes formulation and design for wastewater treatment, Water 11 (2019) 2043.

[65] A.M. Awad, R. Jalab, J. Minier-Matar, S. Adham, M.S. Nasser, S.J. Judd, The status of forward osmosis technology implementation, Desalination 461 (2019) 10–21.

[66] J. Wang, N. Pathak, L. Chekli, S. Phuntsho, Y. Kim, D. Li, et al., Performance of a novel fertilizer-drawn forward osmosis aerobic membrane bioreactor (FDFO-MBR): mitigating salinity build-up by integrating microfiltration, Water 9 (2017) 21.

[67] D.L. Shaffer, J.R. Werber, H. Jaramillo, S. Lin, M. Elimelech, Forward osmosis: where are we now? Desalination 356 (2015) 271–284.

[68] M.L. Stone, C. Rae, F.F. Stewart, A.D. Wilson, Switchable polarity solvents as draw solutes for forward osmosis, Desalination 312 (2013) 124–129.

[69] J.R. McCutcheon, R.L. McGinnis, M. Elimelech, A novel ammonia—carbon dioxide forward (direct) osmosis desalination process, Desalination 174 (2005) 1–11.

[70] S.K. Yen, F.N. Mehnas Haja, M. Su, K.Y. Wang, T.-S. Chung, Study of draw solutes using 2-methylimidazole-based compounds in forward osmosis, J. Membr. Sci. 364 (2010) 242–252.

[71] G. Gwak, D.I. Kim, J. Kim, M. Zhan, S. Hong, An integrated system for CO_2 capture and water treatment by forward osmosis driven by an amine-based draw solution, J. Membr. Sci. 581 (2019) 9–17.

[72] Q. Ge, G.L. Amy, T.-S. Chung, Forward osmosis for oily wastewater reclamation: multi-charged oxalic acid complexes as draw solutes, Water Res. 122 (2017) 580–590.

[73] M.M. Ling, T.-S. Chung, Desalination process using super hydrophilic nanoparticles via forward osmosis integrated with ultrafiltration regeneration, Desalination 278 (2011) 194–202.

[74] Q. Long, L. Shen, R. Chen, J. Huang, S. Xiong, Y. Wang, Synthesis and application of organic phosphonate salts as draw solutes in forward osmosis for oil–water separation, Environ. Sci. Technol. 50 (2016) 12022–12029.

[75] Q. Ge, J. Su, G.L. Amy, T.-S. Chung, Exploration of polyelectrolytes as draw solutes in forward osmosis processes, Water Res. 46 (2012) 1318–1326.

[76] C.H. Tan, H.Y. Ng, A novel hybrid forward osmosis-nanofiltration (FO-NF) process for seawater desalination: draw solution selection and system configuration, Desalin. Water Treat. 13 (2010) 356–361.

[77] R. Alnaizy, A. Aidan, M. Qasim, Copper sulfate as draw solute in forward osmosis desalination, J. Environ. Chem. Eng. 1 (2013) 424–430.

[78] R. Alnaizy, A. Aidan, M. Qasim, Draw solute recovery by metathesis precipitation in forward osmosis desalination, Desalin. Water Treat. 51 (2013) 5516–5525.

[79] M.M. Ling, K.Y. Wang, T.-S. Chung, Highly water-soluble magnetic nanoparticles as novel draw solutes in forward osmosis for water reuse, Ind. Eng. Chem. Res. 49 (2010) 5869–5876.

[80] D. Li, X. Zhang, J. Yao, G.P. Simon, H. Wang, Stimuli-responsive polymer hydrogels as a new class of draw agent for forward osmosis desalination, Chem. Commun. 47 (2011) 1710–1712.

[81] D. Li, X. Zhang, J. Yao, Y. Zeng, G.P. Simon, H. Wang, Composite polymer hydrogels as draw agents in forward osmosis and solar dewatering, Soft Matter 7 (2011) 10048–10056.

[82] Y. Cai, R. Wang, W.B. Krantz, A.G. Fane, X.M. Hu, Exploration of using thermally responsive polyionic liquid hydrogels as draw agents in forward osmosis, RSC Adv. 5 (2015) 97143–97150.

[83] S. Phuntsho, H.K. Shon, S. Hong, S. Lee, S. Vigneswaran, J. Kandasamy, Fertiliser drawn forward osmosis desalination: the concept, performance and limitations for fertigation, Rev. Environ. Sci. Bio/Technol. 11 (2012) 147–168.

[84] S. Phuntsho, H.K. Shon, T. Majeed, I. El Saliby, S. Vigneswaran, J. Kandasamy, et al., Blended fertilizers as draw solutions for fertilizer-drawn forward osmosis desalination, Environ. Sci. Technol. 46 (2012) 4567–4575.

[85] S. Phuntsho, J.E. Kim, M.A.H. Johir, S. Hong, Z. Li, N. Ghaffour, et al., Fertiliser drawn forward osmosis process: pilot-scale desalination of mine impaired water for fertigation, J. Membr. Sci. 508 (2016) 22–31.

[86] Y. Kim, L. Chekli, W.-G. Shim, S. Phuntsho, S. Li, N. Ghaffour, et al., Selection of suitable fertilizer draw solute for a novel fertilizer-drawn forward osmosis–anaerobic membrane bioreactor hybrid system, Bioresour. Technol. 210 (2016) 26–34.

[87] Y. Kim, S. Li, N. Ghaffour, Evaluation of different cleaning strategies for different types of forward osmosis membrane fouling and scaling, J. Membr. Sci. 596 (2020) 117731.

[88] M.A. Shannon, P.W. Bohn, M. Elimelech, J.G. Georgiadis, B.J. Mariñas, A.M. Mayes, Science and technology for water purification in the coming decades, Nature 452 (2008) 337–346.

[89] S. Hong, M. Elimelech, Chemical and physical aspects of natural organic matter (NOM) fouling of nanofiltration membranes, J. Membr. Sci. 132 (1997) 159–181.

[90] Q. She, X. Jin, Q. Li, C.Y. Tang, Relating reverse and forward solute diffusion to membrane fouling in osmotically driven membrane processes, Water Res. 46 (2012) 2478–2486.

[91] M. Xie, E. Bar-Zeev, S.M. Hashmi, L.D. Nghiem, M. Elimelech, Role of reverse divalent cation diffusion in forward osmosis biofouling, Environ. Sci. Technol. 49 (2015) 13222–13229.

[92] J.-Y. Li, Z.-Y. Ni, Z.-Y. Zhou, Y.-X. Hu, X.-H. Xu, L.-H. Cheng, Membrane fouling of forward osmosis in dewatering of soluble algal products: comparison of TFC and CTA membranes, J. Membr. Sci. 552 (2018) 213–221.

[93] Y.-J. Choi, S.-H. Kim, S. Jeong, T.-M. Hwang, Application of ultrasound to mitigate calcium sulfate scaling and colloidal fouling, Desalination 336 (2014) 153–159.

[94] L. Lv, J. Xu, B. Shan, C. Gao, Concentration performance and cleaning strategy for controlling membrane fouling during forward osmosis concentration of actual oily wastewater, J. Membr. Sci. 523 (2017) 15–23.

[95] G. Blandin, H. Vervoort, P. Le-Clech, A.R.D. Verliefde, Fouling and cleaning of high permeability forward osmosis membranes, J. Water Process. Eng. 9 (2016) 161–169.

[96] A. Achilli, T.Y. Cath, E.A. Marchand, A.E. Childress, The forward osmosis membrane bioreactor: a low fouling alternative to MBR processes, Desalination 239 (2009) 10–21.

[97] S. Zou, Y.-N. Wang, F. Wicaksana, T. Aung, P.C.Y. Wong, A.G. Fane, et al., Direct microscopic observation of forward osmosis membrane fouling by microalgae: critical flux and the role of operational conditions, J. Membr. Sci. 436 (2013) 174–185.

[98] C. Boo, M. Elimelech, S. Hong, Fouling control in a forward osmosis process integrating seawater desalination and wastewater reclamation, J. Membr. Sci. 444 (2013) 148–156.

[99] M. Zhan, G. Gwak, D.I. Kim, K. Park, S. Hong, Quantitative analysis of the irreversible membrane fouling of forward osmosis during wastewater reclamation: correlation with the modified fouling index, J. Membr. Sci. 597 (2020) 117757.

[100] B. Mi, M. Elimelech, Chemical and physical aspects of organic fouling of forward osmosis membranes, J. Membr. Sci. 320 (2008) 292–302.

[101] Y. Kim, J. Kim, M. Zhan, D. Min, S. Hong, Optimal flow rate evaluation for low energy, high efficiency cleaning of forward osmosis (FO), Membr. J. 29 (2019) 339–347.

[102] R.W. Holloway, A.E. Childress, K.E. Dennett, T.Y. Cath, Forward osmosis for concentration of anaerobic digester centrate, Water Res. 41 (2007) 4005–4014.

[103] P. Shojaee Nasirabadi, E. Saljoughi, S.M. Mousavi, Membrane processes used for removal of pharmaceuticals, hormones, endocrine disruptors and their metabolites from wastewaters: a review, Desalin. Water Treat. 57 (2016) 24146–24175.

[104] B.K. Pramanik, L. Shu, V. Jegatheesan, A review of the management and treatment of brine solutions, Environ. Sci. Water Res. Technol. 3 (2017) 625–658.

[105] Y.-N. Wang, K. Goh, X. Li, L. Setiawan, R. Wang, Membranes and processes for forward osmosis-based desalination: recent advances and future prospects, Desalination 434 (2018) 81–99.

[106] P. Das, K.K.K. Singh, S. Dutta, Insight into emerging applications of forward osmosis systems, J. Ind. Eng. Chem. 72 (2019) 1–17.

[107] M.R. Bilad, H.A. Arafat, I.F.J. Vankelecom, Membrane technology in microalgae cultivation and harvesting: a review, Biotechnol. Adv. 32 (2014) 1283–1300.

[108] S. Kim, K.H. Chu, Y.A.J. Al-Hamadani, C.M. Park, M. Jang, D.-H. Kim, et al., Removal of contaminants of emerging concern by membranes in water and wastewater: a review, Chem. Eng. J. 335 (2018) 896–914.

[109] M. Xie, H.K. Shon, S.R. Gray, M. Elimelech, Membrane-based processes for wastewater nutrient recovery: technology, challenges, and future direction, Water Res. 89 (2016) 210–221.

[110] B.G. Choi, M. Zhan, K. Shin, S. Lee, S. Hong, Pilot-scale evaluation of FO-RO osmotic dilution process for treating wastewater from coal-fired power plant integrated with seawater desalination, J. Membr. Sci. 540 (2017) 78–87.

[111] M. Zhan, G. Gwak, B.G. Choi, S. Hong, Indexing fouling reversibility in forward osmosis and its implications for sustainable operation of wastewater reclamation, J. Membr. Sci. 574 (2019) 262–269.

[112] I. Chaoui, S. Abderafi, S. Vaudreuil, T. Bounahmidi, Water desalination by forward osmosis: draw solutes and recovery methods – review, Environ. Technol. Rev. 8 (2019) 25–46.

[113] C.A. Nayak, S.S. Valluri, N.K. Rastogi, Effect of high or low molecular weight of components of feed on transmembrane flux during forward osmosis, J. Food Eng. 106 (2011) 48–52.

[114] E.M. Garcia-Castello, J.R. McCutcheon, Dewatering press liquor derived from orange production by forward osmosis, J. Membr. Sci. 372 (2011) 97–101.

[115] D.I. Kim, G. Gwak, M. Zhan, S. Hong, Sustainable dewatering of grapefruit juice through forward osmosis: improving membrane performance, fouling control, and product quality, J. Membr. Sci. 578 (2019) 53–60.

[116] X. An, Y. Hu, N. Wang, Z. Zhou, Z. Liu, Continuous juice concentration by integrating forward osmosis with membrane distillation using potassium sorbate preservative as a draw solute, J. Membr. Sci. 573 (2019) 192–199.

[117] Y.-N. Wang, R. Wang, W. Li, C.Y. Tang, Whey recovery using forward osmosis – evaluating the factors limiting the flux performance, J. Membr. Sci. 533 (2017) 179–189.

[118] P. Pal, J. Nayak, Development and analysis of a sustainable technology in manufacturing acetic acid and whey protein from waste cheese whey, J. Clean. Prod. 112 (2016) 59–70.

[119] H. Song, F. Xie, W. Chen, J. Liu, FO/MD hybrid system for real dairy wastewater recycling, Environ. Technol. 39 (2018) 2411–2421.

[120] A. Haupt, A. Lerch, Forward osmosis treatment of effluents from dairy and automobile industry – results from short-term experiments to show general applicability, Water Sci. Technol. 78 (2018) 467–475.

[121] Y.H. Cho, H.D. Lee, H.B. Park, Integrated membrane processes for separation and purification of organic acid from a biomass fermentation process, Ind. Eng. Chem. Res. 51 (2012) 10207–10219.

[122] L. Wang, H. Chu, B. Dong, Effects on the purification of tannic acid and natural dissolved organic matter by forward osmosis membrane, J. Membr. Sci. 455 (2014) 31–43.

[123] T. Ruprakobkit, L. Ruprakobkit, C. Ratanatamskul, Carboxylic acid concentration by forward osmosis processes: dynamic modeling, experimental validation and simulation, Chem. Eng. J. 306 (2016) 538–549.

[124] K. Jung, J.-D.-R. Choi, D. Lee, C. Seo, J. Lee, S.Y. Lee, et al., Permeation characteristics of volatile fatty acids solution by forward osmosis, Process. Biochem. 50 (2015) 669–677.

[125] D. Mondal, S.K. Nataraj, A.V. Rami Reddy, K.K. Ghara, P. Maiti, S.C. Upadhyay, et al., Four-fold concentration of sucrose in sugarcane juice through energy efficient forward osmosis using sea bittern as draw solution, RSC Adv. 5 (2015) 17872–17878.

[126] E.M. Garcia-Castello, J.R. McCutcheon, M. Elimelech, Performance evaluation of sucrose concentration using forward osmosis, J. Membr. Sci. 338 (2009) 61–66.

[127] M. Madhumala, S. Moulik, T. Sankarshana, S. Sridhar, Forward-osmosis-aided concentration of fructose sugar through hydrophilized polyamide membrane: molecular modeling and economic estimation, Appl. Polym. 134 (2017).

[128] M.P. Marques, V.D. Alves, I.M. Coelhoso, Concentration of tea extracts by osmotic evaporation: optimisation of process parameters and effect on antioxidant activity, Membranes 7 (2017) 1.

[129] J.R. Herron, E.G. Beaudry, C.E. Jochums, L.E. Medina, Osmotic Concentration Apparatus and Method for Direct Osmosis Concentration of Fruit Juices, US Patent 5281,430, 1994.

[130] K.Y. Wang, M.M. Teoh, A. Nugroho, T.-S. Chung, Integrated forward osmosis–membrane distillation (FO–MD) hybrid system for the concentration of protein solutions, Chem. Eng. Sci. 66 (2011) 2421–2430.

[131] Q. Yang, K.Y. Wang, T.-S. Chung, A novel dual-layer forward osmosis membrane for protein enrichment and concentration, Sep. Purif. Technol. 69 (2009) 269–274.

[132] Q. Ge, P. Wang, C. Wan, T.-S. Chung, Polyelectrolyte-promoted forward osmosis–membrane distillation (FO–MD) hybrid process for dye wastewater treatment, Environ. Sci. Technol. 46 (2012) 6236–6243.

[133] J. Kim, J. Kim, J. Lim, S. Hong, Evaluation of ethanol as draw solute for forward osmosis (FO) process of highly saline (waste)water, Desalination 456 (2019) 23–31.

[134] X. Zhang, Z. Ning, D.K. Wang, J.C. Diniz da Costa, A novel ethanol dehydration process by forward osmosis, Chem. Eng. J. 232 (2013) 397–404.

[135] A. Ambrosi, G. Lopes Corrêa, N. Souza de Vargas, L. Martim Gabe, N.S.M. Cardozo, I.C. Tessaro, Impact of osmotic agent on the transport of components using forward osmosis to separate ethanol from aqueous solutions, AIChE J. 63 (2017) 4499–4507.

[136] A. Ambrosi, M. Al-Furaiji, J.R. McCutcheon, N.S.M. Cardozo, I.C. Tessaro, Transport of components in the separation of ethanol from aqueous dilute solutions by forward osmosis, Ind. Eng. Chem. Res. 57 (2018) 2967–2975.

[137] Y. Zhang, K. Nakagawa, M. Shibuya, K. Sasaki, T. Takahashi, T. Shintani, et al., Improved permselectivity of forward osmosis membranes for efficient concentration of pretreated rice straw and bioethanol production, J. Membr. Sci. 566 (2018) 15–24.

[138] C.A. Nayak, N.K. Rastogi, Forward osmosis for the concentration of anthocyanin from *Garcinia indica* Choisy, Sep. Purif. Technol. 71 (2010) 144–151.

[139] M. Ahmed, S. Ahmad, Fayyaz-ul-Hassan, G. Qadir, R. Hayat, F.A. Shaheen, et al., Innovative processes and technologies for nutrient recovery from wastes: a comprehensive review, Sustainability 11 (2019) 4938.

[140] A.J. Ansari, F.I. Hai, W.E. Price, J.E. Drewes, L.D. Nghiem, Forward osmosis as a platform for resource recovery from municipal wastewater—a critical assessment of the literature, J. Membr. Sci. 529 (2017) 195–206.

[141] E.A. Bell, R.W. Holloway, T.Y. Cath, Evaluation of forward osmosis membrane performance and fouling during long-term osmotic membrane bioreactor study, J. Membr. Sci. 517 (2016) 1–13.

[142] R.W. Holloway, J. Regnery, L.D. Nghiem, T.Y. Cath, Removal of trace organic chemicals and performance of a novel hybrid ultrafiltration-osmotic membrane bioreactor, Environ. Sci. Technol. 48 (2014) 10859–10868.

[143] W. Luo, F.I. Hai, W.E. Price, W. Guo, H.H. Ngo, K. Yamamoto, et al., Phosphorus and water recovery by a novel osmotic membrane bioreactor–reverse osmosis system, Bioresour. Technol. 200 (2016) 297–304.

[144] G. Qiu, Y.-P. Ting, Direct phosphorus recovery from municipal wastewater via osmotic membrane bioreactor (OMBR) for wastewater treatment, Bioresour. Technol. 170 (2014) 221–229.

[145] L. Chen, Y. Gu, C. Cao, J. Zhang, J.-W. Ng, C. Tang, Performance of a submerged anaerobic membrane bioreactor with forward osmosis membrane for low-strength wastewater treatment, Water Res. 50 (2014) 114–123.

[146] Y. Gu, L. Chen, J.-W. Ng, C. Lee, V.W.C. Chang, C.Y. Tang, Development of anaerobic osmotic membrane bioreactor for low-strength wastewater treatment at mesophilic condition, J. Membr. Sci. 490 (2015) 197–208.

[147] M.K.Y. Tang, H.Y. Ng, Impacts of different draw solutions on a novel anaerobic forward osmosis membrane bioreactor (AnFOMBR), Water Sci. Technol. 69 (2014) 2036–2042.

[148] A.J. Ansari, F.I. Hai, W. Guo, H.H. Ngo, W.E. Price, L.D. Nghiem, Selection of forward osmosis draw solutes for subsequent integration with anaerobic treatment to facilitate resource recovery from wastewater, Bioresour. Technol. 191 (2015) 30–36.

[149] X. Zhang, Z. Ning, D.K. Wang, J.C. Diniz da Costa, Processing municipal wastewaters by forward osmosis using CTA membrane, J. Membr. Sci. 468 (2014) 269–275.

[150] C.-H. Wei, M. Harb, G. Amy, P.-Y. Hong, T. Leiknes, Sustainable organic loading rate and energy recovery potential of mesophilic anaerobic membrane bioreactor for municipal wastewater treatment, Bioresour. Technol. 166 (2014) 326–334.

[151] S.M. Iskander, B. Brazil, J.T. Novak, Z. He, Resource recovery from landfill leachate using bioelectrochemical systems: opportunities, challenges, and perspectives, Bioresour. Technol. 201 (2016) 347–354.

[152] X. Zhao, C. Liu, Efficient removal of heavy metal ions based on the optimized dissolution-diffusion-flow forward osmosis process, Chem. Eng. J. 334 (2018) 1128–1134.

[153] Y. Cui, Q. Ge, X.-Y. Liu, T.-S. Chung, Novel forward osmosis process to effectively remove heavy metal ions, J. Membr. Sci. 467 (2014) 188–194.

[154] S. You, J. Lu, C.Y. Tang, X. Wang, Rejection of heavy metals in acidic wastewater by a novel thin-film inorganic forward osmosis membrane, Chem. Eng. J. 320 (2017) 532–538.

[155] P. Zhao, B. Gao, Q. Yue, S. Liu, H.K. Shon, The performance of forward osmosis in treating high-salinity wastewater containing heavy metal Ni^{2+}, Chem. Eng. J. 288 (2016) 569–576.

[156] C.-Y. Wu, H. Mouri, S.-S. Chen, D.-Z. Zhang, M. Koga, J. Kobayashi, Removal of trace-amount mercury from wastewater by forward osmosis, J. Water Process. Eng. 14 (2016) 108–116.

[157] M. Larronde-Larretche, X. Jin, Microalgal biomass dewatering using forward osmosis membrane: influence of microalgae species and carbohydrates composition, Algal Res. 23 (2017) 12–19.

[158] M. Larronde-Larretche, X. Jin, Microalgae (*Scenedesmus obliquus*) dewatering using forward osmosis membrane: influence of draw solution chemistry, Algal Res. 15 (2016) 1–8.

[159] S. Zou, Y. Gu, D. Xiao, C.Y. Tang, The role of physical and chemical parameters on forward osmosis membrane fouling during algae separation, J. Membr. Sci. 366 (2011) 356–362.

[160] G. Gwak, D.I. Kim, S. Hong, New industrial application of forward osmosis (FO): precious metal recovery from printed circuit board (PCB) plant wastewater, J. Membr. Sci. 552 (2018) 234–242.

[161] M.O. Barbosa, N.F.F. Moreira, A.R. Ribeiro, M.F.R. Pereira, A.M.T. Silva, Occurrence and removal of organic micropollutants: an overview of the watch list of EU Decision 2015/495, Water Res. 94 (2016) 257–279.

[162] J.H. Al-Rifai, H. Khabbaz, A.I. Schäfer, Removal of pharmaceuticals and endocrine disrupting compounds in a water recycling process using reverse osmosis systems, Sep. Purif. Technol. 77 (2011) 60–67.

[163] C. Bellona, J.E. Drewes, P. Xu, G. Amy, Factors affecting the rejection of organic solutes during NF/RO treatment—a literature review, Water Res. 38 (2004) 2795–2809.

[164] J.E. Drewes, C. Bellona, M. Oedekoven, P. Xu, T.-U. Kim, G. Amy, Rejection of wastewater-derived micropollutants in high-pressure membrane applications leading to indirect potable reuse, Environ. Prog. 24 (2005) 400–409.

[165] T.-U. Kim, J.E. Drewes, R. Scott Summers, G.L. Amy, Solute transport model for trace organic neutral and charged compounds through nanofiltration and reverse osmosis membranes, Water Res. 41 (2007) 3977–3988.

[166] A. D'Haese, P. Le-Clech, S. Van Nevel, K. Verbeken, E.R. Cornelissen, S.J. Khan, et al., Trace organic solutes in closed-loop forward osmosis applications: influence of membrane fouling and modeling of solute build-up, Water Res. 47 (2013) 5232–5244.

[167] Y. Jin, H. Lee, M. Zhan, S. Hong, UV radiation pretreatment for reverse osmosis (RO) process in ultrapure water (UPW) production, Desalination 439 (2018) 138–146.

[168] P. Westerhoff, Y. Yoon, S. Snyder, E. Wert, Fate of endocrine-disruptor, pharmaceutical, and personal care product chemicals during simulated drinking water treatment processes, Environ. Sci. Technol. 39 (2005) 6649–6663.

[169] N. Bolong, A.F. Ismail, M.R. Salim, T. Matsuura, A review of the effects of emerging contaminants in wastewater and options for their removal, Desalination 239 (2009) 229–246.

[170] N.T. Hancock, P. Xu, D.M. Heil, C. Bellona, T.Y. Cath, Comprehensive bench- and pilot-scale investigation of trace organic compounds rejection by forward osmosis, Environ. Sci. Technol. 45 (2011) 8483–8490.

[171] B.D. Coday, B.G.M. Yaffe, P. Xu, T.Y. Cath, Rejection of trace organic compounds by forward osmosis membranes: a literature review, Environ. Sci. Technol. 48 (2014) 3612–3624.

[172] M. Yasukawa, T. Suzuki, M. Higa, Chapter 1—Salinity gradient processes: thermodynamics, applications, and future prospects, in: S. Sarp, N. Hilal (Eds.), Membrane-Based Salinity Gradient Processes for Water Treatment and Power Generation, Elsevier, 2018, pp. 3–56.

[173] S. Phuntsho, H.K. Shon, S. Hong, S. Lee, S. Vigneswaran, A novel low energy fertilizer driven forward osmosis desalination for direct fertigation: evaluating the performance of fertilizer draw solutions, J. Membr. Sci. 375 (2011) 172–181.

[174] J. Wang, Z. Yang, K. Chen, D. Zhou, Practices of detecting and removing nuisance alarms for alarm overloading in thermal power plants, Control. Eng. Pract. 67 (2017) 21–30.

[175] S. Zhao, L. Zou, D. Mulcahy, Effects of membrane orientation on process performance in forward osmosis applications, J. Membr. Sci. 382 (2011) 308–315.

[176] G.T. Gray, J.R. McCutcheon, M. Elimelech, Internal concentration polarization in forward osmosis: role of membrane orientation, Desalination 197 (2006) 1–8.

[177] Z.F. Cui, Y. Jiang, R.W. Field, Chapter 1—Fundamentals of pressure-driven membrane separation processes, in: Z.F. Cui, H.S. Muralidhara (Eds.), Membrane Technology, Butterworth-Heinemann, Oxford, 2010, pp. 1–18.

[178] V. Yangali-Quintanilla, Z. Li, R. Valladares, Q. Li, G. Amy, Indirect desalination of Red Sea water with forward osmosis and low pressure reverse osmosis for water reuse, Desalination 280 (2011) 160–166.

[179] T.Y. Cath, N.T. Hancock, C.D. Lundin, C. Hoppe-Jones, J.E. Drewes, A multi-barrier osmotic dilution process for simultaneous desalination and purification of impaired water, J. Membr. Sci. 362 (2010) 417–426.

[180] N.T. Hau, S.-S. Chen, N.C. Nguyen, K.Z. Huang, H.H. Ngo, W. Guo, Exploration of EDTA sodium salt as novel draw solution in forward osmosis process for dewatering of high nutrient sludge, J. Membr. Sci. 455 (2014) 305–311.

[181] Q. Ge, F. Fu, T.-S. Chung, Ferric and cobaltous hydroacid complexes for forward osmosis (FO) processes, Water Res. 58 (2014) 230–238.

[182] R.F. Rafique, Z. Min, G. Son, S.H. Lee, Removal of cadmium ion using micellar-enhanced ultrafiltration (MEUF) and activated carbon fiber (ACF) hybrid processes: adsorption isotherm study for micelle onto ACF, Desalin. Water Treat. 57 (2016) 7780–7788.

[183] H. Lee, J. Lim, M. Zhan, S. Hong, UV-LED/PMS preoxidation to control fouling caused by harmful marine algae in the UF pretreatment of seawater desalination, Desalination 467 (2019) 219–228.

[184] T. Alejo, M. Arruebo, V. Carcelen, V.M. Monsalvo, V. Sebastian, Advances in draw solutes for forward osmosis: hybrid organic-inorganic nanoparticles and conventional solutes, Chem. Eng. J. 309 (2017) 738–752.

[185] J.R. McCutcheon, R.L. McGinnis, M. Elimelech, Desalination by ammonia–carbon dioxide forward osmosis: influence of draw and feed solution concentrations on process performance, J. Membr. Sci. 278 (2006) 114–123.

[186] S. Adham, A. Hussain, J.M. Matar, R. Dores, A. Janson, Application of membrane distillation for desalting brines from thermal desalination plants, Desalination 314 (2013) 101–108.

[187] P. Wang, T.-S. Chung, A conceptual demonstration of freeze desalination–membrane distillation (FD–MD) hybrid desalination process utilizing liquefied natural gas (LNG) cold energy, Water Res. 46 (2012) 4037–4052.

7

Principle and theoretical background of pressure-retarded osmosis process

Sarper Sarp[1], Daniel Johnson[2], Nidal Hilal[2], Wafa Suwaileh[3]

[1]CENTRE FOR WATER ADVANCED TECHNOLOGIES AND ENVIRONMENTAL RESEARCH (CWATER), COLLEGE OF ENGINEERING, SWANSEA UNIVERSITY, SWANSEA, UNITED KINGDOM [2]NYUAD WATER RESEARCH CENTER, NEW YORK UNIVERSITY ABU DHABI, ABU DHABI, UNITED ARAB EMIRATES [3]RESEARCH AND DEVELOPMENT, QATAR FOUNDATION, DOHA, QATAR

7.1 Introduction

Fossil fuels are the primary energy source responsible for the emission of gases leading to anthropogenic climate change [1]. As a result, it is necessary to explore green power sources [2]. Recently, solar and wind energies are being increasingly used in many countries. One emerging green energy source, pressure-retarded osmosis (PRO), has gained much recent interest in both academia and industrial development [3]. PRO uses a semipermeable membrane to control the osmotic mixing to generate renewable osmotic power from a salinity gradient. The main goal of a PRO system is to economically provide an appropriate amount of green energy. Furthermore, one of the key factors for a successful PRO system is the selection of a suitable draw solution that possesses a greater osmotic pressure than that of the feed solute to drag water through the membrane. Therefore quantifying the osmotic pressure and the extractable mixing energy is required. In this chapter, we discuss the general principles of PRO, before addressing modeling of the osmotic pressure associated with the osmotic power generation.

7.2 Theory and modeling of osmotic pressure

The PRO process depends on the flow of water from a feed with low solute concentration through a membrane module that separates it from a pressurized concentrated solution [1,4], with the concentration gradient providing the driving force (Fig. 7−1). This causes the dilution of the draw solution and concentration of the feed solution. To harness energy, the passage of water within the membrane is retarded by introducing hydraulic pressure.

Osmosis Engineering. DOI: https://doi.org/10.1016/B978-0-12-821016-1.00004-8

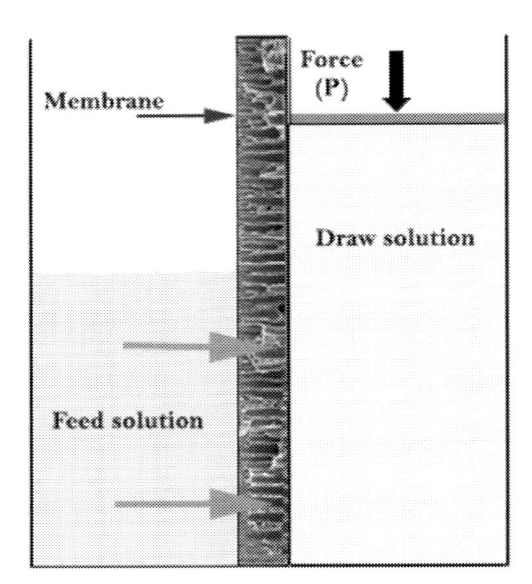

FIGURE 7–1 Illustrating the concept of PRO process. The driving force is the osmotic pressure difference between the feed solution and the draw solution which draws pure water from the feed side to the draw solution side across a semipermeable membrane. PRO describes the area between osmosis and osmotic equilibrium in which $\Delta P < \Delta\pi$. *PRO*, Pressure-retarded osmosis.

As the energy that can be extracted, it is important to accurately estimate the osmotic pressure of the relevant solutions. As a result, several models have been developed for determining the solution osmotic pressure to allow estimation of the energy that can be recovered from solutions using PRO.

7.2.1 Pitzer model for osmotic pressure

Pitzer parameters have been found to be useful for estimating solution osmotic pressure [5]. To overcome the limitations of van't Hoff equation, which is only strictly valid for ideal solutions, the bulk osmotic pressure can be modeled by using the water activity (Eq. 7.14). The molar volume and water activity can be quantified according to Pitzer equations for high salinity solutions. The water activity (a_w) can be calculated by the following [5]:

$$a_w = \exp\left(-0.01802\phi \sum_i M_i\right) \tag{7.1}$$

where M_i is the molality of the solute in moles of solute per kg of solvent. For a solution containing two different ions, M and X, the osmotic coefficient, Φ, can be modeled using the following equation [5]:

$$\Phi - 1 = Z_m Z_x F + 2m\left(\frac{\nu_m \nu_x}{\upsilon}\right)B_{mx} + 2m^2\left[\frac{(\nu_m \nu_m)^2}{\upsilon}\right]C_{mx} \tag{7.2}$$

where Z_x is the charge of X ions while Z_m is the charge of M ions, and υ_x and υ_m describe the respective stoichiometric coefficients of the ions. The constant, F, can be calculated from [5]:

$$F = -\frac{0.3921^{0.5}}{1 + 1.2I^{0.5}} \tag{7.3}$$

where the parameter I reflects the distance at which the electrostatic energy for singly charged ions is equivalent to the thermal energy and is given by [5]:

$$I = 0.5\sum_i m_i z_i^2 \tag{7.4}$$

The charges can be quantified as [5]:

$$|Z_m Z_x| = \frac{\sum_i m_i z_i^2}{\sum_i m_i} \tag{7.5}$$

The Pitzer constants for each solute are expressed by B_{mx} (0), B_{mx} (1), and C_{mx}. These constants at given absolute temperature, T, and diffusion coefficient, D, are analogous to virial coefficients and refer to the net effect of various short-range forces between the M and X ions [6]. There is no difference between straightforward interactions of solute ions at short distances and changes in solvation with concentration. This is because both phenomena may impact the effective interionic potential of average force leading to the second and third virial coefficients [6]. It can be calculated using the following [5]:

$$B_{mx} = B_{mx}(0) + B_{mx}\exp(-2.0I^{0.5}) \tag{7.6}$$

The basic Pitzer expression is applicable to determine the Gibbs energy of an individual electrolyte (G^E) as follows [7]:

$$\frac{G^E}{w_s RT} = f(I) + 2m_M m_X[B_{MX} + Z_M m_M C_{MX}] \tag{7.7}$$

where w_s and $f(I)$ are, respectively, the solvent weight and the ionic strength function form of the Debye–Hückel expression. B_{MX}, Z_M, m, and C_{MX} describe second virial-type coefficient, cation valence, molality, and third virial-type coefficient, respectively. Subscripts M, X, and MX refer to cation, anion, and neutral electrolyte, respectively. Next, by taking into account the description of the activity coefficient, the mean activity coefficient can be quantified as [7]:

$$\ln\gamma_{MX}^{\pm} = \left(\gamma_M^{\upsilon M}\gamma_X^{\upsilon X}\right)^{1/\upsilon_{MX}} = |Z_M Z_X| f^{\gamma} + \left(\frac{2\upsilon_M\upsilon_X}{\upsilon_{MX}}\right) m_{MX} B_{MX}^{\gamma} + \frac{2(\upsilon_M\upsilon_X)^{3/2}}{\upsilon_{MX}} {m_{MX}}^2 C_{MX}^{\gamma} \tag{7.8}$$

where the functions $f(I)$, $f\phi$, and $f\gamma$ reflect the electrostatic interactions and can be calculated using the following equations:

$$f(I) = -\frac{4A_{\varnothing}I}{b_{MX}}\ln(1 + b_{MX}I^{1/2}) \tag{7.9}$$

$$f^{\varnothing} = \frac{4A_{\varnothing}I^{1/2}}{1 + b_{MX}I^{1/2}} \tag{7.10}$$

$$f^{\gamma} = -A_{\varnothing}\left(\frac{I^{1/2}}{1 + b_{MX}I^{1/2}} + \frac{2}{b_{MX}}\ln(1 + b_{MX}I^{1/2})\right) \tag{7.11}$$

To express the second virial coefficient, B_{MX} and its derivatives with respect to the ionic strength can be written as:

$$B_{MX}(I) = \beta_{MX}^{(0)} \frac{2\beta_{MX}^{(1)}}{\alpha_1^2 I}[1 - e^{-\alpha I^{1/2}(1+\alpha I^{1/2})}] + \frac{2\beta_{MX}^{(2)}}{\alpha_1^2 I}[1 - e^{-\alpha_2 I^{1/2}(1+\alpha_2 I^{1/2})}] \tag{7.12}$$

$$B_{MX}^{\varnothing}(I) = \beta_{MX}^{(0)} + \beta_{MX^{e}}^{(1)} + \beta_{MX^{e^-}}^{(2)}\alpha_2 I^{1/2} \tag{7.13}$$

$$fB_{MX}^{\gamma}(I) = 2\beta_{MX}^{(0)} + \frac{2\beta_{MX}^{(1)}}{\alpha_2 I}\left[1 - \frac{e^{-\alpha I^{1/2}}}{2}(1 + \alpha_2 I^{1/2} - \alpha_2{}^2 I)\right] + \frac{2\beta_{MX}^{(2)}}{\alpha_2 I}\left[1 - \left(\frac{e}{2}\right)^{-\alpha I^{1/2}}(1 + \alpha_2 I^{1/2} - \alpha_2{}^2 I)\right] \tag{7.14}$$

The third virial coefficient with respect to the ionic strength can be modeled as:

$$C_{MX}(I) = C_{MX}^{(0)} \frac{4C_{MX}^{(1)}}{\alpha_3^4 I^2}\left[6 - e^{-\alpha I^{1/2}}\left(6 + 6\alpha_3 I^{1/2} + 3\alpha_3^2 I + \alpha_3^3 I^{3/2}\right)\right] \tag{7.15}$$

$$C_{MX}^{\varnothing}(I) = 2|Z_M Z_X|^{1/2}\left[C_{MX}^{(0)} + C_{MX}^{(1)} e^{-\alpha_3 I^{1/2}}\right] \tag{7.16}$$

$$C_{MX}^{\gamma}(I) = 3|Z_M Z_X|^{1/2}\left\{C_{MX}^{(0)} \frac{4C_{MX}^{(1)}}{\alpha_3^4 I^2}\left[6 - e^{-\alpha_3 I^{1/2}}\left(6 + 6e^{-\alpha_3 I^{1/2}} + 3\alpha_3^2 I + \alpha_3^3 I^{3/2} - \frac{\alpha_3^2 I^2}{2}\right)\right]\right\} \tag{7.17}$$

Recently, the elementary Pitzer model has been modified to calculate the activity of solutions with high ionic strength [8]. This model is capable of measuring the activity coefficients for a wide range of solution salinities.

7.2.2 Van Laar's model for osmotic pressure

To avoid the complexity of the previous models, osmotic pressure values can be derived from air humidity osmometry using the water activity equation based on the relative humidity data of organic electrolytes [9]. When we consider the osmotic behavior of nonideal concentrated solutions, the water activity and relative humidity can be quantified using Eq. (7.14).

It was observed that the behavior of the osmotic pressure became higher upon increasing the solution molality [9]. Using this model, a good prediction of the osmotic pressure was obtained which helped in selecting a draw solution with the highest osmotic pressure for forward osmosis (FO) system.

7.2.3 Water and solute activities

To calculate the activity coefficient of water, feed, and draw solutions with high ionic strength, the molal-based Pitzer equation can be utilized. Subsequently, the activity coefficient of ions can be expressed based on the first derivative of the osmotic coefficient as follows [8]:

$$\begin{aligned}\ln(\gamma_{\pm}) = &-|Z_M Z_X|A_{\varnothing}\left[\frac{\sqrt{I}}{1+b\sqrt{I}}+\frac{2}{b}\ln(1+b\sqrt{I})\right]\\ &+mo_s\cdot\frac{2v_M v_X}{v}\left\{2\beta^0_{MX}+2\frac{\beta^1_{MX}}{\alpha^2 I}\left[1-\left(1+\alpha\sqrt{I}-\frac{\alpha^2 I}{2}\right)^{e^{-\alpha\sqrt{I}}}\right]\right\}\\ &+\frac{3}{2}mo_s^2\left[\frac{2(v_M v_X)^{3/2}}{v}C^{\varphi}_{MX}\right]\end{aligned} \tag{7.18}$$

It is worth noting that the sign " $\pm$ " indicates the activity coefficient for the oppositely charged ions. By incorporating the charges of ions term ($|Z_m Z_X|$) and Pitzer constants (B_{mx}, B_{mx}, and C_{mx}) into Eq. (7.2), the final model for calculating the water activity based on the osmotic coefficient can be written as:

$$\varphi - 1 = -|Z_m Z_X|A_{\varphi} + mo_s\cdot\frac{2V_M V_X}{V}\left(\beta^0_{MX}+\beta^0_{MXe^{-\alpha\sqrt{I}}}\right)+mo_s^2\frac{2(V_M V_X)^{3/2}}{V}C^{\varphi}_{MX} \tag{7.19}$$

$$\ln(\alpha_w) = -\varphi\cdot v\cdot\frac{n_s}{n_w} \tag{7.20}$$

where the electrolyte (MX) includes species of v_M and v_X with charges of Z_M and Z_X. Both b and α have constant values of 1.2 and 2.0 $kg^{1/2}/mol^{1/2}$ for different electrolytes. β^0, β^1, and C^{φ} are temperature and pressure dependent and specific for each ion. I and A_{φ} are defined as the ionic strength and the Debye–Hückel slope for the osmotic coefficient.

The water activity coefficient (γ) integrating the Gibbs–Duhem theorem can be used to determine the solute activities for binary solutions:

$$\ln\gamma = \int_0^m\left(\frac{\varphi-1}{m}\right)dm+\varphi-1 \tag{7.21}$$

By combining the water activity with the solute molality, the osmotic coefficient (φ) can be calculated by:

$$\varphi = -55.5\ln\frac{\alpha_w}{m} \tag{7.21}$$

7.2.4 Newton–Raphson method for osmotic pressure

Additional thermodynamic models to determine the osmotic pressure have also been proposed by Yokozeki [10] who established a new equation based on the Newton–Raphson method using an entrepreneurial operating system (EOS). It is a helpful approach to explain the thermodynamic behavior of a real compound that depends on a geometrical factor [11]. Here, the basic equation of an expanded van der Waals formula for liquids was written as:

$$P = \frac{RT}{V-b}\left(\frac{V-d}{V-c}\right)^k - \left(\frac{a}{V^2 + qbV + rb^2}\right) \tag{7.22}$$

where P and V denote the pressure and the volume, respectively, b is the effective molar excluded volume, equal to $(2\pi/3)\alpha^3$, where α describes the hard-sphere diameter. The first part of this equation was used to directly solve the osmotic pressure for sucrose, polyethylene glycol, bovine serum albumin with salt, and hen–egg-white lysozyme with salt solutions, which were expressed as:

$$P = \frac{RT}{V-b} \tag{7.23}$$

By a large number of steps, Eq. (7.23) was reduced to the final model to calculate the osmotic pressure that can be given by:

$$\pi = \frac{x_2^0 RT}{V_0 - b} \approx \frac{x_2^0 RT}{V_0} \approx \frac{x_2^0}{V_0} RT \tag{7.24}$$

It is clear that this final model represents the general van't Hoff equation. Although the EOS method facilitated deriving osmotic equilibria and osmotic pressure model for liquid solutions, it had some limitations restricting its usage [10,11], namely, the complexity of the process.

7.3 Osmotic power generation

One of the methods to conceptually understand the osmotic gradient process is through the osmotic pressure of the solutions. To harness green power from a PRO system a low concentration feedwater is drawn across a semipermeable membrane to a pressurized high concentration feed. Since the pressurized concentrated solution gains the potential energy, a generator or hydroturbine can be used to convert this energy into a mechanical energy which can be also converted to electricity [2]. For instance, the energy that can be generated using a system of river water and seawater with high salinity is approximately 1 kWh/m^3. This means that this energy is equal to that produced from a waterfall with a height of over 200 m. Several theoretical models are available in the literature to quantify the osmotic power [12,13]. Both the specific energy (SE) and power density are effective ways of normalizing the extractable power of the PRO process ($\Delta Q\Delta P$). This SE is equal to power output divided by the sum of the initial feed flow rate $Q_{F,0}$ and initial draw flow rate $Q_{D,0}$.

$$\mathrm{SE} = \frac{\Delta P \Delta Q}{Q_{F,0} + Q_{D,0}} \tag{7.25}$$

where ΔP and ΔQ are the hydraulic pressure difference and the volumetric flow rate through the membrane, respectively. Since PRO is a near isobaric process, the pressure of the draw solution remains constant during operation. So, the PRO system is able to transform the osmotic pressure to hydraulic pressure when the water flux flows from the feed to the draw solution at almost constant pressure. In this respect, the increase of the water flux volume leads to extracting a large amount of energy from the system. It should be noted that the energy that can be obtained from this reversible process is identical to the Gibbs free energy of mixing, ΔG [14]. Many modeling approaches have been proposed to estimate the osmotic energy capable of being extracted from a particular PRO system.

7.3.1 Van't Hoff model for Gibbs free energy

Generally, the osmotic power depends on the change in the free Gibbs energy. A simple equation can be used to predict this energy ΔG_{mix} [12]:

$$\Delta G_{mix} = RT \ln(x_w) \tag{7.26}$$

where x_w is the mole fraction of water in the draw solution, most often seawater. Another equation has been developed including the salt activity of a hypersaline water mixture, the molal concentration (m_o), and a mass-based mixing ratio (φ_m). Typically, the osmotic pressure of concentrated draw solution produced from a PRO process can be transformed into hydraulic pressure by using the Gibbs free energy and volumetric expansion [13]. The primary formula of Gibbs free energy of mixing (ΔG_{mix}) of two different salinity solutions in an ideal system can be written as [13]:

$$-\mathrm{d}(\Delta G_{mix}) = -RT \ln a_s dn_s = \pi_{ss} \overline{V}_s dn_s \tag{7.27}$$

where T, R, a_s, n_s, π_{ss}, and $\overline{V}_s$ describe the absolute temperature, the gas constant, the activity coefficient of the product solution, the number of moles of the product solution, the osmotic pressure of the salty solution, and the molar volume of the final solution, respectively. The Gibbs free energy can be used to calculate this increase in the energy as shown in the following [13,15]:

$$-\frac{\Delta G_{mix,V^o_{LC}}}{\upsilon RT} \approx \frac{C_M}{\phi} \ln C_M - C_{LC} \ln C_{LC} - \frac{1-\phi}{\phi} C_{HC} \ln C_{HC} \tag{7.28}$$

where υ, c, and V^o represent the number of dissociated ions for a single electrolyte molecule, the molar salt concentration, and the specific volume at the reference rate, respectively, whereas LC, HC, M, and Φ, respectively, indicate low concentration, high concentration, mixed solution, and the ratio of the total moles in a solution.

Recent modeling work suggests that the extractable energy from PRO is equal to the molar Gibbs free energy of mixing [14,16]. It quantifies how much energy is released per mole from mixing two solutions having different salt concentrations in an isothermal and isobaric process. A simple expression to compute this energy is expressed as:

$$\frac{-\Delta G_{mix}}{RT} = \sum_i X_{i,M} \ln(\gamma_{i,M} X_{i,M}) - \frac{n_F}{n_M} \sum_i X_{i,F} \ln(\gamma_{i,F} X_{i,F}) - \frac{n_D}{n_M} \sum_i X_{i,D} \ln(\gamma_{i,D} X_{i,D}) \tag{7.29}$$

where n_M, n_F, and n_D are the moles of ions in the mixed and separate feed and draw solutions respectively, while the mole fractions of ions (i) in the solutions are described by $X_{i,M}$, $X_{i,F}$, and $X_{i,D}$. The activity coefficients of the ions (i) are $\gamma_{i,M}$, $\gamma_{i,F}$, and $\gamma_{i,D}$. An expanded model has been developed for saline solutions with different mixing ratios [12]. In this model, it is presumed that both the solutions are homogenous and the volume remained constant. The activity of solute (γ_s) was 1, so could be ignored. The new model is described by the following relation [13,17]:

$$\frac{-\Delta G_{mix,V_A}}{vRT} \approx \frac{C_{s,M}}{\varphi} \ln(\gamma_{s,M} C_{s,M}) - C_{s,A} \ln(\gamma_{s,A} C_{s,A}) - \frac{1-\varnothing}{\varnothing} \cdot C_{s,B} \ln(\gamma_{s,B} C_{s,B}) \tag{7.30}$$

where v, c, and ϕ are the dissociation constant which equals 2 for NaCl solution, the molar concentration, and the volumetric mixing ratio. The subscripts M, A, and B denote the mixed fluid, the freshwater feed, the draw solution, and the mass fraction of water, respectively. However, this model is not applicable for hypersaline solutions and therefore the model was modified to accurately predict the mixing energy for hypersaline solutions.

Here, the molal concentrations, M_0, and a mass-based mixing ratio, ϕ_m, were incorporated into the model, which can be expressed as follows [12]:

$$\begin{aligned}\frac{-\Delta G_{mix,M_A}}{RT} = {} & \frac{Mo_{w,M}}{\varphi} \ln(\gamma_{w,M} X_{w,M}) + \frac{v \cdot mo_{s,M}}{\varphi} C_{s,A} \ln(\gamma_{s,M} X_{s,M}) - mo_{w,A} \ln(\gamma_{w,A} X_{w,A}) \\ & - v \cdot mo_{s,A} \ln(\gamma_{s,A} X_{s,A}) - \frac{mo_{w,B \cdot (1-\varphi)}}{\varphi m} \ln(\gamma_{w,B} X_{w,B}) \\ & - \frac{v \cdot mo_{s,B \cdot (1-\varphi_m)}}{\varphi m} \ln(\gamma_{s,B} X_{s,B})\end{aligned} \tag{7.31}$$

where s, v, c, γ, and ϕ are the salt, the dissociation constant, the molar concentration, the activity coefficient of ions, and the volumetric mixing ratio of the NaCl solution, respectively. It should be noted that this model assumes that increasing the concentration of NaCl solution results in higher energy production. Fig. 7–2 shows the mixing energy generated from a highly concentrated solution per cubic meter. It was observed that there was a strong correlation between the salinity of the draw solution and the free energy. When using a draw solution with higher salinity, an increase in the free energy was noticeable. In contrast, the above model is not practical due to difficulty in solving the activity constants.

By assuming that the solute does not change the volume of solution and that the activity is equal to the mole fraction, that is, solutions behave ideally, the ratio of the total moles in the feed

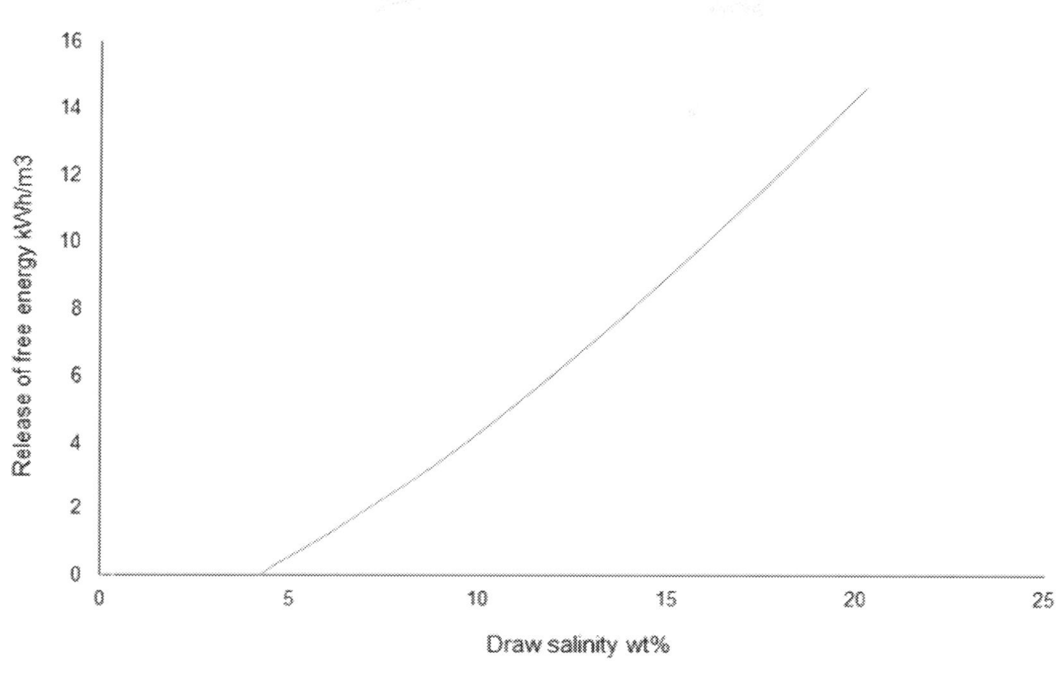

FIGURE 7–2 Showing a linear relationship between the concentration of the draw solution and the mixing energy produced per cubic meter. *Adapted with permission from N. Bajraktari, C. Hélix-Nielsen, H.T. Madsen, Pressure retarded osmosis from hypersaline sources—a review, Desalination 413 (2017) 65–85.*

and draw solutions to the total moles in the mixed solution is identical to the volume fractions ($n_F/n_M \approx \varphi$, $n_D/n_M \approx 1-\varphi$), the Gibbs free energy of mixing can be described by [14]:

$$\frac{-\Delta G_{Vm}}{\upsilon RT} = C_M \ln(C_M) - \varphi C_F \ln(C_F) - (1-\varphi)C_D \ln C_D \tag{7.32}$$

where υ is the van't Hoff factor for strong electrolytes; ΔG_{Vm} is the specific mixing energy per volume of the total mixture that varies according to the molar concentration of solutions; and C_M, C_F, and C_D are the molar concentrations of the feed solution, the draw solution, and the mixture, respectively. Because the maximum Gibbs free energy of mixing depends on the initial concentration of the feed and draw solutions, an expanded equation for calculating this energy can be expressed as:

$$\frac{-\Delta G_{max}}{\upsilon RT} = \frac{C_D C_F}{C_D - C_F}(\ln(C_D) - \ln(C_F)) - \exp\left(\frac{C_D \ln(C_D) - C_F \ln(C_F)}{C_D - C_F} - 1\right) \tag{7.33}$$

where c_M, c_F, and c_D describe the molar concentration of a mixture, the feed, and the draw solution, respectively. Straub et al. reported that the extractable SE from a seawater draw solution is 0.26 kWh/m^3, while it was 2.52 kWh/m^3 for concentrated draw solution from the Dead Sea [14]. This means that the SE was higher upon increasing the salinity of the draw solution. By rewriting the molal-based Pitzer expression, including the activity coefficient of ions (γ) in a hypersaline solution, the expanded model can become [18]:

$$\log(\gamma\pm) = -|Z_m Z_X| A_\varphi \left[\frac{\sqrt{I}}{1-b\sqrt{I}} + \frac{2}{b}\ln(1+b\sqrt{I})\right] + mo_s \cdot \frac{2V_M V_X}{V}\left\{2\beta^0_{MX} + 2\frac{\beta^1_{MX}}{\alpha^2 I}\left[1-\left(1+\alpha\sqrt{I}-\frac{\alpha^2 I}{2}\right)^{e^{-\alpha\sqrt{I}}}\right]\right\} + \frac{3}{2}mo_s\left[\frac{2(V_M V_X)^{3/2}}{V}C^{\varphi}_{MX}\right] \tag{7.34}$$

where A_φ and mo_s describe the Debye–Hückel slope for the osmotic coefficient and the molal concentration of the salt solution. The constants b and α are equal to 1.2 and 2.0 $kg^{1/2}/mol^{1/2}$, respectively, for all 1–1, 2–1, and 3–1 ion pairs. n_M and n_X refer to the number of the cation and anion in a solution while Z_M and Z_X are their charges, b_0, b_1, and C_j describe the temperature and pressure constants for the MX ion pair. β^0 and β^1 are the temperature and the pressure of each species, respectively.

Another study highlighted that the extractable mixing energy from concentrated fertilizers can be calculated as a function of the osmotic pressure (π) of solutions. By assuming validity of the van't Hoff equation (i.e., an ideal solution) and using Eq. (7.35), $d(\Delta G)/d\phi = 0$ was solved to give the following maximum obtainable SE [19]:

$$\Delta G_{V,max} = \frac{\pi_D \pi_F}{\pi_D - \pi_F}(\ln(\pi_D) - \ln(\pi_F)) - \exp\left(\frac{\pi_D \ln(\pi_D) - \pi_F \ln(\pi_F)}{\pi_D - \pi_F} - 1\right) \tag{7.35}$$

7.3.2 Piston model for Gibbs energy and energy density

Wilson and Stewart used Morse's modified van't Hoff equation, which uses molality instead of molarity, to predict the mixing energy [20]. In comparison with the van't Hoff equation which is workable only for diluted solution, Morse equation can be applied for relatively concentrated solutions [21]. At the beginning of the PRO process, it is assumed that the hydraulic pressure is equivalent to the osmotic pressure [14,22]. It is known that the hydraulic pressure is introduced to the concentrated solution resulting in an infinitesimal amount of permeable water. In all expressions the water (ΔV) transported is correlated to the difference in properties between the initial concentrated (V_{HI}) and dilute (V_{LI}) feed solutions.

According to the Morse equation, there is a good correlation between the osmotic pressure and concentration (molality) of various draw solutions used in the PRO system (see Fig. 7–3) [20]. Another important parameter is the initial mass fraction of water, W_{HI}. The elementary equation of the piston model in equilibrium condition involving freshwater feed and as a function of hydraulic pressure can be written as [22]:

$$P_{eq} = \frac{\pi_{HI}}{(V_{HI} + \Delta V / w_{HI})/V_{HI}} \tag{7.36}$$

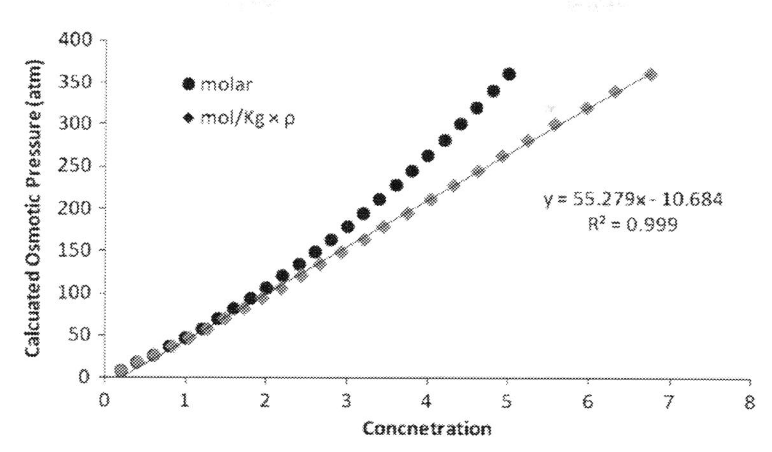

FIGURE 7–3 Correlation between molarity and molality density product (mol/kg × ρ) versus calculated osmotic pressure. *Adapted with permission from A.D. Wilson, F.F. Stewart, Deriving osmotic pressures of draw solutes used in osmotically driven membrane processes, J. Membr. Sci. 431 (2013) 205–211.*

If this equation is integrated as a function of the volume transported, we obtain the work (W) which is created by the reversible piston PRO process. This new formula represents the ideal energy ($\Delta V > 0$) and the minimum energy needed to treat a volume of water producing highly pure water ($\Delta V < 0$). W can be expressed as [22]:

$$W = \int_0^{\Delta V} \frac{\pi_{HI}}{((V_{HI} + \Delta V / w_{HI})/V_{HI})} d\Delta V = V_{HI}\pi_{HI}w_{HI} \ln\left(1 + \frac{\Delta V}{V_{HI}w_{HI}}\right) \quad (7.37)$$

This model is simple and easy to calculate the mixing energy due to the absence of assumptions, approximations, and activity coefficients involved in the classical Gibbs free energy models [22]. It is workable for various types of solution containing different solute species. The applicability of this model can be reduced if the saline solution reaches supersaturation because of the underlying osmotic correlations. In this study, it was reported that the PRO process can be operated using cocurrent or counter-current flow. In comparison to the cocurrent mode the counter-current mode can generate the highest degree of mixing at constant pressure that depicts an ideal mass exchange system providing infinite dimension. When using feed solution with zero osmotic pressure and the hydraulic pressure is not defined, Eq. (7.36) can be expressed as:

$$P_{eq} = \frac{\pi_{HI}}{(V_{HI} + \Delta V / w_{HI})/V_{HI}} - \pi_{LI} \quad (7.38)$$

Also, W from Eq. (7.37) can be modified as:

$$W = \int_0^{\Delta V} \frac{\pi_{HI}}{((V_{HI} + \Delta V / w_{HI})/V_{HI})} \pi_{LI} d\Delta V = V_{HI}\pi_{HI}w_{HI} \ln\left(1 + \frac{\Delta V}{V_{HI}w_{HI}}\right) - \pi_{LI} d\Delta V \quad (7.39)$$

However, when the process reached an equilibrium, the ratio of inlet volume to outlet volume can be quantified using the below equation. The term "$\Delta\pi = \pi_{HI} - \pi_{LI}$" refers to the osmotic pressure gradient between the concentrated inlet and the diluted solution [22]. Both the ratios nominate the hydraulic pressure and the volumes as an independent variable and dependent equilibrium values, respectively. Some restriction was shown when varying the dilution factor and hydraulic pressure. The new equation involving the ratio of the total volume to permeated volume is given by:

$$\frac{V_{HI} + V_{LI}}{\Delta V} = \frac{\pi_{HI} w_{HI} + \pi_{HI} w_{HI} + P_{op}(w_{LI} w_{HI})}{w_{LI} w_{HI}(\Delta\pi - P_{op})} \tag{7.40}$$

When the minimal volume ratios are divided by the hydraulic pressure, the minimal inverse SE density (m^3/kWh) is established. $P_{op}\Delta V$ refers to the work done by the process. This relationship can be expressed as:

$$\frac{V_{HI} + V_{LI}}{P_{op}\Delta V} = \frac{\pi_{HI} w_{HI} + \pi_{LI} w_{LI} + P_{op}(w_{LI} w_{HI})}{w_{HI} w_{LI}(\Delta\pi - P_{op})} \tag{7.41}$$

However, the SE density equation can be changed if the draw solution becomes very diluted as compared to the solution with high salinity ($\pi_{LI} \ll \pi_{HI}$). If we expect that $w_{LI} \approx 1$, $\pi_{LI} \approx 0$, and $\Delta\pi \approx \pi_{HI}$, the final model can be reduced to:

$$\frac{P_{op}\Delta V}{V_{HI} + V_{LI}} \approx \frac{w_{HI} P_{op} + (\pi_{HI} - P_{op})}{\pi_{HI} w_{HI} + P_{op}(1 - w_{HI})} \tag{7.42}$$

It must be noted that the addition or extraction of freshwater can change the osmotic pressure. This developed osmotic pressure was predicted as applied pressure, (π_{HI}), the corresponding mass fraction (w_{HI}), and relative volume of water added $V_{LI}/\Delta V$ and $V_{HI}/\Delta V$. This is a workable model and easier than other models for estimating the Gibbs free energy. One of the drawbacks of other models is involving many dilute-solution assumptions and they are often difficult to modify [16].

In addition, if a large amount of dilute solution is utilized to further dilute the high salinity solution, the maximum energy density ($\mathcal{U}^{H}_{max}$) is obtained as the transmembrane pressure (P_{op}) reaches its minimum [22]. It can be defined as the SE extracted from a unit volume of solution that is diluted when the applied pressure is kept constant [23]. This facilitates a comparison between various draw solutions of known osmotic pressure and water mass fraction. Then, the maximum energy density can be calculated when $P_{op} = 0$ using:

$$\mathcal{U}^{H}_{max}(\mathrm{m}^3/\mathrm{kWh}) = \frac{P_{op}\Delta V}{V_{HI}} = \frac{P_{op} w_{HI}(\pi_{HI} - \pi_{LI})}{\pi_{LI}} \approx \pi_{HI} w_{HI} \tag{7.43}$$

All the terms in this model rely mainly on the initial volume of the high salinity solution V_{HI}. The energy density can also be given as a function of low osmotic pressure and hence the maximum power is exploited if $\Delta V = V_{LI}$. It approaches its peak if $P_{op} = \Delta\pi \approx \pi_{HI}$ for a very dilute solution. To achieve the highest energy density value a unit volume of dilute solution should be mixed with an infinite volume of high salinity solution. The energy density of the final dilute solution can be calculated by [22]:

$$\mathcal{U}^{L}_{P_{op}}(\mathrm{m}^3/\mathrm{kWh}) = \frac{P_{op}\Delta V}{V_{LI}} \approx P_{op}w_{LI} \approx P_{op}\frac{\Delta V}{\Delta V} \tag{7.44}$$

In brief, the estimation of energy density of a solution operated in the PRO system depends on the osmotic pressure and module applied pressure. The attainable power from counter-current PRO operation can be determined from models including mixing energy expression [22]. These models predict the mole quantity of solute needed to generate a kWh unit of power in PRO process relying on the effective performance of the system.

7.4 Dual- and multistage pressure-retarded osmosis process

A new concept of a dual-stage PRO process has been developed for power extraction [24]. In principle, after completing the first PRO operation, the pressurized seawater as a draw solution is transferred to two pathways. Most of the solution flow enters a pressure exchanger. The rest of the solution flow is equivalent to the permeate flow rate in the first stage (V1), which transfers to the second stage of the operations. Some of the merits are that various types of membrane can be utilized in the two stages, a diverse range of draw solutions can be used, the feed salinity has negligible influence on the process performance, and high extractable energy is obtained. A recent study improved the conventional dual-stage process and proposed a new design [25]. In this process the whole solution flow obtained from the first stage enters the second stage to produce drinking water. After that, part of the seawater flow at the end of the membrane treatment in the second stage goes back to the pressure exchanger to increase the solution pressure and returns to the first stage. The other part of the flow is transported to the turbine to generate natural power. The total power generation (P_{wt}, kW/day) for both classical and new systems was estimated by the following model [25]:

$$P_{wt} = \Delta P * (Q_{P1} + Q_{P2}) \tag{7.45}$$

where Q_{P1} and Q_{P2} describe the permeate flow rate (m^3/h) of the first and second stages, respectively. The permeate flow rate can be quantified from the mass balance as follows:

$$\mathrm{Q}_{pt} = Q_{f\text{-}in1} * \%R_{e1} + Q_{f\text{-}in2} * \%R_{e2} \tag{7.46}$$

where $\%R_{e1}$ and $\%R_{e2}$ denote the recovery rates of the first and second steps, respectively. $Q_{f\text{-}in1}$ and $Q_{f\text{-}in2}$ represent the flow rate of draw solution. With respect to the ratio of the draw solution flow rate ($Q_{ds\text{-}in}$) to the feed flow rate ($Q_{f\text{-}in}$), the equation can be rearranged as:

$$\mathrm{Q}_{pt} = \left(\frac{Q_{ds\text{-}in1}}{Q_{ds\text{-}in1}/Q_{f\text{-}in1}}\right)\%R_{e1} + \left(\frac{Q_{ds\text{-}in2}}{Q_{ds\text{-}in2}/Q_{f\text{-}in2}}\right)\%R_{e2} \tag{7.47}$$

where $Q_{ds\text{-}in2}$ equals the sum of permeate and flow rate of the draw solution in the first step. It can be given as:

$$Q_{ds\text{-}in2} = Q_{ds\text{-}in1} + Q_{f\text{-}in1} * \%R_{e1} \tag{7.48}$$

In the single stage system the draw solution flow rate ($Q_{ds\text{-}in1}$) was returned to the pressure exchanger and therefore the ratio of $Q_{ds\text{-}in}/Q_{f\text{-}in}$ equals 1. The formula can be written as:

$$Q_{pt} = \left(\frac{Q_{ds\text{-}in1}}{Q_{ds\text{-}in1}/Q_{f\text{-}in1}}\right)\%R_{e1} + \left(\frac{Q_{ds\text{-}in1} * \%R_{e1}}{Q_{ds\text{-}in2}/Q_{f\text{-}in2}}\right)\%R_{e2} \tag{7.49}$$

In the new design, to determine the total flow rate, the new model is derived by substituting Eq. (7.48) into (7.47):

$$Q_{pt} = \left(\frac{Q_{ds\text{-}in1}}{Q_{ds\text{-}in1}/Q_{f\text{-}in1}}\right)\%R_{e1} + \left(\frac{Q_{ds\text{-}in1} + (Q_{f\text{-}in1} * \%R_{e1})}{Q_{ds\text{-}in2}/Q_{f\text{-}in2}}\right)\%R_{e2} \tag{7.50}$$

In addition, the total permeate flow rate in both systems can be calculated if we assume $Q_{ds\text{-}in}/Q_{f\text{-}in}$ expression is equivalent to n. Then, substituting in Eq. (7.50) into (7.49) yields the final models, which can be rephrased as:

$$Q_{pt} = Q_{ds\text{-}in1}\left(\frac{\%R_{e1}}{n_1} + \frac{(1 + \%R_{e1})}{n_2}\%R_{e2}\right) \tag{7.51}$$

$$Q_{pt} = Q_{ds\text{-}in1}\left(\frac{\%R_{e1}}{n_1} + \frac{\%R_{e1}}{n_2}\%R_{e2}\right) \tag{7.52}$$

The efficiency of the new design could be much better than the traditional design as the power density can be grown by 17.4% upon increasing the concentration of the draw solution (DS) [25]. In the new process, a lower membrane area is needed which minimized the cost of the membrane. The energy expenditure was comparable as the flow rate was increased in the second stage.

Until very recently, a new design of PRO process was proposed to reduce the salinity of the diluted draw solution and get a higher amount of the pressurized solution [15]. This PRO system involves four different steps and the hydraulic pressure was adjusted in each step to obtain high water permeation and power generation. First, the net pressure difference among both the draw (D) and feed (F) solutions was determined using:

$$(\pi_D - P_D) - (\pi_F - P_F) \tag{7.53}$$

The mass balance formula for this multistage PRO (MPRO) at constant feed pressure can be written as:

$$P_{in,draw} \cdot V_{in,draw} + \Delta G_{mix} \cdot M_{in,draw} \cdot r_{mix} = P_{out,draw} \cdot V_{out,draw} + E_{loss} \tag{7.54}$$

where ΔG_{mix} reflects the theoretical Gibb's free energy of mixing. M_{in}, r_{mix}, E_{loss} are the molar flow rate of the draw solution at the inlet, the mixing ratio of both feed and draw solution, and the amount of energy cutoff, respectively. P_{in} and P_{out} refer to the hydraulic pressure of the initial and final draw solutions, respectively. V_{in} and V_{out} are defined as the volumetric flow rate of the initial and final draw solutions, respectively. The energy balance model for this MPRO with different feed pressures was improved as follows:

$$P_{in,draw} \cdot V_{in,draw} + \Delta G_{mix} \cdot M_{in,draw} \cdot r_{mix} + E_{pump} = P_{out,draw} \cdot V_{out,draw} + E_{loss} \tag{7.55}$$

These theoretical equations helped in calculating the extractable energy from PRO with different hydrostatic pressure for each step. It has been found that the MPRO system is a promising process for generating hydraulic energy from the osmotic pressure of hypersaline solution. The extractable energy increased by approximately 20% as compared to MPRO with constant hydraulic pressure. Despite that, in the MPRO, the process under variable pressures can be approximated by testing modules at minimum osmotic pressure [22]. In this respect the energy density in each stage can decrease resulting in reduced power generation per capital cost. Overall, the PRO process is one of the potential technologies for harvesting green power from salinity concentration gradients. So far, indirect alternatives to produce the osmotic energy, such as osmotic power recovery in seawater desalination, have shown increased scientific interest.

Acknowledgment

The authors would like to thank the Royal Society for funding this work through a Royal Society International Collaboration Award (IC160133).

References

[1] F. Helfer, C. Lemckert, Y.G. Anissimov, Osmotic power with pressure retarded osmosis: theory, performance and trends—a review, J. Membr. Sci. 453 (2014) 337–358.

[2] S.H. Chae, Y.M. Kim, H. Park, J. Seo, S.J. Lim, J.H. Kim, Modeling and simulation studies analyzing the pressure-retarded osmosis (PRO) and PRO-hybridized processes, Energies 12 (2019) 1–35.

[3] Z.L. Cheng, T. Chung, Mass transport of various membrane configurations in pressure retarded osmosis (PRO), J. Membr. Sci. 537 (2017) 160–176.

[4] Y. Chen, A.A. Alanezi, J. Zhou, A. Altaee, Hasan, M. Shaheed, Optimization of module pressure retarded osmosis membrane for maximum energy extraction, J. Water Process. Eng. 32 (2019) 100935.

[5] M. Khraisheh, N. Dawas, M.S. Nasser, M.J. Al-Marri, M.A. Atieh, S. Adham, et al., Osmotic pressure estimation using the Pitzer equation for forward osmosis modelling, Environ. Technol. 41 (2019). Available from: https://doi.org/10.1080/09593330.2019.1575476.

[6] K.S. Pitzer, Thermodynamics of electrolytes. I. Theoretical basis and general equations, The J. Phys. Chem. 77 (1973) 268–277.

[7] F. Perez-Villasenor, S. Carro-Sanchez, Comparison among Pitzer-type models for the osmotic and activity coefficients of strong electrolyte solutions at 298.15 K, Ind. Eng. Chem. Res. 50 (2011) 10894–10901.

[8] H.T. Madsen, S.S. Nissen, E.G. Søgaard, Theoretical framework for energy analysis of hypersaline pressure retarded osmosis, Chem. Eng. Sci. 139 (2016) 211–220.

[9] M. Hamdan, A.O. Sharif, G. Derwish, S. Al-Aibi, A. Altaee, Draw solutions for forward osmosis process: osmotic pressure of binary and ternary aqueous solutions of magnesium chloride, sodium chloride, sucrose and maltose, J. Food Eng. 155 (2015) 10–15.

[10] A. Yokozeki, Osmotic pressures studied using a simple equation-of-state and its applications, Appl. Energy 83 (2006) 15–41.

[11] A. Yokozeki, Phase behaviors of binary hard-sphere mixtures using simple analytical equations of state, Int. J. Thermophys. 25 (2004) 643–668.

[12] N. Bajraktari, C. Hélix-Nielsen, H.T. Madsen, Pressure retarded osmosis from hypersaline sources—a review, Desalination 413 (2017) 65–85.

[13] S. Sarp, Z. Li, J. Saththasivam, Pressure retarded osmosis (PRO): past experiences, current developments, and future prospects, Desalination 389 (2016) 2–14.

[14] A.P. Straub, A. Deshmukh, M. Elimelech, Pressure-retarded osmosis for power generation from salinity gradients: is it viable? Energy Environ. Sci. 9 (2015) 31–48.

[15] S. Sarp, N. Hilal, Thermodynamic optimization of multistage pressure retarded osmosis (MPRO) with variable feed pressures for hypersaline solutions, Desalination 477 (2020) 114–245.

[16] S. Lin, A.P. Straub, M. Elimelech, Thermodynamic limits of extractable energy by pressure retarded osmosis, Energy Environ. Sci. 7 (2014) 2706–2714.

[17] N.Y. Yip, M. Elimelech, Thermodynamic and energy efficiency analysis of power generation from natural salinity gradients by pressure retarded osmosis, Environ. Sci. Technol. 46 (2012) 5230–5239.

[18] H. Sharifan, H.T. Madsen, A. Morse, High performance in power generation by pressure-retarded osmosis (PRO) from hypersalinity gradient: case study of hypersaline Lake of Urmia, Iran, Desalination Water Treat. 71 (2017) 302–311.

[19] F. Volpin, R.R. Gonzales, S. Lim, N. Pathak, S. Phuntsho, H.K. Shon, et al., A novel fertiliser-driven osmotic power generation process for fertigation, Desalination 447 (2018) 158–166.

[20] A.D. Wilson, F.F. Stewart, Deriving osmotic pressures of draw solutes used in osmotically driven membrane processes, J. Membr. Sci. 431 (2013) 205–211.

[21] Y. Hartanto, S. Yun, B. Jin, S. Sheng, Functionalized thermo-responsive microgels for high performance forward osmosis desalination, Water Res. 70 (2015) 385–393.

[22] K.K. Reimund, J.R. McCutcheon, A.D. Wilson, Thermodynamic analysis of energy density in pressure retarded osmosis: the impact of solution volumes and costs, J. Membr. Sci. 487 (2015) 240–248.

[23] B.E. Logan, M. Elimelech, Membrane-based processes for sustainable power generation using water, Nature 488 (2012) 313–319.

[24] A. Altaee, N. Hilal, Dual-stage forward osmosis/pressure retarded osmosis process for hypersaline solutions and fracking wastewater treatment, Desalination 350 (2014) 79–85.

[25] A. Altaee, N. Hilal, Design optimization of high performance dual stage pressure retarded osmosis, Desalination 355 (2015) 217–224.

8

Application of PRO process for seawater and wastewater treatment: assessment of membrane performance

Wafa Suwaileh[1], Daniel Johnson[2], Nidal Hilal[2]

[1]RESEARCH AND DEVELOPMENT, QATAR FOUNDATION, DOHA, QATAR [2]NYUAD WATER RESEARCH CENTER, NEW YORK UNIVERSITY ABU DHABI, ABU DHABI, UNITED ARAB EMIRATES

8.1 Introduction

A major challenge is the provision of both sustainable energy and freshwater supplies. A potential direction is in generating green power from mixing solutions with different salinities. The freshwater runoff has been estimated at a rate of 1.1×10^6 m^3/s, and the amount of power that can be exploited in the form of a salinity gradient is around 2.6×10^{12} W [1]. A promising technology for harvesting this energy is pressure-retarded osmosis (PRO). PRO is an emerging membrane separations technology, which extracts the mixing energy from the salt gradient and converts it to electricity [2]. To produce the extractable power a semipermeable membrane is used. PRO membranes should possess high selectivity producing high water permeation and very low reverse solute flux. There are many types of membranes that have been tested for the PRO process. In prior studies, flat sheet reverse osmosis (RO) membranes were the most widely used type but showed poor performance due to a thick porous support layer causing serious internal concentration polarization (CP) [3,4].

Another membrane was a flat sheet cellulosic membrane designed by Hydration Technologies Inc. (HTI). The performance was better than RO membranes. However, the water permeation and power density were low because of membrane deterioration under high pressure. After that, another configuration was developed, which can withstand higher hydraulic pressure in the PRO process. It was found that self-support hollow-fiber membranes could be used without the need for spacers, which eliminated the compaction problem. This allowed the thin-film composite (TFC) hollow-fiber membrane to produce a maximum power output

Osmosis Engineering. DOI: https://doi.org/10.1016/B978-0-12-821016-1.00012-7

of around 10.6 W/m^2 upon using synthetic seawater brine (1.0 mol/L NaCl) and wastewater brine (40.0 mmol/L NaCl) [4].

In contrast, the membrane durability should be enhanced to boost the power density at high hydraulic pressure. The spiral-wound configuration suffers significantly from biofouling governed by the presence of spacers, which reduced the membrane efficiency [5]. To further advance the PRO membranes, most studies have been focused on optimizing the fabrication conditions [6], improving antifouling properties [7], and increasing the mechanical strength [4]. At the same time, fewer works have highlighted modeling of membrane performance [8]. A breakthrough in membrane materials and membrane fabrication is required to increase water flux, lower fouling, control CP, and maximize power output in the PRO process.

Mathematical approaches are a powerful tool to compute and simulate the performance of both membrane configurations under different operating variables in the stand-alone and hybrid processes. Several models were developed to accurately determine the performance of flat sheet and hollow-fiber membranes in terms of water flux, reverse solute flux, and transport parameters for a feasibility assessment of PRO membranes. One particular drawback for such theoretical modeling is that the majority of models deal with a perfect membrane under ideal conditions. Although the stand-alone PRO system has many advantages, its efficiency is restricted due to many challenges at present. Because the selective layer is against the draw solution (DS), foulants can easily penetrate the support layer and deposit under the selective layer causing a huge drop in water flux and power density. Thus next-generation PRO technology has been explored not only as individual pilot plants but also as pretreatment and posttreatment for seawater and wastewater in hybrid systems when coupled with pressure-driven membrane processes such as RO, ultrafiltration (UF), nanofiltration (NF), and membrane distillation (MD) [9]. The main purpose is to obtain drinking water, extractable energy, minimum energy consumption, whilst reducing the disposal of brine. On the other hand, this approach involves extra pumping costs, chemical consumption, and other operating costs [10].

This chapter, therefore, aims to delve into numerical analysis relevant to membrane performance in the PRO process. The derived models for flat sheet membranes are systematically compared to those of hollow-fiber membranes. An overview of recent trends in the development of different PRO membranes and their performances in independent and hybrid systems is extensively discussed. Lastly, the main requirements and research directions to promote the feasibility of the PRO process are outlined.

8.2 Modeling pressure-retarded osmosis process

8.2.1 Water flux and extractable power

The water flow in a PRO process is from the feed side to the draw side due to the difference in the osmotic pressure between the solutions. The osmotic pressure is created due to the difference in the feed solution (FS) and the DS concentrations and hence the water flux obeys the Fick's law [11,12]. According to Fick's law, the chemical potential difference ($\Delta\mu$) between the feed and DSs is given by [11]:

$$\Delta\mu = \mu_f - \mu_d = (\Delta\pi - \Delta P)\overline{V} \tag{8.1}$$

where μ_f and μ_d are the chemical potential of the feed and DSs respectively, $\Delta\pi$ and ΔP are the osmotic pressure and the applied pressure and $\overline{V}$ is the molar volume of water. The water flux through a semipermeable membrane can be transferred from either the feed side to the draw side due to osmotic gradient or from the draw side to the feed side due to the applied hydrostatic pressure. Therefore the water flux can be measured based on the most commonly used solution–diffusion model because the membrane is nonporous. The following equation can be used to calculate the water flux [13]:

$$J_w = -D_w \frac{dc_w}{dx} \tag{8.2}$$

where D_w, c_w, and x denote the diffusion coefficient of water, the concentration and the axis perpendicular to the membrane surface respectively. By combining the chemical potential expression, the gas constant (R), and the absolute temperature (T) based on Henry's law, Eq. (8.2) can be rephrased as:

$$J_w = \frac{D_w c_w}{RT}\frac{d\mu_w}{dx} \approx \frac{D_w c_w}{RT}\frac{\Delta\mu_w}{\Delta x} \tag{8.3}$$

where $\Delta\mu_w$ is the transmembrane difference at a constant temperature. Taking into account the chemical potential definition, the transmembrane difference can be calculated using:

$$\Delta\mu_w = RT \ln \Delta\alpha_w + \overline{V}_w\Delta P \tag{8.4}$$

where α_w and $\overline{V}_w$ are the chemical activity and the partial molar volume of water. As shown in Chapter 1, Basic Principles of Osmosis and Osmotic Pressure, osmotic pressure can be derived in terms of activity, when $RT \ln \Delta\alpha_w$ is replaced by $\overline{V}_w \Delta\pi$, Eq. (1.14) becomes:

$$\Delta\mu_w = -\overline{V}_w\Delta\pi + \overline{V}_w\Delta P = \overline{V}_w(\Delta P - \Delta\pi) \tag{8.5}$$

This indicates that the molar volume of water is independent of the applied pressure. By combining Eq. (8.3) and (1.14), the final expression for quantifying the water flux J_w is given by:

$$J_w = \frac{D_w C_w \overline{V}_w}{RT\Delta x}(\Delta P - \Delta\pi) = A(\Delta P - \Delta\pi) \tag{8.6}$$

where A (L/m^2/h/bar) is the water permeability coefficient. The negative sign became positive because the osmotic pressure difference is higher than the applied hydraulic pressure. In the isothermal equilibrium, the chemical potential difference between the two solutions is dimensioned. Therefore the transmembrane pressure difference is equal to the osmotic pressure gradient ($\Delta P = \Delta\pi$) [12]. As generally accepted, the water flux transported across the membrane (J_w) is correlated to the osmotic pressure difference ($\Delta\pi$), and the applied pressure difference (ΔP). Also, the term "$D_w C_w \overline{V}_w / RT\Delta x$" is equal to the water permeability coefficient. Thus the general water flux in PRO can be expressed as [11,14,15]:

$$J_w = A(\Delta\pi - \Delta P) \tag{8.7}$$

It is noteworthy that this expression can be applied only for an ideal membrane. As the membrane is composed of three layers, the total diffusion resistance of all layers of the membrane (r) should be considered when measuring the water flux within each layer. Subsequently, the water flux in each layer ($J_{f\to d}$) is the same as there is no leakage of water molecules and this can be expressed as [11]:

$$J_{f\to d} = \frac{\Delta\pi}{RTr} \tag{8.8}$$

The water flux in the reverse direction can be modeled as:

$$J_{d\to f} = \frac{\Delta P}{RTr} \tag{8.9}$$

To calculate the total water flux, Eq. (8.6) can be written as:

$$J_w = J_{f\to d} - J_{d\to f} = \frac{1}{RTr}(\Delta\pi - \Delta P) = A(\Delta\pi - \Delta P) \tag{8.10}$$

Generally, the power generated from the PRO process depends on the transmembrane pressure of the DS and the water flux [16]. The amount of this power depends mainly on the concentration of the DS. In a closed system, when using a pretreatment step and introducing more pumping in the PRO process, the net power density harnessed is about 0.2 kWh/m^3 [8]. In the PRO operation, the power density, W (W/m^2), equation is applicable for both flat sheet and hollow-fiber membranes. Generally, it can be estimated by multiplying the water flux by the hydrostatic pressure difference [8,17,18]:

$$W = J_w \Delta P \tag{8.11}$$

By substituting Eq. (8.7) into Eq. (8.11), the following expression of power density can be derived [8,19]:

$$W = A(\Delta\pi - \Delta P)\Delta P \tag{8.12}$$

It must be mentioned that the power density indicates how much osmotic energy obtained per membrane area. Ignoring the internal CP (ICP), the power density approaches its maximum value when $\Delta P = \Delta\pi/2$ [8]. It has been found that an increase in the pressure difference yielded a low water flux and therefore the theoretical power value is equal to half the osmotic pressure gradient. The optimal ΔP can deviate from this expression due to CP and other process variables [10]. It can also be used to evaluate the performance of the system because we know the membrane surface area thereby estimating the size of an industrial system needed for a specific power production capacity [16]. The generation of power depends on the membrane type as a forward osmosis (FO) membrane was shown to have excellent energy production as compared to an NF membrane [20].

The experimental water flux produced from an actual membrane has been found to be lower than the theoretical water flux [21]. This can be attributed to the membrane deformation when a high hydraulic pressure (30 bar) was applied to the DS. This hydraulic pressure can be adjusted based on the osmotic pressure of the DS [8]. On the other hand, the PRO membrane had a small surface area that restricts the completion of the mixing process across the membrane [22]. It is of paramount importance to understand the effect of reverse solute flux on the performance of the PRO process.

8.2.2 Reverse solute flux

One of the technical hurdles for PRO is the reverse solute flux due to diffusion of salt from the higher concentration draw to the FS. The reverse solute flux influences the performance of the PRO membrane and the ultimate efficiency of power plants. First, we have to analyze the mathematical equations of the reverse solute diffusion due to its effects on CP. The PRO membrane should produce a high salt rejection, and minimal reverse solute flux and ICP [23]. It is necessary to study the reverse solute flux to understand its effect on the system dynamics and CP in the PRO membrane. The analysis starts with modeling the reverse solute flux which can then be integrated with respect to the CP into final equation of the water flux. The mass transfer of the draw solute passing through the support layer at each boundary layer is equivalent to the sum of the salinity displacement. This occurs due to the convection and diffusion transport of the solute within the support layer governed by the concentration gradient [14]. So, the reverse solute flux of both flat sheet and hollow-fiber membranes can be also given by Fick's law [12,14,24,25]:

$$D_{s.l}\frac{dC(x)}{dx} - J_w C(x) = J_s \tag{8.13}$$

where $C(x)$ is the solute concentration at area x. $D_{s.l}$ denotes the diffusion coefficient at the sublayer which can be quantified using:

$$D_{s.l} = \frac{\varepsilon D}{\tau} \tag{8.14}$$

where D, ε, and τ denote the diffusion coefficient, the porosity, and the tortuosity of the support layer respectively. However, the water and solute fluxes in hollow-fiber membranes depend on the radial coordinate (r) with forward flow because the flow area varies across the radial position. Therefore Eq. (8.13) can be rearranged considering linear water and solute flux terms $\xi_w = 2\pi r J_w\ (r)$ and $\xi_s = 2\pi r J_s\ (r)$. For internal-selective hollow-fiber with minus direction and for an external-selective hollow-fiber with positive direction, Eq. (8.13) becomes [25]:

$$\xi_s = \pm 2\pi r D_s \frac{dC(r)}{dr} - \xi_w C(r) \tag{8.15}$$

Since both ξ_w and ξ_s do not change and are independent of the radial coordinate (r), the solute flux profile for a flat sheet membrane (see Fig. 8–1B) can be also applied to hollow-fiber membranes. Considering the solute concentration gradient (ΔC_s) through the membrane, the solute transport expression is provided by [12]:

$$J_s = -D_s \frac{dC_s}{dx} \simeq D_s \frac{\Delta C_s}{\delta_m} \tag{8.16}$$

where C_s, D_s, and δ_m are the concentration, diffusivity of draw solute within the membrane, and the membrane thickness, respectively. Assuming ΔC_s is proportional to the concentration between membrane surfaces, ΔC_m yields an expression to determine the solute resistivity through the support layer or the partition coefficient (K) as follows:

$$K = \frac{\Delta C_s}{\Delta C_m} (<1) \tag{8.17}$$

where ΔC_m is quantified as the difference between the concentration of the FS C_f and permeate solution C_p. Substituting Eq. (8.14) into Eq. (8.13) gives a realistic model to calculate the solute flux as:

$$J_s = B\Delta C_m \tag{8.18}$$

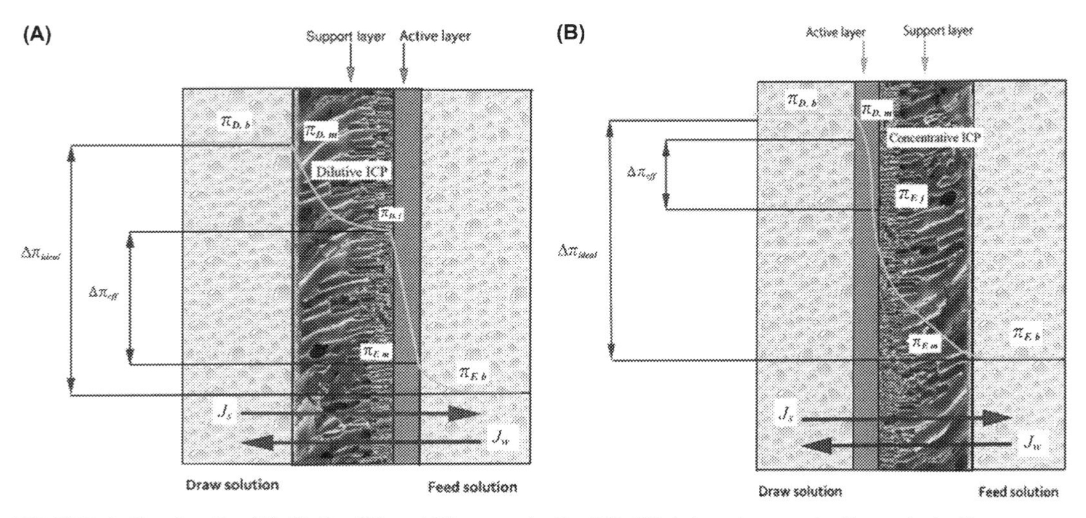

FIGURE 8–1 Showing the (A) dilutive ICP and (B) concentrative ICP. *ICP*, Internal concentration polarization.

where B is the solute permeability coefficient that can be determined using:

$$B = \frac{D_s K}{\delta_m} \tag{8.19}$$

Also, K can be computed using the following formula [24]:

$$K = \frac{\tau t_s}{\varepsilon D} = \frac{S}{D} \tag{8.20}$$

where τ, ε, and λ refer to the tortuosity, porosity, and wall thickness of the membrane, respectively. As noted previously, D is the diffusion coefficient and S is the structural parameter of the membrane support layer. Note that this is identical in principle to the treatment of solute flux in forward osmosis (FO) membranes as described in Chapter 5, Principles of Forward Osmosis.

Subsequently, researchers derived a model considering the operating parameters of the system [24]. To provide a more reliable model for evaluation of the membrane performance, the salt flux and solute permeability were included in the new model. This model expressed the solute flux model as:

$$J_s = B\left[\alpha\left(C_{D,b} + \frac{1}{\varnothing}\left(1 - \frac{A\Delta P}{J_s}\right)^{-1}\right)\exp\left(-\varnothing\frac{J_s}{k_D}\right) - \alpha'\alpha''\left(C_{F,b} + \frac{1}{\varnothing}\left(1 - \frac{A\Delta P}{J_s}\right)^{-1}\right)\exp(\varnothing J_s K)\exp\left(\varnothing\frac{J_s}{k_F}\right)\right] \tag{8.21}$$

where B is the solute permeability coefficient. $C_{D,b}$ and $C_{F,b}$ are the concentrations of the draw and feed bulk solutions respectively, α, α', and α'' denote constants controlled by the operating parameters and can be quantified by:

$$\varnothing = \frac{A}{B}\beta RT \tag{8.22}$$

$$\alpha = \exp\left(\varnothing\frac{A\Delta P}{k_D}\right) \tag{8.23}$$

$$\alpha' = \exp\left(\varnothing\frac{A\Delta P}{k_F}\right) \tag{8.24}$$

$$\alpha'' = \exp(-K\varnothing A\Delta P) \tag{8.25}$$

where β is the van't Hoff coefficient and k is the mass transfer coefficient for the feed and draw sides (subscripts F and D, respectively). The solute permeability for a flat sheet membrane can be calculated using [25]:

$$B = J_w\left(\frac{1-R}{R}\right)\exp\left(\frac{-J_w}{k}\right) \tag{8.26}$$

$R = R_s$ refers to the salt rejection rate and can be calculated from Ref. [26]. Based on the solution–diffusion model, Eq. (8.27) can be used to quantify the B parameter for the flat sheet and the hollow-fiber membrane as follows [27–30]:

$$B = A\frac{(1 - R_s)(\Delta P - \Delta\pi)}{R_s} \tag{8.27}$$

A simplified equation for measuring the solute flux in a hollow-fiber membrane is provided by [31]:

$$J_s = \frac{J_w + A\Delta P_{DS-FS}}{(A/B)\cdot \beta RT} \tag{8.28}$$

This can be reduced to [32,33]:

$$J_s = \frac{B}{\beta RT}\left(\frac{J_w}{A} + W\right) \tag{8.29}$$

This expression is derived according to flat sheet geometry and could be used for both membrane configurations [10]. It is useful for hollow-fiber membranes if an identical fiber thickness in the hollow-fiber membrane is utilized to measure the S value. Both the water and solute fluxes in Eqs. (8.7) and (8.29) could be explained by the mass flow rate across a unit area of the active layer [25].

Another important element in the PRO process is the specific salt flux that refers to the ratio of the salt flux to the water flux J_s/J_w [4], which is dependent on the hydraulic pressure gradient (ΔP) across the membrane, the water permeation (J_w), and the membrane transport characteristics (A and B). The specific salt flux expression for both membrane configurations can be derived as:

$$\frac{J_s}{J_w} = \frac{B}{AiRT}\left(1 + \frac{A\Delta P}{J_w}\right) \tag{8.30}$$

Generally, diminished power production could be also attributed to CP phenomena in the membrane. The system suffers CP influenced by reverse salt flux, which decreases performance. This is due to poor selectivity of the available PRO membrane resulting in high reverse solute flux to the feed stream. This may aggravate CP, thereby diminishing the driving force and ultimately reducing water permeation. The solute flux is not desirable because it causes the accumulation of salts in the FS. When the quantity of salts increases, the concentration of the FS can be elevated causing severe ICP and fouling, as well as reducing the osmotic pressure difference between the bulk solutions. Salt transport through the membrane in the reverse direction can be reduced by increasing the hydraulic pressure on the feed stream [34].

8.2.3 Concentration polarization

CP is a major challenge to the efficiency of the PRO process. It can be classified into ICP and external CP (ECP) similarly to in FO (see Chapter 5: Principles of Forward Osmosis). Much research has been carried out to prevent CP effects in PRO processes and the majority of these has developed membranes with reduced ICP effects [35,36]. An analysis of both polarization phenomena mathematically for PRO membranes is demonstrated in the following sections.

8.2.3.1 Internal concentration polarization

When the salt flux fails to pass through the selective layer to reach the feed stream, the salts can be deposited heavily in the porous support layer [13], changing the osmotic pressure of solution in that layer. This phenomenon reflects the ICP, as it severely influences the performance of the membrane and can lead to reduction of water flux by more than 80% [37]. A diagram of the ICP profile is illustrated in Fig. 8–1. If the solute flux is composed of different species, the effect of ICP magnifies. Since the composition of the solute is more complicated, the interaction between different salt components accelerates the ICP and impacts the membrane productivity. Hence, the solute reverse flux (Eq. 8.13) can be rearranged to include the effect of ICP, considering the boundary conditions as follows [24]:

$$C_{ICP} = \left(C_{F,m} + \frac{J_s}{J_w}\right)\exp(J_w k) - \frac{J_s}{J_w} \tag{8.31}$$

where

$$\begin{cases} C(x=0) = C_{F,m} \\ C(x=t_s) = C_{ICP} \end{cases} \tag{8.32}$$

where x, C_{ICP}, $C_{F,m}$, and t_s denote the distance from the interface between the support and the active layers, the solute concentration at this position, the FS concentration, and the support layer thickness, respectively. It can be observed in Fig. 8–1 that the driving force can be determined by the osmotic pressure gradient across the selective layer. This means that the driving force should be amended for the ICP taking place in the support layer as follows [38]:

$$\Delta\pi_{eff} = \pi_c - \pi_d \exp(J_w k) \tag{8.33}$$

where $\Delta\pi_{eff}$ describes the effective osmotic pressure gradient across the selective layer and k denotes the mass transfer resistance to salt in the support layer. The term "$\exp(J_w k)$" reflects the ICP. By incorporating the ICP term and the mass transfer coefficients on the draw side (k_D) and on the feed side (k_F) into Eq. (8.7), the water flux through the semipermeable membrane can be written as [23,34]:

$$J_w = A\left[\pi_{D,b}\exp\left(\frac{-J_w}{k_D}\right) - \pi_{F,b}\exp\left(\frac{J_w}{k_F}\right)\right] \tag{8.34}$$

The osmotic pressure of the DS and FSs described as $\pi_{D,b}$ and $\pi_{F,b}$, respectively, k_D and k_F are the mass transfer coefficients and can be quantified by:

$$k = \frac{ShD}{d_h} \tag{8.35}$$

where Sh and d_h represent the Sherwood number and the membrane channel diameter. If the hydraulic pressure is applied to the FS, the osmotic driving force minimizes and the ICP aggravates. Thus Eq. (8.13) can be modified as [34]:

$$J_w = A\left(\Delta P + \pi_{D,b} \exp\left(-\frac{J_w}{k}\right)\right) \tag{8.36}$$

where $\pi_{D,b}$ is the bulk osmotic pressure of the DS and k represents the mass transfer coefficient in the boundary layer at the selective layer. Considering the solute resistivity or the partition factor (K), the concentrations of the feed ($C_{F,b}$) and DSs ($C_{D,b}$), and ICP moduli, the water flux model was modified as [37]:

$$J_w = A\left[\pi_{D,b}\frac{1-(C_{F,b}/C_{D,b})\exp(J_wK)}{1+(B/J_w)[\exp(J_wK)-1]} - \Delta P\right] \tag{8.37}$$

Recently, scientists developed another model assuming complete rejection of salt while the concentrations of the feed and DSs were replaced by the osmotic pressures of feed stream ($\pi_{F,b}$) and draw stream ($\pi_{D,m}$) [7,35]. The water flux model is given by:

$$J_w = \frac{1}{K}\ln\frac{A\pi_{D,m} - J_w + B}{A\pi_{F,b} + B} \tag{8.38}$$

where K is the solute resistivity or the partition factor. It should be mentioned that the osmotic pressure of the feed is neglected here, assuming a pure water feed.

The ICP issue can be avoided by enhancing the diffusion permeability of the porous support layer. This could be achieved by manufacturing a thin and loose support layer, but it may influence the mechanical strength of the membrane during the PRO operation.

8.2.3.2 External concentration polarization

Two ECPs take place on the membrane based on the position where it occurs. Dilutive ECP is created at the interface between the membrane and the DS whilst concentrative ECP exists at the interface between the membrane and the FS [14,37]. A schematic of the ECP profile is shown in Fig. 8–2. The ECP moduli have been derived depending on the interaction with boundary conditions. For the dilutive ECP, this can be expressed as [14]:

$$C_{D,m} = \left(C_{D,b} + \frac{J_s}{J_w}\right)\exp\left(-\frac{J_w}{k_D}\right) - \frac{J_s}{J_w} \tag{8.39}$$

where

$$\begin{aligned} C(x=0) &= C_{D,m} \\ C(x=\delta_D) &= C_{D,b} \end{aligned} \tag{8.40}$$

where $C_{D,m}$ and $C_{D,b}$ denote the solute concentration at the interface of the selective layer and the draw concentration, respectively, and x and δ_D are the distance from the interface

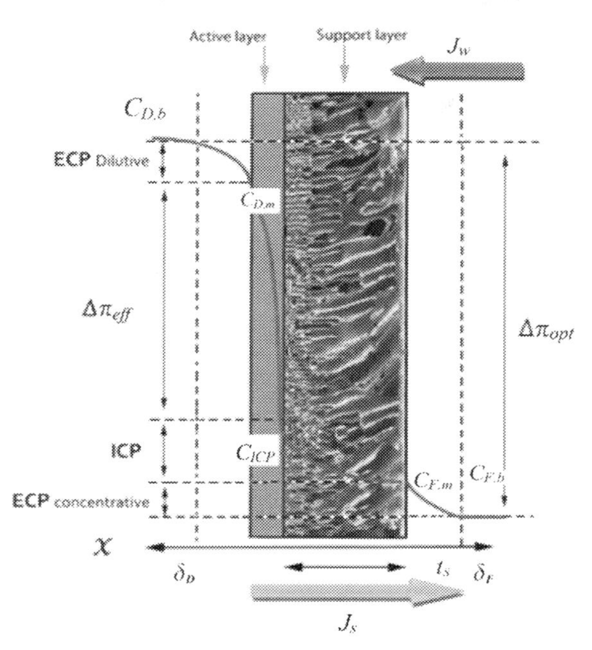

FIGURE 8–2 Concentration polarization involving dilutive and concentrative ECP along with ICP. *ECP*, External concentration polarization; *ICP*, Internal concentration polarization.

between the selective and support layers and the thickness of the draw boundary layer. By incorporating the dilutive ECP in the water flux model, it becomes:

$$J_w = A\left[\pi_{D,b}\frac{1-(C_{F,b}/C_{D,b})\exp(J_wK)}{1+(B/J_w)[\exp(J_wK)-1]} - \Delta P\right] \tag{8.41}$$

For the concentrative ECP the concentration on the feed side of the membrane is [14]:

$$C_{F,m} = \left(C_{F,b} + \frac{J_s}{J_w}\right)\exp\left(-\frac{J_w}{k_F}\right) - \frac{J_s}{J_w} \tag{8.42}$$

where

$$\begin{aligned} C(x=0) &= C_{F,b} \\ C(x=\delta_F) &= C_{F,m} \end{aligned} \tag{8.43}$$

It is important to point out that the concentrative EPC terms, $(\exp(J_w/k))/(R+(1-R)\exp(J_w/k))$, are less significant for the PRO process. As the membrane separation is not usually 100%, the water flux model can be rewritten as [37]:

$$J_w = \left(\frac{1}{K}\right)\left[\ln\frac{B + A\pi_{D,b}\exp(-(J_w/k))\times(R+(1-R)\exp(J_w/k)) - J_w}{B + A\pi_{F,b}((\exp(J_w/k))/(R+(1-R)\exp(J_w/k)))}\right] \tag{8.44}$$

The water flux equation, with the combination of ICP, dilutive ECP, and the reverse solute diffusion on the position of dilutive ECP can be transformed to [14]:

$$C_{D,m} - C_{ICP} = \frac{C_{D,b}\exp(-(J_w/k_D)) - C_{F,b}\exp(J_wS/D)}{1 + (B/J_w)\{\exp(J_wS/D) - \exp(-(J_w/k_D))\}} \tag{8.45}$$

The water flux model has been further modified by reflecting the existence of both ECP and ICP terms and the reverse solute flux [14]. By combining the van't Hoff expression, Eqs. (8.7) and (8.45), a final water flux model for flat sheet membrane can be rephrased appropriately as [13,15,23]:

$$J_w = A\left[\frac{\pi_{D,b}\exp(-(J_w/k)) - \pi_{F,b}\exp(J_wK)}{1 + (B/J_w)\{\exp(J_wK) - \exp(-(J_w/k))\}} - \Delta P\right] \tag{8.46}$$

A final salt flux expression can be written as [15,39]:

$$J_s = A\left[\frac{C_{D,b}\exp(-(J_w/k)) - C_{F,b}\exp(J_wK)}{1 + (B/J_w)\{\exp(J_wK) - \exp(-(J_w/k))\}}\right] \tag{8.47}$$

Subsequently, another model was developed which combined the friction CP (FCP) theory and PRO theory [31]. The FCP reflects the CP effects and a fiber-bore pressure drop in the membrane. According to the solution–diffusion model, the new water flux model for a hollow-fiber membrane can be described as:

$$J_w = A\left[\frac{\pi_{DS,b}\exp(-(J_w/k)) - \pi_{FS,b}\exp(J_wS/D_{diff})}{1 + (B/J_w)\{\exp(J_wS/D_{diff}) - \exp(-(J_w/k))\}} - \Delta P_{DS-FS}\right] \tag{8.48}$$

where $\pi_{DS,b}$ and $\pi_{FS,b}$ are the osmotic pressures in the bulk draw and feed solutions, respectively. ΔP_{DS-FS} (bar) is the hydrostatic pressure gradient between the shell side DS and bore side FS of the membrane. The diffusion and the mass transfer coefficients of the DS are D_{diff} (μL/m/h) and k (L/m^2/h), respectively. More recently, Altaee and Cipolina [40] expanded the water flux equation and included the effect of both CPs and external resistance across the PRO module as follows:

$$J_{w,x} = A\left[\frac{nRTC_{Di,x}(1 + (Q_{Di,x}/Q_{Do,x})/2)\exp(-J_{w,x}/k_d) - nRTC_{Fi,x}(1 - (Q_{Fi,x}/Q_{Fo,x}))\exp(J_{w,x}K + (J_{w,x}/k_f))}{1 + (B/J_{w,x})(\exp(J_{w,x}(J_{w,x}/k_f)) - \exp(-J_{w,x}/k_d)} - \Delta P\right] \tag{8.49}$$

where $x = 0.1$ m, n is the number of moles in the solution, A and B have the value of 1.23 L/h m^2 bar and 2.6 kg/h m^2, respectively. The flow rate of both the feed (k_d) and the DS (k_f) were fixed at 0.18 m/h and $K = 31$ h/m. $C_{Di,x}$ and $C_{Fi,x}$ are the inlet concentration of the DS and FSs at the distance x across the membrane. $Q_{Di,x}$ and $Q_{Do,x}$ denote the flow rate of the inlet and outlet DS and $Q_{Fi,x}$ and $Q_{Fo,x}$ are the flow rate of the inlet and outlet of the FS at distance x over the membrane, respectively.

More recent research was reported by Cheng and Chung [25] to derive expanded models involving the geometry of the hollow-fiber membrane, which included the radius to the outer skin of a single fiber (r_o) (Eq. 8.50) and the radius to the inner skin (lumen) of a single fiber (r_i) (Eq. 8.51):

$$J_w = A\left[\frac{\pi_{D,b}(r_i/(r_i-(D/k_D)))^{-J_w r_i/D} - \pi_{F,b}((r_o+(D/k_F))/r_o)^{J_w r_i/D}(r_o/r_i)^{J_w r_i \tau/D\varepsilon}}{1+(B/J_w)[((r_o+(D/k_F))/r_o)^{J_w r_i/D}(r_o/r_i)^{J_w r_i \tau/D\varepsilon} - (r_i/(r_i-(D/k_D)))^{-J_w r_i/D}]} - \Delta P\right] \tag{8.50}$$

$$J_w = A\left[\frac{\pi_{D,b}((r_o+(D/k_D))/r_o)^{-J_w r_o/D} - \pi_{F,b}(r_i/(r_i-(D/k_F)))^{J_w r_o/D}(r_o/r_i)^{J_w r_o \tau/D\varepsilon}}{1+(B/J_w)[(r_i/(r_i-(D/k_F)))^{J_w r_o/D}(r_o/r_i)^{J_w r_o \tau/D\varepsilon} - ((r_o+(D/k_D))/r_o)^{-J_w r_o/D}]} - \Delta P\right] \tag{8.51}$$

Both the above models can be used for membranes with a double selective layer, one at the inner skin and the other at the outer skin. The difference between the two models depends on the length of the hollow-fiber membrane. In fact, the ECP exerts less impact than ICP in the hollow-fiber membrane during the PRO operations. Likewise, ECP at the bore side can be neglected due to the high structural parameter of the studied membrane. However, the water flux obtained from the flat sheet membrane can be enhanced by designing thinner support layer to minimize the CP impacts. If the structural parameter was reduced, both the ICP ($\exp(J_W S/D) - 1$) and the reverse solute flux ($B/J_W\,[\exp(J_W S/D) - \exp(-J_W/k)]$) were minimized [41]. It is also interesting to note that the ECP became more pronounced upon reducing the hydrodynamic shear force or rapid water flux rate [42]. This means that the ECP can be enhanced when increasing the applied hydrostatic pressure and the water permeation at a low flow rate.

The power density can be also computed based on the empirical parameters. By combining the water flux (Eq. 8.44) into Eq. (8.11) and considering the ECP, the ICP, and the solute flux, the actual power density (W_w) formula is given by [43]:

$$W_w = A\left[\frac{\pi_{D,b}\exp(-(J_w/k)) - \pi_{F,b}\exp(J_w K)}{1+(J_w/B)\{\exp(J_w K) - \exp(-(J_w/k))\}} - \Delta P\right] \times \Delta P \tag{8.52}$$

There is a high correlation between the water permeability (A) and the solute permeability (B). The extractable power is a function of these membrane parameters plus the structural parameter (S). Therefore a balance between both the A and B value is necessary to obtain the greatest power density.

8.3 Membrane development

The most important parameter for the successful process is membrane development. The PRO membrane should achieve high water permeability and salt rejection. A wide range of membranes was used in the PRO process, such as flat sheet and hollow-fiber membranes. Flat sheet PRO membranes can be classified based on their fabrication material into cellulose acetate (CA) membranes, developed by Loeb in the 1960s, and commercial cellulose triacetate (CTA) membranes, by Hydration Technology Innovations (HTI, Albany, OR) [44].

The structure of this latter membrane consists of an embedded support layer and a dense selective layer with a thickness of 10–20 μm designed on the top of the support layer. The TFC membranes are composed of a selective layer deposited on a porous support layer. It has been found that these membranes perform well in the PRO process due to high rejection of salt by the selective layer and great water permeation governed by the highly porous support layer [45]. The development of TFC-PRO membranes is continuing due to their advantages over CTA membranes. As mentioned previously, the support layer structure includes high porosity and performs as a diffusive boundary layer, which reduces the ICP impacts on the PRO process, compared with more dense support layers.

The properties of the selective layer can be controlled by the structure of the support layer. The passage of water takes place at the inner barrier of the selective layer, and the morphology of the particular layer had a less significant effect on the water permeation [46]. The performance of this membrane can be characterized by the membrane transport properties water permeability coefficient (A) (Eq. 8.44 and Eq. 8.46, 8.47, 8.48, 8.49, 8.50 and 8.51), solute permeability coefficient (B) (Eq. 8.26 and Eq. 8.27), and structural parameter for the support layer (S) (Eq. 8.20). The selection of a highly permeable membrane is the most important factor in further increasing the power output in the PRO membrane. The power output can be measured from the applicable equations (Eqs. 8.11 and 8.12) reported previously. For example, recent research was reported on altering the polyamide (PA) selective layer to enhance the power production in the PRO system, where the developed membrane exhibited an increase in the power density of about 12.0 W/m^2 upon using de-ionised (DI) water feed, 1.0 mol/L NaCl as DS, and hydraulic pressure of 15.0 bar [47].

Besides, hollow-fiber membranes were manufactured for the PRO process and attracted more attention in industrial applications due to high packing density and easy scale-up [16]. These membranes were formed in a tubular shape and fabricated by using a nonsolvent-induced phase inversion spinning process. The current obstacles found in these membranes are the poor mechanical strength to withstand applied pressure in the absence of a fabric backing layer [13]. Similarly, to examine the potential of these membranes in the PRO system, the membrane transport characteristics can be determined from the above models. Moreover, adjusting the fabrication parameters and process conditions plays an important role in the PRO membrane performance. For instance, it was found that a highly asymmetric polyethersulfone (PES)-TFC hollow-fiber membrane involving a highly porous support layer and small pore size distribution achieved greater power density. The power density was estimated at 24.3 W/m^2 when using a DI water feed and 1.0 mol/L NaCl as DS at 20.0 bar [29]. In this section the representative milestone of the PRO membrane developments is divided into flat sheet membranes and hollow-fiber membranes.

8.3.1 Performance of reverse osmosis flat sheet membranes

Membranes originally developed for use in RO are commonly used in PRO processes, both in academic research and industrial development. The first RO membrane used in the PRO system was purchased from DuPont company [48]. This membrane was capable of withstanding hydraulic pressure of up to 91.2 bar. However, the attainable power density from the PRO process was very low, below 1.74 W/m^2 when using a highly concentrated DS and hydraulic pressure of 31 bar [49]. However, when the same membrane was operated at 50 bar the extractable power density approached 4.89 W/m^2 [50]. Although this membrane tolerated high pressure and had high salt rejection, the membrane structure minimized the extractable energy. The membrane structure is asymmetric, including a thick and hydrophobic polysulfone (PSF) support layer embedded in a backing fabric layer to tolerate high hydraulic pressure [51]. However, the challenges that remained include the ECP on the selective layer and more severe ICP in the support layer leading to very low water flux and poor economy of energy production [52]. The presence of severe ICP meant that the adverse effects of reverse salt flux and both CPs decreased the effective osmotic pressure across the selective layer yielding a lowered effective osmotic pressure difference [18]. This, in turn, resulted in reduced power density governed by the membrane structure and solute characteristics. To maximize the power density, tailoring the material, structure, and morphology of the membrane is necessary. One viable strategy is to redesign the support layer of the RO membrane, which becomes thinner and more open, thereby reducing the value of S [7].

Cellulosic RO membranes have also been used in PRO. One study reported that CA was used in the prototype osmotic power plant constructed in Statkraft, Norway. The power density generated from this membrane as low as 1.3 W/m^2 when using seawater as a DS and brackish water as an FS [47], which was very low compared to the target power density of 5.0 W/m^2. This was due to a thick, tight support layer, which compressed under high pressure. Jellinek and Masuda [53] explored the performance of a CTA membrane. This membrane generated a power density of about 1.62 W/m^2 at 17.2 bar. In comparison with RO membranes, cellulosic membranes are more hydrophilic, which is preferable to hinder the ICP and enhance the water permeation. Furthermore, these membranes showed low fouling propensity and can be cleaned easy [44]. Due to the poor power density obtained from both types of membranes at high hydraulic pressure, they are not practical for PRO process.

8.3.2 Performance of forward osmosis flat sheet membranes

Since the same characteristics are needed for both FO and PRO systems, CTA/CA-FO membranes with improved performance have also been used for PRO [23,54,55]. However, the mechanical weakness of these membranes is the major hurdle for PRO application. For PRO using seawater FS (0.5 mol/L NaCl) and 2.0 mol/L NaCl as a DS, Kim and Elimelech [56] investigated the performance of a commercial CTA-FO membrane provided by HTI, Albany, OR. This membrane was embedded with a polyester mesh and had a thickness of about 93 μm. This thinner membrane had a lower structural parameter than is typical for RO membranes. The membrane had high water permeation of about 13.9 LMH, while the power density was near the target power density at 4.7 W/m^2 at 12.5 bar. This was attributed to the high salinity of the DS leading to the high osmotic pressure gradient across the membrane and high permeate flow rate.

Achilli et al. [57] used the HTI membrane using 2.5 and 5.0 g/L NaCl as FSs. An improvement in the power density from 2.73 to 5.1 W/m^2 was observed when increasing the salinity of the DS from 35.0 to 60.0 g/L NaCl at 9.7 bar. It was suggested that the water permeability correlated to the increase in the power density. It was found that the influence of solute mass transfer and solute resistivity on the power density was high due to high water flux. Thus the ICP was reduced due to small structural parameters and low solute resistivity in the support layer.

She et al. [42] explored the performance of three different commercial CTA membranes. The main difference was the fabric type, as the first CTA-W contained a woven fabric; the second CTA-NW included a nonwoven fabric and woven-supported CTA-P membrane. In PRO tests utilizing 1.0-mol/L NaCl DS and 10-mmol/L NaCl FS at 12.0 bar, the greatest power density of around ~4.5 W/m^2 was assigned to the CTA-P membrane followed by the CTA-W membrane (~3.8 W/m^2) and CTA-NW membrane (~3.2 W/m^2). The CTA-P membrane showed the best performance in terms of A value achieving 2.08×10^{-12} m/s Pa (0.75 LMH), but the B value was high at $17.57 \pm 0.47 \times 10^{-12}$ m/s. The maximum power density was still below the ideal value for the PRO system due to increased reverse solute flux upon raising the hydraulic pressure. This resulted in severe ICP and low water permeation, thereby reducing the power output. The membrane was compressed during the experiment when increasing the hydraulic pressure above 15.0 bar, which hampered the transport of the FS. It was concluded that improving the membrane structure. And optimizing the feed spacer design is important to overcome these technical limitations.

Kim et al. [58] tested a CTA membrane in a PRO system using 0.72 mol/L NaCl DS including 200 mg/L organics and DI water feed at 15.0 bar. It was observed that the water permeability and solute permeability were better at about 0.98 LMH/bar and 0.82 LMH when using DS including xanthan as compared to that for a DS including alginate. This is because the former had a higher molecular weight and viscosity, while xanthan had lower charge density. An increase in the molecular weight and viscosity may induce the dilutive ECP yielding little concentration on the outer surface. Regarding power density, the maximum value was 8.3 W/m^2 when utilizing 2.0 mol/L NaCl as a DS and DI water feed at 15.0 bar. However, the membrane could not withstand a hydraulic pressure above 15.0 bar.

To avoid the poor power density obtained by the cellulosic membranes, advances in robust TFC membranes have been achieved. Han et al. developed the first TFC-FO membrane to tolerate high applied pressure in PRO and improved its intrinsic transport properties via novel postfabrication treatment. When the membrane was tested utilizing synthetic seawater brine (1.0 M NaCl) as the DS and DI water feed, the highest power density approached 12.0 W/m^2 at 15.0 bar. After treatment, the membrane became highly permeable, and a balanced water and solute permeability are important to increase the power output. It was observed that this membrane possessed an excellent water permeability of about 5.3 LMH/bar and moderate solute permeability around 2.0 LMH. This is the first membrane capable of withstanding hydraulic pressure above 13.5 bar due to good mechanical strength. It can also generate maximum power density as a result of great water permeation governed by low structural parameters and a highly permeable selective layer.

Overall, adding a high concentration of polymer to increase the membrane thickness can enhance the mechanical stability of the membrane at the expense of water permeation and power output. In this respect, new membranes examined under high salinity for different processes should have improved physicochemical characteristics in the wet environment, considering any change under compression and bending stress [47].

8.3.3 Performance of thin-film composite flat sheet membranes

TFC membranes can be fabricated through a conventional phase inversion process for the support layer followed interfacial polymerization (IP) for the selective layer. The structure of these membranes is asymmetric with a porous support layer providing mechanical stability and an active layer providing good permselectivity. One of the merits of using IP to prepare the TFC membrane is that both the structure and characteristics of the support and selective layers can be separately altered and optimized to acquire the best water permeation and salt rejection. Recent trends in TFC flat sheet membranes for PRO are summarized in Table 8–1. Extensive efforts have been devoted to designing a superior membrane with a relatively low structural parameter, balanced water and solute permeabilities, and maximizing extractable power. The support layer structure can be tailored by changing the phase inversion conditions and polymer concentration giving a smaller S value. Improving other aspects can be done by applying posttreatment modification using new monomers [66] and by incorporating various nanomaterials [13] or surfactants [63] into the selective layer. For example, Hickenbottom et al. [67] studied the influence of different spacer configurations and cross-flow velocities on mass transfer, the pressure drop in the membrane cell, and power output of a commercial TFC membrane manufactured by HTI. The membrane achieved a water permeability of about 1.63 LMH/bar and a moderate solute permeability of about 1.42 LMH. The special design and orientation of the spacers for the feed and draw streams reduced the mass transfer resistance across the membrane yielding a power output of 22.0 W/m^2, when using DI water feed and 3.0 mol/L NaCl DS at 35.0 bar.

Li et al. [62] created a TFC membrane by casting a polymer layer created by IP on a Torlon polyamide-imide support layer. The substrate was altered via the polydopamine (PDA) cross-linking method. Then, the selective layer was deposited on the PDA cushion layer adhered to the substrate. The PRO test was run using synthetic seawater (3.5% NaCl) as a DS and DI water feed at 6.0 bar. Although a good water permeability was obtained of 2.39 ± 0.25 LMH/bar and solute permeability of 0.66 ± 0.12 LMH, the power density remained poor of about 2.84 W/m^2. The membrane achieved better performance than the pristine membrane as the water flux increased 3-fold, and the power output 16-fold.

Huang et al. [61] invented a new procedure to form a two-zone support structure. This utilized a "slot-die" device to cast a nylon 6,6 layer on a nonwoven mesh, followed by a second nylon 6,6 layer to ensure full coverage of the mesh and for fine pores. After that, the selective layer was designed on the top of the support layer (see Fig. 8–3). It was observed that the support layer structure contained spongy pores and microvoids forming a "truss

Table 8–1 Comparison between a wide range of pressure-retarded osmosis flat sheet membranes reported in the literature.

Membrane category	Materials Selective layer/support layer	Process parameter			Water Permeability ($L/m^2/h/bar$)	Solute permeability ($L/m^2/h$)	Power density (W/m^2)	Reference
		Draw solution	Feed solution	Pressure (bar)				
Flat sheet	HTI-CTA	1.0 M NaCl	10 mM NaCl	9.72	1.83	1.2	4.5	[43]
	HTI-CTA	1.0 M NaCl	10 mM NaCl	15.0	3.3 ± 0.2	0.31	3.7	[59]
	HTI-CTA	Synthetic seawater (0.4 M)	Synthetic wastewater effluent	15.0	2.540 μm/s bar	0.187 μm/s	5.55	[60]
	CA-modified PEG	0.6 M NaCl	DI water	2.75	7.25×10^{-3} ($cm^3/cm^2/s/atm$)	1.57×10^{-2} ($cm^3/cm^2/s$)	3.1	[30]
	PA-coated PDA PAN	3.5 wt.% NaCl	DI water	10.0	4.0	1.0	2.6	[47]
	PA/PSf-TFC	0.5 M NaCl	Brackish water (40 mM NaCl)	13.0	5.81	0.88	10.0	[45]
	PA-PAN-TFC	Seawater brine (1.0 M NaCl)	DI water	15.0	7.2	–	12.0	[33]
	PA-PEI-TFC	1.0 M NaCl	DI water	17.2	2.09 ± 0.28	1.03 ± 0.082	14.56	[59]
	PA-PEI-TFC	1.0 M NaCl	10 mM NaCl	17.2	–	–	12.76	[59]
	PA/PSf incorporated PVP-TFC	1.0 M NaCl	10 mM NaCl	22.5	1.4	–	6.6	[6]
	HTI-PA/PSf-TFC	1.0 M NaCl	DI water	16.5	2.0	3.422 ± 0.898	14.88 ± 2.08	[18]
	PA/nylon 6,6-TFC-1	0.5 M NaCl	DI water	12.0	1.41 ± 0.30	0.22 ± 0.08	2.3	[61]
	PA/nylon 6,6-TFC-2	0.5 M NaCl	DI water	12.0	2.69	0.94	2.0	[61]
	PA-PAI-altered PDA-TFC	1.0 M NaCl	DI water	6.0	2.39 ± 0.25	0.66 ± 0.12	2.84	[62]
	PA-altered SDS/PI-TFC	1.0 M NaCl	DI water	22.0	–	–	18.09	[63]
	PA/PES-incorporated CNTs-TFN	0.5 M NaCl	DI water	6.0	1.8	–	1.67	[64]
	PA/PAN-pTFN nanofiber	0.5 M NaCl	DI water	9.0	5.30 ± 0.94	4.97 ± 0.62	6.2	[65]
	PA/PAN-mTFN nanofiber	0.5 M NaCl	DI water	10.2	2.83 ± 0.56	0.44 ± 0.30	8.0	[65]
	PA/modified PAN Nanofiber-TNC	1.06 M NaCl (seawater brine)	80 mM NaCl (synthetic brackish)	15.2	4.1	1.74	15.2	[41]
	PA/modified PAN Nanofiber-TFC	1.06 M NaCl (seawater brine)	0.9 mM NaCl (synthetic river) (synthetic brackish)	15.2	4.1	1.74	21.3	[41]

CA, Cellulose acetate; *CNT*, carbon nanotube; *CTA*, cellulose triacetate; *HTI*, Hydration Technologies Inc.; *mTFN*, *m*-phenylene diamine thin film nanocomposite; *pTFN*, polyethyleneimine thin film nanocomposite; *PA*, polyacrylomide; *PAI*, polyamide-imide; *PAN*, polyacrylonitrile; *PDA*, polydopamine; *PEG*, poly ethylene glycol; *PEI*, polyetherimide; *PES*, polyethersulfone; *PI*, polyimide; *PSf*, polysulfone; *PVP*, polyvinylpyrrolidone; *TNC*, thin-film nanofibre composite; *SDS*, sodium dodecylsulfate; *TFC*, thin-film composite; *TFN*, thin film nanocomposite.

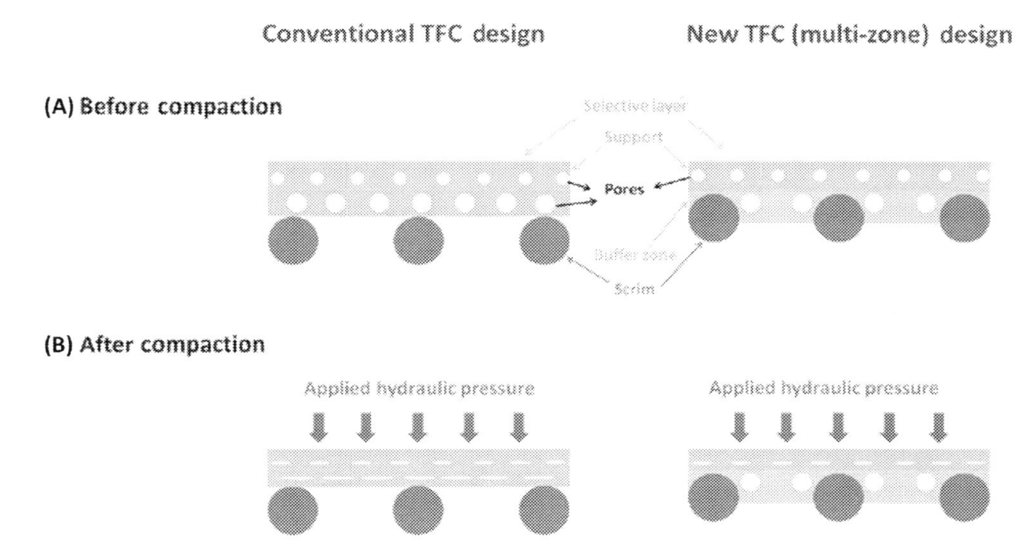

FIGURE 8–3 Structure of both the classical and the newly designed TFC membranes. (A) The classical TFC membrane and multizoned nylon 6,6 supported TFC membrane before pressure compaction. (B) The classical TFC membrane and multizoned nylon 6,6 supported TFC membrane after pressure compaction. *TFC*, Thin-film composite. *Adapted with permission from L. Huang, J.T. Arena, M.T. Meyering, T.J. Hamlin, J.R. McCutcheon, Tailored multi-zoned nylon 6,6 supported thin film composite membranes for pressure retarded osmosis, Desalination 399 (2016) 96–104.*

network." The thickness was reduced by 50% to 520 ± 52 μm. This increased the water permeability to 2.69 ± 0.48 LMH/bar. At a hydrostatic pressure of 15.5 bar and using 0.5 mol/L NaCl as the DS and DI water, the available power was only 2.3 W/m^2. This might be due to the stretching of the polymer material in the support layer at this pressure causing severe reverse solute flux. This is, in turn, induced ICP effects.

Wei et al. [68] developed a support layer through blending both polyvinylpyrrolidone (PVP) into the PSF dope solutions and cast this onto a nonwoven fabric. The ratio of PVP/PSF was optimized to achieve the best membrane productivity. Using 1.0 mol/L NaCl DS and pure water feed at 22.0 bar, a power output of about 12.9 W/m^2 was achieved. The results suggested that the PVP dope solution enhanced the porosity and pore connectivity in the support layer. Therefore the mass transfer of the solution in the support layer improved, which mitigated the impact of the ICP. The high viscosity of the polysulfone (PSf) solution facilitated the formation of a defect-free substrate adhered well to the selective layer.

8.3.4 Performance of nanofiber supported flat sheet membranes

It has been shown that the occurrence of ICP can be hindered by nanofiber supported membrane, which showed interconnected pore structure and high porosity [69,70]. Electrospinning technology has been used to design a very thin and porous support layer embedded into the nonwoven fabric backing layer. This structure led to low tortuosity and acceptable mechanical strength of the support layer [23]. Hoover et al. [71] fabricated a PSF support layer and cast on top of polyethylene terephthalate 340 nm diameter nanofiber mesh to boost the mechanical strength of the membrane under high hydrostatic pressure and cross-flow velocities. A selective layer was then fabricated on the top of the support to achieve high permselectivity. The performance was evaluated using a DI water FS and a 1.0 M NaCl draw. It was reported that the membrane produced low water permeability of 1.13 LMH/bar and salt permeability of 0.23 LMH, which was because of a high mass transfer resistance of the solute in the support layer as the structural parameter was also high at 651 μm. However, the membrane showed enhanced resistance to delamination when increasing solution shear flow, which reduced the deterioration of the membrane at high hydraulic pressure.

Recently, a sulfonated PSF support layer was fabricated by El Khaldi et al. [72], using 247-nm nanofibers with a scaffold-like structure. When the membrane was tested utilizing a 1.0-mol/L NaCl as a DS and DI water FS, very high water flux corresponding 313 LMH was observed. The membrane produced a favorable low reverse solute flux of about 5.3 $g/m^2/h$ while the selectivity ratio was maintained high around 58.8 L/g.

Bui and McCutcheon examined a TFC membrane with a support layer covered with nanofibers of various diameters to improve mechanical stability [65]. The structure of this support layer with a small structural parameter (150 μm) had a role in obtaining increased power density. The results indicated that the membrane tolerated an applied a pressure of 11.5 bar to achieve a power output approximately 8.0 W/m^2 when a 0.5 mol/L NaCl and DI water was used as the DS and FS, respectively. More importantly, this membrane showed relatively good water permeability of about 2.83 ± 0.56 LMH/bar and very low solute permeability of about 0.44 ± 0.3 LMH. The reason behind the lower power output might be the effect of compressible spacers upon raising the hydraulic pressure. This may cause the blockage of the channel and hindrance of the water flow over the membrane. Another reason is the nanofiber support could be compacted at high applied pressure yielding greater structure parameter and ICP impacts.

Song et al. [41] developed a nanofiber supported high porosity TFC membrane with low tortuosity and structural parameter (150 μm). This allowed improved mixing of solute and the diluted FS and hence alleviating the effects of ICP. Due to the unique microstructure of the support layer, the membrane achieved superior water permeability of 4.1 LMH/bar and acceptable solute permeability of 1.74 LMH. As suggested by earlier work, increased water permeation enables increased power density. Accordingly, this robust membrane exhibited excellent power output of 15.2 W/m^2 when employing synthetic brackish feed water (0.08 mol/L NaCl) and seawater brine (1.06 mol/L NaCl) as a DS at 15.2 bar. This can be ascribed to the mass transfer, which allows a further beneficial decrease of ECP impacts, and therefore the power output was enhanced. However, increasing the mass transfer of the solute can induce the pumping cost, so it is preferable to achieve high power output at a minimum mass transfer factor.

Lastly, Kim et al. [73] reported the highest power density for a novel thermally rearranged TFC membrane. This membrane was made of a thermally rearranged nanofibrous polymer support

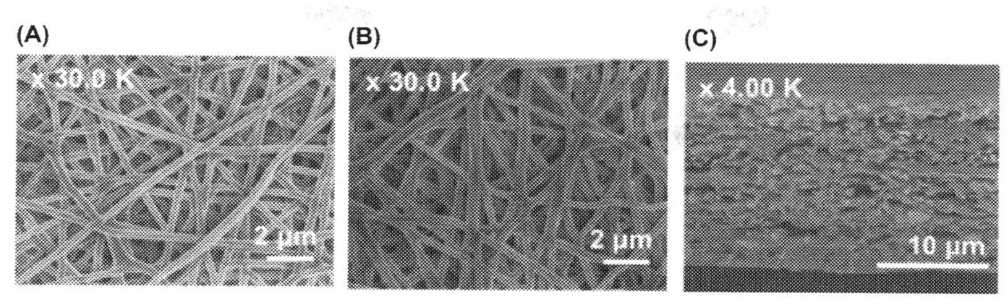

FIGURE 8–4 SEM images illustrating the difference between (A) uncoated-thermally rearranged nanofibrous polymer support, (B) coated-thermally rearranged nanofibrous polymer support coated with polydopamine, (C) a cross section of the uncoated-thermally rearranged nanofibrous polymer support. *Adapted with permission from J.H. Kim, S.J. Moon, S.H. Park, M. Cook, A.G. Livingston, Y.M. Lee, A robust thin film composite membrane incorporating thermally rearranged polymer support for organic solvent nanofiltration and pressure retarded osmosis, J. Membr. Sci. 550 (2018) 322–331.*

layer coated with PDA and an ultrathin PA selective layer (see Fig. 8–4). The newly developed membrane exhibited an improved water permeability of 3.7 LMH/bar and salt rejection of 94% at optimum conditions. Due to this improvement, the power density scored the highest value in the literature at 40 W/m^2 using DI water feed and 3.0 mol/L NaCl DS at 18.0 bar. It was indicated that the growth of the obtainable power was estimated at 1.8-fold as a result of a highly porous, thin, thermally and chemically stable support layer leading to excellent water permeation.

8.3.5 Performance of hollow-fiber membranes

Compared to flat sheet membranes, hollow-fiber membranes have a high surface area, self-mechanical support, high backing density, small footprint and are easy to fabricate in large-scale modules [23]. Furthermore, these modules do not require spacers. Power losses are also less in the feed flow channel than for flat sheet membranes [74]. In 2012 the first hollow-fiber membrane specially for PRO was manufactured and was composed of a PES support and a PA selective layer on the lumen side of the membrane [16]. An overview of recent progress in hollow-fiber membranes is given in Table 8–2.

Higa et al. [86] examined the performance of a commercial CTA hollow-fiber module manufactured by Toyobo Co. Ltd. using a 0.5-mol/L NaCl as the DS and tap water FS. The maximum power density per unit volume of the hollow-fiber membrane was lower around 1.11-kW/m^3 module as compared to the spiral-wound module with 0.75-kW/m^3 module due to larger backing density of the hollow-fiber membrane. The extractable power output per membrane area became 0.14 W/m^2 representing one-sixth of that for the spiral-wound membrane. This is probably due to low water permeability generated from the hollow-fiber membrane corresponding to around 0.0792 LMH/bar, while it was 0.792 LMH/bar for the spiral-wound membrane.

The same membrane was also evaluated by Tanaka et al. [31] but with larger module diameters, similar membrane area, and shorter length. The power efficiency was estimated and compared with Gibbs's free energy of mixing under varying operating conditions. It was

Table 8–2 An overview of various hollow-fiber membranes used for the pressure-retarded osmosis process.

Membrane category	Materials Selective layer/support layer	Process parameter Draw solution	Feed solution	Pressure (bar)	Water Permeability ($L/m^2/h/bar$)	Solute permeability ($L/m^2/h$)	Power density (W/m^2)	Reference
Hollow fiber	PA/PES	1 M NaCl	10 mM NaCl	21.0	3.3 ± 0.2	0.31	24.0	[75]
	PA/PEI-TFC	1.0 M NaCl	10 mM NaCl	15.1	9.22×10^{-12} m/s/Pa	3.86×10^{-8} m/s	20.9	[76]
	PA/PES-TFC	1.0 M NaCl	10 mM NaCl	8.4	1.52 ± 0.3	0.24 ± 0.06	11.0	[4]
	PBI-POSS/PAN	1.0 M NaCl	10 mM NaCl	7.0	0.6	0.1	2.5	[77]
	PA/PI	1.0 M NaCl	DI water	20.0	1.42	0.40	7.6	[28]
	PA/PES	0.6 M NaCl	DI water	7.0	–	–	2.98	[78]
	PA/PI	1.0 M NaCl	10 mM NaCl	15.0	1.9	0.48	14.4	[79]
	PA/PI	1.0 M NaCl	40 mM NaCl	15.0	1.9	0.48	10.6	[79]
	PA/PI	1.0 M Na–Fe–Ca	DI water	12.0	4.3	0.47	16.2	[80]
	PA/P84	1.0 M NaCl 2.0	DI water	21.0	0.918	0.138	12.0	[81]
	PA-P84 copolyimide	1 M NaCl	DI water	6.0	0.918	0.138	13.0	[81]
	PA/PDA-PES	0.6 M NaCl	DI water	7.0	–	–	3.0	[78]
	PA-TBP/PDA-PES	0.6 M NaCl	DI water	8.0	–	–	3.9	[82]
	PA/PES	1.0 M NaCl	DI water	20.0	3.3 ± 0.2	0.31	24.3	[29]
	PA/PES	0.6 M NaCl	DI water	6.0	–	–	1.6	[83]
	PA/PEI	1.0 M NaCl	DI water	20.0	4.0	0.5	28.0	[84]
	PA/PEI	1 M NaCl	Tap water	15.0	–	–	8.9	[85]

PA, polyamide; *PAN*, polyacrylonitrile; *PBI*, polybenzimidazole; *PDA*, polydopamine; *PEI*, polyetherimide; *PES*, polyethersulfone; *PI*, polyimine; *POSS*, polyhedral oligomeric silsesquioxane; *TBP*, tributylphosphate; *TFC*, thin-film composite.

suggested that the new membrane module achieved a lower pressure drop and enhanced the water permeation in some regions inside the module. As a result, the power density was improved significantly, reaching 10.7%–15.7% as compared to that for RO seawater desalination process with 9.2%–14.2%. Interestingly, the water and solute permeabilities of this membrane were decreased to 0.27 LMH/bar and 0.035 LMH, respectively, as compared to the original membrane ($A = 0.60$ LMH/bar, $B = 0.072$ LMH), which may be attributed to the lower hydraulic pressure of 10 bar for this membrane due to its tolerance to lower pressure than the original membrane (20.0 bar).

Another study reported the modification of the hollow-fiber membrane via cross-linking of the selective layer to achieve improved water flux and power density. Li et al. [17] stated that the modification procedure consisted of immersing the membrane in a 0.06 mol/L PAH solution at a temperature of 70°C for 1 h. After rinsing the membrane with water, it was immersed in a 0.5 wt.% GA cross-linker solution for 30 min. The pore size of the selective layer was smaller, leading to an acceptable rejection rate of 72% ± 4% and 92% ± 3% for 500 ppm NaCl and $MgCl_2$ solutions, respectively, at 2.0 bar. Water permeability was 2.0 ± 0.3 LMH/bar, which could be due to the sponge-like structure and highly selective active layer having tight pores in the internal surface. Consequently, the membrane generated a power output around 4.3 W/m^2 at 12.0–13.0 bar in AL-FS orientation and utilizing a real wastewater RO retentate as a feed and 1.0 mol/L NaCl as a DS. The selection of AL-FS mode was beneficial, as the membrane showed excellent antifouling propensity, and there was no need for a pretreatment step for the complex FS.

Cheng et al. [27] designed a highly selective PA outer active layer on top of a PES support layer. The best performing membrane had a selective layer structure of "flake-like" shape with no defects, large pores, and a large pore size distribution. As a consequence, the fast migration of the MPD solution was induced, leading to the occurrence of the reaction over a larger area. The water permeability was enhanced, at 1.420 ± 0.096 LMH/bar, while the solute permeability was minimal around 0.028 ± 0.010 LMH. The developed membrane exhibited a good power density of about 7.81 W/m^2 when using 1.0 mol/L NaCl and DI water as a DS and FS, respectively, at 20.0 bar. A small increase in the feed pressure buildup to 0.17 bar caused an increase in the water permeation and power density to 10.06 W/m^2. The reverse solute diffusion remained almost the same under the same conditions, while the ratio between reverse salt flux and water flux was low around 0.4 g/L. Finally, excellent mechanical stability was reported for the developed membrane under high hydrostatic applied pressure.

To further maximize the power density of the hollow-fiber membranes, an innovative modification procedure was employed associated with optimizing the spinning conditions by Wan et al. [87] who proposed the incorporation of inorganic salts into a PES dope polymer to form the support layer, followed by fabricating inner and outer selective layers. The support layer of the resultant membrane was composed of a spongy structure and a finger-like structure with a low S value of around 430 ± 11 μm. This provided significant mechanical strength to the membrane under high hydraulic pressure (30.0 bar). These alterations to the support increased the water flow within the support layer and reduced the salt flux and

ICP effects. The phase inversion was also tailored by impregnating optimum concentration (1.0%) of $CaCl_2$ into the support layer. It was observed that the mean pore size was small (6.05 nm) due to improved hydrophilicity and viscosity. The coupled modification strategies resulted in membranes with high water permeability of 3.8 ± 0.20 LMH/bar and minimum solute permeability of 0.44 ± 0.03 LMH. These results contributed to the growth of the power density of 38.0 W/m^2 at 30.0 bar in utilizing 1.2 mol/L NaCl and DI water as the DS and FSs.

A novel double-skinned hollow-fiber membrane was manufactured by Han et al. [88] for energy production. The newly developed membrane consisted of an internal polyamide layer as a fouling barrier and an external polyamide protective layer. Experimental findings showed that the water flux was reduced by only 29% at a maximum water recovery rate around 80% when using actual wastewater, including inorganic salts and organic components, with a DS of 1.0 mol/L NaCl. This indicated the superior selectivity of the external active layer, where the foulants could not penetrate through the active layer achieving a good rejection rate of 94.2%. The membrane exhibited high water and negligible solute permeabilities of about 1.5 ± 0.2 LMH/bar and 0.02 LMH, respectively. To estimate the extractable power from the PRO system, wastewater brine as the FS and 1.0 mol/L NaCl as the DS was used. The initial power output reached 9.8 W/m^2 but declined to 8.9 W/m^2 after 12 h of operations at 15.0 bar. The additional internal-selective layer contributed significantly to the antifouling property of the developed membrane resulting in maximum water permeation, low foulant deposition, and high power density.

8.4 Applications in seawater and wastewater treatment

It may be noted that lessons learned from past experiences in the use of membrane technology to separate a wide range of solutions are now being utilized to generate power by the PRO system. The continuous development in the PRO process to harness power can help reduce energy demands in the future. Despite that, PRO has suffered from some technical obstacles, such as the requirement of a pretreatment stage for complex FS, high cost of pretreatment, relatively low attainable power, energy consumption, and environmental impacts of disposing of the concentrated solution [43,89]. Thus integrating the PRO system with other pressure-driven processes is promising as a water and electricity cogeneration process. It shows some advantages involving further fouling mitigation as a pretreatment stage, obtaining freshwater, and maximizing the water permeability and power production [90]. It has been suggested that optimizing the materials and membrane structure to maximize A and lower the structural parameter would provide a power density of higher than 10 W/m^2. This could be obtained if seawater and freshwater were utilized as the DS and FS, respectively, in a bench-scale PRO system.

Moreover, the performance of the hybrid system depends strongly on the concentration of the solutions. In case a pretreatment system can mitigate membrane fouling, wastewater effluent can be a suitable FS instead of brackish water. The salinity of wastewater effluent is between 500 and 3000 mg/L, while that for brackish water is between 1000 and 5000 mg/L [56].

To produce great power output, brine from seawater RO desalination plants with a salinity of 70,000 ppm for 50% recovery could be a good option as a DS. To date, significant improvement of the hybrid process has been achieved based on system integration, which can successfully alleviate the increased need for water and energy [13,91]. Applications of PRO can be classified into the stand-alone process and hybrid processes. The hybrid process can be categorized based on the power efficiency of an open- and a closed-loop PRO process discussing the current state of the PRO system at the prepilot scale and industrial scale.

8.4.1 Individual pressure-retarded osmosis pilot plant

Statkraft launched the first prototype pilot plant in the world in Tofte, Norway, in 2009 [13,92]. It is important to identify the viability of this pilot plant and, subsequently, to analyze the main practical restrictions. In this PRO plant, traditional CA spiral-wound membrane module having an effective membrane area of 2000 m^2 from Nitto Denko/Hydranautics was used. The system was operated utilizing river water and seawater as the FS and DS, respectively, under hydraulic pressure of 10.0–15.0 bar (see Fig. 8–5). Unfortunately, the estimated power output was only 3.0 W/m^2, which is far below the target of 5.0 W/m^2 to make the system economically practical.

In comparison with the current TFC-RO membrane in spiral-wound modules, the estimated power density of about 2.7–3.0 W/m^2 was based on bench-scale operation due to the altered support layer [93]. In the next year the plant was improved, and the power capacity

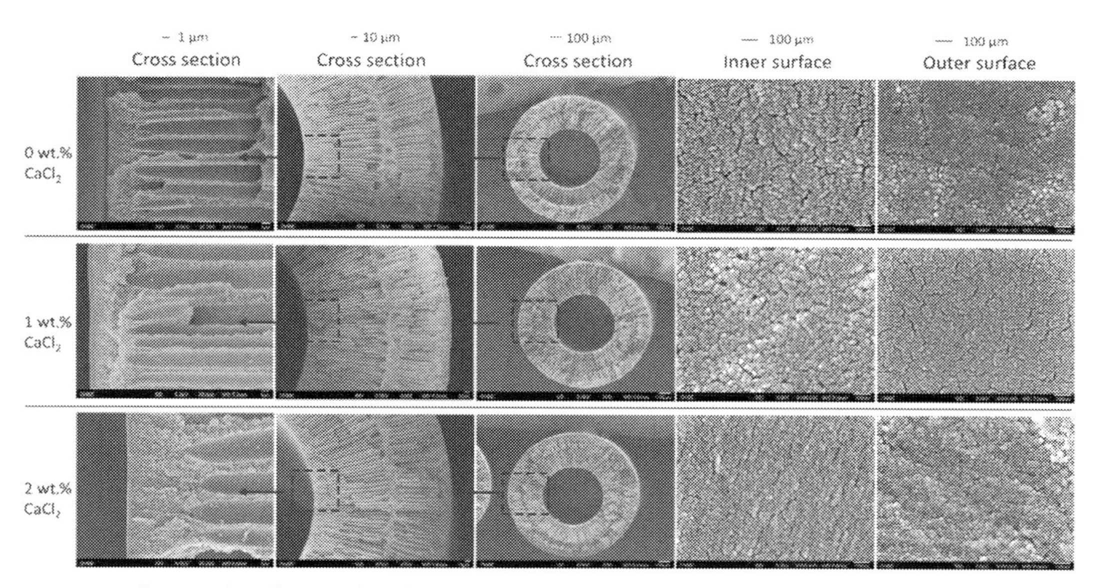

FIGURE 8–5 Showing the influence of $CaCl_2$ concentration on the morphology and structure of different PES hollow-fiber substrates. *PES*, Polyethersulfone. *Adapted with permission from C.F. Wan, T. Yang, W. Gai, Y.D. Lee, T.-S. Chung, Thin-film composite hollow fiber membrane with inorganic salt additives for high mechanical strength and high power density for pressure-retarded osmosis, J. Membr. Sci., 555 (2018) 388–397.*

was expected to reach a peak of 10.0 kW. Another pilot plant facility was installed in Sunndalsøra, Norway, expecting a power output of about 2.0 MW [94]. In 2015 Statkraft proposed the implementation of a pilot plant with a full-scale 25.0-MW power output (Fig. 8–6) [92]. However, the plant was terminated in Tofte due to poor power efficiency, which hindered the economic feasibility of the system.

Another pilot plant was established in Fukuoka City, Japan, in 2012. The project was led by Kyowakiden Industry Co. (a Japanese industrial infrastructure, maintenance, and operation provider) in collaboration with the Tokyo Institute of Technology and Nagasaki University [95]. This PRO plant was tested for a year, and the power density was 10.0 W/m^2 [96]. In 2016, Osmosis Energy UK started operating a PRO industrial plant using hollow-fiber membranes. The maximum power density reached 24.0 W/m^2 at 20.0 bar pressure and 38.0 W/m^2 at 30.0 bar [96] using freshwater feed and NaCl DS.

To maximize the power density further, utilizing a DS with higher salinity than seawater and modifying the membrane support layer are crucial factors. When using this concentrated DS, the power density can be augmented by fourfold as compared to that values in the literature. On the other hand, mass transfer boundary layers can be accelerated, leading to negative outcomes. Even though a concentrated DS can promote the osmotic driving force and back pressure leading to a maximum power density, the operation setups should be adjusted. Representative research indicated that using wastewater as an FS for the PRO process can be limited depending on the availability of the source and location of the plant [21]. The process could be affected by fouling on the feed side due to the presence of complex species in the solution, which necessities pretreatment. According to the results from the

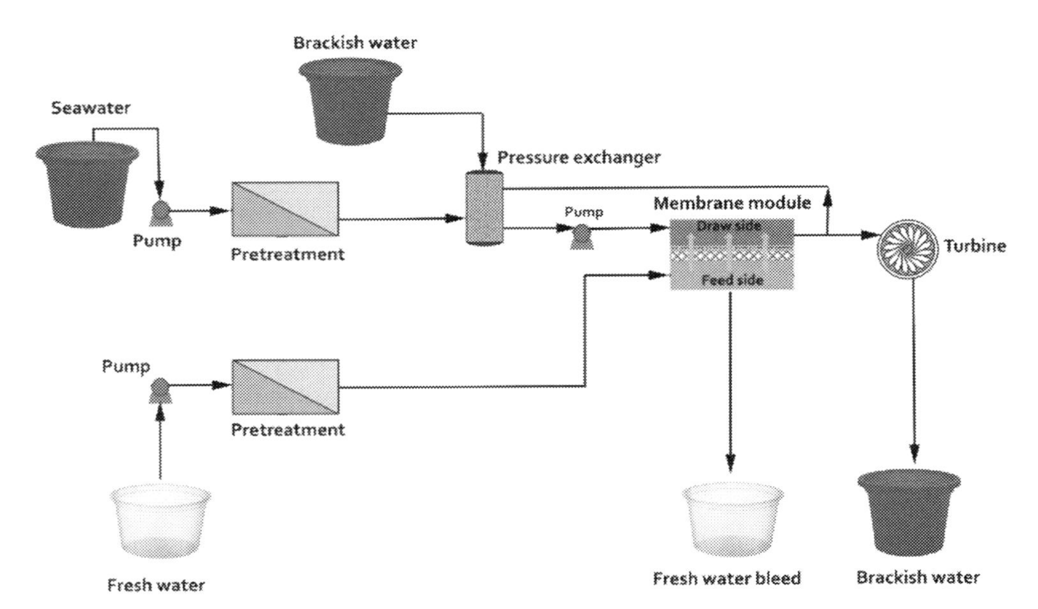

FIGURE 8–6 A diagram illustrates the Statkraft's PRO pilot plant. *PRO*, Pressure-retarded osmosis.

Japanese pilot plant, employing UF-coupled low-pressure RO (LP-RO) as a pretreatment enhanced the power output during the 1-year operation. However, this may impose additional pumping costs, energy consumption, and overall operating cost. Besides, selecting a good fitting system combined PRO system would reduce the capital cost relative to the cost of the PRO stand-alone system. For instance, an RO-coupled PRO system can share the budget for constructing the plant and buying materials. Before designing the hybrid system, the location of building both seawater desalination and water reuse industrial plants should be taken into account. The hybrid system is beneficial because the specific energy consumption of seawater desalination could be decreased to below the minimum theoretical work of separation [97]. Therefore it would be advantageous to hybridize the PRO and a pressure-driven process to expand the water production capacity, the extractable power, improving the quality of the disposed of wastewater, and eliminate environmental problems of disposing of brine to the sea.

8.4.2 Hybrid pressure-retarded osmosis processes

8.4.2.1 Reverse osmosis—pressure-retarded osmosis system

The main hurdle for stand-alone PRO is the low extractable power due to low pressure—driven force raised from mixing the seawater and river water. The net energy can be further reduced when considering the energy expenditure of the pretreatment stage. To overcome these issues, a combination of RO with PRO has been explored to increase the available energy. An alternative DS is hypersaline brine produced as a waste stream by RO. As the brine solution would be diluted during the PRO operation, the required cost of posttreatment for discharging the brine is minimum [90]. Some experts in this field have performed research to study the workability of RO—PRO system under different operating conditions involving the type of feed and DS and optimizing the hydraulic pressure in both the systems to promote the efficiency of the hybrid system [13,89,90,98,99].

Kim et al. [90] evaluated the performance of the RO—PRO system through mathematical models for each system to simulate various parameters for four different hybrid configurations. The difference between these four configurations was the arrangement of process steps and the influent concentration. For example, when RO is placed before the PRO system, the RO brine can be used as the DS for the PRO with no need for the pretreatment method (mode-1 and -2). However, when RO is placed after the PRO system, the PRO acts as a pretreatment stage for the RO system (mode-3 and -4). It has been demonstrated that the comparison between the four modes is based on the water and energy return rate (WERR). The important findings from this work were that when the PRO coupled RO system; their power output grew when using hypersaline brine from the RO process as the DS (mode-2). However, the lowest performing configuration was mode-2 because the membrane size for both systems had the same dimension. It was indicated that the WERR value could be lowered due to the small size of the RO plant, while the PRO plant size showed an insignificant effect. Also, increasing the power cost resulted in minimum WERR value, especially upon using seawater feed for the RO process (mode-2 and -4).

A far-reaching power generation project, the "Mega-ton Water System," proposed the establishment of a hybrid SWRO–PRO–WWT plant in Fukuoka (Japan) [13,100]. The water production capacity of the RO plant was around 1,000,000 m^3/day, while that for sewage wastewater reclamation around 100,000 m^3/day. In this large-scale process during a year, a new PRO hollow-fiber membrane module, 420 m^3/day of sewage effluent as the FS, and 460 m^3/day of the concentrated RO brine as the DS were employed. It was reported that the output of this new membrane was measured as 4.4 W/m^2 upon using 4.7% NaCl as the DS. This project aimed to decrease the salinity of seawater by dilution with concentrated brine from wastewater treatment plants such as membrane bioreactor, and the next stage was the RO plant. A good relationship was found between energy expenditure via high-pressure pump and salinity of the FS; therefore the system consumed low energy. Subsequently, the maximum power extracted from the PRO approached 13.0 W/m^2 under hydrostatic pressure of 30.0 bar due to the high osmotic pressure gradient. It seems possible to fulfill the main objectives of this project of lower energy consumption and cost. This hybrid process is a potential solution for water and energy production, and hence building a full-scale industrial plant is the plan for the near future.

Wan and Chung [98] performed a techno-economic evaluation via the mass transfer model for the hybrid RO–PRO system. In this study, there were two different hybrid configurations utilizing PRO, RO-coupled open-loop PRO system and an RO-coupled closed-loop PRO system. The major variation between the two configurations is that in the closed-loop configuration the diluted and pressurized PRO DSs are recycled as a feed to the RO process (see Fig. 8–7). In this case, there was no need to utilize another ERD in the RO-coupled closed-loop PRO system to generate osmotic power. It has been reported that the RO combined open-loop PRO system produced better osmotic energy than that for RO-coupled closed-loop PRO system. It is because in the RO combined closed-loop PRO system, the concentrated brine as the DS was being diluted, and its salinity was the salinity of the seawater. The maximum power density obtained from the former using a 25% recovery RO as the DS and under optimum hydrostatic pressure of 16.8 bar approached 13.4 W/m^2. The extractable power was enhanced to 15.3 W/m^2 when using a 50% recovery RO as the DS at hydraulic pressure of 18.2 bar. According to the techno-economic analysis, if an existing RO system is coupled closed-loop PRO system, this hybrid configuration is the most cost-effective system. It can be ascribed to the low OpEx of RO system while the cost-saving reached up to $3,698,000/year and $2,081,000/year, for RO and closed-loop PRO, respectively.

Blankert et al. [101] investigated the practicability of a facultative hybrid RO–PRO process using the solution–diffusion model with the PRO process conditions and membrane parameters optimized to increase osmotic power. This system is named facultative because of the large difference between electricity value and water value. It can change from generating outlet from wastewater such as RO process to providing energy recovery from seawater desalination such as PRO process. Three different geometries were examined: spiral wound or flat sheet with spacers, transverse flow element, and cross-wound element depicting the external surface of the hollow fiber. The last geometry was depicting the internal surface of

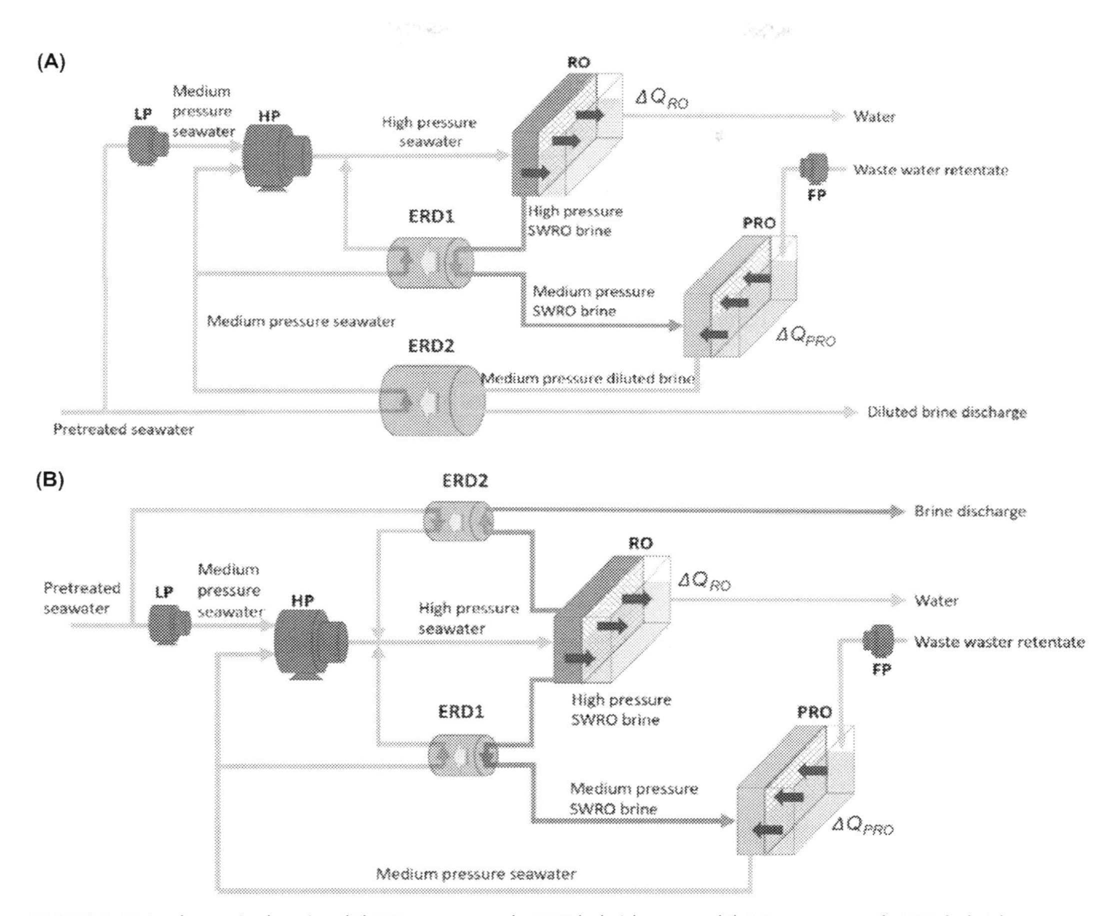

FIGURE 8–7 A schematic showing (A) RO process and oPRO hybrid system. (B) RO process and cPRO hybrid system. *cPRO*, Closed-circuit; *oPRO*, open-circuit pressure-retarded osmosis; *RO*, reverse osmosis. *Adapted with permission from C.F. Wan, T.-S. Chung, Techno-economic evaluation of various RO + PRO and RO + FO integrated processes, Appl. Energy 212 (2018) 1038–1050.*

the hollow fiber. To be feasible economically, the value of permeate should be low enough to maintain the power generation and high enough to shift to water generation. If the highest energy cost is 0.25 USD/kW h, the value of permeate should be below 0.20–0.28 USD/m^3 for the PRO system to satisfy the first assumption. This imposes a high cost of seawater desalination, and therefore using wastewater effluent is desirable. This means that the facultative hybrid RO–PRO principle could be workable if the system is modified, and the treatment of wastewater effluent is needed.

8.4.2.2 Pressure-retarded osmosis–forward osmosis system

Although no hydraulic pressure is used in the FO process, and fouling is insignificant, the technology is not cost-effective for direct seawater desalination [51,102–104]. FO technology also needs an ideal DS with very low reverse solute flux and requires high energy for seawater desalination. Therefore combining FO with PRO has great potential as an emerging hybrid system for water reuse, seawater desalination, and power production. At some point the mixing energy generated experimentally from PRO is lower than the theoretical value raised from the irreversibility of the extraction procedure. Typically, the maximum calculated energy efficiency is roughly 91% of the total energy in the process [105]. To harness more extractable power a similar approach may be applied to the FO–PRO hybrid system, as FO can be used as a pretreatment stage. In this respect, pure water can be driven from the wastewater FS to dilute the seawater DS before it enters the PRO system. Consequently, the diluted seawater from FO can be used as the FS for the PRO system, which can mitigate the fouling of the membrane. It is worth noting that fouling is a severe problem in the PRO membrane because the porous support layer is against the complex FS [103]. Also, fouling mechanism can be more complicated if the reverse solute diffusion is significant.

To investigate the performance of newly developed CA hollow-fiber membranes, a FO/PRO power pilot plant was constructed [106]. In the FO/PRO hybrid system, three hollow-fiber modules were connected in parallel, and their packing density was around 1500 m^2/m^3. In the pilot plant an innovative solution was to install a solenoid valve on the upstream of a hydroturbine machine. In this way the valve was opened, allowing the excess volume from the DS to transfer through a hydroturbine machine. If the valve was closed the hydraulic pressure of the DS could be raised, reaching the same osmotic pressure of the DS, thereby stopping the water permeation. It was outlined that the FO experiment was run using a 0.8 mol/L of $MgSO_4$ DS at a hydraulic pressure of 1.5–3.0 bar, while the intermittent PRO experiment was operated utilizing a 0.8 mol/L of NaCl DS at hydraulic pressure of 11.0 bar. The water and solute fluxes were 33.0 LMH and 2.1 g/L/h in the FO process, while they were 8.0 LMH and 12.7 g/L/h in the PRO process. A better projected power density was achieved of 5.0 W/m^2, which meets the target value for commercialization. The small thickness dimension and large needle-like shape in the structure of the support layer caused excellent water flux, but the mechanical strength was lowered. Therefore the membrane could not produce very high power output due to poor tolerance of the membrane under high applied pressure. Also, the reverse solute flux in the PRO operation was greater, which induced ICP and limited the water permeation and power density.

A case study based on a FO–PRO hybrid process for power generation was reported by Cheng et al. [10]. To obtain more extractable mixing energy, fouling in the PRO process must be alleviated. Here, the FO acts as a pretreatment system to draw pure water from the municipal wastewater feed to pass through the inter-loop solution. As a result, the diluted inter-loop solution can be employed as the FS for the PRO system. It has been confirmed that the power output is higher than 5.0 W/m^2, when the inter-loop solution (0.1 mol/L) and synthetic brine (1.2 mol/L NaCl) as the DS are operated in the PRO process at 23.3 bar. It appears that with an increase in the salinity of the inter-loop FS, the water flux and power output declined sharply, associated with a drop in the hydraulic pressure. The results verified that the PRO-TFC hollow-fiber membrane became more fouling resistance in the fouling test using synthetic wastewater (0.011 mol/L NaCl) and 0.2 mol/L NaCl DS. The thin fouling layer formed on the surface did not suppress remarkably the water flux, and the membrane can be cleaned easily at feed recovery up to 50%. Another interesting observation is that increasing the flow rate (0.64 m/s) in the lumen side of the hollow fiber caused hydrodynamic shear forces that lessen the deposition of foulants on the membrane. The most important point regarding the recovery is that the ultimate water recovery is always adjusted at 50% [97]. However, it was varied for water reuse estimated at 87.5% plus 0.5 m^3/s of outlet due to diluted seawater feed. For example, the FO process can generate water permeation around 25 LMH when using a commercial 8-in. spiral-wound RO module and seawater as the DS. However, the water flux was lower in the PRO process corresponding 18.0 and 15.0 LMH upon using seawater and concentrated brine as the DSs, respectively. Under the same conditions the extractable energy was expected to be 5.0 and 8.8 W/m^2 at hydraulic pressures of 10.0 and 20.0 bar. Spacers for the feed and draw stream sides can be in either the FO system or the PRO system. By using the same membrane with a surface area of 25.9 m^2 and a single spacer in the brine channel, the water recovery approached only 70%. Ultimately, water reuse is preferable over seawater treatment for a water supply because it provides more energy-efficient options.

8.4.2.3 Pressure-retarded osmosis—membrane distillation system

The PRO—MD process offers some appreciable advantages in many studies, although the energy expenditure is high. The most significant merits are that this hybrid system can produce great power density and insignificant fouling [14]. It is generally accepted that the circulated DS contains negligible foulants, which alleviated fouling and minimized the price of the recovery method. By applying the combination strategy to the PRO system the vacuum MD (VMD) system can reduce fouling tendency, increase the recovery rate and the extractable energy [107]. In this study, process variables such as the concentration of both the FS and DS, the water recovery and dilution impact of the DS, the extractable energy, the water permeation, and salt rejection in the MD process, the fouling for both systems were optimized. First, the MD system exhibited negligible fouling as the water flux, and salt rejection reached 63.1 LMH and 99.9% at low operating temperatures 40°C—60°C. Second, the selected commercial PES-TFC hollow-fiber membrane for PRO processes had a superior performance with water and solute permeabilities of 3.3 ± 0.3 LMH/bar and 0.31 ± 0.3 LMH due to reduced S value (450 μm). Alongside this, the membrane withstood hydrostatic pressures greater than 20.0 bar, and hence, a highly concentrated DS can be operated at high applied pressure. As a result, huge growth in the power density was obtained from the PRO—MD hybrid system corresponding to 31.0 and 9.3 W/m^2 when using DI water and real wastewater brine as the FS and the highest DS concentration of 2.0 mol/L NaCl at 20.0 bar. The water and solute fluxes were excellent, at 87.3 LMH and 13.5 $g/m^2/h$ respectively, which minimized the ICP effects. It was observed that increasing the water recovery to 70% caused a rapid decrease in the power density from 7.8 to 2.7 W/m^2. This means that the recovery rate should be reduced to maintain a power output of around 5.0 W/m^2. It should be mentioned that fouling on the feed side was remarkable due to complex wastewater feed in the PRO system. One of the drawbacks is the high energy consumption of the hybrid system, which was imposed by the MD process at great operating temperature.

Lin et al. [108] optimized the performance of the hybrid system through the combination of mass and energy flows for both the processes. The performance and power efficiency of an MD system using low-grade heat sources were investigated. Then, the power density of the PRO system was estimated. In this work the flow rate was correlated strongly to the feed concentration and operating temperature to identify the thermodynamic characteristics of fluids. Thus the flow rate of the hybrid process was optimized to get the greatest power density when using concentrations in the range of 1.0, 2.0, and 4.0 mol/kg NaCl and temperatures varied from 40°C to 80°C. The hybrid process generated a theoretical power efficiency of around 9.8%, which is equal to 81.6% of the Carnot efficiency when using 1.0 M NaCl inlet and temperatures of cold and hot compartments fixed at 20°C and 60°C, respectively. The PRO–MD system could use low-grade heat source <80°C to generate comparable energy to that obtained from binary cycle power plants. It has been explained that the hybrid process produced theoretical efficiency around 9%–10% using a low temperature of about 60°C (75%–83% of Carnot efficiency). However, the experimental power efficiency is assumed lower, and it can be maximized if using greater solution concentration and hybrid pressure. For instance, when using NaCl solutions (1.0, 2.0, and 4.0 mol/kg) in the PRO system at an applied pressure of 46, 100, and 220 bar, the membrane requires modification to improve its mechanical strength.

Lee et al. [109] studied the feasibility of combining multistage VMD (MVMD) with PRO for power production. The closed-loop system operates in MVMD to provide hypersaline brine as the DS for the PRO system and drinking water. A flat sheet HTI-CTA membrane was used for the PRO operation. The PRO membrane had acceptable water permeability of about 1.89×10^{-7} m/s/kPa, solute permeability around 1.11×10^{-7} m/s, and S value of 6.78×10^{-4} m. A small increase in the water permeate occurred when the feed flow rate was kept constant, and the recycling flow was lowered. The concentrated brine from the MVMD-R process was 1.9 mol/L NaCl when adjusting the feed cross velocity and recycling flow at 3.0 kg/min and 90%, respectively.

A new mathematical model was developed to assess the performance of this hybrid system. The results demonstrated that an increase in the water flux and power density was achieved at a higher DS. These values dropped considerably when increasing the FS concentration due to low osmotic pressure gradient and ICP effects. The practical power output produced from the PRO process was 2.8 and 5.1 W/m^2 when using 0.58 and 0.98 mol/L NaCl DSs and DI water feed at 9.7 bar. In comparison with the theoretical data the power outputs were 2.8 and 5.0 W/m^2 under the same conditions. Overall, the greatest power output was 9.7 W/m^2 when utilizing river water feed and brine (1.9 mol/L NaCl) as the DS in the PRO system. The cross velocity of the feed and DS remained constant at 0.5 kg/min, and the hydraulic pressure difference was fixed at 13.0 bar. Since the power density value exceeded the target value of 5.0 W/m^2, this indicated the practicality of the hybrid system.

Thermo-osmotic energy conversion was developed combining, direct contact MD coupled with PRO, which depends on the vapor pressure gradient as the driving force raised from the temperature gradient over the membrane [110]. This process, termed pressure-retarded MD (PRMD), has hydraulic pressure applied to the cold. To produce the required temperature gradient between the pressurized transmembrane water in the hot compartment to a cold compartment, a low-grade heat device can be used. This fluid could be depressurized using a hydroturbine/generator to produce electricity. A commercial nanoporous polytetrafluoroethylene membrane was examined under varied hydraulic pressure from 0 to 10.0 bar and temperature gradient of 40°C. Experimentally, the membrane deformed upon raising the applied pressure resulting in low water flux. When the selective layer was against the hot stream, the water vapor was lost at 2.0 bar. Thus the system was unable to generate power output due to damage of the selective layer structure. When the selective layer was against the cold compartment, the membrane could tolerate hydraulic pressure of 10.0 bar. At this hydraulic pressure the membrane showed poor water permeability and power output of 1.8 LMH and 0.4 W/m^2, respectively, due to membrane deformation. This indicates that the development of a robust membrane with high mechanical stability is necessary for the commercialization of the PRMD process.

8.4.2.4 Nanofiltration–pressure-retarded osmosis system

Few reports have been conducted on pressure-driven processes as a pretreatment strategy for the PRO process. The UF process is cost-effective, but the pore size of the membrane leads to the penetration of salt [32]. NF process was used for wastewater reclamation, but the energy consumption was high due to applying pressure between 10 and 40 bar. The high energy consumption can be avoided by utilizing the low-pressure NF membrane. This type of membrane has an efficient performance in terms of high water permeation and salt rejection rate for multivalent ions and small organic components at low pressure around 2.0 bar [111]. For instance, Chen et al. [112] employed the NF membrane as a pretreatment step before the PRO system. Since real wastewater FS was used in the PRO process, it was necessary to use an NF pretreatment stage to minimize the fouling and enhance water permeation and power density. In this hybrid system a polyetherimide (PEI) PRO hollow-fiber membrane was used due to high water permeability (3.8 LMH/bar), minimum solute permeability (0.25 LMH), and *S* value around 527 μm. When the filtered wastewater and 1.0 mol/L DS were tested in the PRO system, water flux was doubled to 30.5 LMH at 16.0 bar applied pressure, due to high selectivity of NF membrane compared to that for the unfiltered wastewater (9.0 LMH). The PRO membrane could produce high water permeation if the wastewater FS included lower concentration and scaling precursors. At high water flux the membrane needed more cleaning activities to remove foulants that were deposited on the membrane surface. Under the same conditions the greatest power output was achieved at 13.5 W/m^2, while it was only 4.4 W/m^2 for unfiltered FS. Ultimately, both the water permeation and power density could be further maximized in the absence of silica fouling to improve the feasibility of the PRO process.

Yang et al. [113] studied the performance of three different newly fabricated membranes—UF, NF, and LP-RO—as prefiltration methods for wastewater effluent. The pretreated real municipal wastewater solution from each membrane driven process was passed through the PRO system. It was reported that NF and LP-RO were the best pretreatment methods to alleviate fouling and scaling. When using the pretreated wastewater as feed and synthetic brine (0.8 mol/L NaCl) as DS, the PES hollow-fiber membrane was operated in the PRO process at 15.0 bar. The fabricated hollow fiber with a packing density of 2.5% showed good performance. Experimental results revealed that the water flux was the highest, around 20.1 for the pretreated wastewater feed, while it was reduced to 17.6 LMH for that produced from the NF system. Likewise, the PRO system achieved the greatest power output of 8.4 W/m^2 when using pretreated wastewater from the LP-RO process. In comparison with the power output generated from untreated wastewater (2.92 W/m^2), the power output from PRO using treated wastewater via the LP-RO system was increased by fivefold and declined by only 7.0% from the initial value (21.6 W/m^2). The power output from PRO using treated wastewater via the NF system was about 7.3 W/m^2, which was lower than the initial value by 10.7% (19.7 W/m^2). It was concluded that the NF pretreatment strategy is optimal due to higher water permeability (2.78 LMH/bar); less time and energy are required as compared to the LP-RO system (1.51 LMH/bar).

8.5 Conclusion and future research needs

The PRO conception has given a platform to allow academic research and industrial investigation for osmotic power generation. The feasibility of the PRO process for power exploitation was suspended for many years due to the absence of an ideal membrane and incentive to explore other energy sources. Theoretical results highlighted that the trade-off between power density and specific energy is challenging, which restricted the commercialization of the PRO process. The research scope of mathematical modeling should be broadened using advanced computational algorithm methods and machine learning to study mass transport, membrane intrinsic parameters, and fouling.

In recent publications, researchers have addressed significant progress in the advancement of novel PRO membranes. An optimal membrane configuration plays an important role in improving the overall performance, and more tests should be conducted to optimize the transport characteristics and system conditions. It is recognized that the hollow-fiber configuration in the case of using a pressurized DS on the shell side is the best performing membrane up until now. Therefore the self-supporting geometry of the membrane fibers could be used under high pressures. In comparison with flat sheet configuration the hollow-fiber membrane can be manufactured easily to obtain great packing density and improved performance. Hollow-fiber membranes showed excellent intrinsic transport characteristics owing to the breakthroughs in robust fabrication methodologies and new materials.

Past research has highlighted that if the developed membrane is effective for the PRO process, the salinity of seawater can sustain an industrial plant producing power at a density over 5.0 W/m^2. Synthetic feed and DSs, including pure NaCl, were used, which did not depict the actual composition of wastewater effluent, seawater brine, brackish water, and other water sources. Thus more experiments should be carried out using real water sources to understand their effects on the developed membrane during long term operations. By decreasing the S value of the support layer the optimal hydrostatic counter-pressure could be minimized remarkably [114]. In most cases the optimal counter-pressure was higher than the osmotic pressure of the DS governed by the nonlinear ICP during the experiments.

By considering the characteristics of the membrane and solutions, it would be possible to develop the optimal PRO membrane for power exploitation. Future researches should aim at alleviating the fouling issue, which is affected by the type and composition of the feed and DSs, pretreatment of the FS, membrane module properties, and design. Therefore increased efforts are required to set design criteria for membrane modules, such as module configurations, mass transfer phenomena, and membrane spacers, which are rarely discussed. It is important to modify the available membranes via controlling the fabrication conditions of the support layer to reduce the thickness and S value. Meanwhile, impregnating advanced materials such as molecular sieving and aquaporins may provide high selectivity to the PRO membrane and breach the efficiency hurdle for practicality [115]. Also, there are numerous chemicals, hybrid materials, and functional materials that can be incorporated into the support layer to boost the mechanical strength [116,117].

To get a clear insight into fouling mechanism and the suitable cleaning method, more experiments should be carried out using large-scale operation instead of lab-scale operation owing to the variation in the process conditions and membrane cell. It seems that membrane fouling can be minimized by changing the membrane orientation to AL-FS [58]. This would lower the energy expenditure because of no need for the pretreatment method, and it became a cost-effective process. The desired antifouling method should reduce fouling/scaling and keep the operating price and the footprint of the process as low as possible. This may help in advancing the cleaning procedures and antifouling membranes.

To improve power density the draw osmotic pressure should be increased above the pressure at maximum power output. So far, more efforts are required to develop concentrated DSs for maximum power generation. Another promising method is to operate the PRO process based on a multistage setup in which the diluted DS from each stage enters the following stage leading to an increase in extractable energy.

The combination of PRO and pressure-driven membrane processes for water production and energy extraction needs further optimization. Although the NF membrane is a promising pretreatment method providing better water flux and power density than other pressure-driven membranes, fouling is still a major obstacle affecting its performance.

Encouragingly, great advantages on reduced fouling, superior water recovery, and power output have been demonstrated in the MD process. However, the system required high energy demand. Consequently, further experiments are needed to employ low-grade heat from industry, natural energy, and geothermal power sources as thermal energy for the MD system [108]. As the prefiltration of the feed is significant for the hypersaline-based PRO process, the feed spacers are necessary, and the water flux should be maximum. The hybrid process of PRO and RO has great potential for harnessing power, but the major hurdle is also energy consumption. Thus it is of paramount importance to use a large membrane area for any membrane to get the greatest specific energy. On the other side, this may result in low-projected power density. One of the keys to successfully integrating both systems is to use the concentrated brine as the draw for PRO, and the final diluted seawater can be operated as the feed for RO desalination. In this respect, the energy consumption might be decreased, and discharging RO brine to the sea can be eliminated. More work is required to estimate the energy consumption and economics of the pretreatment stage, multistage PRO process, and the hybrid system. An investigation into the capital and operation cost, membrane replacement price, and maintenance for long term pilot plants are recommended. Also, most of the studies were conducted using a bench-scale operation without installing turbines in the PRO system. It is worth mentioning that the obtainable power from the PRO plant depends on not only the membrane properties but also the efficiency of the turbine/ generator and power recovery machine. Finally, it is encouraged to perform more experiments in the pilot plant because the results of the extractable power would be more accurate than that for the lab-scale process and to find out other issues during the operations.

References

[1] G.L. Wick, Power from salinity gradients, Energy 3 (1978) 95–100.

[2] H.T. Madsen, S.S. Nissen, J. Muff, E.G. Søgaard, Pressure retarded osmosis from hypersaline solutions: investigating commercial FO membranes at high pressures, Desalination 420 (2017) 183–190.

[3] H.K. Shon, S. Phuntsho, T.C. Zhang, R.Y. Surampalli, Membrane development for pressure-retarded osmosis, Forward Osmosis: Fundamentals and Applications, American Society of Civil Engineers, 2015, pp. 465–490.

[4] S. Chou, R. Wang, A.G. Fane, Robust and high performance hollow fiber membranes for energy harvesting from salinity gradients by pressure retarded osmosis, J. Membr. Sci. 448 (2013) 44–54.

[5] A. Bogler, S. Lin, E. Bar-Zeeva, Biofouling of membrane distillation, forward osmosis and pressure retarded osmosis: principles, impacts and future directions, J. Membr. Sci. 542 (2017) 378–398.

[6] Q. She, J. Wei, N. Ma, V. Sim, A.G. Fane, R. Wang, et al., Fabrication and characterization of fabric-reinforced pressure retarded osmosis membranes for osmotic power harvesting, J. Membr. Sci. 504 (2016) 75–88.

[7] L. Zhang, Q. She, R. Wang, S. Wongchitphimon, Y. Chen, A.G. Fane, Unique roles of aminosilane in developing anti-fouling thin film composite (TFC) membranes for pressure retarded osmosis (PRO), Desalination 389 (2016) 119–128.

[8] A. Altaee, A. Sharif, Pressure retarded osmosis: advancement in the process applications for power generation and desalination, Desalination 356 (2015) 31–46.

[9] W. He, Y. Wang, A. Sharif, M.H. Shaheed, Thermodynamic analysis of a stand-alone reverse osmosis desalination system powered by pressure retarded osmosis, Desalination 352 (2014) 27–37.

[10] Z.L. Cheng, X. Li, T.S. Chung, The forward osmosis-pressure retarded osmosis (FO-PRO) hybrid system: a new process to mitigate membrane fouling for sustainable osmotic power generation, J. Membr. Sci. 559 (2018) 63–74.

[11] N.M. Bazhin, Water flux in pressure retarded osmosis, Desalination 375 (2015) 21–23.

[12] A.S. Kim, H. Kim, Membrane thermodynamics for osmotic phenomena, InTech 68406 (2017) 1–26. Available from: https://doi.org/10.5772/intechopen.68406.

[13] J. Kim, K. Jeong, M.J. Park, H.K. Shon, J.H. Kim, Recent advances in osmotic energy generation via pressure-retarded osmosis (PRO): a review, Energies 8 (2015) 11821–11845.

[14] S.H. Chae, Y.M. Kim, H. Park, J. Seo, S.J. Lim, J.H. Kim, Modeling and simulation studies analyzing the pressure-retarded osmosis (PRO) and PRO-hybridized processes, Energies 12 (2019) 1–35.

[15] G. O'Toole, L. Jones, C. Coutinho, C. Hayes, M. Napoles, A. Achilli, River-to-sea pressure retarded osmosis: resource utilization in a full-scale facility, Desalination 389 (2016) 39–51.

[16] S. Chou, R. Wang, L. Shi, C. Tang, A.G. Fane, Thin-film composite hollow fiber membranes for pressure retarded osmosis (PRO) process with high power density, J. Membr. Sci. 389 (2012) 25–33.

[17] Y. Li, S. Zhao, L. Setiawan, L. Zhang, R. Wang, Integral hollow fiber membrane with chemical cross-linking for pressure retarded osmosis operated in the orientation of active layer facing feed solution, J. Membr. Sci. 550 (2018) 163–172.

[18] H. Gong, D.D. Anastasio, K. Wang, J.R. McCutcheon, Finding better draw solutes for osmotic heat engines: understanding transport of ions during pressure retarded osmosis, Desalination 421 (2017) 32–39.

[19] J. Lee, S. Kim, Predicting power density of pressure retarded osmosis (PRO) membranes using a new characterization method based on a single PRO test, Desalination 389 (2016) 224–234.

[20] H. Kim, J.-S. Choi, S. Lee, Pressure retarded osmosis for energy production: membrane materials and operating conditions, Water Sci. Technol. 65 (2012) 1789–1794.

[21] N. Bajraktari, C. Hélix-Nielsen, H.T. Madsen, Pressure retarded osmosis from hypersaline sources—a review, Desalination 413 (2017) 65–85.

[22] A.P. Straub, S. Lin, M. Elimelech, Module-scale analysis of pressure retarded osmosis: performance limitations and implications for full-scale operation, Environ. Sci. Technol. 48 (2014) 12435–12444.

[23] S. Sarp, Z. Li, J. Saththasivam, Pressure retarded osmosis (PRO): past experiences, current developments, and future prospects, Desalination 389 (2016) 2–14.

[24] K. Touati, F. Tadeo, Study of the reverse salt diffusion in pressure retarded osmosis: influence on concentration polarization and effect of the operating conditions, Desalination 389 (2016) 171–186.

[25] Z.L. Cheng, T.S. Chung, Mass transport of various membrane configurations in pressure-retarded osmosis (PRO), J. Membr. Sci. 537 (2017) 160–176.

[26] B. Kim, G. Gwak, S.-K. Hong, Review on methodology for determining forward osmosis (FO) membrane characteristics: water permeability (A), solute permeability (B), and structural parameter (S), Desalination 422 (2017) 5–16.

[27] Z.L. Cheng, X. Li, Y.D. Liu, T.S. Chung, Robust outer-selective thin-film composite polyethersulfone hollow fiber membranes with low reverse salt flux for renewable salinity-gradient energy generation, J. Membr. Sci. 506 (2016) 119–129.

[28] S.-P. Sun, T.-S. Chung, Outer-selective pressure-retarded osmosis hollow fiber membranes from vacuum-assisted interfacial polymerization for osmotic power generation, Environ. Sci. Technol. 47 (2013) 13167–13174.

[29] S. Zhang, P. Sukitpaneenit, T.-S. Chung, Design of robust hollow fiber membranes with high power density for osmotic energy production, Chem. Eng. J. 241 (2014) 457–465.

[30] M. Sharma, P. Mondal, A. Chakraborty, J. Kuttippurath, M. Purkait, Effect of different molecular weight polyethylene glycol on flat sheet cellulose acetate membranes for evaluating power density performance in pressure retarded osmosis study, J. Water Process. Eng. 30 (2019) 100632.

[31] Y. Tanaka, M. Yasukawa, S. Goda, H. Sakurai, M. Shibuya, T. Takahashi, et al., Experimental and simulation studies of two types of 5-inch scale hollow fiber membrane modules for pressure-retarded osmosis, Desalination 447 (2018) 133–146.

[32] C.F. Wan, T. Chung, Osmotic power generation by pressure retarded osmosis using seawater brine as the draw solution and wastewater retentate as the feed, J. Membr. Sci. 479 (2015) 148–158.

[33] G. Han, S. Zhang, X. Li, T. Chung, High performance thin film composite pressure retarded osmosis (PRO) membranes for renewable salinity-gradient energy generation, J. Membr. Sci. 440 (2013) 108–121.

[34] Y. Oh, S. Lee, M. Elimelech, S. Lee, S.K. Hong, Effect of hydraulic pressure and membrane orientation on water flux and reverse solute flux in pressure assisted osmosis, J. Membr. Sci. 465 (2014) 159–166.

[35] S. Zhang, K.Y. Wang, T.S. Chung, H. Chen, Y.C. Jean, G. Amy, Well-constructed cellulose acetate membranes for forward osmosis: minimized internal concentration polarization with an ultra-thin selective layer, J. Membr. Sci. 360 (2010) 522–535.

[36] T.P.N. Nguyen, B.M. Jun, J.H. Lee, Y.N. Kwon, Comparison of integrally asymmetric and thin film composite structures for a desirable fashion of forward osmosis membranes, J. Membr. Sci. 495 (2015) 457–470.

[37] T.P. Nguyen, B. Jun, H.G. Park, S. Han, Y. Kim, H.K. Lee, et al., Concentration polarization effect and preferable membrane configuration at pressure-retarded osmosis operation, Desalination 389 (2016) 58–67.

[38] J.W. Post, J. Veerman, H.V.M. Hamelers, G.J.W. Euverink, S.J. Metz, K. Nymeijer, et al., Salinity-gradient power: evaluation of pressure-retarded osmosis and reverse electrodialysis, J. Membr. Sci. 288 (2007) 218–230.

[39] Y.G. Anissimov, Aspects of mathematical modelling of pressure retarded osmosis, Membranes 6 (2016) 1–11.

[40] A. Altaee, A. Cipolina, Modelling and optimization of modular system for power generation from a salinity gradient, Renew. Energy 141 (2019) 139–147.

[41] X. Song, Z. Liu, D.D. Sun, Energy recovery from concentrated seawater brine by thin-film nanofiber composite pressure retarded osmosis membranes with high power density, Energy Environ. Sci. 6 (2013) 1199–1210.

[42] Q. She, X. Jin, C.Y. Tang, Osmotic power production from salinity gradient resource by pressure retarded osmosis: effects of operating conditions and reverse solute diffusion, J. Membr. Sci. 401–402 (2012) 262–273.

[43] G. Han, S. Zhang, X. Li, T.S. Chung, Progress in pressure retarded osmosis (PRO) membranes for osmotic power generation, Prog. Polym. Sci. 51 (2015) 1–27.

[44] I.L. Alsvik, M.-B. Hägg, Pressure retarded osmosis and forward osmosis membranes: materials and methods, Polymers 5 (2013) 303–327.

[45] N.Y. Yip, A. Tiraferri, W.A. Phillip, J.D. Schiffman, L.A. Hoover, Y.C. Kim, et al., Thin-film composite pressure retarded osmosis membranes for sustainable power generation from salinity gradients, Environ. Sci. Technol. 45 (2011) 4360–4369.

[46] A.K. Ghosh, B.H. Jeong, X. Huang, E.M.V. Hoek, Impacts of reaction and curing conditions on polyamide composite reverse osmosis membrane properties, J. Membr. Sci. 311 (2008) 34–45.

[47] S. Zhang, F. Fu, T.-S. Chung, Substrate modifications and alcohol treatment on thin film composite membranes for osmotic power, Chem. Eng. Sci. 87 (2013) 40–50.

[48] G.D. Mehta, S. Loeb, Performance of permasep B-9 and B-10 membranes in various osmotic regions and at high osmotic pressures, J. Membr. Sci. 4 (3) (1979) 35–49.

[49] S. Loeb, F.V. Hessen, D. Shahaf, Production of energy from concentrated brines by pressure-retarded osmosis: II. Experimental results and projected energy costs, J. Membr. Sci. 1 (2) (1976) 49–69.

[50] G.D. Mehta, S. Loeb, Internal polarization in the porous substructure of a semipermeable membrane under pressure-retarded osmosis, J. Membr. Sci. 4 (26) (1978) 1–5.

[51] W.A. Suwaileh, D.J. Johnson, S. Sarp, N. Hilal, Advances in forward osmosis membranes: altering the sub-layer structure via recent fabrication and chemical modification approaches, Desalination 436 (2018) 176–201.

[52] A. Achilli, A.E. Childress, Pressure retarded osmosis: from the vision of Sidney Loeb to the first prototype installation—review, Desalination 261 (2010) 205–211.

[53] H.H.G. Jellinek, H. Masuda, Osmo-power. Theory and performance of an osmo-power pilot plant, Ocean. Eng. 8 (1981) 103–128.

[54] C. Klaysom, T.Y. Cath, T. Depuydt, I.F.J. Vankelecom, Forward and pressure retarded osmosis: potential solutions for global challenges in energy and water supply, Chem. Soc. Rev. 42 (69) (2013) 59–89.

[55] Y. Xu, X.Y. Peng, C.Y. Tang, Q.S.A. Fu, S.Z. Nie, Effect of draw solution concentration and operating conditions on forward osmosis and pressure retarded osmosis performance in a spiral wound module, J. Membr. Sci. 348 (2010) 298–309.

[56] Y.C. Kim, M. Elimelech, Potential of osmotic power generation by pressure retarded osmosis using seawater as feed solution: analysis and experiments, J. Membr. Sci. 429 (2013) 330–337.

[57] A. Achilli, T.Y. Cath, A.E. Childress, Power generation with pressure retarded osmosis: an experimental and theoretical investigation, J. Membr. Sci. 343 (2009) 42–52.

[58] J. Kim, B. Kim, D.I. Kim, S. Hong, Evaluation of apparent membrane performance parameters in pressure retarded osmosis processes under varying draw pressures and with draw solutions containing organics, J. Membr. Sci. 493 (2015) 636–644.

[59] Y. Li, R. Wang, S. Qi, C. Tang, Structural stability and mass transfer properties of pressure retarded osmosis (PRO) membrane under high operating pressures, J. Membr. Sci. 488 (2015) 143–153.

[60] D.I. Kim, J. Kim, S. Hong, Changing membrane orientation in pressure retarded osmosis for sustainable power generation with low fouling, Desalination 389 (2016) 197–206.

[61] L. Huang, J.T. Arena, M.T. Meyering, T.J. Hamlin, J.R. McCutcheon, Tailored multi-zoned nylon 6,6 supported thin film composite membranes for pressure retarded osmosis, Desalination 399 (2016) 96–104.

[62] X. Li, S. Zhang, F. Fu, T.-S. Chung, Deformation and reinforcement of thin-film composite (TFC) polyamide-imide (PAI) membranes for osmotic power generation, J. Membr. Sci. 434 (2013) 204–217.

[63] Y. Cui, X. Liu, T.S. Chung, Enhanced osmotic energy generation from salinity gradients by modifying thin film composite membranes, Chem. Eng. J. 242 (2014) 195–203.

[64] M. Son, H. Park, L. Liu, H. Choi, J.H. Kim, H. Choi, Thin-film nanocomposite membrane with CNT positioning in support layer for energy harvesting from saline water, Chem. Eng. J. 284 (2016) 68–77.

[65] N.-N. Bui, J.R. McCutcheon, Nanofiber supported thin-film composite membrane for pressure retarded osmosis, Environ. Sci. Technol. 48 (2014) 4129–4136.

[66] X. Li, T.S. Chung, Effects of free volume in thin-film composite membranes on osmotic power generation, AIChE J. 59 (47) (2013) 49–61.

[67] K.L. Hickenbottom, J. Vanneste, M. Elimelech, T.Y. Cath, Assessing the current state of commercially available membranes and spacers for energy production with pressure retarded osmosis, Desalination 389 (2016) 108–118.

[68] J. Wei, Y. Li, L. Setiawan, R. Wang, Influence of macromolecular additive on reinforced flat-sheet thin film composite pressure-retarded osmosis membranes, J. Membr. Sci. 511 (2016) 54–64.

[69] R.R. Gonzales, M.J. Park, L. Tijing, D.S. Han, S. Phuntsho, H.K. Shon, Modification of nanofiber support layer for thin film composite forward osmosis membranes via layer-by-layer polyelectrolyte deposition, Membranes 8 (2018) 1–15.

[70] N.N. Bui, M.L. Lind, E.M.V. Hoek, M.J.R. Cutcheon, Electrospun nanofiber supported thin film composite membranes for engineered osmosis, J. Membr. Sci. 385–386 (2011) 10–19.

[71] L.A. Hoover, J.D. Schiffman, M. Elimelech, Nanofibers in thin-film composite membrane support layers: enabling expanded application of forward and pressure retarded osmosis, Desalination 308 (2013) 73–81.

[72] R.M. El Khaldi, M.E. Pasaoglu, S. Guclu, Y.Z. Menceloglu, R. Ozdogan, M. Celebi, et al., Fabrication of high-performance nanofiber-based FO membranes, Desalin. Water Treat. 147 (2019) 56–72.

[73] J.H. Kim, S.J. Moon, S.H. Park, M. Cook, A.G. Livingston, Y.M. Lee, A robust thin film composite membrane incorporating thermally rearranged polymer support for organic solvent nanofiltration and pressure retarded osmosis, J. Membr. Sci. 550 (2018) 322–331.

[74] E. Sivertsen, T. Holt, W. Thelin, G. Brekke, Pressure retarded osmosis efficiency for different hollow fibre membrane module flow configurations, Desalination 312 (2013) 107–123.

[75] S. Zhang, T.-S. Chung, Minimizing the instant and accumulative effects of salt permeability to sustain ultrahigh osmotic power density, Environ. Sci. Technol. 47 (2013) 10085–10092.

[76] R.W.S. Chou, L. Shia, Q. She, C. Tang, A.G. Fanea, Thin-film composite hollow fiber membranes for pressure retarded osmosis (PRO) process with high power density, J. Membr. Sci. 389 (2012) 25–33.

[77] F.-J. Fu, S. Zhang, S.-P. Sun, K.-Y. Wang, T.-S. Chung, POSS-containing delamination-free dual-layer hollow fiber membranes for forward osmosis and osmotic power generation, J. Membr. Sci. 443 (2013) 144–155.

[78] P.G. Ingole, W. Choi, K.H. Kim, C.H. Park, W.K. Choi, H.K. Lee, Synthesis, characterization and surface modification of PES hollow fiber membrane support with polydopamine and thin film composite for energy generation, Chem. Eng. J. 243 (2014) 137–146.

[79] G. Han, T.-S. Chung, Robust and high performance pressure retarded osmosis hollow fiber membranes for osmotic power generation, AIChE J. 60 (2014) 1107–1119.

[80] G. Han, Q. Ge, T.-S. Chung, Conceptual demonstration of novel closed-loop pressure retarded osmosis process for sustainable osmotic energy generation, Appl. Energy 132 (2014) 383–393.

[81] X. Li, T.S. Chung, Thin-film composite P84 co-polyimide hollow fiber membranes for osmotic power generation, Appl. Energy 114 (2014) 600–610.

[82] P.G. Ingole, K.H. Kim, C.H. Park, W.K. Choi, H.K. Lee, Preparation, modification and characterization of polymeric hollow fiber membranes for pressure-retarded osmosis, RSC Adv. 4 (2014) 51430–51439.

[83] P.G. Ingole, W. Choi, K.-H. Kim, H.-D. Jo, W.-K. Choi, J.-S. Park, et al., Preparation, characterization and performance evaluations of thin film composite hollow fiber membrane for energy generation, Desalination 345 (2014) 136–145.

[84] Y. Chen, L. Setiawan, S. Chou, X. Hub, R. Wang, Identification of safe and stable operation conditions for pressure retarded osmosis with high performance hollow fiber membrane, J. Membr. Sci. 503 (2016) 90–100.

[85] Y. Chen, C.H. Loh, L. Zhang, L. Setiawan, Q. She, Fang, et al., Module scale-up and performance evaluation of thin film composite hollow fiber membranes for pressure retarded osmosis, J. Membr. Sci. 548 (2018) 398–407.

[86] M. Higa, D. Shigefuji, M. Shibuya, S. Izumikawa, Y. Ikebe, M. Yasukawa, et al., Experimental study of a hollow fiber membrane module in pressure-retarded osmosis: module performance comparison with volumetric-based power outputs, Desalination 420 (2017) 45–53.

[87] C.F. Wan, T. Yang, W. Gai, Y.D. Lee, T.-S. Chung, Thin-film composite hollow fiber membrane with inorganic salt additives for high mechanical strength and high power density for pressure-retarded osmosis, J. Membr. Sci. 555 (2018) 388–397.

[88] G. Han, Z.L. Cheng, T.-S. Chung, Thin-film composite (TFC) hollow fiber membrane with double-polyamide active layers for internal concentration polarization and fouling mitigation in osmotic processes, J. Membr. Sci. 523 (2017) 497–504.

[89] J. Kim, J. Lee, J.H. Kim, Overview of pressure-retarded osmosis (PRO) process and hybrid application to sea water reverse osmosis process, Desalin. Water Treat. 43 (2012) 193–200.

[90] J. Kim, M. Park, S.A. Snyder, J.H. Kim, Reverse osmosis (RO) and pressure retarded osmosis (PRO) hybrid processes: model-based scenario study, Desalination 322 (2013) 121–130.

[91] K. Touati, F. Tadeo, Green energy generation by pressure retarded osmosis: state of the art and technical advancement—review, Int. J. Green Energy 14 (2017) 337–360.

[92] J.J. Wu, R.W. Field, On the understanding and feasibility of "Breakthrough" osmosis, Sci. Rep. 9 (2019).

[93] F. Helfer, C. Lemckert, Y.G. Anissimov, Osmotic power with pressure retarded osmosis: theory, performance and trends—a review, J. Membr. Sci. 453 (2014) 337–358.

[94] R. Reidy, IDE Adapts Desal Technology for Osmotic Power Generation, Pump Industry Analyst, Elsevier Limited, Oxford, 2013, p. 3.

[95] K. Saito, M. Irie, S. Zaitsu, H. Sakai, H. Hayashi, A. Tanioka, Power generation with salinity gradient by pressure retarded osmosis using concentrated brine from SWRO system and treated sewage as pure water, Desalin. Water Treat. 41 (2012) 114–121.

[96] M. Perry, F.O. Tech, 2017. <https://www.forwardosmosistech.com/according-to-osmosis-energy-pressure-retarded-osmosis-using-hollow-fibers-is-now-a-viable-means-of-energy-production/>. 2017 (accessed 23.09.20).

[97] V.S.T. Sim, Q. She, T.H. Chong, C.Y. Tang, A.G. Fane, W.B. Krantz, Strategic co-location in a hybrid process involving desalination and pressure retarded osmosis (PRO), Membranes 3 (2013) 98–125.

[98] C.F. Wan, T.-S. Chung, Techno-economic evaluation of various RO + PRO and RO + FO integrated processes, Appl. Energy 212 (2018) 1038–1050.

[99] M. Li, Optimization of multi-stage hybrid RO-PRO membrane processes at the water–energy nexus, Chem. Eng. Res. Des. 137 (2018) 1–9.

[100] M. Kurihara, M. Hanakawa, Mega-ton water system: Japanese national research and development project on seawater desalination and wastewater reclamation, Desalination 308 (2013) 131–137.

[101] B. Blankert, Y. Kim, H. Vrouwenvelder, N. Ghaffour, Facultative hybrid RO-PRO concept to improve economic performance of PRO: feasibility and maximizing efficiency, Desalination 478 (2020) 114–268.

[102] D.J. Johnson, W.A. Suwaileh, A. Mohammed, N. Hilal, Osmotic's potential: an overview of draw solutes for forward osmosis, Desalination 434 (2018) 100–120.

[103] T.-S. Chung, L. Luo, C.F. Wan, Y. Cui, G. Amya, What is next for forward osmosis (FO) and pressure retarded osmosis (PRO), Sep. Purif. Technol. 156 (2015) 856–860.

[104] C. Klaysom, T.Y. Cath, T. Depuydt, I.F.J. Vankelecom, Forward and pressure retarded osmosis: potential solutions for global challenges in energy and water supply, Chem. Soc. Rev. 42 (2013) 6959–6989.

[105] D. Attarde, M. Jain, S.K. Gupta, Modeling of a forward osmosis and a pressure-retarded osmosis spiral wound module using the Spiegler-Kedem model and experimental validation, Sep. Purif. Technol. 164 (2016) 182–197.

[106] N.R.J.-D. Mermier, C.P. Borges, Direct osmosis process for power generation using salinity gradient: FO/PRO pilot plant investigation using hollow fiber modules, Chem. Eng. Process. 103 (2016) 27–36.

[107] G. Han, J. Zuo, C. Wan, T.-S. Chung, Hybrid pressure retarded osmosis–membrane distillation (PRO–MD) process for osmotic power and clean water generation, Environ. Sci. Water Res. Technol. 1 (2015) 507–515.

[108] S. Lin, N.Y. Yip, T.Y. Cath, C.O. Osuji, M. Elimelech, Hybrid pressure retarded osmosis–membrane distillation system for power generation from low-grade heat: thermodynamic analysis and energy efficiency, Environ. Sci. Technol. 48 (2014) 5306–5313.

[109] J.-G. Lee, Y.-D. Kim, S.-M. Shim, B.-G. Ima, W.-S. Kim, Numerical study of a hybrid multi-stage vacuum membrane distillation and pressure-retarded osmosis system, Desalination 363 (2015) 82–91.

[110] Z. Yuan, L. Wei, J.D. Afroze, K. Goh, Y. Chen, Y. Yu, et al., Pressure-retarded membrane distillation for low-grade heat recovery: the critical roles of pressure-induced membrane deformation, J. Membr. Sci. 579 (2019) 90–101.

[111] C. Liu, L. Shi, R. Wang, Enhanced hollow fiber membrane performance via semidynamic layer-by-layer polyelectrolyte inner surface deposition for nanofiltration and forward osmosis applications, React. Funct. Polym. 86 (2015) 154–160.

[112] Y. Chen, C. Liu, L. Setiawan, Y.N. Wang, X. Hub, R. Wang, Enhancing pressure retarded osmosis performance with low-pressure nanofiltration pretreatment: membrane fouling analysis and mitigation, J. Membr. Sci. 543 (2017) 114–122.

[113] T. Yang, C.F. Wan, J.Y. Xiong, T.-S. Chung, Pre-treatment of wastewater retentate to mitigate fouling on the pressure retarded osmosis (PRO) process, Sep. Purif. Technol. 215 (2019) 390–397.

[114] A. Yaroshchuk, Optimal hydrostatic counter-pressure in pressure-retarded osmosis with composite/asymmetric membranes, J. Membr. Sci. 477 (2015) 157–160.

[115] C.F. Wan, Y. Cui, W.X. Gai, Z.L. Cheng, T.-S. Chung, Nanostructured membranes for enhanced forward osmosis and pressure-retarded osmosis, Sustainable Nanoscale Engineering: From Materials Design to Chemical Processing, Elsevier, 2020, pp. 373–394.

[116] S.F. Anis, B.S. Lalia, A.O. Mostafa, R. Hashaikeh, Electrospun nickel–tungsten oxide composite fibers as active electrocatalysts for hydrogen evolution reaction, J. Mater. Sci. 52 (2017) 7269–7281.

[117] S.F. Anisa, R. Hashaikeha, N. Hilal, Functional materials in desalination: a review, Desalination 468 (2019) 114077.

9 Osmotic distillation and osmotic membrane distillation for the treatment of different feed solutions

A.B. Yavuz[1,2], V. Karanikola[3], M.C. García-Payo[1], M. Khayet[1,4]

[1] DEPARTMENT OF STRUCTURE OF MATTER, THERMAL PHYSICS AND ELECTRONICS, FACULTY OF PHYSICS, UNIVERSITY COMPLUTENSE OF MADRID, MADRID, SPAIN [2] PATNOS SULTAN ALPARSLAN NATURAL SCIENCES AND ENGINEERING FACULTY, AGRI IBRAHIM CECEN UNIVERSITY, AGRI, TURKEY [3] CHEMICAL AND ENVIRONMENTAL ENGINEERING, UNIVERSITY OF ARIZONA, TUCSON, AZ, UNITED STATES [4] MADRID INSTITUTE FOR ADVANCED STUDIES OF WATER (IMDEA WATER INSTITUTE), MADRID, SPAIN

9.1 Introduction

The study of the nonisothermal water transport through porous hydrophobic membranes began in the mid-1960s being the involved process "membrane evaporation" [1,2], later known generally as "membrane distillation (MD)" or particularly as "direct contact MD (DCMD)" [3,4]. In the 1980s the same type of membranes was used in the process "osmotic distillation, OD" [5–8]. In this case, water transport was observed under an isothermal condition, but with a concentration difference between both sides of the membrane, originated from an apparent osmotic pressure. Being closely related, OD process was considered a DCMD variant by many researchers, although there are some notable differences between them. For instance, some of the first references to OD appeared in papers dealing with MD experiments performed with different aqueous solutions [9–14]. In these references the effect of the concentration difference between both sides of the membrane pores was studied as a factor affecting the DCMD permeate flux and not as a separate process. In both processes, to establish the thermodynamic force required for water transport, it is necessary to maintain a water vapor pressure difference between both sides of the membrane pores. In fact, the physical origin of such a transmembrane vapor pressure (Δp_v) is quite different. As shown in Fig. 9–1, while in DCMD a temperature difference between both ends of the membrane pores is the driving force (i.e., the feed solution is maintained at a temperature higher than that of the permeate), in OD a difference of the concentration or composition of both phases adjoining the membrane at both sides (ΔC) is the driving force, being higher in the

Osmosis Engineering. DOI: https://doi.org/10.1016/B978-0-12-821016-1.00005-X

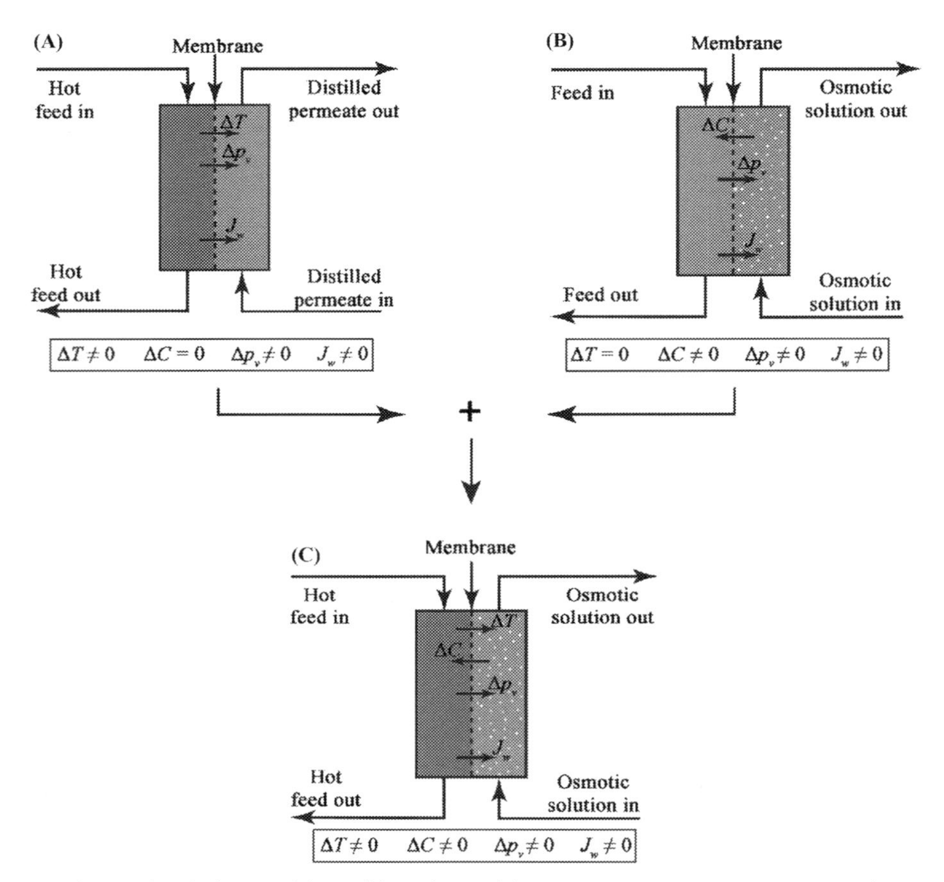

FIGURE 9–1 Schema of typical DCMD (A), OD (B), and OMD (C) processes. *DCMD*, Direct contact membrane distillation; *OD*, osmotic distillation; *OMD*, osmotic membrane distillation.

permeate side. An osmotic solution (i.e., draw solution) with a high osmotic pressure is generally used in the permeate side of the OD membrane module. The osmotic agent in OD is normally a salt aqueous solution. Details on the used osmotic solutions in OD are summarized in Section 9.3.

In MD, when the temperature difference between both sides of the membrane (ΔT) was maintained the same, the permeate flux (J_w) decreased with the increase of the solute concentration, but it took place from the hot to the cold side of the membrane. However, in those MD experiments carried out with high concentration differences, MD permeate fluxes were observed from the cold to the hot side of the membrane following the expected sense of the driving force, which is the vapor pressure difference. When both MD and OD were combined, the process was termed "osmotic MD, OMD" [15–17]. The driving forces in this case are temperature and concentration differences (ΔT and ΔC) contributing both the generation of the vapor pressure difference (Δp_v) (Fig. 9–1).

As illustrated in Fig. 9–2, worldwide research interest on both OD and OMD processes is growing very slightly. This scarce literature may be attributed mainly to their practical applicability such as the concentration of liquid wastes (fruit and vegetable juices, milk, wine, recovery of flavonoids, concentration and recovery of olive mill wastewater, etc.). It seems that researchers are interested more on OD rather than OMD. The number of papers published each year on OD (139 papers published in international refereed journals up to December 2019, WOS source) is greater than that of OMD (33 papers published in international refereed journals up to December 2019). Compared to OD, only 23.7% of papers have been published on OMD. This may be due to the wider application fields of the isothermal

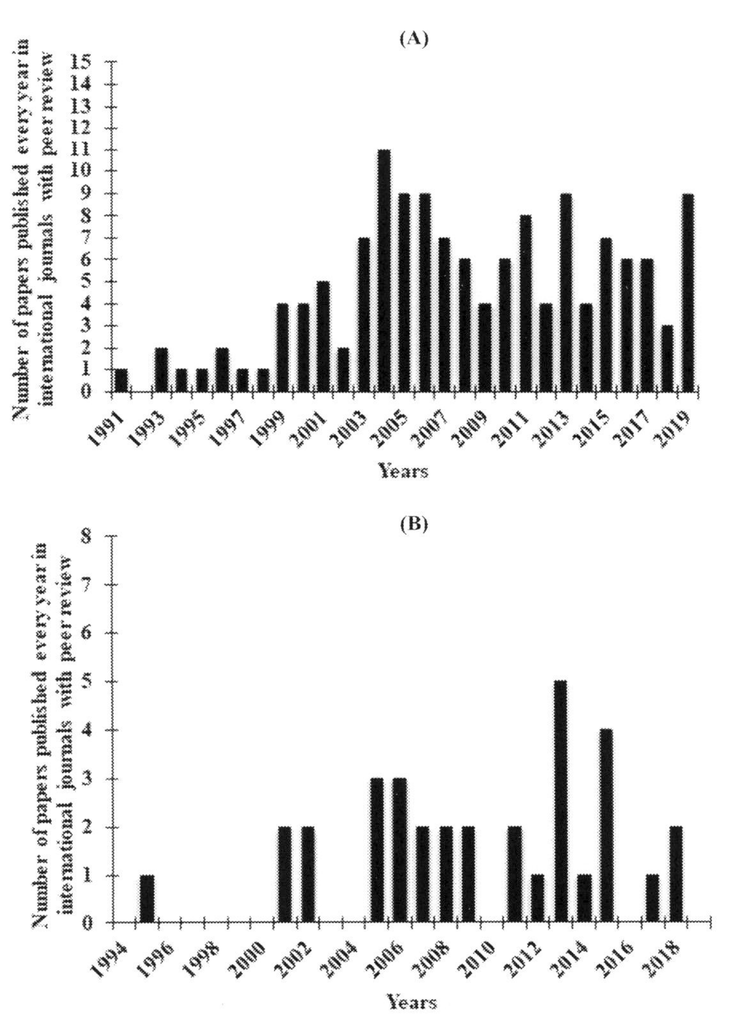

FIGURE 9–2 Number of papers published each year on OD process (A) and OMD (B) up to December 2019. *OD*, Osmotic distillation; *OMD*, osmotic membrane distillation.

OD process as it can be carried out at low temperatures when producing heat-sensitive concentrations preventing therefore the loss of volatile solute(s) and/or denaturalization of others (i.e., below 37°C) and to the subsequent lower energy consumption of OD. It must be pointed out that very few studies have been focused on osmotic membrane engineering (i.e., membranes for OD and OMD). Among the 139 published papers on OD, only 11 were dedicated to the preparation and modification of membranes (i.e., 7.9%). In fact, much more membranes have been developed for MD, and these can be proposed for OD and/or OMD.

The present chapter is focused on the abovementioned two processes, OD and OMD, covering different types of membranes, osmotic solutions, theoretical modeling, experimental effects of some significant operating parameters and applications.

9.2 Membranes used in osmotic distillation and osmotic membrane distillation processes

Similar to DCMD, the membrane required for OD and OMD must be porous and hydrophobic. Tables 9–1 and 9–2 summarize different types of flat sheet and capillary commercial membranes used in OD and OMD processes. The membranes commonly used in OD and OMD are fabricated from the hydrophobic polymers polypropylene, polyvinylidene fluoride, and polytetrafluoroethylene (PTFE). However, other polymers such as ethylene chlorotrifluoroethylene, ultrahigh molecular weight polyethylene, and polyether-block-amid are rarely considered. For the OD process, capillary membranes have been used more than flat sheet membranes and more research studies have been carried out using OD rather than OMD variant. As stated earlier, because of the similarity of the mass transfer mechanism among OD, OMD, and DCMD, most of the desired membrane characteristics of these three processes are very similar. A thorough review of the principal characteristics of an MD membrane was reported by Khayet [3] and Khayet and Matsuura [4]. For instance, OD and OMD membranes should exhibit a porous structure with a narrow pore size distribution and high porosity (i.e., void volume fraction). From Tables 9–1 and 9–2, it can be seen that the pore size of the used flat sheet membranes varies between 0.1 and 1.2 μm, while that of capillary membranes varies between 0.03 and 0.8 μm.

The membrane for OD and OMD can be a single hydrophobic layer or composite double- or triple-layered membrane prepared with different materials and structures. At least one layer must be made of a hydrophobic material and must be as thin as possible provided that the permeate flux is inversely proportional to the thickness of the hydrophobic layer, as reported in Section 9.4. For supported membranes, the backing material must be designed to guarantee low internal concentration and temperature polarization effects other than a low mass transfer resistance. It must be pointed out that the requirement of low thermal conductivity coefficient is less necessary for an OD membrane because the process is isothermal, although heat transfers from the permeate membrane side to the feed side (see Section 9.4).

Table 9–1 Commercial membranes used in osmotic distillation process.

Flat.sheet membranes						
Material	**Commercial Code**	**Manufacturer**	**δ (μm)**	**ε (%)**	**d_p (μm)**	**References**
PVDF	VVSP	Millipore, United States	120	65	0.1	[18,19]
	GVSP		120	80	0.2	[18]
	VVHP		125	75	0.1	[20]
	GVHP		125	75	0.22	[19–21]
	HVHP		125	70	0.45	[19,20]
	Unspecified	TS Filter Membranes, China	–	–	0.20	[22]
PP	Celgard 2400	Hoechst Celanese, United States	25	41	0.012×0.04	[18,19]
	Celgard 2500		25	55	0.05, 0.12 0.05×0.19 0.075×0.25	[18,19,23–25]
	Accurel 1E-PP	Enka, Germany	90	75	0.1	[18,19]
	Accurel 2E-PP		150	75	0.2	[18,19]
	Accurel 2E-HF	Membrana GmbH, Germany	165	60–75	0.2	[26]
	Unspecified	3M, Maplewood, MN, United States	110	79	0.2	[27,28]
PTFE	Goretex 10387	Gore, United States	8.5	78	0.2	[29]
	Goretex L31189		50	–	0.2	[18]
	Poreflon 020–40	Sumitomo Electric, Japan	40	–	0.2	[18,19]
	Poreflon 045-40		40	70	0.45	[18,19]
	TF200[a]	Pall Gelman, United States	178	80	0.2	[21,30–33]
	TF450[a]		178	80	0.45	[20,21,30,31]
	TF1000[a]		178	80	1	[21,30]
	FHLP	Millipore, United States	175	70	0.2	[15,21]
	TefSep PTFE laminated[a,b]	Micron Separations, United States	175	60	0.22	[34]
	Unspecified	TS Filter Membranes, China	–	–	0.2	[22]
UHMWPE	UPVP	Millipore, United States	90	80	0.2	[23]
ECTFE	Unspecified	3M, Maplewood, MN, United States	82	71	0.2	[27,28]
Acrylic fluorourethane copolymer	11104/2TPR	Gelman Sciences Pty. Ltd., Australia	9.0	–	0.2	[29]

Capillary membranes							
Material	Commercial Code	Manufacturer	D_o (μm)	δ (μm)	ε (%)	d_p (μm)	References
PP	Accurel PP Q3/2	Membrana GmbH, Germany	1000	200	70	0.2	[26,66,75]
	Accurel PP S6/2		2600	450	70	0.2	[66]
	Self-made membrane module	Unspecified	420	31	40	0.03	[16]
	2.5 × 8 Extra-Flow Module Liqui-Cel®, Membrane Contactor	Celgard LLC, Charlotte, United States	300	30	40	0.03	[71,76,83, 85,87,91, 100,111, 113,122]
	2.5 × 8 Extra-Flow Module Liqui-Cel®, Membrane Contactor		300	–	–	0.2	[103,115]
	2.5 × 8 Extra-Flow Module Liqui-Cel®, Membrane Contactor		300	40	40	0.04	[107,123]
	2.5 × 8 Extra-Flow Module Liqui-Cel®, Membrane Contactor		–	–	–	–	[80]
	4 × 28 Extra-Flow Liqui-Cel®, Membrane Contactor	Membrana GmbH, Wuppertal, Germany	–	–	–	–	[118]
	1 × 5.5 Minimodule Liqui-Cel®, Membrane Contactor	Celgard LLC, Charlotte, United States	300	–	40	0.3	[93]
	1 × 5.5 Minimodule Liqui-Cel®, Membrane Contactor		300	42	40	0.03	[108,109,119, 120,132]
	1.7 × 5.5 Minimodule Liqui-Cel®, Membrane Contactor		300	–	40	0.3	[93]
	1.7 × 5.5 Minimodule Liqui-Cel®, Membrane Contactor		300	–	40	0.03	[125]
	1.7 × 5.5 Minimodule Liqui-Cel®, Membrane Contactor	Membrana GmbH, Wuppertal, Germany	–	–	–	–	[114,116, 128,135]
	1.7 × 5.5 Minimodule Liqui-Cel®, Membrane Contactor	Membrana, Charlotte, United States	300	30	40	0.04	[134]
	0.5 × 1 Liqui-Cel® Micromodule	Membrana, Polypore Company, United States	300	–	40	0.06	[108]
	Liqui-Cel® Minimodule (model X-50)	Celgard LLC, Charlotte, United States	300	–	40	–	[124]
	Celgard Liqui-Cel® G542 Minimodule		300	–	–	0.2	[106]
	Unspecified	Unspecified	426	45	30	0.05	[92]
	Unspecified	Hangzhou Hualu Membrane Engineering Co. Ltd., China	450	60	35	0.05–0.2	[81]

(Continued)

Table 9–1 (Continued)

Capillary membranes							
Material	Commercial Code	Manufacturer	D_o (μm)	δ (μm)	ε (%)	d_p (μm)	References
	MD 150 CS 2N Module	Microdyn, Germany	2800	–	–	0.2	[86]
	MD 020 CP 2N Module		2800	500	–	0.2	[84,86,99,101, 110,112,117, 121,126,127, 130,133]
	Unspecified		–	1500	70	0.2	[50]
	Model E06 Module	COGIA Company, Palaiseau, France	–	–	–	0.2	[70]
	COGIA Module	COGIA Company, Palaiseau, France	2600	400[c]	–	0.2	[69]
	Unspecified	Unspecified	–	–	–	0.2	[77,78]
	PP375	Memcor Australia	625	–	–	0.2	[82]
	Unspecified	Unspecified	–	–	–	0.2	[90]
	Unspecified	JU. CLA.S. LTD, Verona, Italy	1000	200	–	–	[131]
PVDF	PV375	Memcor Australia	625	125	75	0.2	[74,79,82]
	PV660		1000	170	64	0.2	[74,79,82,88,94]
	Unspecified		1000	180	75	0.2	[89,96]
PTFE	Unspecified	Unspecified	–	–	–	0.4	[102]
Ceramic	Unspecified	Pall Corporation, Exekia Division,	–	–	–	0.2	[73]
	Unspecified	United States	–	–	–	0.8	[73]

ECTFE, Ethylene chlorotrifluoroethylene; *PP*, polypropylene; *PTFE*, polytetrafluoroethylene; *PVDF*, polyvinylidene fluoride; *UHMWPE*, ultrahigh molecular weight polyethylene.

[a]Membrane on polypropylene support (thickness values include support material).

[b]It is stated that the membrane properties are quite close to those of TF200 membrane and the porosity was estimated to be on the order of 60%.

[c]Calculated based on outer diameter D_o/inner diameter D_{in} = 2600/1800 μm.

Table 9–2 Commercial membranes used in osmotic membrane distillation process.

Material	Commercial Code	Manufacturer	δ (μm)	ε (%)	d_p (μm)	References
Flat sheet membranes						
PVDF	Durapore	Millipore, United States	–	77	0.45	[35]
	Unspecified	Durapore Merck, Germany	125	75	0.45	[36]
PP	Accurel	Enka, Germany	150	75	0.05	[37–41]
	Accurel		150	75	0.2	[37–44]
	Unspecified	Celgard, United States	170	75	0.1	[36]
PTFE	Unspecified	AS, Ltd., Denmark	–	–	0.025	[37]
	TefSep PTFE laminated	Micron Separations, Inc., MA, United States	175	60	0.22	[45]
	11807	Sartorius Stedim, Germany	65	62	0.2	[36,46]
	11806		80	80	0.45	[36,46–48]
	11802		–	–	1.2	[46]
	Unspecified	Unspecified	–	–	0.22	[49]
	Unspecified	Unspecified	–	–	0.45	[49]
	TF200[a]	Gelman, United States	178	80	0.2	[30]
	TF450[a]		178	80	0.45	[30]
	TF1000[a]		178	80	1	[30]
Polypropylene capillary membranes						
PP (D_o = 420 μm)	Self-made membrane module	Unspecified	31	40	0.03	[16,17]
PP	–	Microdyn, Germany	1500	70	0.2	[50]
PP	–		100	70	0.2	[51]
PP (D_o = 2700 μm)	MD020CP2N Module	Microdyn-Nadir, Germany	450	70	0.2	[52]
PP (D_o = 800 μm)	Unspecified	Microdyn, Germany	200	70	0.2	[53,54]
PP (D_o = 300 μm)	2.5 × 8 Extra-Flow Module Liqui-Cel®, Membrane Contactor X-50	Membrana, Charlotte, United States	40	40	0.04	[55–58]
PP (D_o = 2600 μm)	Accurel PP S6/2	Membrana GmbH, Germany	450	73	0.2	[59]

PP, Polypropylene; *PTFE*, polytetrafluoroethylene; *PVDF*, polyvinylidene fluoride.
[a]Membrane on polypropylene support (thickness values include support material).

It is worth noting that all commercial membranes used in OD and OMD were not originally designed for these processes. Because of their low liquid entry pressure and the wide field of application of these processes, especially in the concentration of liquid wastes (fruit and vegetable juices, milk, wine, olive mill wastewater, etc.), the pores of these membranes were easily wet reducing the membrane performance and the efficiency of the process as a consequence. As can be seen in Table 9–3, very few research studies have been carried out in the field of OD and OMD membrane engineering. To reduce the risk of pore wetting, both flat sheet and capillary membranes were surface modified by coating hydrophilic polymers such polyvinyl alcohol, alginate, alginic acid–silica, alginate–carrageenan, and chitosan [23,60–63]. It must be pointed out that only one research study has been devoted so far to the preparation of a membrane for OD process [65].

Table 9–3 Modified or prepared membranes for osmotic distillation process.

Material	Commercial Code	Manufacturer	δ (μm)	ε (%)	d_p (μm)	Reference
Flat sheet–modified membranes						
PVA-coated PVDF	GVSP	Millipore	120	80	0.2	[23]
PVA-coated PP	Celgard 2500	Hoechst Celanese, United States	25	55	0.05 0.12 0.05 × 0.19 0.075 × 0.25	[23]
PVA-coated UHMWPE	UPVP	Millipore, United States	90	80	0.2	[23]
PHEMA-coated PVDF	GVSP		120	80	0.2	[23]
Sodium alginate–hydrogel-coated PTFE	Poreflon 020–40	Sumitomo Electric, Japan	40	–	0.2	[60]
Alginic acid–silica hydrogel-coated PTFE	Desal type KS K 150	Desalination Systems Inc., Vista, CA, United States	–	–	0.1	[61]
Sodium alginate–carrageenan-coated PTFE	Desal type KS K 150	Desalination Systems Inc., Vista, CA, United States	–	–	0.1	[62]
Capillary-modified membranes						
Chitosan-coated PVDF $D_o = 1300$ μm	Unspecified	Memcor, Australia	–	–	0.2	[63]
Grafted ceramic	Unspecified	Membralox Pall Company, United States	–	–	0.2	[64]
	Unspecified		–	–	0.5	[64]
	Unspecified		–	–	0.8	[64]
Prepared nanofibrous membrane						
PEBA	Lab made		–	70.9	0.836	[65]

PEBA, Polyether-block-amid; *PHEMA*, polyhydroxy 2-ethyl methacrylate; *PP*, polypropylene; *PTFE*, polytetrafluoroethylene; *PVA*, polyvinyl alcohol; *PVDF*, polyvinylidene fluoride; *UHMWPE*, ultrahigh molecular weight polyethylene.

9.3 Osmotic solutions used in osmotic distillation and osmotic membrane distillation processes

Taking into account that the driving force in OD is the concentration difference between both sides of the porous hydrophobic membrane, aqueous osmotic solutions also called stripping solutions have been used considering different osmotic agents. In general, these solutions are mainly applied to reduce the chemical potential of water in the permeate side of the membrane. The osmotic agent should exhibit high solubility in water, nontoxicity, high superficial tension to avoid wetting of the pores of the membrane from its permeate side, low volatility, low viscosity, and no interaction with the membrane material or with any component of the food components transported through the membrane.

Details of both the feed and osmotic solutions considered in OD and OMD processes are summarized in Tables 9–4 and 9–5, respectively. It is worth noting that NaCl and $CaCl_2$ are the most commonly used osmotic agents in OD and OMD research studies because they are economical. More particularly, $CaCl_2$ is used in most studies of OD as it is cheap and produces higher osmotic pressure thus results in higher driving force for the process [30,38,43,50,51,65,76,93,106,116,126,138]. However, these two osmotic agents have some drawbacks. NaCl can provide a limited driving force due to its low solubility in water. Unlike NaCl, $CaCl_2$ has high solubility in water.

Although it is an effective osmotic agent recommended by many researchers, highly concentrated $CaCl_2$ solutions can also cause side problems such as a high corrosion risk. Another issue associated with $CaCl_2$ is the temperature dependence of its solubility. In the case of temperature drop, crystallization may occur causing blockages in the system [34,75]. Other osmotic agents have been proposed in other research studies such as NaOH, $MgCl_2$, CH_3COOK, H_2SO_4, K_2HPO_4, glycol, propylene glycol solutions or glycerol–NaCl–water, K_2HPO_4–KH_2PO_4, $K_4P_2O_7$–$H_4P_2O_7$, K_2HPO_4–H_3PO_4 mixtures.

A comprehensive study on osmotic agents used in OD has been reported in Ref. [25] who compared K_2HPO_4, $K_4P_2O_7$, and CH_3COOK with the conventional osmotic agents NaCl and $CaCl_2$. KH_2PO_4 was not used alone because it has lower solubility compared to the other osmotic agents. To increase the osmotic activity, KH_2PO_4/K_2HPO_4, $K_4P_2O_7/H_4P_2O_7$, and K_2HPO_4/H_3PO_4 blended agents were used. It was concluded that phosphate-based agents KH_2PO_4/K_2HPO_4, $K_4P_2O_7/H_4P_2O_7$, K_2HPO_4/H_3PO_4, and $K_4P_2O_7$ were more suitable than NaCl and $CaCl_2$ in OD applications. It was stated that NaCl and $CaCl_2$ resulted in lower permeate fluxes and significantly higher corrosion on stainless steel, and CH_3COOK was comparable to phosphate agents inducing generally low corrosion.

In dealcoholization processes, OD process has been applied instead of OMD, and pure water or dilute alcohol solutions have been employed instead of salt solutions as stripping or osmotic solutions[105,108,109,135]. For wine dealcoholization, pure water was used as an osmotic solution. However, glycerol solutions were proposed in Ref. [123] to avoid watering of wine. To preserve sulfur dioxide contained in wine under OD treatment, potassium metabisulfite had been added to the stripping solution [131].

Table 9–4 Feed and osmotic solutions considered in osmotic distillation process.

Feed solution	Osmotic solution	Reference
Pure water	0.5–5 M NaCl solution	[21]
Tomato puree	28 wt.% NaCl brine	[29]
Pure water	1–5 mol/L NaCl solution	[15]
Aqueous solutions (1.00 L) containing mixtures of ethanol (200 μL), 3-methylbutanal (0.01 μL), ethyl acetate (1.0 μL), alpha-pinene (0.01 μL), beta-myrcene (1.0 μL), ethyl hexanoate (0.01 μL), and limonene (1.0 μL)	45 wt.% $CaCl_2$ solution	[18]
Gordo grape juice	45 wt.% $CaCl_2$ solution	
Valencia orange juice	45 wt.% $CaCl_2$ solution	
Pure water	24.5 wt.% $CaCl_2$ solution	[23]
Different concentrations of sucrose solutions (40, 55, 60, 65, and 68 wt.%)	36 wt.% $CaCl_2$ solution	
Concentrated feed (Gordo grape juice 68 wt.% sucrose)	36 wt.% $CaCl_2$ solution	
Different oil (limonene) dispersions (0.2, 0.5, and 1 wt.% in water and in sucrose solution)	36 wt.% $CaCl_2$ solution	
Pure water	25 wt.% NaCl solution	[66]
Pure water	30 wt.% $MgCl_2$ solution	
Sugar solution (0, 35, 45, 55, 60, and 65 wt.% sucrose solution)	45 wt.% $CaCl_2$ initial concentration	[31]
Pure water	$CaCl_2$ solution (concentration varied from 32.2 to 45.5 wt.%)	[67]
Gordo grape juice	40 wt.% $CaCl_2$ initial concentration	[68]
Gordo grape juice permeate from UF through different membranes (KW K500, KS K150, EW E500, PW)	40 wt.% $CaCl_2$ initial concentration	
Deionized water	1–5 mol/kg brine	[16]
Tap water	5.1–5.6 M $CaCl_2$ solution	[69]
Clarified passion fruit juice	5.1–5.6 M $CaCl_2$ solution	
Clarified orange juice	4.6 M $CaCl_2$ solution	[70]
Clarified passion fruit juice	4.6 M $CaCl_2$ solution	
Permeate of PV–MF processes for ethanol–water extracts of the Echinacea plant	40 wt.% $CaCl_2$ solution	[24]
Retentate of UF–RO processes for freshly squeezed orange, lemon, and carrot juices	4.1–4.5 M (60–66 wt.%) $CaCl_2 \cdot 2H_2O$ solution	[71]
Solution (160 g/L of sucrose, 40 g/L of citric acid, 4–6 mg/kg for each of hexyl acetate, ethyl butyrate, hexanol, and benzaldehyde)	3.5–5.1 (± 0.2) mol/L $CaCl_2$ solution	[72]
Pure water	50 wt.% $CaCl_2$ solution	[73]
Deionized water	1–5 M. $CaCl_2 \cdot 2H_2O$ solution	[20]
40–60 wt.% glucose solution	45 wt.% $CaCl_2$ solution	[74]
Orange oil–water mixtures (deionized water, 0.2, 0.4, and 0.8 wt.%)	40 wt.% $CaCl_2$ initial concentration	[60]
Pure water	15–40 wt.% $CaCl_2$ solution	[75]

(Continued)

Table 9–4 (Continued)

Feed solution	Osmotic solution	Reference
Pure water	Glycerol–NaCl–water mixtures (0.34 g salt/g glycerol)	
Pure water	30–70 wt.% glycerol solution	
Pure water	35–75 wt.% propylene glycol solution	
Clarified camu–camu juice	4.0–5.2 mol/L $CaCl_2$ solution	[32]
Kiwifruit juice clarified by UF process	60 wt.% $CaCl_2 \cdot 2H_2O$ solution	[76]
Orange oil–water mixtures (deionized water, 0.2, 0.4, 0.8, and 1.2 wt.%)	40 wt.% $CaCl_2$ solution	[61]
Sodium DBS detergent solutions (0.1, 0.3, and 0.5 wt.%)	40 wt.% $CaCl_2$ solution	
Clarified orange juice	5.5 mol/L $CaCl_2$ solution	[77]
MF permeate of melon juice	5.3–5.6 mol/L $CaCl_2$ solution	[78]
Sodium DBS detergent solutions (deionized water, 0.1, 0.3, and 0.5 wt.%)	40 wt.% $CaCl_2$ solution	[62]
Orange oil–water mixtures (0.2, 0.4, and 0.8 wt.%)	40 wt.% $CaCl_2$ solution	
Fresh full-cream milk	40 wt.% $CaCl_2$ solution	
Glucose solution (30%–60% solids)	Saturated $CaCl_2$ solution	[79]
10.78 $\pm$ 0.09 and 13.02 $\pm$ 0.12 vol.% ethanol solutions	Water	[80]
Synthetic wine	Water	
Wine	Water	
Different cyanide containing wastewaters (0.019–0.058 mol/L acrylonitrile, 0.115–0.154 mol/L benzonitrile, 0.019–0.077 mol/L caffeine, and 0.077–0.135 mol/L praziquantal)	10 wt.% NaOH solution	[81]
Pure water	24 wt.% NaCl solution	[34]
Pure water	40 wt.% $CaCl_2$ solution	
Pure water	70 wt.% glycerol solution	
Pure water	Mixture of glycerol (42.74 wt.%) and NaCl (14.52 wt.%)	
Clarified apple juice	42 wt.% $CaCl_2$ solution	[82]
UF treated grape juice	42 wt.% $CaCl_2$ solution	
FC grape juice semiconcentrate	42 wt.% $CaCl_2$ solution	
Kiwifruit juice clarified by UF process	60 wt.% $CaCl_2$ solution	[83]
Diluted must (10°Brix)	60 wt.% $CaCl_2 \cdot 2H_2O$ solution	[84]
Model solution of sugar (10–20 wt.% sucrose solution)	60 wt.% $CaCl_2 \cdot 2H_2O$ solution	
Distilled water	3.5 M $CaCl_2$ solution	[50]
20% Sucrose solution	3.5 M $CaCl_2$ solution	
45% Sucrose solution	6 M $CaCl_2$ solution	
Apple juice	6 M $CaCl_2$ solution	
12°Brix sucrose solution	4.9 M $CaCl_2 \cdot 2H_2O$ solution	[26]
Citral and ethyl butyrate solution	3 M $CaCl_2$ solution	[26]

(Continued)

Table 9–4 (Continued)

Feed solution	Osmotic solution	Reference
Model juice solution (sucrose 45 wt.%, citral 18 mg/L, and ethyl butyrate 18 mg/L)	3 M $CaCl_2$ solution	
Orange juice	4.9 M $CaCl_2$ solution	
Cactus pear juice clarified by UF process	60 wt.% $CaCl_2 \cdot 2H_2O$ solution	[85]
Pretreated black currant juice	$CaCl_2$ brine solution	[86]
Kiwifruit juice clarified by UF process	60 wt.% $CaCl_2 \cdot 2H_2O$ solution	[87]
Deionized water	73.8 wt.% potassium acetate (CH_3COOK) solution	[25]
Deionized water	70.3 wt.% dipotassium phosphate (K_2HPO_4) solution	
Deionized water	67.3 wt.% potassium pyrophosphate ($K_4P_2O_7$) solution	
Deionized water	69.3/1.4 wt.% K_2HPO_4/KH_2PO_4 (48:1 wt./wt.) solution	
Deionized water	64.1/6.4 wt.% $K_4P_2O_7/H_4P_2O_7$ (10:1 wt./wt.) solution	
Deionized water	65.8/5.5 wt.% K_2HPO_4/H_3PO_4 (12:1 wt./wt.) solution	
Deionized water	25.4 wt.% NaCl solution	
Deionized water	44.84 wt.% $CaCl_2$ solution	
Chardonnay grape juice clarified by UF process	43 wt.% $CaCl_2 \cdot 2H_2O$ solution	[88]
Fructose solutions (35–55 wt.%)	43 wt.% $CaCl_2 \cdot 2H_2O$ solution	
Distilled water	44 wt.% $CaCl_2 \cdot 2H_2O$ solution	[89]
Clarified pineapple juice	5.5–6.0 mol/L $CaCl_2$ solution	[33]
Water	5.5–6.0 mol/L $CaCl_2$ solution	[90]
Sucrose solution	5.5–6.0 mol/L $CaCl_2$ solution	
Roselle extract	5.5–6.0 mol/L $CaCl_2$ solution	
Apple juice	5.5–6.0 mol/L $CaCl_2$ solution	
Grape juice	5.5–6.0 mol/L $CaCl_2$ solution	
Retentate of UF–RO processes for blood orange juice	60 wt.% $CaCl_2 \cdot 2H_2O$ solution	[91]
Deionized water	1.0–4.0 M NaCl solution	[92]
Pure water	(2.0, 4.0, and 6.0 mol/kg) $CaCl_2$ solutions	[93]
Noni juice	(2.0, 4.0 and 6.0 mol/kg) $CaCl_2$ solutions	
Red wine	Pure water	[94]
13 vol.% ethanol solution	Pure water	
13 vol.% ethanol solution	50 wt.% glycerol solution	
13 vol.% ethanol solution	40 wt.% $CaCl_2$ solution	
50 mg/L ethyl acetate solution	Pure water	
400 mg/L isoamyl alcohol solution	Pure water	
Reaction mixture [90% yield isoamyl acetate, unreacted acetic acid, 2.8 wt.% water, excess of isoamyl alcohol (starting from 1:7 acid:alcohol molar ratio), internal standard toluene (0.5 vol.%)]	6 M $CaCl_2$ solution	[95]
Deionized water	44 wt.% $CaCl_2 \cdot 2H_2O$ solution	[96]

(Continued)

Table 9–4 (Continued)

Feed solution	Osmotic solution	Reference
"Oblachinska" sour cherry juice	43 wt.% $CaCl_2$ solution	[97]
Tomato juice	1–5 M $CaCl_2 \cdot 6H_2O$	[98]
Tomato juice	1–5 M $MgCl_2$ $6H_2O$	
Apple juice clarified by UF process	65 wt.% $CaCl_2 \cdot 2H_2O$ solution	[99]
Mixture of ethyl acetate, ethyl hexanoate, limonene and water (concentrations of ethyl acetate and ethyl hexanoate were 500 mg/L, limonene concentrations varied in the range 0–500 mg/L)	45 wt.% $CaCl_2 \cdot 2H_2O$ solution	[63]
Permeate of MF–NF processes for OMW	60 wt.% $CaCl_2 \cdot 2H_2O$ solution	[100]
Apple juice	0, 32.5 and 65 wt.% $CaCl_2$ solution	[101]
Pure water	60 wt.% $K_4P_2O_7$	[64]
Model sucrose solution	60 wt.% $K_4P_2O_7$	
Clarified cane sugar juice	60 wt.% $K_4P_2O_7$	
Ammonia (NH_3) solutions (250 and 1000 mg/L)	10 wt.% sulfuric acid (H_2SO_4) solution	[102]
Pomegranate juice (*Punica granatum* L.) clarified by UF process	10.2 mol/L $CaCl_2$ solution	[103]
Retentate of MF–RO processes for raspberry juice	65–70 wt.% $CaCl_2$ solution	[104]
Hydroalcoholic solutions (10, 12.5, 15 vol.% ethanol)	Distilled water	[105]
Red wine	Distilled water	
Pure water	40 wt.% $CaCl_2$ solution	[106]
10 wt.% sucrose solution	40 wt.% $CaCl_2$ solution	
Simulated fruit juice in an aqueous solution with aroma compounds (ethyl acetate, ethanol, butanol, acetaldehyde, 500 ppm for each compound)	30, 50 wt.% $CaCl_2$ solution	
NF retentate of orange press liquor	10.2 mol/L $CaCl_2 \cdot 2H_2O$ solution	[107]
Red wine	Pure water	[108]
Hydroalcoholic solutions (0.7, 1.2, 3.0, 6.0, 13.0 vol.% ethanol)	Pure water	
Hydroalcoholic solutions (0.8–5 vol.%)	Pure water or aqueous solutions with low-alcohol concentrations (0.08, 0.17, 0.41, and 0.77 vol.% ethanol concentration)	[109]
Pomegranate juice clarified by UF process	65 wt.% $CaCl_2 \cdot 2H_2O$ solution	[110]
Blood orange juice clarified by UF process	11.2 mol/L $CaCl_2 \cdot 2H_2O$ solution	[111]
Distilled water	0–5 M $CaCl_2$ solution	[30]
Crude olive mill wastewater	5 M $CaCl_2$ solution	
Preconcentrated sage extract with 4.4 wt.% TSS	42 wt.% $CaCl_2$ solution	[112]
Wine	Water	[113]
Alcoholic beers of five different commercial brands (containing 3.5–5 vol.% alcohol)	DW	[114]
Alcoholic beers of five different commercial brands (containing 3.5–5 vol.% alcohol)	CW	
The blood orange juice clarified by UF process	10.2 mol/L $CaCl_2$ solution	[115]
Diluted commercial concentrated cranberry juice	Concentrated $CaCl_2$ solution (30, 40, 50 wt.%)	[116]
Tomato juice	65 wt.% $CaCl_2$ solution	[117]

(Continued)

Table 9–4 (Continued)

Feed solution	Osmotic solution	Reference
Beer samples were produced at the Italian Brewing Research Centre with a 100 L pilot plant (CERB, University of Perugia, Italy)	DW	[118]
Beer samples were produced at the Italian Brewing Research Centre with a 110 L pilot plant (CERB, University of Perugia, Italy)	Dealcoholized beer (0.77, 0.41, 0.17, and 0.08 vol.% alcohol concentration)	[119]
Beer samples were produced at the Italian Brewing Research Centre with a 110 L pilot plant (CERB, University of Perugia, Italy)	Aqueous solution made by diluting original beer (0.5, 0.4, 0.2, 0.1, and 0.1 vol.% alcohol concentration)	[120]
Clarified black mulberry juice	65 wt.% $CaCl_2 \cdot 2H_2O$ solution	[121]
NF retentate of OMW	60 wt.% $CaCl_2 \cdot 2H_2O$ solution	[122]
Water coming from must concentration plants + ethanol (14 vol.% ethanol)	Demineralized local tap water	[123]
Red wine (14.28 vol.% ethanol)	Demineralized local tap water	
Water coming from must concentration plants (0 vol.% ethanol)	Demineralized local tap water	
Water coming from must concentration plants + 20.7 wt.% glycerol	Demineralized local tap water	
Water coming from must concentration plants + ethanol (14 vol.% ethanol)	15.5 wt.% glycerol solution	
Red wine (14.3 vol.% ethanol)	20.7 wt.% glycerol solution	
Clarified apple juice	5 M $CaCl_2$ solution	[124]
Clarified orange juice	5 M $CaCl_2$ solution	
135 g TSS/kg sucrose solution	3.5, 5, and 6 M $CaCl_2$ solution	
Beer samples were produced at the Italian Brewing Research Centre with a 30 L pilot plant (CERB, University of Perugia, Italy)	Carbonated beer-diluted solution (0.5 and 0.8 vol.% alcohol concentration)	[125]
Orange juice	Saturated solution of potassium pyrophosphate ($K_4P_2O_7$)	[126]
Surrogate solution (30 μL/L ethyl acetate, 70 μL/L butyl acetate, 30 μL/L hexanal, 80 μL/L linalool, 260 μL/L limonene, 10 μL/L myrcene, 300 mg/L ascorbic acid, and 160 g/L sucrose)	1291 g/L potassium pyrophosphate ($K_4P_2O_7$) solution	
Surrogate solution (30 μL/L ethyl acetate, 70 μL/L butyl acetate, 30 μL/L hexanal, 80 μL/L linalool, 260 μL/L limonene, 10 μL/L myrcene, 300 mg/L ascorbic acid, and 160 g/L sucrose)	316 g/L NaCl solution	
Broccoli juice clarified by UF process	65 wt.% $CaCl_2 \cdot 2H_2O$ solution	[127]
Pomegranate juice	3.62, 4.62, and 5.43 mol/L $CaCl_2$ solution	[65]
Red wine	Pure water	[128]
RO-treated red wine	Pure water	
Deionized water	2, 4, and 5 M NaCl solution	[27]
Muscadine pomace extract	4 M NaCl solution	
Red beet juice	4.3 M $CaCl_2$ solution	[129]

(Continued)

Table 9–4 (Continued)

Feed solution	Osmotic solution	Reference
Preconcentrated (UF–RO) pomegranate juice	65 wt.% $CaCl_2 \cdot 2H_2O$ solution	[130]
Clarified pomegranate juice	6 M $CaCl_2 \cdot 2H_2O$ solution	[22]
Three different wines	Recirculating microfiltered (1 μm) water containing 0.05 g/L of potassium metabisulfite	[131]
White wine	Distilled water	[132]
Commercial pasteurized clear grape juice	65 wt.% $CaCl_2 \cdot 2H_2O$ solution	[133]
Commercial pasteurized clear pomegranate juice	65 wt.% $CaCl_2 \cdot 2H_2O$ solution	
Black carrot juice	65 wt.% $CaCl_2 \cdot 2H_2O$ solution	
Clarified pomegranate juice	6 M $CaCl_2$ solution	[134]
Red wine	Pure water	[135]

CW, Carbonated water; *DBS*, dodecylbenzene sulfonate; *DW*, deaerated water; *FC*, freeze concentrated; *MF*, microfiltration; *NF*, nanofiltration; *OMW*, olive mill wastewater; *PV*, pervaporation; *RO*, reverse osmosis; *TSS*, total suspended solids; *UF*, ultrafiltration.

Table 9–5 Feed and osmotic solutions considered in osmotic membrane distillation process.

Feed solution	Osmotic solution	Reference
Fluosilicic acid solutions of various initial concentrations (2–35 wt.%)	Saturated NaCl solution	[35]
Deionized water	1–5 M brine	[16]
Deionized water	1–5 M brine	[17]
Pure water	2–5 M NaCl solution	[37]
Pure water	5 M $CaCl_2 \cdot 2H_2O$ solution	
Sugarcane juice	5 M $CaCl_2 \cdot 2H_2O$ solution	
Sugarcane juice	5 M NaCl solution	
Sugarcane juice	4 M K_2HPO_4 solution	
Pure water	24 wt.% NaCl solution	[45]
Phycocyanin solution	2–6 M NaCl solutions	[38]
Phycocyanin solution	2–10 M $CaCl_2 \cdot 2H_2O$ solutions	
Sweet-lime juice	2–6 M NaCl solutions	
Sweet-lime juice	2–10 $CaCl_2$ solutions	
Pineapple juice	2–6 M NaCl solutions	[39]
Pineapple juice	2–14 M $CaCl_2 \cdot 2H_2O$ solutions	
Sweet-lime juice	2–6 M NaCl solutions	
Sweet-lime juice	2–14 M $CaCl_2 \cdot 2H_2O$ solutions	
Distilled water	3.5 M $CaCl_2$ solution	[50]
20% sucrose solution	3.5 M $CaCl_2$ solution	
45% sucrose solution	6 M $CaCl_2$ solution	
20% sucrose solution	6 M $CaCl_2$ solution	
Apple juice	6 M $CaCl_2$ solution	
30–60 wt.% sucrose solutions	24.6 wt.% NaCl solution	[47]

(Continued)

Table 9–5 (Continued)

Feed solution	Osmotic solution	Reference
30–60 wt.% sucrose solutions	50 wt.% $CaCl_2$ solution	
Red natural colorant, namely, anthocyanin from red radish	14 M NaCl solution	[42]
Red natural colorant, namely, anthocyanin from red radish	14 M K_2HPO_4 solution	
Chokeberry juice	6 M $CaCl_2 \cdot 2H_2O$ solution	[53]
Redcurrant juice	6 M $CaCl_2 \cdot 2H_2O$ solution	
Cherry juice	6 M $CaCl_2 \cdot 2H_2O$ solution	
Pineapple juice	2–10 M $CaCl_2 \cdot 2H_2O$ solutions	[43]
30–60 wt.% sucrose solutions	50 wt.% $CaCl_2$ solution	[48]
Dealcoholized anthocyanin extract	14 M $CaCl_2 \cdot 2H_2O$ solution	[40]
Anthocyanin extract	1–6 M NaCl solution	[136]
Cornelian cherry juice clarified by UF process	6 M $CaCl_2 \cdot 2H_2O$ solution	[54]
Bergamot juice clarified by UF process	10.2 mol/kg calcium chloride dehydrate	[55]
Deionized water and different concentrations of red grape juice (5, 10, and 20°Brix)	50 wt.% $CaCl_2$ solution	[46]
Distilled water	1 mol/L NaCl solution	[49]
OMW	5 M $CaCl_2$ solution	[30]
Anthocyanin solution	$CaCl_2 \cdot 2H_2O$ solution	[44]
Anthocyanin solution	12 M $CaCl_2 \cdot 2H_2O$ solution	[41]
Apple juice	25 wt.% NaCl solution	[36]
Apple juice	50 wt.% $CaCl_2$ solution	
Beet juice	25 wt.% NaCl solution	
Beet juice	50 wt.% $CaCl_2$ solution	
0.05, 0.2, 0.3, 0.4, 0.5 M sucrose solutions	6 M $CaCl_2$ solution	[51]
0.2 M sucrose solution	2, 4, 6 M $CaCl_2$ solutions	
Whitebeam juice	$CaCl_2$ solution	
Blackthorn juice	$CaCl_2$ solution	
Cornelian cherry juice	$CaCl_2$ solution	
Elderberry juice	$CaCl_2$ solution	
LiCl salt solution	4.5 M $CaCl_2 \cdot 2H_2O$ solution	[52]
Na_2CO_3 solution	NaCl solution	[57]
Pure water	43 wt.% $CaCl_2$ solution	[137]
Sucrose solution	43 wt.% $CaCl_2$ solution	
Fructose solution	43 wt.% $CaCl_2$ solution	
Glucose solution	43 wt.% $CaCl_2$ solution	
Filtered cactus pear juice	43 wt.% $CaCl_2$ solution	
Unfiltered cactus pear juice	43 wt.% $CaCl_2$ solution	
Aqueous solution of pure Na_2CO_3, Na_2SO_4, KNO_3, or mixtures	200–300 g/L NaCl solution	[58]
Distilled water	Saturated NaCl solution	[59]
UF–RO retentate of Nagpur mandarin (*Citrus reticulata* Blanco) fruit juice	55–60 wt.% $CaCl_2 \cdot 2H_2O$ solution	[56]

OMW, Olive mill wastewater; *UF*, ultrafiltration; *RO*, reverse osmosis.

For the dealcoholization of beers, which are highly susceptible to oxidation in presence of oxygen, deaerated water (DW), carbonated water (CW), or beers with low-alcohol levels were used as osmotic solutions. Oxygen content is an important factor to take into consideration for beer stability [118]. The amount of alcohol removed is the same when CW or DW is used as a stripping solution. However, there is no significant difference between CW and DW for the most important volatile components (i.e., similar losses). The biggest difference between the two stripping solutions could be seen in the loss of carbon dioxide. When CW was used as a stripping solution, the loss of carbon dioxide in the low-alcohol beer is less. Nevertheless, the amount of available carbon dioxide in both processes is still low and it is necessary to add carbon dioxide before packaging in a commercial process [114].

In general, although the choice of osmotic solution plays a significant role in OD and OMD performance as most studies suggest, there are still very few studies that explore other osmotic agents. A possible inspiration would be the wider possible osmotic solutions considered in forward osmosis (FO) as reviewed in Refs. [139–141].

9.4 Mechanism of transport in OD and OMD: Temperature polarization, concentration polarization, and theoretical models

9.4.1 Mass transfer through the membrane

In OD and OMD separation processes the mass transfer through a porous and hydrophobic membrane is similar to that in DCMD, except with some slight differences related mainly with the way the driving force "transmembrane vapor pressure difference" is generated. DCMD is a thermally driven separation process; OD is an osmotically driven separation process, in which an osmotic agent is used to lower the water vapor pressure of the permeate relative to that of the feed; and OMD is a combination of both DCMD and OD. In the three processes liquid/vapor interfaces are formed at both sides of the membrane pore and water molecules and/or volatile components evaporate at the formed feed liquid/vapor interface, cross the pore in vapor phase, and finally condense at the formed permeate liquid/vapor interface. The developed theoretical models are generally based on the kinetic theory of gases to predict the membrane performance [3,4], and the pore size together with the applied temperature and hydrostatic pressure plays important roles to establish the mechanism of mass transport through the membrane pore. In general, air is present in the void volume fraction of the and vapor transport is governed by the combined Knudsen/ordinary molecular diffusion model quantified by Bosanquet equation that considers an equal probability of Knudsen diffusion (D_k) and molecular diffusion (D_M) as follows [142]:

$$D_e = \left(\frac{1}{D_K} + \frac{1}{D_M}\right)^{-1} = \left(\frac{3}{2}\frac{\delta\tau}{\varepsilon d_p}\left(\frac{\pi RT}{8M_w}\right)^{1/2} + \frac{RT}{M_w}\frac{\delta\tau}{\varepsilon}\frac{P_a}{PD_{w/a}}\right)^{-1} \tag{9.1}$$

where D_e is the effective diffusion coefficient, M_w is the molecular weight of water, R is the gas constant, δ is the membrane thickness, d_p is the mean size of the pore, ε is the void volume fraction, τ is the pore tortuosity, P_a is the air pressure in the pore, P is the total pressure (atmospheric pressure in this case), and $D_{w/a}$ is the ordinary diffusion of water vapor in air. $PD_{w/a}$ (Pa m^2/s) can be calculated from the following equation [143].

$$PD_{w/a} = 1.89510^{-5}T^{2.072} \tag{9.2}$$

It is worth noting that the mass transfer associated to viscous or Poiseuille type of flow is negligible in DCMD, OD, and OMD because the hydrostatic pressures of the feed and permeate are normally maintained at atmospheric pressure [3,4,142–144]. In this case, Eq. (9.1) is a limiting case of dusty gas model, which combines all transport mechanisms through the membrane, namely, Knudsen diffusion, molecular diffusion, and viscous flow [145]. Moreover, a membrane of uniform and cylindrical pores is generally considered although the used membranes in OD and OMD exhibit pore size distributions. This indicates that different mechanisms may occur simultaneously and the physical nature of mass transport may be different when using membranes under different OD and OMD operating conditions.

In DCMD, a three-dimensional network of interconnected cylindrical pores with a pore size distribution was considered for Monte Carlo simulation and would be also of great interest for OD and OMD [144].

Actually, there are different contributions to mass transfer through the membrane pore of each diffusion type and Eq. (9.1) was revised as [142]:

$$D_e = \left(\frac{\alpha}{D_K} + \frac{1-\alpha}{D_M}\right)^{-1} \tag{9.3}$$

where α is the contribution of Knudsen type of diffusion.

The used quantity to determine the operative mechanism in a membrane subjected to a given experimental condition is Knudsen number ($K_n = \lambda/d_p$) being λ the mean free path of the transported molecules through the membrane pore. Briefly, when K_n is higher than unity, the probability of collisions between molecules, and pore wall is greater than that between molecule and molecule and therefore Knudsen type of flow is responsible for mass transport. As stated previously, in OD and OMD, air is present in the pore and the pressure is close to atmospheric pressure. In this case, when K_n is lower than 0.01, molecular diffusion is considered responsible for mass transport through the membrane pore. When K_n lies in the range 0.01 and 1, the effective diffusion through the pore is the combined Knudsen/ordinary-diffusion mechanism [143–146].

9.4.2 Heat transfer in osmotic distillation and osmotic membrane distillation

In OMD separation process, as can be seen in Fig. 9–3, the heat transfer through the membrane is due to the latent heat accompanying mass permeate flux (J_w) and the heat transferred by conduction across both the membrane matrix and the gas-filled membrane pores (Q_m) as described by the following equation [3,4,38]:

$$Q_m = J_w \Delta H_v + \frac{k}{\delta}\left(T_{m,f} - T_{m,p}\right) \tag{9.4}$$

where ΔH_v is the involved heat of vaporization, $T_{m,f}$ is the feed temperature at the membrane surface, $T_{m,p}$ is the permeate temperature at the membrane surface, and k is the thermal conductivity of the membrane that can be calculated as [3,4,38]:

$$k = \left(\frac{\varepsilon}{k_g} + \frac{1-\varepsilon}{k_m}\right)^{-1} \tag{9.5}$$

where k_g and k_m are the thermal conductivity of the gas-filled pores and that of the membrane matrix, respectively.

Although OD is an isothermal separation process, a temperature gradient occurs due to the evaporation at the feed/membrane interface and the subsequent condensation at the permeate/membrane interface (Fig. 9–3). This causes a heat transfer through the membrane, but in contrast to OMD, Q_m is smaller and reversed in the opposite direction of the permeate flux (J_w) (Fig. 9–3). The occurred transmembrane temperature difference was estimated to be within the range 0.5–0.8K for permeate fluxes $18\text{–}28.8 \cdot 10^{-3}$ kg/m^2 h resulting

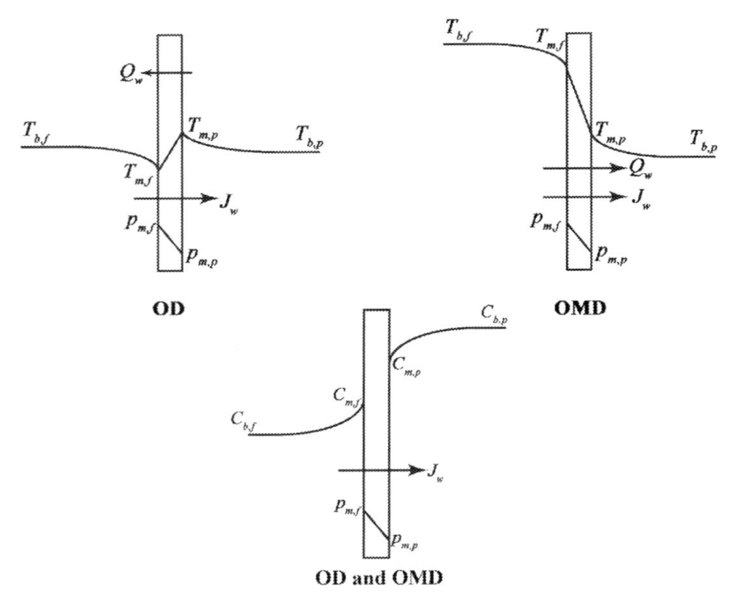

FIGURE 9–3 Temperature and concentration profiles in OD and OMD processes. *OD*, Osmotic distillation; *OMD*, osmotic membrane distillation.

therefore in a negligible reduction of the permeate flux [21]. However, the thermal effect observed in OD due to the evaporation and condensation occurred at both surfaces of the membrane increases with the increase of the permeate flux due to high concentration gradients. A permeate flux of 12 kg/m^2 h resulted in about 2K transmembrane temperature difference (i.e., 30% decline of the driving force) [31,147].

9.4.3 Heat and mass transfer boundary layers: Temperature and concentration polarization effects in OD and OMD

In OD and OMD studies the permeate flux, J_w, is linearly related to the applied transmembrane partial pressure ($\Delta p_v = p_{m,f} - p_{m,p}$) either by using an osmotic agent in the permeate side of the membrane and/or the temperature gradient [21].

$$J_w = D_e\left(p_{m,f} - p_{m,p}\right) \tag{9.6}$$

where the partial vapor pressure of water, $p_{m,f}$ and $p_{m,p}$ can be determined from the following equation [3,147]:

$$p = p^0(T)a(x) \tag{9.7}$$

where a is the activity of the water, x is the water mole fraction in the solution, and p^0 is the vapor pressure of pure water determined as [3]:

$$p^0(T) = e^{(23.20-(3846.44/(T-46.13)))} \tag{9.8}$$

where p^0 is in Pa and T in K.

Both $p_{m,f}$ and $p_{m,p}$ depend on the temperature and concentration at the feed and permeate membrane surfaces, respectively. As shown in Fig. 9–3, these are different from those at bulk liquid solutions due to the simultaneous heat and mass transfer through the membrane. These phenomena are called temperature polarization and concentration polarization resulting in a reduction of the driving force and the permeate flux as a consequence. The concentration polarization in the feed side refers to the accumulation of solutes at the membrane surface due to the evaporation phenomenon, whereas that in the permeate refers to the dilution of the osmotic solution at the permeate membrane surface. The overall OD and OMD processes are controlled by both the heat and mass transfer through both the membrane and the liquid phases adjoining the membrane.

At steady-state conditions, the heat transfer through the feed boundary layer (Q_f) becomes equal to that through the membrane (Q_m, Eq. 9.4) and that of the permeate (Q_p) [3,4,38]:

$$Q_f = h_f\left(T_{b,f} - T_{m,f}\right) = Q_m = Q_p = h_p\left(T_{b,p} - T_{m,p}\right) = H\left(T_{b,f} - T_{b,p}\right) \tag{9.9}$$

where H is the global heat transfer coefficient of the process, and h_f and h_p are the heat transfer coefficients of the feed and permeate, respectively.

Compared to OD, the temperature polarization coefficient is more significant in OMD and it is defined as:

$$\theta = \frac{T_{m,f} - T_{m,p}}{T_{b,f} - T_{b,p}} \tag{9.10}$$

The temperatures at the membrane surfaces can be estimated taking into account the heat transfer coefficients h_f and h_p as well as the permeate flux J_w as described thoroughly for DCMD process by Khayet and coworkers [3,4].

On the contrary, the concentration polarization in both OD and OMD is important specifically on the side of the osmotic solution. The dilution of the osmotic solution at the permeate membrane surface can significantly reduce the water activity and the necessary driving force. The effect of the concentration polarization is more pronounced for greater permeate fluxes.

The concentration polarization coefficients can be defined separately for the feed and permeate as follows:

$$\psi_f = \frac{C_{m,f}}{C_{b,f}} \tag{9.11}$$

$$\psi_p = \frac{C_{m,p}}{C_{b,p}} \tag{9.12}$$

where ψ_f is greater than 1 whereas ψ_p is lower than 1.

It is worth quoting that the solute concentration at the membrane surface (C_m) was predicted by the film theory model as [21,31]:

$$C_m = C_b e^{J_w/k_s} \tag{9.13}$$

where C_b is the solute concentration at the bulk aqueous solution and k_s is the solute mass transfer coefficient for the diffusive mass transfer through the boundary layer.

Both the concentration and temperature polarization effects can be evaluated by means of a joined coefficient defined as the fraction of the applied driving force that contributes to mass transfer:

$$\phi = \frac{p_{m,f} - p_{m,p}}{p_{b,f} - p_{b,p}} \tag{9.14}$$

The permeate flux can finally be rewritten as a function of the water vapor pressure of the bulk feed and permeate solutions:

$$J_w = D_e \phi (p_{b,f} - p_{b,p}) = D'_e (p_{b,f} - p_{b,p}) \tag{9.15}$$

9.5 OD and OMD applications and effects of different involved operating parameters

As discussed earlier, OD is an isothermal evaporative membrane process that involves two liquid streams separated by a hydrophobic porous membrane. The driving force for mass transport is the concentration difference across the membrane. This concentration difference results in a vapor pressure gradient across the membrane. Given this thermodynamic trait of this process, OD should be suitable for concentrating solutions at a high degree that are sensitive to high temperatures and pressures.

OMD is the same as OD with the only difference that it is a nonisothermal evaporative membrane process that involves two liquid streams separated by a hydrophobic porous membrane. This process is very similar to DCMD. However, instead of distilled water on the permeate side the carrier is a solution of higher osmotic concentration than the feed solution thus water transport occurs from the feed side (concentrating the feed solution) to the permeate side (diluting the osmotic solution) [31,59,147]. The temperature of the feed side is usually low enough that it does not affect the integrity of heat-sensitive feed solutions, such as juices. As shown in Tables 9–4 and 9–5, some studies have tested this process and were mostly focused on bench-scale studies of sucrose, wastewater, and juices.

Many of the current applications of OD and OMD are fundamental experimental studies and most of them are focused on Juices concentration. Thermal processes are often not preferred or even suitable when the solution to be concentrated is sensitive to heat as it compromises and degrades the quality of the concentrate (e.g., juices). OD and OMD have several advantages over other membrane processes for concentrating juices and sensitive solutions. The most prominent advantage is the utilization of low pressure and temperature. Since high pressures and temperatures are not required, the cost and sustainability of the processes are highly favorable. The other big subsequent advantage of OD and OMD related to pressure utilization is that they can achieve high concentration factors of the feed solutions without being limited by the osmotic pressure limitations of a typical pressure-driven process. However, both OD and OMD have yet to be utilized beyond the pilot plant scale. Lack of membranes and modules availability with sufficient surface area for industrial implementation is one of the main reasons. Most of the studies presented in Tables 9–4 and 9–5 utilize PTFE membranes, which cannot be fabricated in hollow-fiber form with mechanical stability characteristics. Furthermore, the fluxes produced from the processes are pretty low at about 1 kg/m^{2}h; however, some studies have shown that optimizing the operating variables could increase the performance of the process [31].

The most notable studied operational variables are temperature of the feed solution, its circulation flowrate, osmotic solution, and its concentration. Studies have shown improvements in an order of magnitude by varying the experimental parameters mentioned previously. For instance, it was stated that varying the experimental conditions the permeate flux could be improved from 0.5 to 23 kg/m^{2}h [31]. In this chapter, we discuss the effect of the flowrate, concentration, temperature, and membrane characteristics under the perspective of each application published to date.

From Tables 9–4 and 9–5, it is evident that the majority of the studies conducted both for OD and OMD are on fruit juices with the exception of very few studies for wastewater applications and mostly for urea concentration. It should be noted, at least a couple of studies focused on the preservation of the quality of the juice by studying the aroma compounds or loss of volatile components. Shaw et al. showed that it is possible as concentrating fruit juices (either with only OD or combined with other processes) aroma loss can occur up to 35% on average [70]. Other studies have measured volatiles transport through OD as a surrogate for aroma and the results vary based on the operational variables selected of the process [72,106,128]. When OD is compared with vacuum MD, OD outperforms in terms of aroma losses [72]. In all studies examining aroma and volatile compounds while concentrating solutions, it is crucial to optimize operational variables to ensure that sensitive solutions are not compromised in terms of quality.

Multiple fruit juice types have been studied from traditional orange and apple juice to several exotic types such as kiwi and pomegranate. The main reason for juice concentration in the industry is the reduction of costs and space. As stated previously in this chapter the main advantage of OD is that it can provide acceptable degree of concentration without compromising the quality of the product. For that reason OD has been studied in the last couple of decades to concentrate multiple juices such as orange juice [26,38,43, 71,77,91,111,117], apple juice [50,90,99], pomegranate juice [65,130,134,148], tomato juice [29,98,117], noni juice [93], kiwi juice [76,83,87], lime juice [38], melon juice [78], passion fruit juice [69,70], and cranberry juice [116].

One of the principal objectives of this chapter is to discuss the advancement of the OD and OMD application studies (listed in Tables 9–4 and 9–5) and discuss the effect of the operating parameters of the processes for improvement of the performance of the process.

9.5.1 Temperature effect

OD is an isothermal process; however, studies have shown that a significant enhancement of the initial permeate flux from the feed to the brine side can be achieved with increasing temperature of both the feed and the brine [33,124]. El-Abbassi et al. showed that an increase of temperature from 20°C to 40°C could improve transport of about 40% [30]. Rehman et al. showed increasing temperature of a sucrose solution from 23°C to 35°C raised the average flux by 25% [124]. This is due to the increase of the water vapor pressure following an Arrhenius type of dependence with temperature, which increases the driving force although the OD process is isothermal [30]. However, the extent of mass transfer increase with temperature depends on solute content. More specifically, Courel et al. showed that as feed temperature increases from 20°C to 35°C the vapor transfer increases by 120% for a 35 wt.% sucrose solution but only by 32% at 60wt.% solute content [31].

The benefit of combining a temperature and a concentration gradient (i.e., OMD), to increase the performance of OD, has been examined by several studies. The application of a temperature gradient between pure water and brine has a significant effect on mass transfer. However, when the feed solution is not pure water and varies in concentration, the enhancement of flux is not as straightforward. Courel et al. [31] reported that the increase of the permeate flux, when high concentration feed solution (55–60 w/w% sucrose) was subjected to a temperature gradient of 10°C–15°C, was limited to 14%. The authors also reported that for even higher concentration solution (65 wt.% sucrose), the flux enhancement did not exceed a 19% improvement [31]. Even though the increased temperature gradient is favoring mass transport from the feed side, the increased dilution cause on the brine side is reducing the osmotic driving force; thus the two driving forces are not additive beyond a certain point. Ongaratto et al. showed that they could achieve a typical permeate flux of 1.46 kg/m^2h by imposing a 15°C temperature gradient between the feed and brine solutions [126].

Bélafi-Bakó and Koroknai showed that both osmotic and temperature gradient driving forces have an input in the effectiveness of OD process for water, sucrose solutions, and apple juice [50]. Increasing the feed temperature resulted in permeate flux increase due to an increased driving force; however, at a feed temperature of 45°C ($\Delta T = 20°C$), the flux became insensitive to temperature due to the conductive heat loss. It should be noted that in any case higher temperatures are prone to compromise the integrity of the feed solution (e.g., juice).

Due to the sensitive nature of certain solutions to temperature, several studies have shown that even a less than 10 degrees temperature difference can have an effect in the flux of the feed to the brine without degrading the solution. In their study, Vaillant et al. showed that when the feed and osmotic solutions were maintained at 26°C and 31°C, respectively, the very sensitive to temperature melon juice did not lose its integrity while maintaining a good permeate flux [78]. On the other hand, Onsekizoglu et al. showed that the quality of apple juice did not degrade after OMD application considering a temperature difference of 20°C between the feed and brine side [99].

When the treated feed stream is not sensitive to higher temperatures, such as treating olive mill wastewater, then using OMD with a temperature difference of 20°C results in better permeate fluxes [30] than conventional OD by a small fraction. Another study of extracting the *Echinacea* genus from *E. purpurea* showed that keeping the feed and brine temperatures at 40°C and 24°C, respectively, has a positive effect without compromising the final product [24].

9.5.2 Flowrate effect

Most of the studies, which varied the circulation flowrate either of the feed or the brine stream or both simultaneously, demonstrate that high feed flowrates is advantageous in terms of the increase in permeate flux and concentration of the feed solution [8,116,129]. However, Babu et al. showed that the effect of feed flowrate was found prominent only at higher concentration of feed solutions [38]. This observation is expected as the concentration polarization is added to the resistance to mass transport due to the increasing concentration on the feed steam. Since the driving force to mass transport in OD and OMD is osmotic gradient, this is directly correlated to the concentration of the feed solution and any operational variable can change the boundary layer and have an important effect on the performance of the process.

To further strengthen this effect, several studies conducted on pure water at a constant temperature (~25°C) showed that permeate flux from the feed to the brine side was insensitive to flowrate on both sides of the membrane [50,93].

When high temperature is applied similar to MD (even with pure water), the process shows improvement of flux with increasing flowrate. This is an indication that for OMD the two boundary layers corresponding to temperature and concentration profiles in Fig. 9–3, the feed flowrate will have an impact of both resistances to mass transport [106].

The degree of each volatile type extracted from the OD feed can be feed flowrate dependent. Some investigations showed that by increasing the residence time on the feed side (by lowering the feed flowrate), it increases the degree of volatiles stripped to the brine side [80,106]. This is particularly important when the aromas need to be preserved or removed from the feed solution.

On the other hand, the dependence on brine flowrate appears to be insignificant compared to the other operational variables, that is, feed-side mass transfer resistance was shown to be the dominant hydrodynamic resistance controlling volatiles mass transfer [106]. Courel et al. showed that increasing the brine flowrate by 10% it only increases flux between 2% and 66% [31].

9.5.3 Osmotic solution effect

Several studies on OD and OMD have observed that the most important parameter affecting the process is the osmotic agent concentration. Increasing the concentration of the osmotic agent increases the pressure gradient across the membrane [8,26,27,30,31,36, 43,46,68,93,110,111,116,117,147,149]. As stated in Section 9.3, the osmotic solution has to meet certain properties to provide high driving force for mass transfer such as high solubility in water, low volatility, low viscosity, and high surface tension [147]. It is important to maintain the concentration of the osmotic solution close to the saturation to achieve a high yield of the OD and OMD separation processes. The main disadvantage of near-saturated salt solutions is scaling and precipitation. The formation of deposit hinders the mixing of condensed vapor with the osmotic solution (i.e., brine dilution) and increases the resistance of heat transfer, which enhances a temperature of the interfacial layer: membrane (condensed vapor)/brine. These phenomena decrease a value of the mass driving force, and thus they cause a decline of the permeate flux [59,150]. Organic osmotic solutions are less popular and the few studies have used propylene glycol, glycerol, and glycerol/sodium chloride [75]. Organic solutions offer the main advantage of high solubility compared to salt solutions, thus avoiding scaling and corrosion while producing the same driving force as salt solutions. However, due to their high cost, they are considered mostly an unviable solution [31]. Interestingly, potassium pyrophosphate as osmotic solution has not been studied in depth, even though it exhibits high osmotic potential and has a higher solubility compared to calcium chloride [126].

Since OD is a process used mostly for the concentration of thermally sensitive solutions such as juices, the aroma quality of the product is important. Few studies have been focused on studying the effect of the osmotic solution and aroma losses from the feed solution. Ongaratto et al. found that when potassium pyrophosphate was used as brine, there was a reduction of the aroma losses from the feed solution compared to other solutions [126]. This is due to the reduced activity of volatile compounds as the concentration of the brine increases, and since potassium pyrophosphate has a higher solubility, it allows for further concentration increase. As the volatile activity reduces, less volatile compound permeation occurs through the membrane [26,126]. Hasanoğlu et al. studied the effect of aroma transport using two different osmotic solutions of 30 and 50 wt.% $CaCl_2$ [106]. It was observed that each aroma compound was affected differently by the osmotic solution concentration with the more volatile compounds being affected the most. Bélafi-Bakó and Koroknai showed that applying a temperature gradient (20°C) while varying the osmotic concentration could indeed have an additive effect on the driving forces and thus higher permeate fluxes could be achieved using $CaCl_2$ solution while concentrating a sucrose solution to a target 45% [50]. However, this could not be applied in all cases as increasing the temperature would compromise the quality of the feed solution.

9.6 Conclusion

Very few studies on membrane engineering for OD have been carried, especially for membrane surface modification to reduce the risk of pore wetting. Only one study was performed on nanofibrous membrane development for OD separation process. No research study has been found on preparation or modification of any membrane for OMD. Instead, sufficient worldwide research papers have been published on theoretical modeling, field of applications as most of the studies reported herein are for the treatment and concentration of fruit juices with the exception of some studies on beverage dealcoholization and wastewater, and the effects of the involved operating parameters on the OD and/or OMD performance.

For a given feed solution the most important effect on the performance of OD under the same conditions is the concentration of the osmotic solution and type of the osmotic agent. The results have shown that the increase in temperature, brine/extraction solution concentration, and feed flowrates tend to enhance OD performance.

Both OD and OMD have yet to be utilized beyond pilot plant scales. The lack of membranes and modules availability with sufficient surface area for industrial implementation is one of the main reasons.

Although the osmotic solution plays a significant role in OD and OMD performance, compared to FO still very few studies explore the use of other possible osmotic agents.

References

[1] M. Findley, Vaporization through porous membranes, Ind. Eng. Chem. Process Des. Dev. 6 (2) (1967) 226–230.

[2] M. Findley, et al., Mass and heat transfer relations in evaporation through porous membranes, AIChE J. 15 (4) (1969) 483–489.

[3] M. Khayet, Membranes and theoretical modeling of membrane distillation: a review, Adv. Colloid Interface Sci. 164 (1–2) (2011) 56–88.

[4] M. Khayet, T. Matsuura, Membrane Distillation: Principles and Applications, Elsevier, 2011.

[5] R. Johnson, R. Valks, M. Léfèbvre, Osmotic distillation, in: Workshop on Membrane Distillation, Rome. 1986.

[6] M. Lefebvre, R. Johnson, V. Yip, Theoretical and practical aspects of osmotic distillation, in: International Congress of Membranes, Tokyo, 1987.

[7] R. Johnson, R. Valks, M. Lefebvre, Osmotic distillation—a low temperature concentration technique, Aust. J. Biotechnol. 3 (3) (1989) 206–207.

[8] J. Sheng, R. Johnson, M. Lefebvre, Mass and heat transfer mechanisms in the osmotic distillation process, Desalination 80 (2–3) (1991) 113–121.

[9] G. Sarti, C. Gostoli, S. Matulli, Low energy cost desalination processes using hydrophobic membranes, Desalination 56 (1985) 277–286.

[10] E. Drioli, Y. Wu, V. Calabro, Membrane distillation in the treatment of aqueous solutions, J. Membr. Sci. 33 (3) (1987) 277–284.

[11] G. Sarti, C. Gostoli, Use of hydrophobic membranes in thermal separation of liquid mixtures: theory and experiments, Membranes and Membrane Processes, Springer, 1986, pp. 349–360.

[12] Z. Honda, et al., Nonisothermal mass transport of organic aqueous solution in hydrophobic porous membrane, Membranes and Membrane Processes, Springer, 1986, pp. 587–594.

[13] K. Schneider, et al., Membranes and modules for transmembrane distillation, J. Membr. Sci. 39 (1) (1988) 25–42.

[14] R. Schofield, A. Fane, C. Fell, Heat and mass transfer in membrane distillation, J. Membr. Sci. 33 (3) (1987) 299–313.

[15] M. Godino, et al., Coupled phenomena membrane distillation and osmotic distillation through a porous hydrophobic membrane, Sep. Sci. Technol. 30 (6) (1995) 993–1011.

[16] Z. Wang, F. Zheng, S. Wang, Experimental study of membrane distillation with brine circulated in the cold side, J. Membr. Sci. 183 (2) (2001) 171–179.

[17] Z. Wang, et al., Membrane osmotic distillation and its mathematical simulation, Desalination 139 (1–3) (2001) 423–428.

[18] A. Barbe, et al., Retention of volatile organic flavour/fragrance components in the concentration of liquid foods by osmotic distillation, J. Membr. Sci. 145 (1) (1998) 67–75.

[19] R.A. Johnson, M.H. Nguyen, Understanding Membrane Distillation and Osmotic Distillation, John Wiley & Sons, 2017.

[20] V. Alves, I. Coelhoso, Effect of membrane characteristics on mass and heat transfer in the osmotic evaporation process, J. Membr. Sci. 228 (2) (2004) 159–167.

[21] J.I. Mengual, L. Pen, A. Vela, Osmotic distillation through porous hydrophobic membranes, J. Membr. Sci. 82 (1–2) (1993) 129–140.

[22] W.-U. Rehman, et al., Effect of membrane wetting on the performance of PVDF and PTFE membranes in the concentration of pomegranate juice through osmotic distillation, J. Membr. Sci. 584 (2019) 66–78.

[23] J. Mansouri, A. Fane, Osmotic distillation of oily feeds, J. Membr. Sci. 153 (1) (1999) 103–120.

[24] R. Johnson, J. Sun, J. Sun, A pervaporation–microfiltration–osmotic distillation hybrid process for the concentration of ethanol–water extracts of the Echinacea plant, J. Membr. Sci. 209 (1) (2002) 221–232.

[25] C.H. Shin, R. Johnson, Identification of an appropriate osmotic agent for use in osmotic distillation, J. Ind. Eng. Chem. 13 (6) (2007) 926–931.

[26] V. Alves, I. Coelhoso, Orange juice concentration by osmotic evaporation and membrane distillation: a comparative study, J. Food Eng. 74 (1) (2006) 125–133.

[27] Z. Anari, et al., Combined osmotic and membrane distillation for concentration of anthocyanin from muscadine pomace, J. Food Sci. 84 (8) (2019) 2199–2208.

[28] C. Mai, Concentration of Hot Water Extracts of Anthocyanins Obtained From Muscadine Grape Pomace Using Membrane-Osmotic Distillation. Food Science Undergraduate Honors Theses. University of Arkansas, 2019. https://scholarworks.uark.edu/fdscuht/9/.

[29] R. Durham, M. Nguyen, Hydrophobic membrane evaluation and cleaning for osmotic distillation of tomato puree, J. Membr. Sci. 87 (1–2) (1994) 181–189.

[30] A. El-Abbassi, et al., Treatment of crude olive mill wastewaters by osmotic distillation and osmotic membrane distillation, Sep. Purif. Technol. 104 (2013) 327–332.

[31] M. Courel, et al., Effect of operating conditions on water transport during the concentration of sucrose solutions by osmotic distillation, J. Membr. Sci. 170 (2) (2000) 281–289.

[32] R.B. Rodrigues, et al., Evaluation of reverse osmosis and osmotic evaporation to concentrate camu–camu juice (*Myrciaria dubia*), J. Food Eng. 63 (1) (2004) 97–102.

[33] C. Hongvaleerat, et al., Concentration of pineapple juice by osmotic evaporation, J. Food Eng. 88 (4) (2008) 548–552.

[34] M. Celere, C. Gostoli, Heat and mass transfer in osmotic distillation with brines, glycerol and glycerol–salt mixtures, J. Membr. Sci. 257 (1–2) (2005) 99–110.

[35] M. Tomaszewska, Concentration and purification of fluosilicic acid by membrane distillation, Ind. Eng. Chem. Res. 39 (8) (2000) 3038–3041.

[36] J. Kujawa, et al., Raw juice concentration by osmotic membrane distillation process with hydrophobic polymeric membranes, Food Bioprocess Technol. 8 (10) (2015) 2146–2158.

[37] A. Narayan, et al., Acoustic field-assisted osmotic membrane distillation, Desalination 147 (1) (2002) 149–156.

[38] B.R. Babu, N. Rastogi, K. Raghavarao, Mass transfer in osmotic membrane distillation of phycocyanin colorant and sweet-lime juice, J. Membr. Sci. 272 (1–2) (2006) 58–69.

[39] N. Nagaraj, et al., Mass transfer in osmotic membrane distillation, J. Membr. Sci. 268 (1) (2006) 48–56.

[40] G. Patil, et al., Extraction, dealcoholization and concentration of anthocyanin from red radish, Chem. Eng. Process.: Process. Intensif. 48 (1) (2009) 364–369.

[41] C. Jampani, K. Raghavarao, Process integration for purification and concentration of red cabbage (*Brassica oleracea* L.) anthocyanins, Sep. Purif. Technol. 141 (2015) 10–16.

[42] G. Patil, K. Raghavarao, Integrated membrane process for the concentration of anthocyanin, J. Food Eng. 78 (4) (2007) 1233–1239.

[43] B.R. Babu, N. Rastogi, K. Raghavarao, Concentration and temperature polarization effects during osmotic membrane distillation, J. Membr. Sci. 322 (1) (2008) 146–153.

[44] J. Chandrasekhar, K. Raghavarao, Separation and concentration of anthocyanins from jamun: an integrated process, Chem. Eng. Commun. 202 (10) (2015) 1368–1379.

[45] M. Celere, C. Gostoli, The heat and mass transfer phenomena in osmotic membrane distillation, Desalination 147 (1–3) (2002) 133–138.

[46] W. Kujawski, et al., Application of osmotic membrane distillation process in red grape juice concentration, J. Food Eng. 116 (4) (2013) 801–808.

[47] J. Warczok, et al., Application of osmotic membrane distillation for reconcentration of sugar solutions from osmotic dehydration, Sep. Purif. Technol. 57 (3) (2007) 425–429.

[48] W. Kujawski, et al., Application of pervaporation and osmotic membrane distillation to the regeneration of spent solutions from the osmotic food dehydration, Pol. J. Chem. Technol. 11 (2) (2009) 41–45.

[49] G. Freiberg, Experimental and theoretical study on the effectiveness of vacuum in osmotic membrane distillation, Pet. Chem. 53 (8) (2013) 572–577.

[50] K. Bélafi-Bakó, B. Koroknai, Enhanced water flux in fruit juice concentration: coupled operation of osmotic evaporation and membrane distillation, J. Membr. Sci. 269 (1–2) (2006) 187–193.

[51] A. Boór, K. Bélafi-Bakó, N. Nemestóthy, Concentration of colourful wild berry fruit juices by membrane osmotic distillation via cascade model systems, J. Membr. Sci. Res. 2 (4) (2016) 201–206.

[52] C.A. Quist-Jensen, et al., A study of membrane distillation and crystallization for lithium recovery from high-concentrated aqueous solutions, J. Membr. Sci. 505 (2016) 167–173.

[53] B. Koroknai, et al., Preservation of antioxidant capacity and flux enhancement in concentration of red fruit juices by membrane processes, Desalination 228 (1–3) (2008) 295–301.

[54] K. Bélafi-Bakó, A. Boór, Concentration of cornelian cherry fruit juice by membrane osmotic distillation, Desalin. Water Treat. 35 (1–3) (2011) 271–274.

[55] A. Cassano, C. Conidi, E. Drioli, A membrane-based process for the valorization of the bergamot juice, Sep. Sci. Technol. 48 (4) (2013) 537–546.

[56] D. Kumar, et al., Osmotic membrane distillation for retention of antioxidant potential in Nagpur mandarin (*Citrus reticulata* Blanco) fruit juice concentrate, J. Food Process. Eng. 43 (1) (2020) e13096.

[57] I.R. Salmón, R. Janssens, P. Luis, *Mass and heat transfer study in osmotic membrane distillation-crystallization for CO_2 valorization as sodium carbonate*, Sep. Purif. Technol. 176 (2017) 173–183.

[58] I.R. Salmón, et al., Salt recovery from wastewater using membrane distillation–crystallization, Cryst. Growth Des. 18 (12) (2018) 7275–7285.

[59] M. Gryta, The long-term studies of osmotic membrane distillation, Chem. Pap. 72 (1) (2018) 99–107.

[60] J. Xu, et al., Alginate-coated microporous PTFE membranes for use in the osmotic distillation of oily feeds, J. Membr. Sci. 240 (1–2) (2004) 81–89.

[61] J. Xu, et al., Alginic acid–silica hydrogel coatings for the protection of osmotic distillation membranes against wet-out by surface-active agents, J. Membr. Sci. 260 (1–2) (2005) 19–25.

[62] J. Xu, J. Bartley, R. Johnson, Application of sodium alginate-carrageenan coatings to PTFE membranes for protection against wet-out by surface-active agents, Sep. Sci. Technol. 40 (5) (2005) 1067–1081.

[63] A. Chanachai, K. Meksup, R. Jiraratananon, Coating of hydrophobic hollow fiber PVDF membrane with chitosan for protection against wetting and flavor loss in osmotic distillation process, Sep. Purif. Technol. 72 (2) (2010) 217–224.

[64] A. Vargas-Garcia, et al., Effect of grafting on microstructure, composition and surface and transport properties of ceramic membranes for osmotic evaporation, Sep. Purif. Technol. 80 (3) (2011) 473–481.

[65] A. Roozitalab, A. Raisi, A. Aroujalian, A comparative study on pomegranate juice concentration by osmotic distillation and thermal evaporation processes, Korean J. Chem. Eng. 36 (9) (2019) 1474–1481.

[66] C. Gostoli, Thermal effects in osmotic distillation, J. Membr. Sci. 163 (1) (1999) 75–91.

[67] M. Courel, et al., Modelling of water transport in osmotic distillation using asymmetric membrane, J. Membr. Sci. 173 (1) (2000) 107–122.

[68] A. Bailey, et al., The effect of ultrafiltration on the subsequent concentration of grape juice by osmotic distillation, J. Membr. Sci. 164 (1–2) (2000) 195–204.

[69] F. Vaillant, et al., Concentration of passion fruit juice on an industrial pilot scale using osmotic evaporation, J. Food Eng. 47 (3) (2001) 195–202.

[70] P.E. Shaw, et al., Evaluation of concentrated orange and passion fruit juices prepared by osmotic evaporation, LWT—Food Sci. Technol. 34 (2) (2001) 60–65.

[71] A. Cassano, et al., Clarification and concentration of citrus and carrot juices by integrated membrane processes, J. Food Eng. 57 (2) (2003) 153–163.

[72] F. Ali, et al., Evaluating transfers of aroma compounds during the concentration of sucrose solutions by osmotic distillation in a batch-type pilot plant, J. Food Eng. 60 (1) (2003) 1–8.

[73] F. Brodard, et al., New hydrophobic membranes for osmotic evaporation process, Sep. Purif. Technol. 32 (1–3) (2003) 3–7.

[74] V.A. Bui, M.H. Nguyen, J. Muller, A laboratory study on glucose concentration by osmotic distillation in hollow fibre module, J. Food Eng. 63 (2) (2004) 237–245.

[75] M. Celere, C. Gostoli, Osmotic distillation with propylene glycol, glycerol and glycerol–salt mixtures, J. Membr. Sci. 229 (1–2) (2004) 159–170.

[76] A. Cassano, B. Jiao, E. Drioli, Production of concentrated kiwifruit juice by integrated membrane process, Food Res. Int. 37 (2) (2004) 139–148.

[77] M. Cisse, et al., The quality of orange juice processed by coupling crossflow microfiltration and osmotic evaporation, Int. J. Food Sci. Technol. 40 (1) (2005) 105–116.

[78] F. Vaillant, et al., Clarification and concentration of melon juice using membrane processes, Innovative Food Sci. Emerg. Technol. 6 (2) (2005) 213–220.

[79] A. Bui, H. Nguyen, M. Joachim, Characterisation of the polarisations in osmotic distillation of glucose solutions in hollow fibre module, J. Food Eng. 68 (3) (2005) 391–402.

[80] N. Diban, et al., Ethanol and aroma compounds transfer study for partial dealcoholization of wine using membrane contactor, J. Membr. Sci. 311 (1–2) (2008) 136–146.

[81] B. Han, Z. Shen, S.R. Wickramasinghe, Cyanide removal from industrial wastewaters using gas membranes, J. Membr. Sci. 257 (1–2) (2005) 171–181.

[82] A. Bui, H. Nguyen, Scaling up of osmotic distillation from laboratory to pilot plant for concentration of fruit juices, Int. J. Food Eng. 1 (2) (2005).

[83] A. Cassano, et al., Integrated membrane process for the production of highly nutritional kiwifruit juice, Desalination 189 (1–3) (2006) 21–30.

[84] A. Rektor, G. Vatai, E. Békássy-Molnár, Multi-step membrane processes for the concentration of grape juice, Desalination 191 (1–3) (2006) 446–453.

[85] A. Cassano, et al., A membrane-based process for the clarification and the concentration of the cactus pear juice, J. Food Eng. 80 (3) (2007) 914–921.

[86] Á. Kozák, et al., Comparison of integrated large scale and laboratory scale membrane processes for the production of black currant juice concentrate, Chem. Eng.Process.: Process. Intensif. 47 (7) (2008) 1171–1177.

[87] A. Cassano, E. Drioli, Concentration of clarified kiwifruit juice by osmotic distillation, J. Food Eng. 79 (4) (2007) 1397–1404.

[88] R. Thanedgunbaworn, R. Jiraratananon, M. Nguyen, Mass and heat transfer analysis in fructose concentration by osmotic distillation process using hollow fibre module, J. Food Eng. 78 (1) (2007) 126–135.

[89] R. Thanedgunbaworn, R. Jiraratananon, M.H. Nguyen, Shell-side mass transfer of hollow fibre modules in osmotic distillation process, J. Membr. Sci. 290 (1–2) (2007) 105–113.

[90] M. Cissé, et al., Athermal concentration by osmotic evaporation of roselle extract, apple and grape juices and impact on quality, Innovative Food Sci. Emerg. Technol. 12 (3) (2011) 352–360.

[91] G. Galaverna, et al., A new integrated membrane process for the production of concentrated blood orange juice: effect on bioactive compounds and antioxidant activity, Food Chem. 106 (3) (2008) 1021–1030.

[92] Z. Shen, et al., Suppression of osmotic distillation in gas membrane processes, Sep. Sci. Technol. 43 (15) (2008) 3813–3825.

[93] H. Valdés, et al., Concentration of noni juice by means of osmotic distillation, J. Membr. Sci. 330 (1–2) (2009) 205–213.

[94] S. Varavuth, R. Jiraratananon, S. Atchariyawut, Experimental study on dealcoholization of wine by osmotic distillation process, Sep. Purif. Technol. 66 (2) (2009) 313–321.

[95] E. Fehér, et al., Semi-continuous enzymatic production and membrane assisted separation of isoamyl acetate in alcohol—ionic liquid biphasic system, Desalination 241 (1–3) (2009) 8–13.

[96] R. Thanedgunbaworn, R. Jiraratananon, M.H. Nguyen, Vapour transport mechanism in osmotic distillation process, Int. J. Food Eng. 5 (5) (2009).

[97] G. Racz, et al., Concentration of 'Oblachinska' sour cherry juice using osmotic distillation, Int. J. Hortic. Sci. 18 (1) (2012) 31–34.

[98] T.N. Musa, *Osmotic evaporation of tomato juice by using* $CaCl_2 \cdot 6H_2O$ *and* $MgCl_2 \cdot 6H_2O$ *as osmotic solution*, Iraqi J. Agric. Sci. 41 (6) (2010) 130–144.

[99] P. Onsekizoglu, K.S. Bahceci, M.J. Acar, Clarification and the concentration of apple juice using membrane processes: a comparative quality assessment, J. Membr. Sci. 352 (1–2) (2010) 160–165.

[100] E. Garcia-Castello, et al., Recovery and concentration of polyphenols from olive mill wastewaters by integrated membrane system, Water Res. 44 (13) (2010) 3883–3892.

[101] P. Onsekizoglu, K.S. Bahceci, J. Acar, The use of factorial design for modeling membrane distillation, J. Membr. Sci. 349 (1–2) (2010) 225–230.

[102] Y. Ahn, Y.-H. Hwang, H.-S. Shin, Application of PTFE membrane for ammonia removal in a membrane contactor, Water Sci. Technol. 63 (12) (2011) 2944–2948.

[103] A. Cassano, C. Conidi, E. Drioli, Clarification and concentration of pomegranate juice (*Punica granatum* L.) using membrane processes, J. Food Eng. 107 (3–4) (2011) 366–373.

[104] Z. Molnár, et al., Concentration of raspberry (*Rubus idaeus* L.) juice using membrane processes, Acta Aliment. 41 (Suppl. 1) (2012) 147–159.

[105] L. Liguori, et al., Effect of process parameters on partial dealcoholization of wine by osmotic distillation, Food Bioprocess Technol. 6 (9) (2013) 2514–2524.

[106] A. Hasanoğlu, et al., Effect of the operating variables on the extraction and recovery of aroma compounds in an osmotic distillation process coupled to a vacuum membrane distillation system, J. Food Eng. 111 (4) (2012) 632–641.

[107] A. Cassano, C. Conidi, R. Ruby-Figueroa, Recovery of flavonoids from orange press liquor by an integrated membrane process, Membranes 4 (3) (2014) 509–524.

[108] L. Liguori, et al., Evolution of quality parameters during red wine dealcoholization by osmotic distillation, Food Chem. 140 (1–2) (2013) 68–75.

[109] P. Russo, et al., Investigation of osmotic distillation technique for beer dealcoholization, Chem. Eng. (2013) 32.

[110] P. Onsekizoglu, Production of high quality clarified pomegranate juice concentrate by membrane processes, J. Membr. Sci. 442 (2013) 264–271.

[111] F. Destani, et al., Recovery and concentration of phenolic compounds in blood orange juice by membrane operations, J. Food Eng. 117 (3) (2013) 263–271.

[112] M. Torun, et al., Concentration of sage (*Salvia fruticosa* Miller) extract by using integrated membrane process, Sep. Purif. Technol. 132 (2014) 244–251.

[113] M. Schmitt, M. Murgo, S. Prieto. Does osmotic distillation change the isotopic relation of wines?, in: BIO Web of Conferences 3, 02006 (2014), EDP Sciences. https://doi.org/10.1051/bioconf/20140302006.

[114] G. De Francesco, et al., Effects of operating conditions during low-alcohol beer production by osmotic distillation, J. Agric. Food Chem. 62 (14) (2014) 3279–3286.

[115] C.O. Rossi, A. Cassano, F. Destani, Rheological behavior of blood orange juice concentrated by osmotic distillation and thermal evaporation, Appl. Rheol. 24 (6) (2014) 20–25.

[116] C. Zambra, et al., Concentration of cranberry juice by osmotic distillation process, J. Food Eng. 144 (2015) 58–65.

[117] K.S. Bahçeci, H.G. Akıllıoğlu, V. Gökmen, Osmotic and membrane distillation for the concentration of tomato juice: effects on quality and safety characteristics, Innovative Food Sci. Emerg. Technol. 31 (2015) 131–138.

[118] G. De Francesco, et al., Pilot plant production of low-alcohol beer by osmotic distillation, J. Am. Soc. Brew. Chem. 73 (1) (2015) 41–48.

[119] L. Liguori, et al., Production and characterization of alcohol-free beer by membrane process, Food Bioprod. Process. 94 (2015) 158–168.

[120] L. Liguori, et al., Quality attributes of low-alcohol top-fermented beers produced by membrane contactor, Food Bioprocess Technol. 9 (1) (2016) 191–200.

[121] C. Dincer, I. Tontul, A. Topuz, A comparative study of black mulberry juice concentrates by thermal evaporation and osmotic distillation as influenced by storage, Innovative Food Sci. Emerg. Technol. 38 (2016) 57–64.

[122] F. Bazzarelli, et al., Advances in membrane operations for water purification and biophenols recovery/valorization from OMWWs, J. Membr. Sci. 497 (2016) 402–409.

[123] R. Ferrarini, et al., Variation of oxygen isotopic ratio during wine dealcoholization by membrane contactors: experiments and modelling, J. Membr. Sci. 498 (2016) 385–394.

[124] W.U. Rehman, et al., Osmotic distillation and quality evaluation of sucrose, apple and orange juices in hollow fiber membrane contactor, Chem. Ind. Chem. Eng. Q. 23 (2) (2017) 217–227.

[125] L. Loredana, et al., Impact of osmotic distillation on the sensory properties and quality of low alcohol beer, J. Food Qual. 2018 (2018).

[126] R.S. Ongaratto, et al., Osmotic distillation applying potassium pyrophosphate as brine, J. Food Eng. 228 (2018) 69–78.

[127] E. Yilmaz, P.O. Bagci, Production of phytotherapeutics from broccoli juice by integrated membrane processes, Food Chem. 242 (2018) 264–271.

[128] P. Russo, et al., Combined membrane process for dealcoholization of wines: osmotic distillation and reverse osmosis, Chem. Eng. Trans. 75 (2019) 7–12.

[129] M. Mahdavi, et al., Concentrating red beet juice using osmotic distillation: effects of device structure, Nutr. Food Sci. Res. 6 (2) (2019) 37–43.

[130] P.O. Bagci, et al., Coupling reverse osmosis and osmotic distillation for clarified pomegranate juice concentration: use of plasma modified reverse osmosis membranes for improved performance, Innovative Food Sci. Emerg. Technol. 52 (2019) 213–220.

[131] S. Motta, et al., Comparison of the physicochemical and volatile composition of wine fractions obtained by two different dealcoholization techniques, Food Chem. 221 (2017) 1–10.

[132] L. Liguori, et al., Impact of dealcoholization on quality properties in white wine at various alcohol content levels, J. Food Sci. Technol. 56 (8) (2019) 3707–3720.

[133] C. Dinçer, et al., Mathematical modeling of concentrations of grape, pomegranate and black carrot juices by various methods, Gıda 44 (6) (2019) 1092–1105.

[134] W.U. Rehman, et al., Pomegranate juice concentration using osmotic distillation with membrane contactor, Sep. Purif. Technol. 224 (2019) 481–489.

[135] O. Corona, et al., Quality and volatile compounds in red wine at different degrees of dealcoholization by membrane process, Eur. Food Res. Technol. 245 (11) (2019) 2601–2611.

[136] C.A. Nayak, N.K. Rastogi, Comparison of osmotic membrane distillation and forward osmosis membrane processes for concentration of anthocyanin, Desalin. Water Treat. 16 (1–3) (2010) 134–145.

[137] L. Terki, et al., Implementation of osmotic membrane distillation with various hydrophobic porous membranes for concentration of sugars solutions and preservation of the quality of cactus pear juice, J. Food Eng. 230 (2018) 28–38.

[138] J. Warczok, et al., Reconcentration of spent solutions from osmotic dehydration using direct osmosis in two configurations, J. Food Eng. 80 (1) (2007) 317–326.

[139] A. Achilli, T.Y. Cath, A.E. Childress, Selection of inorganic-based draw solutions for forward osmosis applications, J. Membr. Sci. 364 (1–2) (2010) 233–241.

[140] N. Akther, et al., Recent advancements in forward osmosis desalination: a review, Chem. Eng. J. 281 (2015) 502–522.

[141] K. Lutchmiah, et al., Forward osmosis for application in wastewater treatment: a review, Water Res. 58 (2014) 179–197.

[142] M. Essalhi, M. Khayet, Self-sustained webs of polyvinylidene fluoride electrospun nanofibers at different electrospinning times: 2. Theoretical analysis, polarization effects and thermal efficiency, J. Membr. Sci. 433 (2013) 180–191.

[143] J. Phattaranawik, R. Jiraratananon, A. Fane, Effect of pore size distribution and air flux on mass transport in direct contact membrane distillation, J. Membr. Sci. 215 (1–2) (2003) 75–85.

[144] M. Khayet, A. Imdakm, T. Matsuura, Monte Carlo simulation and experimental heat and mass transfer in direct contact membrane distillation, Int. J. Heat Mass Transfer 53 (7–8) (2010) 1249–1259.

[145] E.A. Mason, A.P. Malinauskas, Gas Transport in Porous Media: The Dusty Gas Model, Chemical Engineering Monographs, 17, Elsevier, Amsterdam, 1983.

[146] M. Khayet, A. Velázquez, J.I. Mengual, Modelling mass transport through a porous partition: effect of pore size distribution, J. Non-equilibr. Thermodyn. 29 (3) (2004) 279–299.

[147] M. Gryta, Osmotic MD and other membrane distillation variants, J. Membr. Sci. 246 (2) (2005) 145–156.

[148] W. Zou, K.R. Davey, An integrated two-step Fr 13 synthesis-demonstrated with membrane fouling in combined ultrafiltration-osmotic distillation (UF-OD) for concentrated juice, Chem. Eng. Sci. 152 (2016) 213–226.

[149] T.Y. Cath, et al., Membrane contactor processes for wastewater reclamation in space: Part I. Direct osmotic concentration as pretreatment for reverse osmosis, J. Membr. Sci. 257 (1–2) (2005) 85–98.

[150] L. Wang, J. Min, Modeling and analyses of membrane osmotic distillation using non-equilibrium thermodynamics, J. Membr. Sci. 378 (1–2) (2011) 462–470.

10

Thermo-osmosis

M. Essalhi[1,2], N.T. Hassan Kiadeh[2], M.C. García-Payo[1], M. Khayet[1,3]

[1]DEPARTMENT OF STRUCTURE OF MATTER, THERMAL PHYSICS AND ELECTRONICS, FACULTY OF PHYSICS, UNIVERSITY COMPLUTENSE OF MADRID, MADRID, SPAIN

[2]DEPARTMENT OF CHEMISTRY, UMEÅ UNIVERSITY, UMEÅ, SWEDEN

[3]MADRID INSTITUTE FOR ADVANCED STUDIES OF WATER (IMDEA WATER INSTITUTE), MADRID, SPAIN

10.1 Introduction and a brief historical review

During recent years, membrane separation processes and related transport phenomena have acquired considerable importance at high speed, due to their wide scope of application in both academia and industry, and to their high separation efficiency attributed mainly to the growing development of new generation membranes and modules.

The transmembrane nonequilibrium thermodynamics through any type of membrane, either porous or dense, hydrophilic or hydrophobic as well as a mixed matrix or composite membrane, is responsible for the transport of matter and/or energy. This can be induced by a pressure difference, a temperature difference, a difference in concentration, a difference in electrical potential or, in general, by a difference in chemical potential between the two phases in contact with the two membrane sides. Depending on this driving force and the type of membrane, different separation processes can be distinguished, as outlined in Fig. 10–1.

In general, transmembrane nonisothermal transport phenomenon has received much less attention than the corresponding isothermal phenomenon, such as those caused by a pressure gradient in reverse osmosis (RO), nanofiltration, ultrafiltration. For instance, RO is nowadays, without a doubt, the most implemented membrane process at industrial level especially in water desalination, producing drinking water with a cost of less than 0.4 US$/m^3. This is due to the continuous development of appropriate membranes since in 1959 Reid and Breton [1] discovered that cellulose acetate (CA) was an ideal material for RO membrane preparation. The pioneering research works of Loeb and Sourirajan [2] were also key in RO development, since exhaustive and systematic research studies with this type of membranes were carried out, improving their selectivity and water production.

The study of nonisothermal fluxes of matter dates back to 1873 when Feddersen [3] published some qualitative results about air flow through different porous media (platinum, sponge, gypsum, etc.) caused solely by a temperature difference while the pressure was uniform. It was detected that gas flux always occurred in the direction of low to high temperature.

Osmosis Engineering. DOI: https://doi.org/10.1016/B978-0-12-821016-1.00001-2

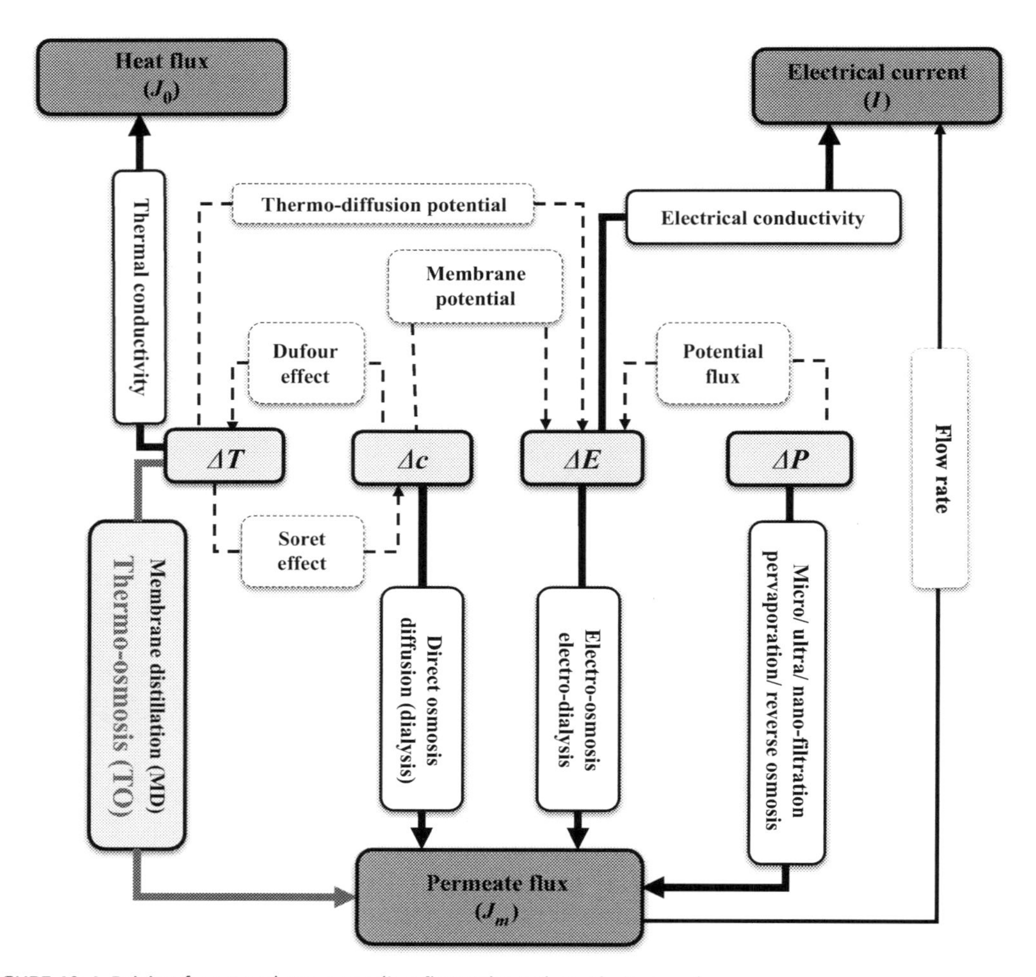

FIGURE 10–1 Driving forces and corresponding fluxes through synthetic membranes.

Feddersen baptized this phenomenon with the name thermal diffusion. Today, the nomenclature "Thermodiffusion" is used to refer to the movement of particles in continuous mixtures subjected to constant temperature gradients, constant pressure and concentration, and in the absence of membranes.

The existence of nonisothermal transport of liquids through a gelatin membrane was first described by Lippmann [4] in 1907, and 5 years later, it was investigated by Aubert [5] in more detail using membranes of gelatin pig's bladder, parchment paper, and viscose. In this phenomenon, there is no liquid/vapor phase transition and it is known as thermo-osmosis (TO) or thermal osmosis (TO). Lippmann [4] also observed TO in air, obviously without prior knowledge of the work of Feddersen [3] and Reynolds [6]. However, at that time the lack of theory on irreversible processes prevented the progress towards understanding this phenomenon.

In 1926 Eastman [7] developed a very intuitive theory relating the fluid flow with the heat flow transported through a given membrane, but the theory was not satisfactory. Not long afterward, in 1931, Onsager [8] proposed the famous reciprocal relations, which through the work of various later theorists formed the basis of the irreversible processes of thermodynamics (IPT).

Lippmann and Aubert [4,5] observed that the permeate flux sometimes occurred in the direction of hot to cold and others in the opposite direction or even no flux could be detected for some materials. Their conclusion was that TO was electrical in origin and it was therefore closely linked to the electro-osmosis (EO) process. In other words, TO was only possible to be detected using electrically charged membranes and electrolyte solutions, which led Freundlich [9] to relate TO to an anomalous osmosis. However, in 1941 Derjaguin and Sidorenkov [10] published TO data tests using pure water and other nonelectrolyte liquids through synthetic glass membranes. Later on, in 1948, these results were questions by Hutchison et al. [11] who repeated the previous tests and concluded that practically all the observed phenomenon was due to the thermal expansion of the used liquids and the experimental device. Subsequently, some not very systematic studies on TO [12,13] were carried out, always suggesting that nonisothermal transport is possible only with electrolyte solutions and with electrically charged and activated membranes. In 1952 the understanding of the IPT had progressed to the point that Denbigh and Raumann [14] were able to formulate a viable theory of TO and interpret obtained quantitative data within the framework of this theory. Rastogi and his co-workers [15–19], who reported on TO of water and methanol for a "DuPont 600" cellophane membrane, indicated that the direction of the TO permeate flux was from the hot to the cold side of the membrane. The same authors also studied TO of binary mixtures of water with other alcohols such as isopropanol and methanol with *n*-butanol.

Voellmy and Lauger [20] carried out a TO experimental study using polystyrene membrane and found that the TO permeate flux of toluene was from the cold to the hot membrane side. This permeate flux was $6 \cdot 10^{-8}$ cm/s agreeing with the thermodiffusion coefficient of polystyrene molecule in toluene. In the case of phenolsulfonate membrane a change in the sign of the TO permeate transport was observed at a temperature of 38°C when using a 10^{-3} mol/dm^3 KCl aqueous solution as an external medium. In other words, for temperatures below 38°C, the TO permeate flux went from the hot to the cold side of the membrane, whereas for temperatures greater than 38°C up to 60°C, the direction of TO transport was reversed. Dirksen [21] used compacted saturated clay membranes in TO and found that the water flux occurred from the warm to the cold side of these membranes.

The few experimental studies on TO that have been done in the past were recapitulated by Carr and Sollner [13] who reviewed older researchers and reported experiments with different types of membranes and electrolyte solutions. However, no detailed considerations about the dependence of TO permeate flux on temperature difference were given. It was also concluded that TO was an anomalous osmosis closely related to EO and it likely occurred only for charged membrane and in the presence of electrolyte solution.

It was the work of Haase and his collaborators [22–24] that demonstrated the existence of TO in nonelectrolyte solutions through uncharged membranes. Later, in the first volumes of the Journal of Membrane Science, Vink and Chishti [25] in 1976 and Mengual et al. [26] in 1978 corroborated this result with pure water and CA membranes. It is worth quoting that until the 1960s, it was claimed that the nonisothermal transport could only occur in dense membranes [17,27]. On the contrary, in subsequent research studies [28–31], it was observed when using mixtures that liquid-filled membrane pores subjected to a temperature gradient behaved like a microscopic Soret cell. This phenomenon was termed thermo-dialysis.

Other than the advancements carried out on TO membranes, porous and hydrophobic membranes with pores that stay dry during vapor transport have been developed. The said membranes are maintained in contact between two liquid solutions kept at different temperatures and liquid/vapor interfaces are formed at each side of the membrane pores. The transport will be in the vapor phase from the hot interface where evaporation takes place toward the cold side where the condensation is carried out. It should be noted that, although the phenomenology is similar, the transport mechanism is completely different from TO. The new process was called membrane distillation (MD). A historical review on MD may be found in Ref. [32]. Briefly, in 1963, Bodel [33] filled the first patent on MD and the first paper on MD was published 4 years later in the Industrial & Engineering Chemistry-Process Design & Development [34]. However, the real interest in MD occurred after 1980 when the first porous Teflon membrane (Gore-Tex®) was developed and marketed [32]. Since then, the MD permeate flux turned out to be orders of magnitude higher than that of TO, which surprised the researchers of that time and generated some debates. Today, in contrast to TO, MD has made its way to its industrial implementation thanks to the progress made in recent years in the development of novel membranes, modules, and autonomous pilot plants coupling renewable energies.

From 1872 up to December 2019, as can be seen in Fig. 10–2, only a total of 169 papers were published in international refereed journals (Web of Science). Most of the studies were focused on the theoretical part of TO (41.9%), while those researches carried out on only experimental studies or theoretical and experimental studies were less, 28.8% and 29.4%, respectively. Recently, TO has gained increasing attention in energetic applications such as energy conversion and in the treatment and desalination of water, motivating therefore researchers to study different mechanisms of liquid transport in nano-channels. There is a great need to improve the "microscopic" understanding of transport that occurs in the presence of interfaces, formulating theoretical models capable of predicting the TO flow as well as the energy harvest rate of TO process. This explains the emerging interest on TO during last year (2019) with a total of 12 published papers.

Compared to other membrane separation processes, there is a relative lack of publications on TO, as can be seen in Fig. 10–2, which is due to several causes. The most significant one is the low interest that this process induces from an industrial point of view. In addition, since the TO permeate flux is very small, the involved experimental tests are complicated as they are long and laborious. Therefore it is not surprising that there are discrepancies, both qualitative and quantitative, between the results obtained by different authors, even when similar set-ups are used. Part of these discrepancies has already been explained by the well-known temperature polarization

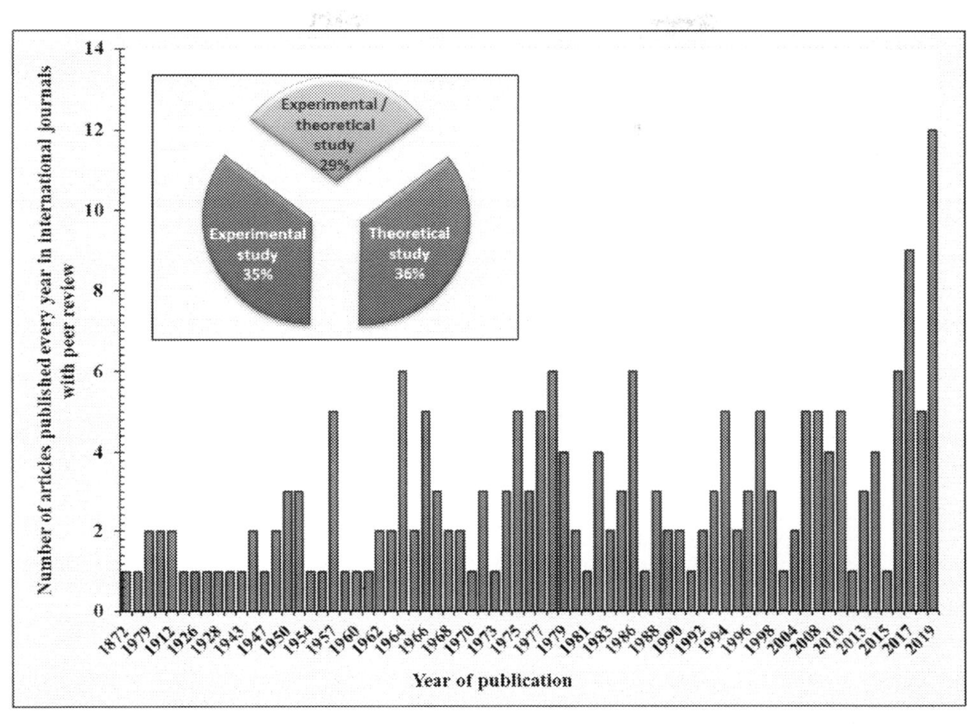

FIGURE 10–2 Growth of TO research activity presented as number of papers published in refereed journals per year up to December 2019. *TO*, Thermo-osmosis.

phenomenon, which refers to the establishment of liquid layers that adhere to both sides of the membrane and form additional barriers for the transport of both energy and matter. Moreover, the precise nature of the thermo-osmotic flow is not yet fully understood. Several different transfer mechanisms and modes have been proposed. Most of the research studies, both theoretical and experimental, have been focused on inert, partially saturated systems. The nature of the thermo-osmotic process becomes more complicated when the involved porous medium is charged or the permeant is an electrolyte. In this case the applied thermal gradient gives rise to a secondary electrical effect that must be taken into account.

10.2 Membranes for thermo-osmosis

Different types of membranes prepared from different materials were used in TO as summarized in Table 10–1 for water, electrolyte solutions, and some other pure liquids. The considered membranes are either hydrophobic (i.e., polytetrafluoroethyelene, membranes) or hydrophilic, uncharged or neutral (Cellophane, Millipore), cation charged (Nafion®), anion-charged (Neosepta®) or weakly charged (i.e., CA membrane), dense or porous. Depending on the membrane, different TO permeate fluxes were reported because of its dependence on

Table 10–1 Separation media, commercial and synthetic membranes used in thermo-osmosis (δ: thickness, d_p: pore size, ε: porosity).

Year	Membrane	Observations	Manufacturer	Properties	Reference
1952	Dipped latex sheet	Sulfur content < 1%.	–	$75 < \delta < 85$ μm	[47]
	Vulcanized dipped latex sheet			$300 < \delta < 320$ μm	
	Smoked rubber sheet (rolled)	Sulfur content = 2.5%.		$\delta = 220$ μm	
1956	Natural rubber membrane	Sulfur content = 2%.	Laboratories of the Firestone Tire and Rubber Company, Ohio, United States	$\delta = 254$ μm	[48]
1962	Nonoxidized collodion (prepared) Cellophane	Nonactivated membranes with solutions of electrolytes.	–	Porosity of conventional dialyzing membranes	[13]
	Oxidized collodion (prepared) Protamine collodion	Activated membranes with solutions of electrolytes.			
1964	DuPont 600 cellophane	Soaked in 10% NaOH solution for 15 min at 25°C.	–	–	[15]
		Washed first for a few minutes in progressive dilutions of NaOH and then with dilute HCl.			
1965	DuPont 600 cellophane	–	Kalle A.G., Wiesbaden-Biebrich, Germany	–	[23]
1966	Type-A	Pure gum rubber material.	Consolidated Electrodynamics Corporation, California, United States	$325 < \delta < 350$ μm	[49]
	Type-B			$360 < \delta < 440$ μm	
				$270 < \delta < 280$ μm	
				$275 < \delta < 285$ μm	
				$580 < \delta < 590$ μm	
1967	DuPont 600 cellophane	Treated with 10% NaOH solution: 50 min.	–	–	[18]
		Kept successively in progressive dilution of NaOH and then in dilute HCl.			
1970		–	Dialysierzwecke der Fa. Kalle Fa. AG. Wiesbaden-Biebrich; Germany	–	[50]
1971		–	Dialysierzwecke der Fa. Kalle AG., Wiesbaden-Biebrich, Germany	–	[24]
1971	Unglazed porous porcelain	–	–	$\delta = 4000$ μm	[51]
1973	Carbolac® (L)	–	G.L. Cabot Ltd., Massachusetts, United States	$\varepsilon = 48.4\%$	[52]
	Graphon® (M)			$\varepsilon = 47.2\%$	
	Graphon® (N)			$\varepsilon = 42.3\%$	

Year	Membrane	Code	Preparation/characteristics	Manufacturer	Thickness/pore size	Ref.
1975	Dense cellulose acetate (prepared)		Dried in a vacuum oven at 85°C for 4 h. 20 g CA in 80 mL acetone Evaporation of the solution spread on a glass plate with a coating knife.	Eastman Chemical Company, Tennessee, United States	$\delta = 26\ \mu m$ $\delta = 74\ \mu m$ $\delta = 77\ \mu m$	[27]
	Cation-exchange membrane (h-m)		Prepared from Amberlite XE-69 60% and poly(viny1 chloride) 40%.	Katayama Chemical Co., Osaka, Japan	–	[36]
	Gel membrane (g-m)		Made by placing 95% gel of sodium polyacrylate between two inert cellophane membranes.			
	Oxidized collodion membranes	(0-m-1) (0-m-2)	–			
	Untreated collodion membrane (c-m)		–			
1976	Cellophane		–	Union Carbide Corp., Chicago, IL, United States	$\delta = 76\ \mu m$	[25]
	CA		CA membrane, obtained by acetylating.of cellophane membrane: in a 1:1 mixture of acetic anhydride in pyridine, at 55°C for 15 h.		$\delta = 104\ \mu m$	
1977	CA		Sulfur content = 0.1%.	Arthur H. Thomas Company, Swedesboro, United States	$\delta = 100\ \mu m$ $d_p = 4.8$ nm	[53]
	Pyrex sinter membranes (G4)		Impregnated with copper ferrocyanide to reduce pore size.	–	$\delta = 250\ \mu m$ $d_p \sim 10$ nm	[54]
	Porous glass		–	–	–	[55]
1977	Cellulose membrane		Regenerated cellulose (type SM 11536).	Sartorius Membranfilter GmbH, Göttingen, Germany	$\delta = 90\ \mu m$ $5 < d_p < 10$ nm	[56]
1977	Oxidized collodion membranes reinforced with wide-mesh cotton gauze	o-m-1 o-m-2 o-m-3	Prepared by controlling the drying and oxidation times, have similar transport numbers but have different thicknesses.	Katayama Chemical Co., Osaka, Japan	$\delta = 1200\ \mu m$ $\delta = 530\ \mu m$ $\delta = 630\ \mu m$	[37]
		c-m	Untreated collodion membrane and, therefore, it has little fixed charges.		$\delta = 530\ \mu m$	
		i-m-1 i-m-2	Collodion-sulfonated polystyrene interpolymer membranes.		$\delta = 1000\ \mu m$ $\delta = 300\ \mu m$	

(Continued)

Table 10–1 (Continued)

Year	Membrane	Observations	Manufacturer	Properties	Reference
1978	CA (2.7 degrees of acetylation)	150 mg of CA in 25 cm^3 of acetone. Then 8 h evaporation at 30°C.	–	$\delta = 130\ \mu m$	[26]
	Unglazed porous porcelain	–	–	$\delta = 4000\ \mu m$	[57]
	Cation-exchange membrane C-2	Has sulfonic acid group.	Asahi Chemical Industry Co. Ltd., Tokyo, Japan	$\delta = 1250\ \mu m$	[58]
	Anion-exchange membrane A-1	60% of ion exchange resin powder and 40% polyvinyl chloride (5 wt. % of vinyl acetate content) as a binder.		$\delta = 990\ \mu m$	
	MCE (code manufacturer HAWP)	Mixed (CA) ester filters.	Millipore Corp., Bedford, MA, United States	$d_p = 0.45\ \mu m$	[28]
	Glass fiber AP Millipore (AP-20)	–		–	
	Nuclepore®	–	Nuclepore Cop., Pleasanton, CA, United States	$30 < d_p < 8000$ nm	
	Metricel-GA4	Made of CA.	Gelman Co., Ann Arbor, MI, United States	–	
	Versapore-epoxy	–		–	
1979	Unglazed porcelain membrane	–	–	$\delta = 4000\ \mu m$	[59]
1980	AP-20 Millipore	–	–	–	[31]
1982	AP-20 Millipore	Consisting of glass microfibers kept together by an acrylic glue.	–	$\delta = 320\ \mu m$	[60]
1983	Sintered stainless steel plug	–	–	–	[61]
	TF1000 Gelman	PTFE supported on polypropylene net.	Gelman Co., Ann Arbor, MI, United States	$\delta = 150\ \mu m$ $d_p = 1.0\ \mu m$; $\varepsilon = 80\%$	[62]
	TF200 Gelman			$\delta = 175\ \mu m$ $d_p = 0.2\ \mu m$; $\varepsilon = 80\%$	
	PTFE Membrane (Manufacturer code FGLP)	PTFE supported on polypropylene.	Millipore Corp., Bedford, MA, United States	$\delta = 130\ \mu m$ $d_p = 0.2\ \mu m$; $\varepsilon = 70\%$	
	PTFE Membrane (Manufacturer code LCWP)	PTFE unsupported.		$\delta = 125\ \mu m$ $d_p = 1.0\ \mu m$; $\varepsilon = 68\%$	
	AP-20 Millipore	Microfibers of borosilicated glass kept together by an acrylic glue.		$\delta = 300\ \mu m$; $\varepsilon = 50\%$	
1984	Anion-exchange membrane (A1)	Prepared by adsorption of a polycation on a highly porous collodion membrane. Keeping in 2% polycation (*I*) for 7 days. Dried and compressed at 200 kg/cm^2 at 60°C.	–	$\delta = 690\ \mu m$	[38]

1986	Membrane no. 1	150 mg CA In 60 cm^3 of acetone.	–	$\delta = 12.7\ \mu m$	[63]
	Membrane no. 2	80 mg CA 8 h evaporation time at 30 °C.		$\delta = 9.3\ \mu m$	
	Collodion membrane (c-m)	A collodion was poured onto a glass plate and dried for about 30 min at room temperature.	Wako Pure Chemical Industries, Ltd., Osaka, Japan	$\delta = 420\ \mu m$	[64]
	Poly(styrenesulfonic acid)-collodion interpolymer membrane (i-m)	10 wt.% poly(styrenesulfonic acid) and prepared in the same way.		$\delta = 370\ \mu m$	
	Nafion® 417 Neosepta® AFN(P) Mixed ion exchange membranes (M-8/2, M-4/6, M-3/7)	Cation exchange.	–	–	[65]
1987	TF1000 Gelman	PTFE supported on polypropylene net.	Gelman Co., Ann Arbor, MI, United States	$\delta = 150\ \mu m$ $d_p = 1.0\ \mu m$; $\varepsilon = 80\%$	[66]
	TF450 Gelman			$\delta = 175\ \mu m$ $d_p = 0.45\ \mu m$; $\varepsilon = 80\%$	
	TF200 Gelman			$\delta = 175\ \mu m$ $d_p = 0.2\ \mu m$; $\varepsilon = 80\%$	
	PTFE Membrane (Manufacturer code FGLP)	PTFE supported on polyethylene.	Millipore Corp., Bedford, MA, United States	$\delta = 130\ \mu m$ $d_p = 0.2\ \mu m$; $\varepsilon = 70\%$	
	Celgard 2400	Polypropylene film.	Celgard, LLC., United States	$\delta = 25\ \mu m$ $d_p = 0.04\ \mu m$; $\varepsilon = 38\%$	
	Celgard 2402			$\delta = 50\ \mu m$ $d_p = 0.04\ \mu m$; $\varepsilon = 38\%$	
1988	Cellophane 500P Cellophane 600P	Both membranes were washed twice in distilled and deionized water for 72 h.	Cellophane Española S.A., Burgos, Spain	$\delta = 51\ \mu m$; $\varepsilon = 74\%$ $\delta = 62\ \mu m$; $\varepsilon = 75\%$	[67]
	Membrane no. 1 Membrane no. 2 Membrane no. 3	200 mg CA 250 mg CA 300 mg CA 8 h evaporation time at 30 °C. In 60 cm^3 of acetone.	–	$\delta = 14.5\ \mu m$ $\delta = 23.75\ \mu m$ $\delta = 26.55\ \mu m$	[68]
	Membrane no. 1 Membrane no. 2	150 mg CA 80 mg CA 8 h evaporation time at 30 °C. In 60 cm^3 of acetone.	–	$\delta = 12.7\ \mu m$ $\delta = 9.3\ \mu m$	[69]

(Continued)

Table 10–1 (Continued)

Year	Membrane	Observations	Manufacturer	Properties	Reference
1989	Membrane no. 1	200 mg CA In 60 cm^3 of acetone.	–	$\delta = 14.5\ \mu m$	[70]
	Membrane no. 2	250 mg CA 8 h evaporation time at		$\delta = 23.75\ \mu m$	
	Membrane no. 3	300 mg CA 30°C.		$\delta = 26.5\ \mu m$	
1990	Gelman TF200	PTFE supported on polypropylene net.	Gelman Sciences Co., Ann Arbor, MI, United States	$\delta = 178\ \mu m$ $d_p = 0.2\ \mu m$; $\varepsilon = 80\%$	[71]
	Gelman TF450			$\delta = 178\ \mu m$ $d_p = 0.45\ \mu m$; $\varepsilon = 80\%$	
	Gelman TF1000			$\delta = 178\ \mu m$ $d_p = 1.0\ \mu m$; $\varepsilon = 80\%$	
	PTFE Membrane (Manufacturer code FGLP)	PTFE supported on polyethylene net.	Millipore Corp., Bedford, MA, United States	$\delta = 130\ \mu m$ $d_p = 0.9\ \mu m$; $\varepsilon = 70\%$	
	PTFE Membrane (Manufacturer code FHLP)			$\delta = 130\ \mu m$ $d_p = 0.4\ \mu m$; $\varepsilon = 70\%$	
	Collodion membranes (prepared): CM-1 to CM-7	Noncharged membrane: dissolving dried collodion in methanol, casting on a glass plate and drying it for 5–90 min.	–	$230 < \delta_{CM} < 1100\ \mu m$	[72]
	Poly(vinylidene fluoride) (Yumicron®): Y-9205	Noncharged membrane: reinforced with a nonwoven fabric of polyester and treated with a hydrophilic agent.	Yuasa Ionics Co., Ltd., Osaka, Japan	$\delta = 230\ \mu m$	
	Poly (tetrafluoroethylene) (Fluoropore®): FX-030, FX-050, and FX-500	Noncharged membrane.	Sumitomo Electric Industry Co., Ltd., Osaka, Japan	$80 < \delta < 300\ \mu m$ $300 < d_p < 5000$ nm	
	PE membrane (Hipore®): (1100, 1200, 2100, 2200, and 3050)		Asahi Chemical Industry Co., Ltd., Tokyo, Japan	$50 < \delta < 200\ \mu m$ $50 < d_p < 500$ nm	
	PVC hydrophobic membrane (Yumicron®): MF-40B, MF-60B, MF-90, and MF-250B.	Noncharged membrane: reinforced with a nonwoven fabric of polyester.	Yuasa Ionics Co., Ltd., Takatsuki, Japan	$90 < \delta < 170\ \mu m$ $400 < d_p < 2500$ nm	
	Nafion® 417	Cation-exchange membrane	DuPont Cop., United States	–	
	Neosepta® AFN	Anion-exchange membrane.	Tokuyama Soda Co., Ltd., Tokyo, Japan	–	

1992	Nafion® 417		Cation-exchange membranes: perfluorosulfonic acid-type.	Tokuyama Soda Co., Ltd. and Asahi Glass Co., Ltd., Tokyo, Japan	$\delta = 430\ \mu m$	[42]
	Flemion® AR1 25		Cation-exchange membranes: perfluorocarboxyic acid-type.		$\delta = 190\ \mu m$	
	Neosepta®	C66-10F	Cation-exchange membranes: hydrocarbonsulfonic acid-type.		$\delta = 0.360\ \mu m$	
		C66-5F			$\delta = 0.140\ \mu m$	
		CM-1			$\delta = 0.140\ \mu m$	
		CM-2			$\delta = 0.120\ \mu m$	
	Neosepta AFN		Anion-exchange membranes: porous AFN(P) membrane prepared by treating the AFN membrane with NaOH solutions.	Tokuyama Soda Co., Ltd., Tokyo, Japan	$\delta = 160\ \mu m$	[73]
	Neosepta AFN(P)				$\delta = 150\ \mu m$	
1993	Membrane-N		Carbolac membranes-type.	G.L. Cabot Ltd., Massachusetts, United States	$\varepsilon = 42.3\%$	[74]
	Membrane-O				$\varepsilon = 43.6\%$	
	Membrane-T		Graphon membrane-type.		$\varepsilon = 51.2\%$	
	Membrane-U				$\varepsilon = 50.2\%$	
	Aciplex® A-201		Anion-exchange membranes.	Asahi Chemical Industry Co., Ltd. and Tokuyama Soda Co., Ltd., Tokyo, Japan	$\delta = 218\ \mu m$	[75]
	Aciplex® A-211				$\delta = 454\ \mu m$	
	Neosepta® AM-1				$\delta = 123\ \mu m$	
1994	Flemion S		Cation-exchange membrane: perfluorosulfonic acid-type.	Asahi Glass Engineering Co., Ltd., Tokyo, Japan	$\delta = 210\ \mu m$	[76]
	MPS HPS		Anion-exchange membranes: prepared by quaternization of poly (4-vilylpyridine-*co*-styrene) with methyl iodide and 1,6-dibromo-*n*-hexane, respectively.	–	–	[77]
	Neosepta®: AM-1		Anion-exchange membranes: hydrocarbon-type.	Asahi Chemical Industry Co., Ltd., Tokyo, Japan	$\delta = 123\ \mu m$	[78]
	Aciplex®	A-201			$\delta = 218\ \mu m$	
		A-221			$\delta = 124\ \mu m$	
	STA-1 to STA-5		Anion-exchange membranes: fluorocarbone-type.		$90 < \delta < 118\ \mu m$	
	Tosflex® IE-DF 17				$\delta = 230\ \mu m$	
	Flemion S		Cation-exchange membrane: perflorosulfonic acid-type.	Asahi Glass Engineering Co., Ltd., Tokyo, Japan	$\delta = 210\ \mu m$	[79]
1995	Aciplex® K-181		Cation-exchange membranes: perfluorosulfonic acid-type membranes.	Asahi Chemical Industry Co., Ltd. Tokuyama Soda Co., Ltd., Tokyo, Japan	$\delta = 94\ \mu m$	[80]
	Aciplex® K-182		Made by treating the surface of K-181 with polyelectrolyte.		$\delta = 93\ \mu m$	
	Neosepta® C66-5T		Prepared by the paste method.		$\delta = 140\ \mu m$	

(*Continued*)

Table 10–1 (Continued)

Year	Membrane		Observations	Manufacturer	Properties	Reference
1996	Cellulose nitride (SM 11311)		–	Sartorius, Gottingen, Germany	$\delta = 90\ \mu m$ $d_p = 0.24\ \mu m$; $\varepsilon = 71\%$	[81]
	Cellulose nitride	PH70	–	Schleicher & Schuell, Dassel, Germany	$\delta = 105\ \mu m$ $d_p = 0.38\ \mu m$; $\varepsilon = 30\%$	
		PH75			$\delta = 105\ \mu m$ $d_p = 0.64\ \mu m$; $\varepsilon = 40\%$	
		PH79			$\delta = 105\ \mu m$ $d_p = 0.92\ \mu m$; $\varepsilon = 39\%$	
		PH83			$\delta = 140\ \mu m$ $d_p = 1.34\ \mu m$; $\varepsilon = 42\%$	
		PH85			$\delta = 135\ \mu m$ $d_p = 1.94\ \mu m$; $\varepsilon = 37\%$	
		PH90			$\delta = 150\ \mu m$ $d_p = 2.54\ \mu m$; $\varepsilon = 40\%$	
1997	Membrane-T Membrane-U		Graphon membrane-type.	G.L. Cabot Ltd., Massachusetts, United States	$\varepsilon = 51.2\%$ $\varepsilon = 50.2\%$	[82]
2004	Nafion® 115 Nafion® 112		Proton-conducting membrane.	El Cell AB, Sweden	$\delta = 55\ \mu m$ $\delta = 127\ \mu m$	[83]
2006	Nafion® 117		–	–	$\delta = 183.1\ \mu m$	[43]
	Gore 5510 MEA		15-μm-thick reinforced perfluorosulfonic acid membrane coated on both sides with 0.4 mg_{Pt}/cm^2 (Pt/C) catalyst.	GORE® Fuel Cell Technologies, United States	–	[84]
	Catalyzed Gore-select Teflon-reinforced Nafion® Nafion112 Catalyzed Nafion® 112		–	W. L. Gore & Associates, Inc., United States	–	[85]

2008	SGL10AA		–	SGL	$\delta = 390$ mm	[86]
	Nafion® 112B			Carbon Group, United States	$\delta = 50\ \mu m$	
	Nafion® 112 reinforced	A			$\delta = 50\ \mu m$	
		B			$\delta = 18\ \mu m$	
	SGL10AA		–		$\delta = 390$ mm	[44]
	Nafion® 112				$\delta = 50\ \mu m$	
	Flemion® SH50				$\delta = 50\ \mu m$	
	GOR-Select				$\delta = 18\ \mu m$	
	Nafion® 117		Cation-exchange membrane: perfluorinated polyethylene with pendant ether-linked side chains terminated with sulfonated groups.	DuPont Cop., United States	$\delta = 183\ \mu m$	[87]
	MK-40		Cation-exchange membrane: sulfonated groups chemically attached to polystyrene/ divinylbenzene copolymer chains dispersed in the polyethylene matrix.	Ionics Inc., United States	$\delta = 510\ \mu m$	
	CR61-CZL-412		Cation-exchange membrane: poly (styreneco-divinylbenzene) with sulfonic acid functional groups.	Ionics Inc., United States	$\delta = 570\ \mu m$	
2009	Nafion® 117		Cation-exchange membrane: perfluorinated polyethylene with pendant ether-linked side chains terminated with sulfonated groups.	DuPont de Nemours, United States	$\delta = 183\ \mu m$	[88]
	MK-40		Cation-exchange membrane: sulfonated groups chemically attached to polystyrene/ divinylbenzene copolymer chains dispersed in the polyethylene matrix.	Schekinazot, Russia.	$\delta = 480\ \mu m$	
	CR61-AZL-412		Cation-exchange membrane: heterogeneous crosslinked sulfonated copolymer of vinyl compounds cast in homogeneous films on synthetic reinforced fabrics.	Ionics Inc., United States	$\delta = 550\ \mu m$	
	Nafion® 112		–	SGL Carbon Group, United States	$\delta = 51\ \mu m$	[89]
	Gore-Select membrane (R-PEM)				$\delta = 18\ \mu m$	

(*Continued*)

Table 10–1 (Continued)

Year	Membrane		Observations	Manufacturer	Properties	Reference
	Gore-Primea 5710	SGL 10AA			$\delta = 390\ \mu m$	
		SGL 10BA			$\delta = 400\ \mu m$	
		SGL 10BB			$\delta = 420\ \mu m$	
	SGL10AA		–	SGL Carbon Group, United States	$\delta = 390\ \mu m$	[44]
	Nafion® 112				$\delta = 50\ \mu m$	
	Flemion® SH50				$\delta = 50\ \mu m$	
	GOR-Select				$\delta = 18\ \mu m$	
2013	t-BNNT (transmembrane nanotube)		Single BNNT inserted into a hole in a silicon nitride (SiN) membrane.	–	–	[90]
2014	Hydrophobic-modified Gamma-alumina membrane		Alumina membrane annealed at 1000°C. The membranes were placed on a droplet (0.7 μL) of the photoresist for 2 h to fill the nanopores and then baked on a hot plate at 100°C for 50 min to evaporate the solvent. Then, washed with 5 wt.% phosphoric acid for 1 min, dried with nitrogen and exposed to vapor of perfluorodecyltrichlorosilane (Gelest) overnight in a vacuum desiccator.	Synkera Technologies Inc., United States	–	[91]
2015	Nafion® 117		–	DuPont de Nemours, United States	$\delta = 183\ \mu m$	[92]
	Three sandstone cores	Banksmeadow	–	Hawkesbury sandstones, New South Wales, Australia	$\delta = 49.5 \cdot 10^3\ \mu m$; $\varepsilon = 13\%$	[93]
		Newcastle			$\delta = 47.7 \cdot 10^3\ \mu m$; $\varepsilon = 15\%$	
		Somersby			$\delta = 50.2 \cdot 10^3\ \mu m$; $\varepsilon = 21\%$	
2016	CTA		–	–	–	[94]
	CA					
	TFC					
	DM		Nonwoven carbon fiber paper.	W. L. Gore and Associates, Newark, Delaware, United States	$\delta = 165\ \mu m$; $\varepsilon = 70.5\%$	[95]
	MEA Freestanding MPLs	GORE MPL1	PTFE-carbon black composite web with continuous, highly uniform (crack-free), and conformable structure.	Ion-Power Inc., NewClaste, Delaware, United States	$\delta = 70\ \mu m$; $\varepsilon = 83\%$	
		GORE MPL2			$\delta = 45\ \mu m$; $\varepsilon = 62\%$	
		GORE MPL3			$\delta = 45\ \mu m$; $\varepsilon = 83\%$	
	PTFE		–	–	$\delta = 136\ \mu m$; $d_p = 20$ nm; $\varepsilon = 77\%$	[96]
	2017	ZIF-25 membrane	–	–	–	[97]

2018	PVDF	PVDF (2 g) with graphene (2, 10, 20, and 30 mg), and PEG (80 mg) in DMF (10 mL). Casting by a 400-μm knife on glass. Then, the membranes were placed in the oven kept at 60°C for 6 h.	PVDF: Solvay Company, Brussels, Belgium Graphene: Tanfeng Tech, Inc., Suzhou, China	–	[98]
2019	SNC/PET hybrid nano-channel membrane	SNCs of 2.3 nm in diameter positioned on the PET membrane with track-etched nano-channels.	–	–	[99]
	Flat-sheet ultrafiltration MCE membrane	Lower membrane surface energy was obtained by vapor deposition treatment using heptadecafluoro-1,1,2,2-tetrahydrodecyltrichlorosilane.	Millipore Sigma, United States	$\delta = 100\ \mu m$ $25 < d_p < 50$ nm $72 < \varepsilon < 78\%$	[45]
	MXene/Kevlar nanofiber composite membrane	$Ti_3C_2T_x$ (MXene) prepared by selectively etching of Al layer from Ti_3AlC_2 (MAX) using HF acid. ANF prepared from Kevlar yarns. MXene and ANF suspension in DMSO. MXene/ANF reassembled by vacuum-assisted filtration method to form a black gray, paper-like flexible thin membrane.	Forsman Scientific (Beijing) Co., Ltd., China DuPont Corp., United States	$2 < \delta < 15\ \mu m$	[46]

ANF, Aramid nanofiber; *BNNT*, boron nitride nanotubes; *CA*, cellulose acetate; *CTA*, cellulose triacetate; *DM*, diffusion layer media; *DMF*, dimethylformamide; *HPS*, Anion exchange membranes obtained by quaternization of polysulfone (PS) membrane with 1,6 dibromo-n-hexane; *MCE*, mixed cellulose ester; *MEA*, membrane electrode assemblies; *MPLs*, microporous layer; *MPS*, Anion exchange membranes obtained by quaternization of polysulfone (PS) membrane with methyl iodide; *PET*, poly (ethylene terephthalate); *PTFE*, polytetrafluoroethylene; *PVC*, poly(vinyl chloride); *PVDF*, polyvinylidene fluoride; *SNC*, ultra-small silica nanochannels; *TFC*, thin-film composite; *ZIF*, zeolitic imidazolate frameworks.

not only the temperature difference but also on the average temperature and its small magnitude. For instance, the TO coefficient for water varies in five orders of magnitude (i.e., $10^{-8}-10^{-13}$ kg/(msK)) and it can be both negative or positive [35]. It is to point out that still it is not very clear why water transport through charged membranes takes place from the cold to the hot side of the membrane [27,36–38], while for uncharged or weakly charged membranes, it is directed from hot to cold side [22,25–28,36,37,39,40]. Moreover, the direction of the TO permeate flux changes with the temperature of the used solutions [23,24], the pore size of the membrane [28,30], the type of solvent [25], etc. As a consequence, the conclusions drawn in the literature are contradictory and, in some studies, confusing. Another issue is related to the void volume fraction of the membrane. The first studies claimed that the nonisothermal transport occurred only with adequate dense membranes [17,27]. However, in several research studies [28–31,41], it has been confirmed that the nonisothermal transport can also occur through porous membranes. Similar to other membrane separation processes, another parameter that affects the TO permeate flux is the membrane thickness. For the same average temperature the TO coefficients seem to increase with the increase in the thickness [35,42–44]. Because of all these concerns, very few research studies have been carried out on the preparation or modification of specific membranes for TO [45,46].

10.3 Electrolyte solutions used in thermo-osmosis

TO of electrolyte solutions was first reported by Carr and Sollner [13]. The electrolyte solutions were KCl aqueous solutions and the TO permeate flux took place from the hot to the cold side of oxidized collodion membranes concluding that the TO of electrolyte solutions through charged membranes was an electro-kinetic process related to EO. In other words, it was concluded that the TO phenomenon occurred only through electrically charged membranes and not through uncharged ones. The TO permeate flux depended on the transmembrane temperature difference, and its direction was governed by both the sign of the membrane charge and the concentration of the electrolyte solution. Later on, based on a theoretical study, Kobatake and Fujita [39] claimed a dependence of the electro-osmotic coefficient on the salt concentration of the used electrolyte solution through a porous membrane. Then, interesting studies on TO in electrolyte solutions were published by the research group of Tasaka [38,42,64,65,73,75,76,78–80,100,101].

It was observed for anion-exchange membranes in the presence of different counter-ions, the direction of the TO permeate flux occurred from the cold side to the hot side of the membranes regardless of the electrolyte type, its concentration or the applied temperature [38,73,75,78]. In contrast, for cation-exchange membranes the direction of the TO permeate flux and its magnitude changed with the type of the counter-ions [42,80].

Tables 10–2 and 10–3 show the commonly used electrolytes solutions together with the corresponding membranes and the observed direction of the TO permeate flux. As can be seen, the hydrophobicity and charge of the membrane, as well as the electrolyte solution, have no clear influence on the direction of the TO permeate flux. For instance, Mita et al.

Table 10–2 Electrolyte solutions and corresponding membranes used in thermo-osmosis-inducing permeate fluxes from the cold to the hot membrane side.

	Membrane		Electrolyte solution		
Year	**Code**	**Type**	**Type**	**Concentration**	**Reference**
1964	Cellophane	Dense	KCl	$1.0 \cdot 10^{-1}$ M	[102]
1975	(h-m)	Cation exchange	LiCl, NaCl, KCl,	$1.0 \cdot 10^{-3}$–$5.0 \cdot 10^{-2}$ M	[36]
	(g-m)	Gel membrane	KF, KBr, KI		
	(0-m-1)	Oxidized collodion			
	(0-m-2)	membranes	KCl	0.02 mol/kg	[36]
	MC 3470		NaCl, $(CH_3))_4NCl$	0.1 M	[103]
1978	C-2	Cation exchange	KCl, $MgCl_2$, and $BaCl_2$	0.1–0.02 mol/kg	[58]
	A-1	Anion exchange	K_2SO_4 and K_2CO_3		
1984	A1	Anion exchange	HCl, LiCl, KCl, KIO_3	0.01 mol/kg	[38]
1992	Neosepta® AFN		KF, KCl, KNO_3, KIO_3, HCOONa	0.01 mol/kg	[73]
1992	Neosepta® C66-10F	Cation exchange	HCl, KCl	0.001 mol/kg	[42]
1993	Aciplex® A-201	Anion exchange	KCl, KIO_3	$1.0 \cdot 10^{-3}$–$1.0 \cdot 10^{-1}$ M	[75]
1993	Aciplex® A-211		KCl	$1.0 \cdot 10^{-3}$–$1.0 \cdot 10^{-2}$ M	[75]
1994	Neosepta® AM-1	Anion exchange	LiCl, KCl	0.001–5 mol/kg	[78]
1994	Aciplex® A-201		KF, KCl, KNO_3, KIO_3, HCOONa	0.01 mol/kg	[78]
1994	MPS-10		KCl	$1.0 \cdot 10^{-3}$–$1.0 \cdot 10^{-1}$ M	[77]
1994	HPS-2		KCl	$1.0 \cdot 10^{-3}$–$1.0 \cdot 10^{-1}$ M	[77]
1994	Flemion® S	Cation exchange	HCl, NaCl	0.1 mol/kg	[79]
1995	Aciplex® K-181		HCl, NaCl	0.1 mol/kg	[80]
1998	M-1(4.5)	Anion exchange	KCl, KIO_3	$1.0 \cdot 10^{-3}$–$1.0 \cdot 10^{-1}$ M	[104]
1998	M-1(8)		KCl	$1.0 \cdot 10^{-2}$ M	[104]

[60] found that the direction of the TO permeate flux through AP-20 Millipore membranes in the presence of alkali chloride solution was similar to that of pure water, from the hot to the cold side of the membrane. For hydrophobic membranes (Fluoropore®, Hipore®, Yumicron®), Tasaka et al. [72] obtained TO permeate fluxes from the hot to cold side and did not observe any dependence on the electrolyte concentration. For charged membranes the TO permeate flux increased with increasing the electrolyte concentration.

Tasaka et al. [38] observed similar TO permeate fluxes through anion-exchange membrane Al with all used electrolytes at different concentrations (i.e., from 10^{-3} to 2 mol/kg of aqueous KCl, LiCl, and NH_4Cl and from 10^{-3} to 0.3 mol/kg of aqueous K_2SO_4) except for the aqueous KIO_3 solution (i.e., 10^{-3}–0.3 mol/kg). The higher TO permeate fluxes obtained for this last electrolyte solution were attributed to the larger ionic radius of IO^{3-} compared to Cl^- inducing greater pore volume fraction of the membrane. A significant effect of the electrolyte type on the TO permeate flux was

Table 10–3 Electrolyte solutions and corresponding membranes used in thermo-osmosis-inducing permeate fluxes from the hot to the cold membrane side.

	Membrane		Electrolyte solution		
Year	Code	Type	Type	Concentration	Ref.
1975	Cellulose Acetate	Dense	KCl	$1.0 \cdot 10^{-3}$–$5.0 \cdot 10^{-2}$ M	[36]
	(h-m)	Cation exchange			
	(g-m)	Gel membrane			
	(0-m-1)	Oxidized collodion membranes			
	(0-m-2)				
1977	o-m-1	Oxidized collodion membranes reinforced with wide-mesh cotton gauze	KCl	$5.0 \cdot 10^{-3}$–$5.0 \cdot 10^{-2}$ M	[37]
	o-m-2				
	o-m-3				
	c-m				
	i-m-1				
	i-m-2				
1982	AP-20 Millipore	Glass microfibers	CH_3COONa, NaCl	0.02 M	[60]
			KCl	0.05 M	
1986	i-m	Poly(styrenesulfonic acid)-collodion interpolymer	KCl	$1.0 \cdot 10^{-3}$ M	[65]
	M-10/0	Cation exchange	KCl	$1.0 \cdot 10^{-3}$ M	
	M-0/10	Anion exchange	KCl	$1.0 \cdot 10^{-3}$ M	
	M-8/2, M-4/6, M-3/7	Mixed ion exchange			
1992	Neosepta® AFN	Anion exchange	LiCl	0.001–10 mol/kg	[42]
	Neosepta® AFN(P)	Anion exchange			
	Nafion® 417	Cation exchange			
	Flemion® AR1 25		LiCl	10 mol/kg	
	Neosepta®				
1994	Flemion® S		HCl, LiCl, NaCl, KCl, NH_4Cl	$1.0 \cdot 10^{-2}$ M	[76]
1995	Nafion® 117		HCl	$1.0 \cdot 10^{-2}$ M	[105]
1995	Flemion® S		LiCl, KCl	0.01 mol/kg	[80]
	Aciplex® K-181		NH_4Cl	0.01 mol/kg	
	Aciplex® K-182		NH_4Cl	0.01 mol/kg	
	Neosepta® C66-5T		NH_4Cl	0.01 mol/kg	
1997	Ionics 61 CZL 386		KCl	$7.5 \cdot 10^{-4}$–$7.5 \cdot 10^{-1}$ M	[106]
			LiCl, NaCl, KCl, CsCl	$1.0 \cdot 10^{-3}$ M	
2013	Nafion® 1110		HCl	N.A.[a]	[107]

[a]Not available.

observed by Suzuki et al. [78] when using hydrocarbon-type and fluorocarbon-type anion-exchange membranes. For Neosepta® membranes in the presence of KCl solutions, the TO permeate flux was maintained the same for low concentrations but it decreased with the increase of the concentration. However, when using LiCl solution, an opposite effect was observed concluding that the TO permeate flux depended on both the water mobility in the membrane and the water content.

10.4 Theoretical studies developed for thermo-osmosis

TO was the first nonequilibrium phenomenon studied, from the experimental and theoretical point of view, for liquids, gases, and gaseous mixtures [10,11,14,15,47,68,108–111]. The process consists of the passage of a fluid through a membrane or media due to a temperature gradient and the slip of the fluid in the boundary layer close to the surface gives rise to momentum transfer. Recently, this velocity slip has gained growing attention as a driving force for micro- or nano-devices (i.e., displacement of colloidal particles under thermal gradients) [112–115]. A coupled TO and mechanical-caloric (i.e., pore pressure gradient on the heat flux) effects have been considered on thermo-hydro-mechanical behavior of saturated porous media and clays [114–118]. TO is, therefore, one of the most fundamental manifestations of thermal forces and eventually particle motion. TO flow is an interfacial phenomenon arising from driving forces by the thermodynamics gradients in a microscopic boundary region, where the properties of the fluid are influenced by the interactions with the surface. This type of system is not under equilibrium since there is a temperature gradient and a flow of heat. Under suitable conditions, it gives rise to a stationary difference of temperature.

10.4.1 Thermo-osmosis and linear irreversible thermodynamics processes

Derjaguin and Sidorenkov [10] formulated a generic description of TO based on the equations of irreversible thermodynamics processes (ITP). Derjaguin et al. [119] used the Onsager reciprocity relations to relate the TO slip due to a temperature gradient and the excess heat flux due to a pressure gradient (i.e., hydrodynamic flow) [120]. Onsager's approach of linear ITP expresses the heat and mass fluxes and permits to obtain equations for determining the rate of the permeate and the pressure ratio at the stationary state.

A system can be driven out of equilibrium by thermodynamic forces ($_i$). The central assumption of linear ITP is that the thermodynamic forces are sufficiently small to validate a first-order Taylor expansion of the thermodynamics fluxes as:

$$J_i = \sum_i L_{ij}\mathcal{F}_i \tag{10.1}$$

being $L_{ij} = (\partial J_i/\partial \mathcal{F}_i)\big|_{eq}$ the Onsager coefficients with $L_{ij} = L_{ji}$. In TO, this driving force is a temperature difference and, in general, a total chemical potential difference (μ_i), which induces a thermodynamic flux (J_i), such as a heat flux (J_q) or volume/mass flux (J_V) [10,11,35,65,80,121]:

$$J_q = -L_{qq}\nabla T - \sum_j L_{qj}\nabla\mu_j \tag{10.2}$$

$$J_i = -L_{iq}\nabla T - \sum_j L_{ij}\nabla\mu_j \tag{10.3}$$

The relationship between chemical potential and pressure can be derived by using the Gibbs–Duhem equation based on the assumption of local isothermal condition as [65,99,122]:

$$\nabla\mu_i = \overline{v}_i\nabla p \tag{10.4}$$

where $\overline{v}_i$ is the partial molar volume and ∇p the pressure gradient along the slit direction (the x-axis). Therefore the heat and volume fluxes can be described from the corresponding flux–force relations and the entropy production rate can be written as [65,99,120]:

$$\sigma = -J_q\frac{\nabla T}{T^2} - J_V\frac{\nabla p}{T} \tag{10.5}$$

If a local nonisothermal condition is considered, the chemical potential must be described in terms of the temperature and pressure gradient $\mu = \mu(T,p)$. From the Gibbs–Helmholtz relation, $\partial(\mu/T)/\partial T|_p = h_B\rho/T^2$, where h_B is the average enthalpy per particle in the bulk (i.e., away from the surface and in the absence of an external field) and ρ is the average density of the fluid [120,123]. If the enthalpy excess is considered, the entropy production will be given by:

$$\sigma = -(J_q - h_B J_V\rho)\frac{\nabla T}{T^2} - J_V\frac{\nabla p}{T} \tag{10.6}$$

From small gradients and taken into account only solvent–solute mixtures and pure fluids, the heat and volume fluxes can be written as [120,123]:

$$J_q - h_B J_V\rho = -L_{qq}\frac{\nabla T}{T^2} - L_{qp}\frac{\nabla p}{T} \tag{10.7}$$

$$J_V = -L_{pq}\frac{\nabla T}{T^2} - L_{pp}\frac{\nabla p}{T} \tag{10.8}$$

The heat flux is proportional to the temperature gradient in the absence of a pressure gradient and L_{qq} coefficient is related to the Fourier type stationary thermal conductivity. The volume flux is proportional to the pressure gradient in the absence of a temperature gradient and the hydraulic permeability K is proportional to L_{pp} ($L_{pp} = K/\eta$) where η is the fluid viscosity [124]. The hydraulic permeability can be determined from the volume flow and the overall pressure difference at an uniform composition [65,124]:

$$K = -\eta\left(\frac{J_V}{\Delta p}\right)_{d\mu=0} \tag{10.9}$$

From general considerations the cross-terms of the transport matrix, namely, the TO coefficient (L_{pq}), describing surface induced flow under a temperature gradient (∇T), and the mechano-caloric coefficient (L_{qp}), describing heat flux due to a pressure gradient (∇p), must be equal to each other (i.e., Onsager reciprocity relation). There are different approaches to determine the off-diagonal Onsager coefficients. Derjaguin and Sidorenkov related the TO coefficient to the interfacial excess enthalpy using linear ITP [10,125]:

$$L_{pq} = L_{qp} = \frac{1}{\eta}\int_0^{\infty} z\delta h(z)dz \tag{10.10}$$

where η is the liquid viscosity, z the distance to the surface ($z = 0$ the position of the interface, $z = \infty$ the bulk liquid region), and $\delta h(z)$ the excess of specific enthalpy as compared to the bulk.

The excess enthalpy cannot fully explain the massive enhancement of L_{qV} at low solid–liquid interface interaction energy (molecular interactions) and hydrodynamic slip. For high interfacial energies, due to the oscillations of the excess enthalpy profiles in the vicinity of the solid–liquid interface, the TO coefficient depends closely on the thickness of the stagnant liquid layer. But, for low interfacial energies, hydrodynamic slip largely amplifies the TO coefficient [114]. Fu et al. [113] introduced a characteristic length d representative of the thickness of the interfacial liquid layer where the enthalpy differs from the bulk:

$$d = \frac{\int_{z_s}^{\infty}(z - z_s)\delta h(z)dz}{\int_{z_s}^{\infty}\delta h(z)dz} \tag{10.11}$$

Eq. (10.10) was then rewritten as:

$$L_{qp} = L_{qp}^{no\ slip}(1 + b/d) \tag{10.12}$$

where $L_{qV}^{no\ slip}$ is the same as Derjaguin's coefficient (in the absence of slip, i.e., $b = 0$) and $(1 + b/d)$ represents the contribution of the hydrodynamic slip to the TO force.

Tasaka et al. [111] assumed that the L_{pq} coefficient is proportional to the electrolyte concentration for TO through charged membranes. Bregulla et al. [125] determined both radial and vertical velocity components of the TO flow in thin films. The excess enthalpy of polyethylene glycol films in water results from the balance of the hydrogen bridging of the oxygen and the opposite hydrophobic effect of polyethylene. The TO coefficient was estimated as:

$$L_{qp} = \frac{\Delta H}{4\pi\eta d} \tag{10.13}$$

This agreed well with the experimental measurements with a numerical value of $d = 3.5$ Å. Bregulla et al. [125] found that the slip velocities were much stronger at those interfaces covered with nonionic block copolymers as compared to charged glass interfaces. It was finally suggested that TO flows along solid–liquid interfaces might contribute considerably to thermophoretic measurements in thin-film geometries and may be harnessed for microfluidic applications.

Proesmans and Frenkel [120] compared three numerical routes to obtain the TO and mechanic–caloric coefficients: (1) by using the Onsager reciprocity relations, (2) by using the appropriate Green–Kubo relation, and (3) via the excess enthalpy. The numerical results were found to be mutually consistent and to agree with the theoretical prediction based on the assumption that hydrodynamics and thermodynamics were locally valid. To determine the Onsager coefficient, they calculated first the heat flux induced by a pressure gradient near the surface and in the bulk. From these pressure gradients, an Onsager coefficient was found to be equal to that obtained from the temperature gradient but in the absence of a bulk pressure gradient. Therefore they could determine the TO coefficients fully in terms of predetermined variables. In addition, they checked that their results were compatible with the TO coefficient by using Derjaguin's method, which was based on the excess enthalpy [120].

Another interesting method to determine the TO coefficients was network thermodynamics, which includes both circuit theory and thermodynamics. This method has been used previously to model biological systems. Imai [126] showed that the TO phenomenon could be modeled with the aid of networks, which represent the interaction between two energy processes in a medium by bond graphs. TO is modeled as a power coupling, which can be represented using a transducer and two resistors, where the resistor relation was assumed to be linear. Since the temperature difference (ΔT) induces a volume flow, a transformer-type transducer was used for the available power conversion. Imai [126] used the two-port resistors, one connected in series at the port of the transducer with the independent flow variable and another resistor connected in parallel at the other port of the transducer. The hydraulic circuit was modeled as a series circuit. Therefore the entropy flow was the sum of the entropy conduction due to the temperature difference and the convective entropy flow due to the volume flow given by:

$$\sigma = \kappa \Delta T^2 + L_p(\Delta p + s_v \Delta T)^2 = \kappa \Delta T^2 + J_V(\Delta p + s_v \Delta T) = J_q \Delta T + J_V \Delta p \tag{10.14}$$

where κ is the entropy conductance and s_v is the transfer ratio, which has an unit of entropy per unit of volume of water in the membrane. The sign of s_v depends on the characteristics of the membrane.

10.4.2 Thermo-osmosis using intermolecular interactions

Onsager's approach cannot be used for a quantitative prediction of the magnitude of TO flows from the knowledge of the intermolecular interactions. Moreover, the validity of continuum hydrodynamics is questionable in the first few molecular layers near a wall/membrane. TO occurs at the interface between a liquid (or gas) and solid surface when exposed to a temperature gradient. The liquid flow along a surface is a result of interaction forces near the solid–liquid interface. This liquid flow is well-known as TO slip. The characteristic length of action for this force is on the order of several molecular radii. The flow of liquid is caused by a temperature-induced longitudinal pressure gradient (parallel to the temperature gradient) established along the surface of the particle or wall. If temperature gradients cause flow near a surface, it is only because a local pressure gradient is induced. Therefore the concept of local thermal equilibrium must be included. At the molecular level, this means that when the particles move across a temperature gradient, their velocity distribution quickly adopts to the local temperature and, therefore, their dynamics must be noninertial [123]. Following this assumption, Farago [123] described the temporal evolution of the mass and surface heat energy densities by means of the corresponding diffusion equations, and that in a steady-state, the divergences of the relevant currents vanish. Statistical–mechanical tools were used to derive equations for the transport cross-coefficients and demonstrated their validity to Derjaguin's method and Onsager reciprocal relations. Farago [123] used those statistical–mechanical tools for two basic models: (1) an incompressible continuum solvent containing noninteracting solute particles and (2) a single-component fluid without thermal expansivity. The only difference for the TO coefficients in both cases was the particle mobility relevant to the situation. In the solvent–solute model a constant single-particle mobility was assumed since each solute particle was treated as an independent Brownian particle. In contrast, the mobility in the single-component fluid model was dominated by the particle–particle interactions considering the collective Fickian mobility.

The same basic assumptions have been recently used by Anzini et al. [127] to develop a field-theoretical approach for TO. However, the focus was not on the Onsager's cross-coefficients, but on the fluid slip velocity, which is a different part of Derjaguin's theory that requires taking into account hydrodynamic considerations (liquids) or kinetic theories (gases).

Ganti et al. [128] proposed three different methods to compute TO slip on the basis of molecular simulations. One method was based on a computation of the thermally induced stress gradient computed using equilibrium simulations and then represented as a body force in nonequilibrium simulations. In the second approach the excess enthalpy density near the wall was computed and a local thermodynamics formalism was used to derive the body force acting on the fluid. In both methods, macroscopic thermodynamics or hydrodynamics did not hold close to an interface. The final approach was based on Onsager's reciprocal relations, which allowed to estimate the TO slip from the excess heat flux due to a pressure gradient. It was found that the TO coefficient obtained for all methods did not differ significantly. However, the TO flow profile for less attractive Lennard–Jones potential, the viscosity and forces were clearly not constant showing a significant discrepancy from Dejaguin's result [121].

Recently, Ganti et al. [128] computed the TO force at simple solid–fluid interfaces as a tensor in the Hamiltonian. They estimated the TO forces based on computing gradients of the stress tensor. This approach is to be useful to obtain a quantitative prediction of the magnitude of TO flows from the knowledge of the intermolecular interactions for the first few molecular layers near the membrane surface.

Semenov and Schimpf [112] used a method based on the Navier–Stokes equation for the flow profile near the solid surface and in the bulk volume assuming that the solid surface could be considered a quasi-planar interface. The one-dimensional liquid flow profile $u_x(z)$ (where z is the coordinate normal to the solid surface and x is the respective tangential coordinate) was written taking into account the Navier–Stokes equation as:

$$\frac{\partial P}{\partial x} = \eta \frac{\partial^2 u_x(z)}{\partial z^2} \tag{10.15}$$

where η is the dynamic viscosity of the liquid and $P(x,z)$ is the pressure distribution. $P(z)$ at the surface layer near the solid–liquid boundary was obtained by mechanical equilibrium conditions as:

$$\frac{\partial P}{\partial z} = -\frac{1}{v_l}\frac{\partial \Phi}{\partial z} \tag{10.16}$$

where v_l is the partial molecular volume of the liquid and Φ is the molecule–solid surface interaction potential. This interaction potential varied in the direction normal to the solid surface and it was not directly caused by temperature gradient. By taking into consideration the boundary conditions ($u_x(\mathrm{z}=0)=0$ and $\partial u_x(\mathrm{z}\rightarrow\infty)/\partial \mathrm{z}=0$), the flow profile within the surface layer was described as the balance of forces within the layer [114]:

$$\eta \frac{\partial^2 u_x(z)}{\partial z^2} = -\frac{\partial(1/v_l)}{\partial x}\Phi(z) \tag{10.17}$$

The interaction potential $\Phi(z)$ depends on the physical forces along the tangential coordinate and the temperature gradient along the surface creates a tangential force. This interaction potential is the type of intermolecular London–van der Waals potential (Lennard–Jones potential) [112], and the slip velocity, u, was obtained as a function of the temperature gradient tangential to the solid–liquid interface as:

$$u = \frac{2-\ln 3}{12}\frac{\alpha_T \sigma_{wl}^2 A_{wl}}{\eta v_l}\frac{\partial T}{\partial z} \tag{10.18}$$

where $\alpha_T = (1/v_l)(\partial v_l/\partial T)$ is the cubic thermal expansion coefficient, σ_{wl}^2 is the minimum molecular approach distance between molecules of the liquid and the surface, and A_{wl} is a constant derived from the molecular-surface interaction potentials in liquid that can be defined by the Hamaker potential [114]. The approach developed by Semenov and Schimpf [114] allowed also to obtain expressions for the angular frequency of the rotor. Therefore the flow rate in the pump and angular velocity of the TO engine could be determined. Consequently, devices may be constructed by a single material, which is a promising difference for TO devices [46,91,114].

10.4.3 Thermo-osmosis for energy conversion

Recently, TO energy conversion (TOEC) has been introduced as a new technology to generate electricity from low-grade heat sources [45,96,129]. In TOEC the permeate stream is pressurized. As water vapor diffuses across a porous hydrophobic membrane and condenses into the permeate stream, the condensed water adds volume to the pressurized permeate reservoir, which can be used to propel a turbine and produce electricity. Shaulsky et al. [45] proposed a one-dimensional finite-element model based on coupled energy and mass balances to determine the performance of the hydrophobic asymmetric membrane for TOEC. These asymmetric membranes exhibit a thicker layer with large-diameter pores, thereby facilitating high vapor permeability, while a thin top layer with much smaller pores was used to prevent wetting of the membrane due to high pressure needed for improved power density and energy-efficient. Several assumptions were carried out in this study: (1) TOEC process was adiabatic, (2) axial dispersion of mass and heat was negligible in the flow direction, and (3) steady-state conditions. Taking into account these assumptions, the maximum hydraulic pressure difference, Δp_h, that can be theoretically generated by a given temperature difference in TO was defined as [129]:

$$\Delta p_{h,max} = \frac{Q}{v_l}\left(1 - \frac{T_C}{T_H}\right) \tag{10.19}$$

where T_C is the temperature on the cold side of the membrane, T_H is the temperature on the hot side of the membrane, v_l is the molar volume of the liquid, and v_l is the heat transferred per mole of fluid.

Chen et al. [99] reported the conversion and storage of TO energy from combined salinity and temperature gradients using ultra-small silica nano-channels (SNC). A hybrid membrane consisting of SNC with nano-channels 2.3 nm in diameter was studied. Finite element simulations based on coupled Poisson–Nernst–Planck and Einstein–Stokes equations were carried out to study the mechanism of energy storage. The authors concluded that the permselectivity was a decisive factor.

10.5 Applications of thermo-osmosis process

TO has widespread applications in processes such as water treatment, charge(s) separation, waste heat recovery and energy conversion, electrochemical systems such as fuel

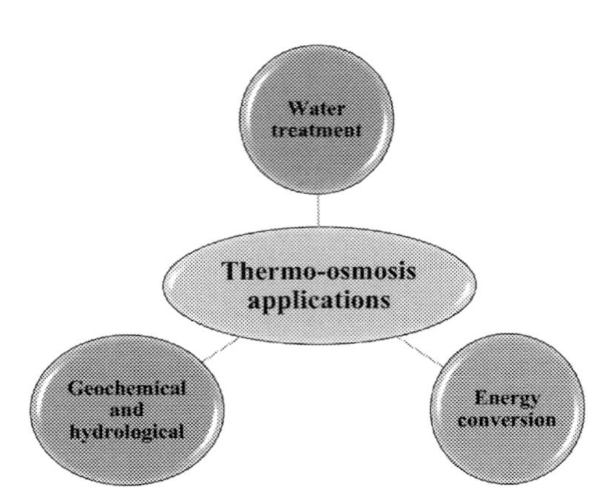

FIGURE 10–3 Some fields of applications of TO process. *TO*, Thermo-osmosis.

cells, reverse electro-dialysis [35,44,45,84–86,89,90,94–97,99,130–141]. Especially, TO has interests in those cases in which it is required to work with a variable temperature and without significant hydraulic pressures. Fig. 10–3 shows some possible fields of application of TO.

The renewed interest in TO separation process marks its technical importance in some interesting technological applications. For example, TO is considered important in fuel cell application [35,44,85,86,89,105,142], water recovery from organic wastewaters [43,56,94,143], and water management [42,43,84,85,91].

The conversion of low-quality waste heat into electrical energy is an attractive opportunity to harvest a sustainable energy resource. As indicated previously in Section 10.4.3, it has been shown that a TOEC process that uses nano-structured membranes converted residual heat into electrical energy from sources at low temperatures. In this context, research on the conversion of low-grade heat into electrical energy is of great importance. The concept of TOEC process capable of converting thermal energy gradients into electricity was thoroughly presented in Refs. [96,131].

TO process is also relevant for a wide variety of applications in geotechnical engineering, including the use of clay barriers designed for waste containment, EO for soil consolidation, highly compacted bentonite buffers for high elimination of radioactive nuclear waste, and electro-kinetics for the removal of soil contaminants, among others [93,118,144].

Table 10–4 shows some membranes used in different TO water treatment applications together with the observed water permeate flux and those used in TOEC process together with the observed energy output.

Table 10–4 Membranes used in different thermo-osmosis water treatment applications together with the corresponding water permeate flux (J_w) (A) and those used in thermal-osmotic energy conversion (TOEC) process together with the observed energy output (P_{Output}) (B).

(A) Water treatment applications

	Membrane		Water treatment					
Year	**Code**	**Type**	**Liquid type**		**ΔT (°C)**	**J_w kg/(m²s)**	**Reference**
1992	Nafion 417	Cation exchange	Water management: PEFC[a]				[42]
			HCl, LiCl, NaCl, KC1, and $MgCl_2$		10	$3.6 \cdot 10^{-5}$	
1998	PA	Polyamide	Draw	Feed	32	$4.2 \cdot 10^{-4}$	[145]
			NaCl (3.9 M)	Tomato juice (0.13 M)			
2006	Nafion 112	PTFE-reinforced	Water management: PEMFC[b]				[84,85]
			Water		1	$4.1 \cdot 10^{-4}$	
	Gore-Select (5510 MEA)	Catalyst-coated	Water management: PEMFC[b]				
			Water		1	$6.9 \cdot 10^{-4}$	
2006	Nafion 117	Cation exchange	Water management: PEFC[a]				[43]
			Mixtures of water/methanol		10	$7.5 \cdot 10^{-6}$	
2008	Nafion 112	Reinforced perfluorinated membranes	Water		0.3	$3.3 \cdot 10^{-4}$	[44]
2011	CT	Cellulose triacetate	Draw	Feed	20	$4.2 \cdot 10^{-3}$	[146]
			Na_2SO_4 (1.5 M)	NaCl (0.6 M)			
2012	CT	Cellulose triacetate	Draw	Feed	20	$5.3 \cdot 10^{-4}$	[141]
			KCl (0.5–3 M)	NaCl (0–0.86 M)			
2012	CT	Cellulose triacetate	Draw	Feed	20	$3.1 \cdot 10^{-3}$	[147]
			NaCl (1 M)	NaCl (0.1 M)			
2013	CT	Cellulose triacetate	NaCl (0.5 M)	Deionized water	–	$1.5 \cdot 10^{-3}$	[148]
	PA	Polyamide			20	$4.7 \cdot 10^{-3}$	
2016	CTA	Cellulose triacetate	Draw	Feed	27	Order of water flux:	[94]
	CA	Cellulose acetate	NaCl	Organic wastewaters:		(CTA > CA > TFC)	
	TFC	Thin-film composite		50 g/L of PTX[c] or BTX[d]			
2017	ZIF-25	Zeolitic imidazolate frameworks	NaCl (0.5 M)		70	15.8	[97]
2019	MCE	Hydrophobic mixed cellulose ester modified with perfluorodecyltrichlorosilane	NaCl (1 M)		39	Permeance: 0.9×10^{-7} kg/(m²sPa)	[45]

(B) TOEC applications

	Membrane			TOEC process			
Year	**Code**	**Type**	**Properties**	**Liquid type**	**ΔT (°C)**	**P_{Output} (W/m²)**	**Reference**
2016	PTFE	Hydrophobic	$\delta = 135\ \mu m$	Water	20	1.55 ± 0.05	[96]
			$d_p = 20$ nm; $\varepsilon = 77\%$		40	3.53 ± 0.29	
2016	Nafion 117	Cation exchange	–	0.02 mol/L (TMACl)[e]	5.4	$0.33 \cdot 10^{-4}$	[130]
					20.5	$5.14 \cdot 10^{-4}$	
2019	SNC/PET	Hybrid nano-channel ultra-small silica positioned on the PET membrane with track-etched nano-channels	$d_p = 2.3$ nm	NaCl	10	1.40	[99]

PET, Poly(ethylene terephthalate); *PTFE*, polytetrafluoroethylene; *SNC*, Ultra-small silica nanochannels; *ZIF*, Zeolitic imidazolate frameworks.
[a]PEFC: Polymer electrolyte fuel cells.
[b]PEMFC: Proton exchange membrane fuel cells.
[c]PTX: Phenol, toluene and xylene mixture.
[d]BTX: Benzene, toluene and xylene mixture.
[e]TMACl: Tetramethylammonium chloride (99% Fluka).

References

[1] C. Reid, E. Breton, Water and ion flow across cellulosic membranes, J. Appl. Polym. Sci. 1 (2) (1959) 133–143.

[2] S. Loeb, S. Sourirajan, Sea water demineralization by means of an osmotic membrane, In Saline Water Conversion—II; Advances in Chemistry, ACS Publications, Washington, DC, 1962, pp. 117–132.

[3] W. Feddersen, Ueber thermodiffusion von gasen, Pogg. Ann. Phys. Chem. 148 (5) (1873) 302–311.

[4] G. Lippmann, Endosmose zwischen zwei Flussigkeiten von gleicher chemischer Zusammensetzung und verschiedener Temperatur, C. R. Acad. Sci. Paris 145 (1907) 105–106.

[5] M. Aubert, Thermo-osmose, Ann. Chim. Phys. 26 (1912) 145–208.

[6] O. Reynolds, On certain dimensional properties of matter in the gaseous state, Philos. Trans. R. Soc. Lond. 170 (B) (1879) 727–845.

[7] E.D. Eastman, Thermodynamics of non-isothermal systems, J. Am. Chem. Soc. 48 (6) (1926) 1482–1493.

[8] L. Onsager, Reciprocal relations in irreversible processes. II, Phys. Rev. 38 (12) (1931) 2265.

[9] H. Freundlich, Kapillarchemie, eine darstellung der chemie der kolloide und verwandter gebiete, Akademische Verlagsgesellschaft, 1922.

[10] B. Derjaguin, G. Sidorenkov, Thermoosmosis at ordinary temperatures and its analogy with the thermo-mechanical effect in helium II, C.R. Acad. Sci. USSR 32 (1941) 622–626.

[11] H.P. Hutchison, I.S. Nixon, K.G. Denbigh, The thermo-osmosis of liquids through porous materials, Discuss. Faraday Soc. 3 (1948) 86–94.

[12] K.F. Alexander, K. Wirtz, Thermoosmose in wässrigen Systemen, Z. Phys. Chem. 195 (1950) 165–174.

[13] C.W. Carr, K. Sollner, New experiments on thermo-osmosis, J. Electrochem. Soc. 109 (7) (1962) 616–622.

[14] K.G. Denbigh, G. Raumann, The thermo-osmosis of gases through a membrane. 1. Theoretical, Proc. R. Soc. London, Ser. A 210 (1102) (1952) 377–387.

[15] R.P. Rastogi, R.K. Agarwal, R.L. Blokhra, Cross-phenomenological coefficients. Part 1. Studies on thermo-osmosis, Trans. Faraday Soc. 60 (5008) (1964) 1386–1390.

[16] R.P. Rastogi, R.K. Agarwal, R.L. Blokhra, Thermo-osmosis through membrane, Indian J. Chem. 2 (4) (1964) 166–172.

[17] R.P. Rastogi, K. Singh, Cross-phenomenological coefficients. Part 5. Thermo-osmosis of liquids through cellophane membrane, Trans. Faraday Soc. 62 (523P) (1966) 1754–1761.

[18] R.P. Rastogi, K. Singh, B.M. Misra, Thermo-osmosis and electro-osmosis of water and electrophoresis of suspensions in water, Desalination 3 (1) (1967) 32–36.

[19] R.P. Rastogi, K. Singh, Cross-phenomenological coefficients. Part 8. Thermo-osmosis of ideal gases, Trans. Faraday Soc. 63 (540P) (1967) 2917–2925.

[20] H. Voellmy, P. Lauger, Untersuchungen uber thermoosmose in flüssigkeiten untersuchungen, Ber. Bunsenges. Phys. Chem. 70 (2) (1966) 165–170.

[21] C. Dirksen, Thermo-osmosis through compacted saturated clay membranes, Soil. Sci. Soc. Am. Proc. 33 (6) (1969) 821–826.

[22] R. Haase, C. Steinert, Thermoosmose in flüssigkeiten. II. Messungen, Z. Phys. Chem. 21 (1959) 270–297.

[23] R. Haase, H.J. Degreiff, Thermoosmose in flüssihkeiten. III. Richtungsumkehr und zeitverlauf, Z. Phys. Chem. 44 (5–6) (1965) 301–313.

[24] R. Haase, H.J. Degreiff, Thermoosmose in flüssigkeiten V. Untersuchungen am system cellophan + methanol, Z. Naturforsch. A 26 (10) (1971) 118–119.

[25] H. Vink, S.A.A. Chishti, Thermal osmosis in liquids, J. Membr. Sci. 1 (2) (1976) 149–164.

[26] J.I. Mengual, J. Aguilar, C. Fernández-Pineda, Thermoosmosis of water through cellulose acetate membranes, J. Membr. Sci. 4 (2) (1978) 209–219.

[27] M.S. Dariel, O. Kedem, Thermoosmosis in semipermeable membranes, J. Phys. Chem. 79 (1971) 1773.

[28] F. Gaeta, D. Mita, Non-isothermal mass transport in porous media, J. Membr. Sci. 3 (2) (1978) 191–214.

[29] F. Belluci, E. Drioli, F.G. Summa, et al., Thermodialysis of non-ideal aqueous solutions: an experimental study, Trans. Farad. Soc. II 75 (1979) 247–260.

[30] F.S. Gaeta, D.G. Mita, Thermal diffusion across porous partitions: the process of thermodialysis, J. Chem. Phys. 83 (1979) 2276–2285.

[31] F. Bellucci, E. Drioli, F.S. Gaeta, et al., Temperature gradient affecting mass-transport in synthetic membranes, J. Membr. Sci. 7 (2) (1980) 169–183.

[32] M. Khayet, T. Matsuura, Membrane Distillation: Principles and Applications, Elsevier, 2011.

[33] B.R. Bodel, Silicone Rubber Vapor Diffusion in Saline Water Distillation. US Patent 285032, 1963.

[34] M.E. Findley, Vaporization through porous membranes, Ind. Eng. Chem. Process. Des. Dev. 6 (2) (1967) 226–230.

[35] V.M. Barragán, S. Kjelstrup, Thermo-osmosis in membrane systems: a review, J. Non-Equilib. Thermodyn. 42 (3) (2017) 217–236.

[36] M. Tasaka, M. Nagasawa, Nonisothermal membrane phenomena through charged membranes, J. Polym. Sci.: Polym. Symp. 49 (1975) 31–42.

[37] M. Tasaka, S. Abe, S. Sugiura, M. Nagasawa, Thermoosmosis through charged membranes, Biophys. Chem. 6 (3) (1977) 271–278.

[38] M. Tasaka, K. Kishi, M. Okita, Thermo-osmosis of various electrolyte-solution through anion-exchange membranes, J. Membr. Sci. 17 (2) (1984) 149–160.

[39] Y. Kobatake, H. Fujita, Osmotic flows in charged membranes. II. Thermoosmosis, J. Chem. Phys. 41 (10) (1964) 2963–2966.

[40] Y. Kobatake, H. Fujita, Flows through charged membranes. I. Flip-flop current vs voltage relation, J. Chem. Phys. 40 (8) (1964) 2212–2218.

[41] D.G. Mita, U. Asprino, A. D'Acunto, et al., Heat-flow induced mass transport through porous partitions, Gazz. Chim. Ital. 109 (1979) 475.

[42] M. Tasaka, T. Hirai, R. Kiyono, Y. Aki, Solvent transport across cation-exchange membranes under a temperature difference and under an osmotic pressure difference, J. Membr. Sci. 71 (1–2) (1992) 151–159.

[43] J.P.G. Villaluenga, B. Seoane, V.M. Barragán, C. Ruiz-Bauzá, Thermo-osmosis of mixtures of water and methanol through a Nafion membrane, J. Membr. Sci. 274 (1–2) (2006) 116–122.

[44] S. Kim, M.M. Mench, Investigation of temperature-driven water transport in polymer electrolyte fuel cell: thermo-osmosis in membranes, J. Membr. Sci. 328 (1–2) (2009) 113–120.

[45] E. Shaulsky, V. Karanikola, A.P. Straub, et al., Asymmetric membranes for membrane distillation and thermo-osmotic energy conversion, Desalination 452 (2019) 141–148.

[46] Z. Zhang, S. Yang, P.P. Zhang, et al., Mechanically strong MXene/Kevlar nanofiber composite membranes as high-performance nanofluidic osmotic power generators, Nat. Commun. (2019) 10.

[47] K.G. Denbigh, G. Raumann, The thermo-osmosis of gases through a membrane. 2. Experimental, Proc. R. Soc. London, Ser. A 210 (1103) (1952) 518–533.

[48] R.J. Bearman, The thermo-osmosis of the rare gases through a rubber membrane, J. Phys. Chem. 61 (6) (1957) 708–713.

[49] M.Y. Bearman, R.J. Bearman, Isothermal permeabilities from thermo-osmosis experiments, J. Appl. Polym. Sci. 10 (5) (1966) 773–786.

[50] R. Haase, H.J. Degreiff, H.J. Buchner, Thermoosmose in flüssigkeiten, Z. Naturforsch. A 25 (7) (1970) 1080–1085.

[51] H.P. Singh, Thermo-osmosis of oxygen sulphide and ethylene through porous unglazed porcelain, Indian J. Chem. 9 (1) (1971) 52–63.

[52] R. Ash, R.M. Barrer, J.H. Clint, R.J. Dolphin, C.L. Murray, Isothermal and thermo-osmotic transport of sorbable gases in microporous carbon membranes, Philos. Trans. R. Soc. A 275 (1249) (1973) 255–307.

[53] L. D'Ilario, M. Canella, Pseudothermoosmosis in a composite membrane, Polym. J. 9 (3) (1977) 253–260.

[54] P.C. Shukla, Studies on thermo-osmosis of NaCl & K_2SO_4 solutions, Indian J. Chem. Sect. A—Inorg. Phys. Theor. Anal. Chem. 15 (4) (1977) 340–342.

[55] P. Voznyi, N. Churaev, Thermo-osmosis flow of water in porous glassed. 2. Experimental results, Colloid J. URSS 39 (3) (1977) 378–383.

[56] S. Duckwitz, H. Moraal, Transport of nonelectrolyte-water mixtures through cellulose membranes, Z. Naturforsch. 32 (1977) 1077–1083.

[57] R.P. Rastogi, B. Mishra, Thermoosmosis of mixtures of oxygen and ethylene, J. Phys. Chem. 82 (21) (1978) 2341–2346.

[58] M. Tasaka, K. Ogawa, T. Yayazaki, Thermal membrane potential across charged membranes in 2-1 and 1-2 electrolyte solutions, Biophys. Chem. 7 (4) (1978) 279–283.

[59] R.P. Rastogi, A.P. Rai, Thermo-osmosis of gaseous mixtures. 4. Thermo-osmotic concentration difference, J. Membr. Sci. 4 (3) (1979) 291–304.

[60] D. Mita, F. Bellucci, M. Cutuli, F. Gaeta, Nonisothermal matter transport in sodium chloride and potassium chloride aqueous solutions. 2. Heterogeneous membrane system (thermodialysis), J. Phys. Chem. 86 (15) (1982) 2975–2982.

[61] T.H.K. Frederking, Darcy law of thermo-osmosis for zero net mass flow at low temperatures, Proc. ASME-JSME Therm. Eng. 2 (1983) 191–197.

[62] N. Pagliuca, D.G. Mita, F.S. Gaeta, Isothermal and non-Isothermal water transport in porous membranes. 1. The power balance, J. Membr. Sci. 14 (1) (1983) 31–57.

[63] J.I. Mengual, F. García-López, C. Fernández-Pineda, Permeation and thermal osmosis of water through cellulose acetate membranes, J. Membr. Sci. 26 (2) (1986) 211–230.

[64] M. Tasaka, H. Futamura, The effect of temperature on thermoosmosis, J. Membr. Sci. 28 (2) (1986) 183–190.

[65] M. Tasaka, Thermal membrane potential and thermoosmosis across charged membranes, Pure Appl. Chem. 58 (12) (1986) 1637–1646.

[66] N. Pagliuca, U. Bencivenga, D. Mita, G. Perna, F. Gaeta, Isothermal and non-isothermal water transport in porous membranes. II. The steady state, J. Membr. Sci. 33 (1) (1987) 1–25.

[67] C. Fernández-Pineda, M.I. Vázquez-González, Differential thermo-osmotic permeability in water cellophane systems, J. Chem. Soc. Faraday Trans. 84 (1988) 647–656.

[68] J.I. Mengual, F. García-López, Thermoosmosis of water, methanol, and ethanol through cellulose acetate membranes, J. Colloid Interface Sci. 125 (1988) 667–678.

[69] J.I. Mengual, F. García-López, C. Fernández-Pineda, Osmosis and thermo-osmosis through cellulose acetate membranes, J. Non-Equilib. Thermodyn. 13 (1988) 177–192.

[70] J.M. Ortiz-Zárate, F. García-López, J.I. Mengual, The effect off unstirred layers on thermoosmosis, J. Non-Equilib. Thermodyn. 14 (1989) 267–278.

[71] J.M. Ortiz de Zárate, F. García-López, J.I. Mengual, Temperature polarization in non-isothermal mass transport through membranes, J. Chem. Soc. Faraday Trans. 86 (16) (1990) 2891–2896.

[72] M. Tasaka, T. Mizuta, O. Sekiguchi, Mass transfer through polymer membranes due to a temperature gradient, J. Membr. Sci. 54 (1–2) (1990) 191–204.

[73] M. Tasaka, T. Urata, R. Kiyono, Y. Aki, Solvent transport across anion-exchange membranes under a temperature difference and an osmotic pressure difference, J. Membr. Sci. 67 (1) (1992) 83–91.

[74] R. Ash, R.M. Barrer, A.V. Edge, T. Foley, C.L. Murray, Thermo-osmosis of sorbable gases in porous media. 3. Single gases, J. Membr. Sci. 76 (1) (1993) 1–26.

[75] K. Hanaoka, R. Kiyono, M. Tasaka, Thermal membrane potential across anion-exchange membranes in KCl and KIO_3 solutions and transported entropy of ions, J. Membr. Sci. 82 (3) (1993) 255–263.

[76] K. Hanaoka, R. Kiyono, M. Tasaka, Non-isothermal membrane phenomena across perfluorosulfonic acid-type membranes, Flemion S: Part II. Thermal membrane potential and transported entropy of ions, Colloid Polym. Sci. 272 (8) (1994) 979–985.

[77] R. Kiyono, A. Kuwashita, Y. Tanaka, O. Sekiguchi, M. Tasaka, Thermal membrane potential across poly (4-vinylpyridine-*co*-styrene) membranes, Bull. Soc. Water Sci. Jpn. 48 (1994) 159–164.

[78] T. Suzuki, R. Kiyono, M. Tasaka, Solvent transport across anion-exchange membranes under a temperature difference and transported entropy of water, J. Membr. Sci. 92 (1) (1994) 85–93.

[79] T. Suzuki, Y. Takahashi, R. Kiyono, M. Tasaka, Non-isothermal membrane phenomena across perfluorosufonic acid-type membranes, Flemion-S: Part 1. Thermoosmosis and transported entropy of water, Colloid Polym. Sci. 272 (8) (1994) 971–978.

[80] T. Suzuki, K. Iwano, R. Kiyono, M. Tasaka, Thermoosmosis and transported entropy of water across hydrocarbonsulfonic acid-type cation-exchange membranes, Bull. Chem. Soc. Jpn. 68 (2) (1995) 493–501.

[81] D. Dedes, D. Woermann, Convective gas flow in plant aeration and thermo-osmosis: a model experiment, Aquat. Bot. 54 (2–3) (1996) 111–120.

[82] R. Ash, R.M. Barrer, A. Vernon, J. Edge, T. Foley, Thermo-osmosis of sorbable gases in porous media. 4. Mixture separation by two procedures, J. Membr. Sci. 125 (1) (1997) 41–59.

[83] P.J. Vie, S. Kjelstrup, Thermal conductivities from temperature profiles in the polymer electrolyte fuel cell, Electrochim. Acta 49 (7) (2004) 1069–1077.

[84] R. Zaffou, S.Y. Jung, H.R. Kunz, J.M. Fenton, Temperature-driven water transport through membrane electrode assembly of proton exchange membrane fuel cells, Electrochem. Solid-State Lett. 9 (9) (2006) A418–A422.

[85] R. Zaffou, H.R. Kunz, J.M. Fenton, Temperature-driven water transport in polymer electrolyte fuel cells, ECS Trans. 3 (1) (2006) 909–913.

[86] S. Kim, M.M. Mench, Temperature gradient induced water transport in polymer electrolyte fuel cells, ECS Trans. 13 (28) (2008) 89–105.

[87] J.P.G. Villaluenga, V.M. Barragán, M.A. Izquierdo-Gil, et al., Comparative study of liquid uptake and permeation characteristics of sulfonated cation-exchange membranes in water and methanol, J. Membr. Sci. 323 (2) (2008) 421–427.

[88] V.M. Barragán, J.P.G. Villaluenga, M.P. Godino, et al., Experimental estimation of equilibrium and transport properties of sulfonated cation-exchange membranes with different morphologies, J. Colloid Interface Sci. 333 (2) (2009) 497–502.

[89] S. Kim, M.M. Mench, Investigation of temperature-driven water transport in polymer electrolyte fuel cell: phase-change-induced flow, J. Electrochem. Soc. 156 (3) (2009) B353–B362.

[90] A. Siria, P. Poncharal, A.L. Biance, et al., Giant osmotic energy conversion measured in a single transmembrane boron nitride nanotube, Nature 494 (7438) (2013) 455–458.

[91] J. Lee, T. Laoui, R. Karnik, Nanofluidic transport governed by the liquid/vapour interface, Nat. Nanotechnol. 9 (4) (2014) 317–323.

[92] V.M. Barragán, S. Muñoz, Influence of a microwave irradiation on the swelling and permeation properties of a Nafion membrane, J. Membr. Sep. Technol. 4 (2) (2015) 32–39.

[93] H. Roshan, M.S. Andersen, R.I. Acworth, Effect of solid-fluid thermal expansion on thermo-osmotic tests: an experimental and analytical study, J. Pet. Sci. Eng. 126 (2015) 222–230.

[94] A.F. Al-Alawy, R.M. Al-Alawy, Thermal osmosis of mixtures of water and organic compounds through different membranes, Iraqi J. Chem. Pet. Eng. 17 (2) (2016) 53–68.

[95] Y.A. Gandomi, M. Edmundson, F. Busby, M.M. Mench, Water management in polymer electrolyte fuel cells through asymmetric thermal and mass transport engineering of the micro-porous layers, J. Electrochem. Soc. 163 (8) (2016) F933–F944.

[96] A.P. Straub, N.Y. Yip, S.H. Lin, J. Lee, M. Elimelech, Harvesting low-grade heat energy using thermo-osmotic vapour transport through nanoporous membranes, Nat. Energy 1 (2016) 1–6.

[97] K.M. Gupta, J.W. Jiang, Water desalination through a zeolitic imidazolate framework membrane by electro- and thermo-osmosis: which could be more efficient? ChemistrySelect 2 (14) (2017) 3981–3986.

[98] Q. Luo, Q. Huang, Z. Chen, et al., Temperature dependence of the pore structure in polyvinylidene fluoride (PVDF)/graphene composite membrane probed by electrochemical impedance spectroscopy, Polymers 10 (10) (2018) 1123.

[99] K.X. Chen, L.N. Yao, F. Yan, et al., Thermo-osmotic energy conversion and storage by nanochannels, J. Mater. Chem. A 7 (44) (2019) 25258–25261.

[100] R. Kiyono, Y. Tanaka, O. Sekiguchi, M. Tasaka, Thermal membrane potential across cation-exchange membranes for various halide solutions, Colloid Polym. Sci. 271 (12) (1993) 1183–1190.

[101] M. Tasaka, T. Suzuki, R. Kiyono, M. Hamada, K. Yoshie, Thermoosmosis and transported entropy of water across poly (4-vinylpyridine/styrene) and poly (*N*-vinyl-2-methylimidazole/styrene) type membranes in electrolyte solutions, J. Phys. Chem. 100 (40) (1996) 16361–16364.

[102] T. Ikeda, M. Tsuchiya, M. Nakano, The thermal membrane potential as a function of the apparent ionic transport number of membranes, Bull. Chem. Soc. Jpn. 37 (10) (1964) 1482–1485.

[103] W.E. Goldstein, F. Verhoff, An investigation of anomalous osmosis and thermoosmosis, AIChE J. 21 (2) (1975) 229–238.

[104] M.S. Huda, R. Kiyono, M. Tasaka, T. Yamaguchi, T. Sata, Thermal membrane potential across anion-exchange membranes with different benzyltrialkylammonium groups, Sep. Purif. Technol. 14 (1–3) (1998) 95–106.

[105] S. Kjelstrup, M. Ottøy, R. Halseid, M. Strømgård, Thermoelectric power relevant for the solid polymer fuel cell, J. Membr. Sci. 107 (1995) 219–228.

[106] V. Barragán, C. Ruiz-Bauzá, Effect of unstirred solution layers on the thermal membrane potential through cation-exchange membranes, J. Membr. Sci. 125 (2) (1997) 219–229.

[107] S. Kjelstrup, P. Vie, L. Akyalcin, et al., The Seebeck coefficient and the Peltier effect in a polymer electrolyte membrane cell with two hydrogen electrodes, Electrochim. Acta 99 (2013) 166–175.

[108] K.G. Denbigh, Thermo-osmosis of gases through a membrane, Nature 163 (4132) (1949) 60.

[109] K.G. Denbigh, G. Raumann, The thermo-osmosis of gases through a membrane, Nature 165 (4188) (1950) 199–200.

[110] C.M. Crowe, The thermo-osmosis of gases through polymers—the transient approach to the steady state, Trans. Faraday Soc. 53 (11) (1957) 1413–1422.

[111] M. Tasaka, S. Abe, S. Sugiura, M. Nagasawa, Thermoosmosis through charged membranes. Theoretical analysis of concentration dependence, Biophys. Chem. 6 (3) (1978) 271–278.

[112] S. Semenov, M. Schimpf, Thermoosmosis as driving mechanism for micro- or nanoscale engine driven by external temperature gradient, J. Phys. Chem. C. 119 (45) (2015) 25628–25633.

[113] L. Fu, S. Merabia, L. Joly, What controls thermo-osmosis? Molecular simulations show the critical role of interfacial hydrodynamics, Phys. Rev. Lett. 119 (21) (2017) 6.

[114] S.N. Semenov, M.E. Schimpf, Thermoosomosis in microfluidic devices containing a temperature gradient normal to the channel walls, Eur. Phys. J. E 42 (11) (2019) 141.

[115] Z. Song, F.Y. Liang, S.L. Chen, Thermo-osmosis and mechano-caloric couplings on THM responses of porous medium under point heat source, Comput. Geotech. 112 (2019) 93–103.

[116] L. Fu, S. Merabia, L. Joly, Understanding fast and robust thermo-osmotic flows through carbon nanotube membranes: thermodynamics meets hydrodynamics, J. Phys. Chem. Lett. 9 (8) (2018) 2086–2092.

[117] V. Salomoni, A mathematical framework for modelling 3D coupled THM phenomena within saturated porous media undergoing finite strains, Composites B 146 (2018) 42–48.

[118] R. Zagorscak, M. Sedighi, H.R. Thomas, Effects of thermo-osmosis on hydraulic behavior of saturated clays, Int. J. Geomech. 17 (3) (2017) 1–10.

[119] B.V. Derjaguin, N.V. Churaev, V.M. Muller, V. Kisin, Surface Forces, Springer, 1987.

[120] K. Proesmans, D. Frenkel, Comparing theory and simulation for thermo-osmosis, J. Chem. Phys. 151 (12) (2019) 6.

[121] B. Derjaguin, G. Sidorenkov, E. Zubashchenko, E. Kiseleva, Kinetic phenomena in the boundary layers of liquids 1. The capillary osmosis, Prog. Surf. Sci. 43 (1–4) (1993) 138–152.

[122] R. Ganti, Y.W. Liu, D. Frenkel, Molecular simulation of thermo-osmotic slip, Phys. Rev. Lett. 119 (3) (2017) 5.

[123] O. Farago, A simple statistical-mechanical interpretation of Onsager reciprocal relations and Derjaguin theory of thermo-osmosis, Eur. Phys. J. E 42 (10) (2019) 8.

[124] S. Kjelstrup, D. Bedeaux, A. Hansen, B. Hafskjold, O. Galteland, Non-isothermal transport of multi-phase fluids in porous media. Constitutive equations, Front. Phys. 6 (2019) 12.

[125] A.P. Bregulla, A. Wurger, K. Gunther, M. Mertig, F. Cichos, Thermo-osmotic flow in thin films, Phys. Rev. Lett. 116 (18) (2016) 5.

[126] Y. Imai, Network representation of thermo-osmotic and thermoelectric phenomena, Int. J. Circuit Theory Appl. 25 (3) (1997) 219–228.

[127] P. Anzini, G.M. Colombo, Z. Filiberti, A. Parola, Thermal forces from a microscopic perspective, Phys. Rev. Lett. 123 (2) (2019) 028002.

[128] R. Ganti, Y.W. Liu, D. Frenkel, Hamiltonian transformation to compute thermo-osmotic forces, Phys. Rev. Lett. 121 (6) (2018).

[129] M. Rahimi, A.P. Straub, F. Zhang, et al., Emerging electrochemical and membrane-based systems to convert low-grade heat to electricity, Energy Environ. Sci. 11 (2) (2018) 276–285.

[130] M. Jokinen, J.A. Manzanares, K. Kontturi, L. Murtomäki, Thermal potential of ion-exchange membranes and its application to thermoelectric power generation, J. Membr. Sci. 499 (2016) 234–244.

[131] M. Elimelech, W.A. Phillip, The future of seawater desalination: energy, technology, and the environment, Science 333 (6043) (2011) 712–717.

[132] S. Chu, A. Majumdar, Opportunities and challenges for a sustainable energy future, Nature 488 (7411) (2012) 294–303.

[133] W.A. Phillip, Thermal-energy conversion: under pressure, Nat. Energy 1 (7) (2016) 1–2.

[134] N.Y. Yip, D. Brogioli, H.V.M. Hamelers, K. Nijmeijer, Salinity gradients for sustainable energy: primer, progress, and prospects, Environ. Sci. Technol. 50 (22) (2016) 12072–12094.

[135] A. Siria, M.L. Bocquet, L. Bocquet, New avenues for the large-scale harvesting of blue energy, Nat. Rev. Chem. 1 (11) (2017) 1–10.

[136] A.P. Straub, M. Elimelech, Energy efficiency and performance limiting effects in thermo-osmotic energy conversion from low-grade heat, Environ. Sci. Technol. 51 (21) (2017) 12925–12937.

[137] R. Long, Z. Luo, Z. Kuang, Z. Liu, W. Liu, Effects of heat transfer and the membrane thermal conductivity on the thermally nanofluidic salinity gradient energy conversion, Nano Energy 67 (2020) 104284.

[138] K. Xiao, L. Chen, L. Jiang, M. Antonietti, Carbon nitride nanotube for ion transport based photo-rechargeable electric energy storage, Nano Energy 67 (2020) 104230.

[139] A.M. Benneker, T. Rijnaarts, R.G. Lammertink, J.A. Wood, Effect of temperature gradients in (reverse) electrodialysis in the Ohmic regime, J. Membr. Sci. 548 (2018) 421–428.

[140] F. Giacalone, F. Vassallo, F. Scargiali, et al., The first operating thermolytic reverse electrodialysis heat engine, J. Membr. Sci. 595 (2020) 117522.

[141] S. Phuntsho, S. Vigneswaran, J. Kandasamy, et al., Influence of temperature and temperature difference in the performance of forward osmosis desalination process, J. Membr. Sci. 415 (2012) 734–744.

[142] M. Khandelwal, M. Mench, An integrated modeling approach for temperature driven water transport in a polymer electrolyte fuel cell stack after shutdown, J. Power Sources 195 (19) (2010) 6549–6558.

[143] E. Nagy, O. Borlai, A. Ujhidy, Membrane permeation of water-alcohol binary mixtures, J. Membr. Sci. 7 (1980) 109–118.

[144] J. Goncalves, G. de Marsily, J. Tremosa, Importance of thermo-osmosis for fluid flow and transport in clay formations hosting a nuclear waste repository, Earth Planet. Sci. Lett. 339 (2012) 1–10.

[145] K.B. Petrotos, P. Quantick, H. Petropakis, A study of the direct osmotic concentration of tomato juice in tubular membrane–module configuration. I. The effect of certain basic process parameters on the process performance, J. Membr. Sci. 150 (1) (1998) 99–110.

[146] S. Zhao, L. Zou, Effects of working temperature on separation performance, membrane scaling and cleaning in forward osmosis desalination, Desalination 278 (1–3) (2011) 157–164.

[147] S.J. You, X.H. Wang, M. Zhong, et al., Temperature as a factor affecting transmembrane water flux in forward osmosis: steady-state modeling and experimental validation, Chem. Eng. J. 198 (2012) 52–60.

[148] M. Xie, W.E. Price, L.D. Nghiem, M. Elimelech, Effects of feed and draw solution temperature and transmembrane temperature difference on the rejection of trace organic contaminants by forward osmosis, J. Membr. Sci. 438 (2013) 57–64.

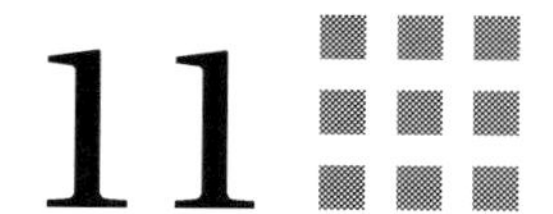

11 The applications of integrated osmosis processes for desalination and wastewater treatment

Nur Diyana Suzaimi[1], Pei Sean Goh[1], Ahmad Fauzi Ismail[2]

[1]ADVANCED MEMBRANE TECHNOLOGY RESEARCH CENTRE (AMTEC), UNIVERSITY OF TECHNOLOGY MALAYSIA, SKUDAI, MALAYSIA [2]ADVANCED MEMBRANE TECHNOLOGY RESEARCH CENTER, SCHOOL OF CHEMICAL AND ENERGY ENGINEERING, FACULTY OF ENGINEERING, UNIVERSITI TEKNOLOGI MALAYSIA, JOHOR, MALAYSIA

11.1 Introduction

Accelerated pace of population growth and uncontrollable anthropogenic activities has produced more wastewater-containing undesirable contaminants and shrunk limited freshwater resources [1,2]. Globalization has brought unprecedented challenges on both water and energy demand. Basically, water and energy are inextricably linked and equally important [3,4]. Concerns over that, cost-effective innovation coupled with technological advancement has revolutionized the water industry with the emergence of several technologies capable of meeting rising demands of clean water from alternative resources, for example, wastewater and seawater. Among them, construction of desalination and/or wastewater treatment using membrane-based technology is gaining worldwide acceptance as an attractive alternative to alleviate water scarcity crises and increase the productivity [5–9]. Assessment on the available literature convinced the membrane separation as the most effective method to attenuate contaminants and salinity of water bodies in real field applications [10–13]. This sustainable and versatile technology is inherently energy saving, offers flexible operation, eco-friendly, and requires small footprint compared to conventional technologies [2,8,14,15]. In addition, the characteristics and properties of the membrane, operational parameter, and feedwater quality dominate the successful implementation of this technology [6].

Membrane technology for water desalination and water/wastewater treatment has expanded considerably in recent times. Nevertheless, they too have shortcomings such as high energy requirement, high maintenance, and high amount of rejected water [6,8,16]. The increasing interest in membranes is driven by the stress from current water distribution system gives impetus to the idea of integration/hybridization of this technology, offering interesting opportunities

Osmosis Engineering. DOI: https://doi.org/10.1016/B978-0-12-821016-1.00013-9

in various aspects. These technologies consume low energy and cost expenditure as well as high quality of final products.

This book chapter compiles the applications of integrated osmosis processes in desalination and water treatment to cater water quality and quantity problems. The applicability of several integrated osmosis systems is expounded. In the first part, the advantages and potential integration of osmosis processes are comprehensively discussed. The second and third part of this chapter cover the performance of membranes in desalination and water treatments. Finally, the challenges, issues, and the opportunities of integration in advancing membrane technology are included for advancing and providing a potential guide for ongoing and future research of osmosis-based membrane technology.

11.2 Osmosis processes

Membrane separation technology that is based on the osmosis process can be mainly divided into reverse osmosis (RO), forward osmosis (FO), and pressure-retarded osmosis (PRO). The schematic diagrams of the processes are depicted in Fig. 11–1. Osmosis refers to the natural movement of water across a semipermeable membrane into a more concentrated solution. As pressure ($\Delta P = 2000-7000$ kPa) is applied to the more concentrated solution, the direction of water flow through the membrane is reversed, hence it is termed RO [17]. The RO membrane preferentially allows water molecules to pass through it while impeding the passage of salts/dissolved elements present in the feed. Ever since seawater/brackish water became an important freshwater source, RO became an important technology for desalination [6]. RO is frequently called into service for water purification, mainly due to its high recovery factor, lower investment, and easy scale up [8,14]. Owing to these benefits, studies showed RO-based membrane desalination technology accounting about 60% of all desalination plants [9].

However, notwithstanding the advantages, RO has significant membrane fouling and energy implications [11]. Potential low-fouling and energy separation process has hence

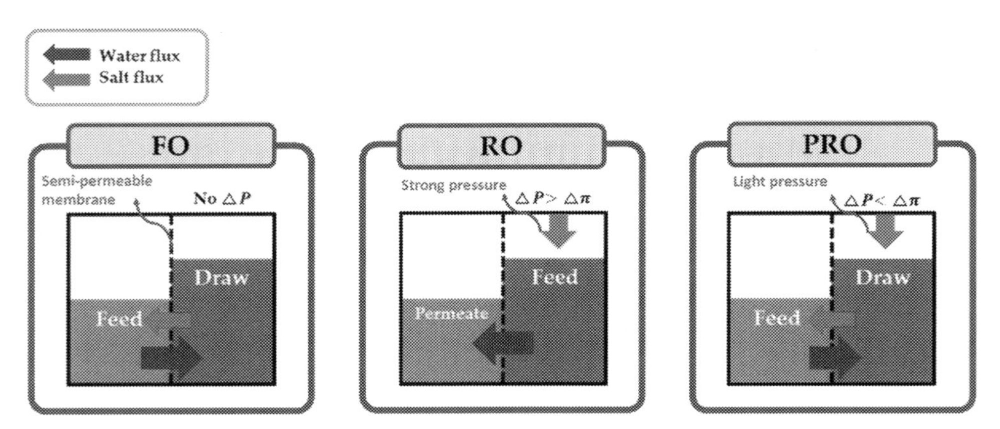

FIGURE 11–1 Schematic of osmosis process [18].

made FO become an eligible alternative. FO has a similar working process with RO but has the opposite movement of water molecules. Unlike the RO process, in FO water is adsorbed through a membrane due to the natural osmotic pressure differences between the feed and the draw solution (DS) [19]. FO is an emerging membrane separation technology that has gained much attention for water treatment applications in the past few years due to high pollutants rejection, high water recovery, and lower membrane fouling propensity than the RO membrane process [11,20]. DS plays a key role in separation performance and energy efficiency of FO. Not only that, compatibility of FO membrane also can significantly affect the performance. Seawater desalination using FO had been proposed in the mid-1960s [21], followed by various areas including food processing, wastewater treatment, desalination, and power generation [2,11,20,21]. In most applications, FO has to be coupled with another separation process in a so-called hybrid FO system to either separate the DS from the final product water or to be used as an advanced pretreatment process to conventional desalination technologies. PRO is a membrane-based energy generation process, employs seawater/concentrated seawater as the DS, while wastewater effluent is used as the feed solution (FS) [9,10]. These technologies are based on the concept of osmotic pressure gradient applications across a semipermeable membrane for power and freshwater supply. PRO potentially supplies water and energy at competitive costs [2–10]. The facts that there are no carbon dioxide emissions and that there is less periodicity due to the weather conditions make this process even more attractive.

11.2.1 Integration of osmosis processes

Along with the continuous research on the water-energy nexus process, the development of new processes has been one of the important factors in keeping resources available in the long-term context, thereby improving their effectiveness in tackling water and energy challenges. Integrated hybridized desalination and water treatment systems may provide a compelling answer to these challenges. Osmosis systems have been integrated into the existing or combined processes to either replace conventional pretreatment technologies or as a posttreatment to reduce the drawbacks and enhance the outputs.

The idea of integration/hybridization expands the role of the osmosis process from only focusing on water-related processes to multifunctional tasks, supporting other types of applications, such as energy (biogas), crystallization, and fertilizer. Such a change of paradigm will offer another perspective to other researchers and industries, where they will be more convincing on the role of RO, FO, and PRO process in nonwater-based application. This will be a healthy development to encourage more researchers and industries to explore membrane technology in various applications. Several potential applications of integrated osmosis would be for agricultural industry, electricity generation, food and beverage industry, chemical industry, pharmaceutical industry, coal processing, microalgae cultivation, textile industry, contaminants of emerging concern, and cooling water treatment [13,22–29].

11.3 Integrated osmosis process for desalination

11.3.1 Integration of reverse osmosis process

Despite the advantages and a wide application of RO processes, there are numbers of limitations encountered that slow down the industrialization and commercialization. The performance of the RO process was improved through its integration with other technologies. Combination of two modules as alternative means were designed to evaluate process efficiency and cut down the RO feed pretreatment cost [19].

11.3.1.1 Reverse osmosis—adsorption and reverse osmosis—nanofiltration

Aiming to produce high-quality water for irrigation and domestic use, hybridization of RO desalination technology and the adsorption (AD) has been utilized to treat brackish water [30]. AD is one of the most effective and economically attractive techniques in industries for practical waste treatment applications [31,32]. In desalination, AD comprises (1) the AD/desorption bed, (2) the evaporator, and (3) condenser producing desalinated water (from the condenser) and cooling (from the evaporator) [27,28]. The evaporator receives sea/brackish water that is evaporated by means of AD. An RO—AD process enhanced its specific energy consumption (SEC) and simultaneously produce a cooling effect that can be exploited for local process cooling or air conditioning, with the AD cycle producing more than 6 m^3/t of drinking water (<15ppm) at 85°C. The results showed that the cost of the combined RO—AD system is lowest, 0.44 £/m^3, compared with other RO configurations. Ali et al. [33] also worked on an innovative combination between RO and AD desalination systems. The proposed system has been studied theoretically and the mathematical model was solved to study the effect of RO brine recycling employing AD desalination on overall system desalinated water recovery. The AD desalination produces dual useful effects: high-quality potable water and a cooling effect. The brine leaving RO system is fed to AD desalination system. The AD desalination is then driven by a low temperature heat source such as solar energy. Results showed that the proposed combination system recovery increased by about 25% compared to a single-stage RO system with low potential permeate salinity. In the case of salt production the brine concentration was about 220,000 ppm. In this case the permeate recovery increased by about 15% compared to the case of 110,000 ppm brine salinity concentration. In addition to system performance improvements, a cooling effect is generated and can be utilized for cooling applications.

Laskowska et al. [34] analyzed an approach of using an RO—NF (nanofiltration) hybrid system to overcome RO limitations resulting from high osmotic pressure of treated solution for concentration of brine. The RO retentate of mine was used as an NF feed, RO retentate pressure as the driving pressure for NF. In such hybrid system, NF produced more concentrated saline water than the RO and showed higher water recovery. NF was regarded as an alternative energy recovery device for RO due to energy consumption reduction (213.2—123.3 kWh/m^3) in the brine concentration process.

11.3.1.2 Microfiltration–reverse osmosis, ultrafiltration–reverse osmosis, nanofiltration–reverse osmosis

Thus far, a well-established pretreatment technique for water desalination was introduced to address the limitations of RO (membrane fouling) and maintaining the integrity. These pressure-driven membrane processes, microfiltration (MF), ultrafiltration (UF), and NF were said to surpass the conventional techniques (i.e., coagulation and granular media filtration) by allowing high permeate flux [35–37]. They are dictated by the membrane's pore size/molecular weight cutoff with different advantages. MF is an appropriate choice for removal of larger particulate matter at higher permeate fluxes, whereas UF, with smaller pore sizes of membrane than MF, appears as an appropriate choice for larger particulates removal. The properties of NF membranes (lies between RO and UF) allows dissolved contaminants as well as particulate and colloidal materials to be removed [24,38].

In the early 2000s, UF was recognized as the pretreatment technology of choice before RO to control the membrane fouling and provide steady-state filtered water [39–41]. Most recently, Gao et al. [42] explored integrated UF–RO desalination through direct UF backwash with RO concentrate. UF backwash enabled enhancement in feedwater quality, demonstrating a measurable impact on both UF fouling rate and effectiveness. As a result, the filtration flux and inline coagulant dose were increased. NF membrane is used especially in water treatment and water hardness removal, due to close similarity to RO. Talaeipour et al. [43] proposed NF process is applied to pretreat feed of brackish water. The salinity rejection became significantly better (78.56%) when NF process was used prior to RO process. The results concluded that the integration of NF–RO can be completed by each other to remove salinity, total dissolved solid (TDS), Cl, and Na. To date, Kaya et al. [44] studied the applicability of using the integrated membrane systems based on NF coupled with RO as a pretreatment stage for desalination of the brackish water RO. NF270 and NF90 were selected as the NF membranes, while the brackish water RO membrane was BW30. The permeates after NF process were collected and used as the feed for RO. Performance in terms of rejection and water recovery was increased significantly and the highest when employing NF90-30 bar + BW30–35 bar system was 51.6% and 98.2%, respectively. Unlike single RO system, the qualities of the product waters showed integrated systems complied with the irrigation standards. Considering the results, freshwater consumption can be greatly reduced by producing irrigation water from seawater.

11.3.1.3 Reverse osmosis–pressure-retarded osmosis

RO coupled with PRO has also been proposed in the context to treat seawater [5]. A performance analysis of PRO–RO over RO–PRO showed its feasibility. RO process takes place before the PRO in which the concentrated RO brine is used either as a DS or as a FS depending on its concentration. Net power consumption for RO–PRO was higher but the estimated cost was 4.3% lower than FO–RO and PRO–RO. Similar results were observed for power density at feed TDS 10 g/L, where RO–PRO permits 4.3 times higher than that in the PRO–RO design as can be seen in Fig. 11–2A. This indicates the

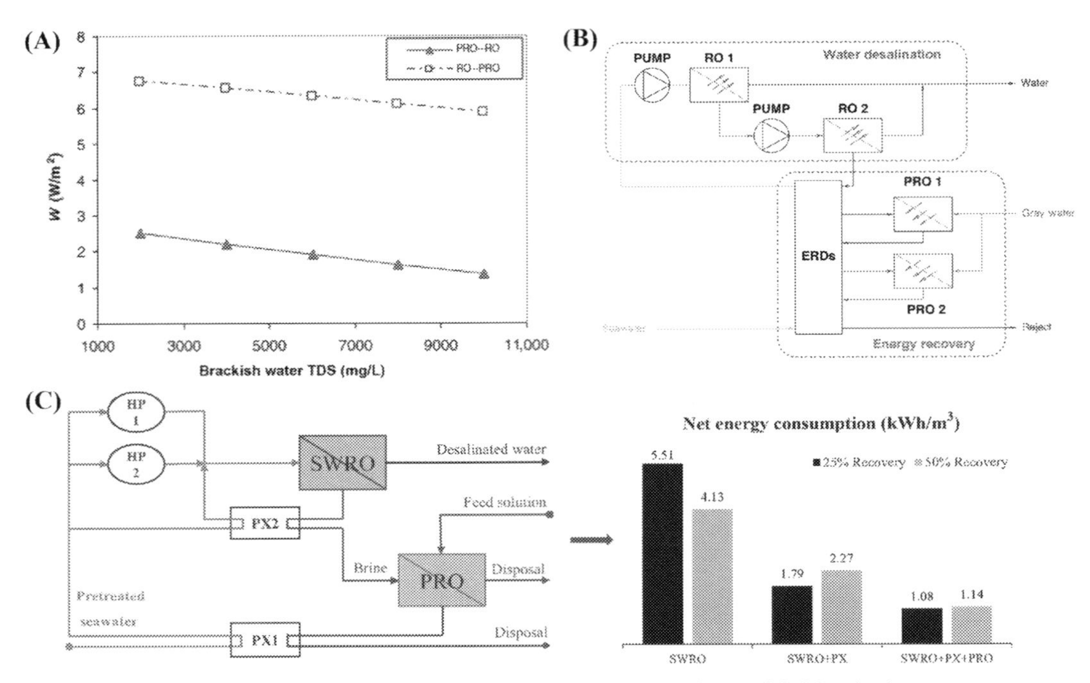

FIGURE 11–2 (A) Comparison of power density in PRO–RO and RO–PRO designs [5]; (B) Hybrid RO–PRO process consisting of two RO stages and two PRO stages connected in series [46]; (C) Schematic drawings and net energy consumption of the integrated RO–PRO processes [47]. *PRO*, Pressure-retarded osmosis; *RO*, reverse osmosis.

RO–PRO design is more efficient than the PRO–RO at higher feed TDS. The integrated system allows a seawater desalination process to save energy and alleviate the problem of disposing of waste RO brine and the environmental effects. He et al. [45] evaluated a hybrid of RO–PRO desalination plant in counter-current PRO configuration to find the optimum SECs. From the results, higher effectiveness in the hybrid RO–PRO was obtained according to the low SEC.

Recently, Wan and Chung [48] determined the overall energy consumption of desalination by developing the first economics-driven designs of RO–PRO. The operating expenditure and capital expenditure of the integrated system is investigated in the cost model to validate the economic feasibilities of the integrations RO and closed-loop (RO–cPRO) offers great advantages, such as high efficiency of osmotic energy recovery without additional energy or generation devices and RO brines can be recycled. Another integrated RO–PRO focuses on systematic optimization and operation of multistage RO–PRO, that is, RO–RO–PRO–PRO [46] as can be seen in Fig. 11–2A. The work focused on energy-efficient seawater desalination which integrated RO for desalination and PRO for power generation into a single process. Multistage design systems allow a stage-dependent applied pressure profile to better match the osmotic pressure of the concentrate stream, thus improving energy efficiency but more stage will require a higher capital cost.

Process of RO–PRO process using hollow fiber membrane has been carried out by Wan et al. [47] to explore and assess the treatment performance and economics. Schematic drawing of the process and the energy consumption was present in Fig. 11–2B. This work presented significant and attractive results with the smallest SEC. SEC reflects a saving in energy cost for seawater desalination. The operation at 25% and 50% recovery and the brines are diluted to the seawater level, the SECs reduced to 1.08 and 1.14 kWh, respectively. Accordingly, incorporating a PRO system into the existing plant leads to an energy saving of 25.6–40.7 million kWh/day.

11.3.2 Integration of forward osmosis process

11.3.2.1 Forward osmosis–reverse osmosis

Stand-alone FO process was operationally combined with various processes because it still suffers from some challenges, mainly membrane fouling that results in reduced permeate quality and quantity as well as increased operational cost. Whilst RO is an efficient process, cost dictated by both energy consumption and membrane replacement is the main concern for RO [4,49]. Therefore RO was proposed to be integrated with FO that has the advantage of minimized membrane fouling and lower energy cost in RO units [1,5,50,51]. FO–RO hybrid system was recognized to be price-competitive relative to an RO system when its recovery rate is very high, the flux and membrane cost of the FO are similar to those of RO, and the electricity cost is expensive [52]. FO and RO can be used as the front end pretreat and post-treatment desalination technologies, respectively.

As early as 1992, Yaeli [53] pioneered an innovative system, combining FO and low-pressure RO (LPRO) process to recover glucose. FO was therefore used as a pretreatment to reduce the fouling propensity of feedwater in the subsequent RO process. His team used a similar system to desalinate seawater [54]. This integration leads to 50% improvement of energy consumption ($\sim$1.5 kWh/m^3), almost double the independent RO desalination (2.5–4 kWh/m^3). Its cost efficiency would surpass RO only if the water flux above 5.5 LMH can be achieved. In a study by Cath et al. [55], wastewater was used as feed while seawater as DS. The diluted seawater after FO providing a low saline solution, subsequently, flows into RO units to desalinate. This hybrid system with impaired sources as feed can produce high quality of drinking water and achieve a favorable economic return with the water recovery rate up to 63% [8]. It can be seen that the hybrid FO–RO process may be a competitive choice for desalination of high saline water compared with RO alone.

A similar study was done for seawater desalination to enhance the energy efficiency. Comprehensive study by Volpin et al. [56] evaluates the performances and fouling tendency in the FO–RO hybrid process for wastewater reclamation and seawater dilution. Real wastewater and real seawater were used as feed and DS to get meaningful results. FO process effectively operates as a pretreatment for RO process by rejecting contaminants while reducing the seawater osmotic pressure. Initial water flux as high as 22.5 LMH was observed when using secondary effluent. On the other hand, flux decline respectively 25% and 50% was observed after 80 h when biologically treated secondary effluent was used. Given a stable FO performance, the high quality of diluted seawater is likely to decrease the cost of the downstream RO filtration for potable water production.

Altaee et al. evaluated the hybrid FO–RO system for desalination purposes and comparing the performance of the system with PRO–RO and RO–PRO [5]. The design consists of both FO and RO system, FO was installed in front of the RO in a parallel connection to develop less energy-intensive systems. Among them, FO–RO was recorded as the most efficient power system FO–RO providing the best product water quality and average FO membrane cost. This design utilized the diluted DS as the RO feed. The results of this study could be used to develop guidelines for the optimal design of the FO–RO hybrid process. FO–RO hybrid processes offer a promising solution not only to lower desalination energy needs but also to increase water reuse efficiency by combining seawater desalination and water reuse. Interestingly, due to the lower fouling propensity compared to pressure-driven membrane system, FO has the potential to treat feedwater of various qualities (potentially even including raw sewage), allowing to lower wastewater treatment costs. FO–RO schemes do require further validation but also radical shift in current consideration of water supply.

11.3.2.2 Forward osmosis–membrane distillation

Membrane distillation (MD) is an efficient separation process based on the principle of vapor–liquid equilibrium allowing only vapors to pass through a porous hydrophobic membrane [4,57,58]. The separation is achieved due to vapor pressure differences between the membrane surfaces. In fact, it has been demonstrated that FO used as a pretreatment process can improve the overall efficiency of conventional desalination processes in applications with challenging feedwaters (i.e., having high salinity, high fouling potential, or containing specific contaminants) [20]. One good example is the coupling of FO with MD to desalinate waters that are usually challenging for MD alone.

Recently, FO–MD has been used as an alternative for simultaneous treatment of domestic wastewater and seawater desalination. Seawater can be utilized as a DS due to its high osmotic pressure. Before desalination, seawater can be diluted before desalination, hence reducing the energy cost of desalination, and simultaneously, contaminants in the wastewater are prevented from migrating into the product water. A comprehensive study by Li et al. [19] recorded high removal efficiency ($>$ 90%) as high molecular weight contaminants were all removed, leaving only a few low molecular weight contaminants permeated through the membrane. By comparison, the MD membrane fouling was mainly induced by inorganic salts and was not as severe as that of the FO membrane. During 120-h continuous operation, the FO–MD integrated system exhibited satisfying performance stability and maintained a high-water yield and high-product water quality. The results indicated the potential of the FO–MD integrated system for wastewater treatment in cities, water purification and desalination since the system could simultaneously treat wastewater treatment and desalinate seawater within the same reactor.

11.3.2.3 Forward osmosis–ultrafiltration and forward osmosis–nanofiltration

Experimental results on the hybrid FO–UF process showed that this hybrid process can achieve high water fluxes with very low reverse solute fluxes. A high PAA–Na rejection (i.e., 499%) was obtained by the UF process for recycling and further reuse by the FO process. The DS recycled by the UF process showed reasonable repeatable performance and indicated no aggregation problems in comparison with MNPs. However, at high concentration, PAA–Na solution has a very high viscosity, limiting the application of this hybrid process in ambient conditions.

Ling and Chung [59] explored the use of surface-dissociated nanoparticles as DS in a hybrid FO desalination process. The hybrid system consisted of FO coupled with an electric field integrated into NF system. The electric field, <70 V, was employed to regenerate the negatively charged nanoparticles, while the NF system was used to recover the product water and reconcentrate the alkaline solution. This concentrated alkaline solution was used to dissolve and dissociate the nanoparticles prior to be reinjected in the FO process. The results of this study showed that the stability of water flux as well as the size of the MNPs after five cycles. However, alkaline solution entering the NF unit might damage the membrane as well as the corrosion of pumps, pipes, and fillings. The energy cost of this hybrid system however has not been evaluated.

The performance of FO–NF system for dewatering brackish water and wastewater was studied by Dutta and Nath [60]. Experimental tests used potassium chloride and monoammonium phosphate as DS for FO. Moderate water flux and low reversed solute flux were achieved for desalination of brackish water and wastewater concentration by FO. Posttreatment by NF at low pressure (5 bar) achieved high rejection rate (<95%) of nutrients (nitrogen, phosphorous, potassium) with permeate contained 117/248/409 mg/L which is suitable for fertigation and hydroponics and subsequently eradicated the eutrophication issue. They also found that NF potentially recycles the excess nutrient (retentate stream) back to the FO process. In the energy aspects and sustainability, energy required in FO process itself is much less compared to the NF process used for recovery of DS.

11.3.3 Integration of pressure-retarded osmosis process

Integrating PRO with other membrane operations is highly useful to offset energy consumption of other system. This concept is currently gaining considerable attention industrially as well. For this process, several combinations of feed and DSs have been tested including river and seawater [28], seawater brine and wastewater retentate, freshwater and synthetic NaCl solution [29], and seawater and municipal wastewater [30]. Synergetic effects can be achieved by introducing PRO into desalination systems in terms of reduction of waste footprint of FS and the dilution of the DS, which can reduce the energy demand of desalination system and can minimize the environmental consequences of brine disposal.

11.3.3.1 Pressure-retarded osmosis–reverse osmosis

Prante et al. [17] have developed the first module-based PRO model for full-scale applications. The authors have also reported experimental data on a RO unit getting a benefit from energy generated by a PRO unit. The system brings energy demand down about 40% with a minimum net energy consumption of 1.2 kWh/m^3 at 50% RO recovery. In another study by Kim et al. [61], PRO–RO was proposed for integrating wastewater reuse and seawater desalination by theoretical SEC and experimental fouling control of PRO membrane. SEC demonstrates low energy use of a PRO–RO process. However, a substantial flux decline was observed owing to the susceptibility of PRO to membrane fouling pronounced within the support layer. Lastly, the antiscaling pretreatment was very effective for lessening calcium phosphate and calcium carbonate scaling within the support layer. This suggests PRO–RO hybrid process could be successfully applied with optimized fouling control strategies in PRO.

11.3.3.2 Pressure-retarded osmosis–membrane distillation

The development of PRO–MD hybrid process gives twofold benefits, prevents the shortcomings of each subsystem, and strengthens their advantages [3,4,57]. As MD is a high cost of energy supply, hence MD–PRO is encouraged to be used instead. Despite lack of literature, many countries have also started to develop PRO-integrated technologies by building and commercializing out pilot plant with various design capacity [18] because of these key benefits: (1) low desalination cost, (2) minimal environmental issue and fouling, (3) high recovery of water [62]. In the system, MD is responsible to produce DS (freshwater and concentrated brine) for the PRO system, while PRO system will be the one that produces energy. Experimental results by Han et al. [3] showed that PRO–MD can generate osmotic energy from freshwater as well as wastewater. Employing 2 M NaCl as the DS and deionized water and wastewater brine as feeds, power densities of 31.0 and 9.3 W/m^2 were obtained by the PRO system. Simultaneously, at 40°C–60°C high-purity water (flux is 32.5–63.1 LMH) can be produced by the MD system without any fouling. In this system, a heat exchanger can further increase net energy recovery in the PRO stage; meanwhile, supplying thermal energy to the MD feed would decrease the amount of energy consumed. PRO–MD provides insightful guidelines for the exploration of alternative green technologies for renewable osmotic power and clean water production.

11.4 Integrated osmosis process for wastewater treatment

As well as desalination, membrane technology is also used for wastewater treatment, particularly in those areas where water is scarce. Various approaches such as AD, MF, and UF have been employed to decontaminate emerging pollutants and trace metals from wastewater. However, they too suffer several constraints not only in terms of cost but also in terms of feasibility, efficiency, practicability, reliability, environmental impact, sludge production, operation difficulty, pretreatment requirements, and by-products that ultimately lead to in unsuccessfully meeting the stringent standards. Consequently, this has prompted the exploration of osmosis integration as alternative to overcome these concerns.

11.4.1 Integration of reverse osmosis process

11.4.1.1 Microfiltration–reverse osmosis, ultrafiltration–reverse osmosis, nanofiltration–reverse osmosis

As seen in most of the studies and applications, the membrane filtrations i.e., MF, UF, and NF commonly serve as to RO seawater desalination. Also, those processes are effective to successfully treat feed of wastewater prior to RO treatment. A pilot study in collaboration with the Inland Empire Utility Agency (IEUA) was conducted in Chino using MF–RO to treat primary effluent for water reuse [63]. MF did not remove dissolved inorganic matters, hence acts as a pretreatment stage to optimize operating conditions for the RO process (e.g., stabilize water and reduce fouling of the RO membranes). Sending primary effluent directly to MF followed by RO could save some energy and bypass the tertiary treatment.

Petrinic et al. [24] applied UF as a pretreatment in combination with RO to reuse industrial wastewater. The treatment highly rejects (>90%) metal elements, organic, and inorganic compounds contaminants, while chemical oxygen demand (COD) and biochemical oxygen demand (BOD) were completely removed from the effluent. This allows the membrane to run at high recovery. Based on the presented results, UF enhanced the productivity and decreased lifetime costs of RO, as UF–RO membrane system is capable to produce 30,000 m^3 per year of treated wastewater.

Low concentration of phenol may be life-threatening to living things; hence, a combined process of UF–RO and UF–NF was applied to treat phenolic-containing wastewater from a paper mill [64]. Among them, UF–RO recorded the best results of rejections COD and TDS, 56.4% and 9.6%, respectively. The resulting permeate was further treated by NF membrane or RO membrane to remove the remaining COD and phenol. RO membrane exhibited superior performance over NF membrane with relatively high permeate flux of 26.4 LMH at 30 bar and rejected 95.5% COD and 94.9% phenol. Consuming less energy and having a good phenol rejection, UF–RO can be considered as a good potential wastewater treatment.

11.4.2 Integration of forward osmosis process

FO usually requires an additional separation process to separate drinkable water from the DS. The schematic diagram of FO hybrid is presented in Fig. 11–3.

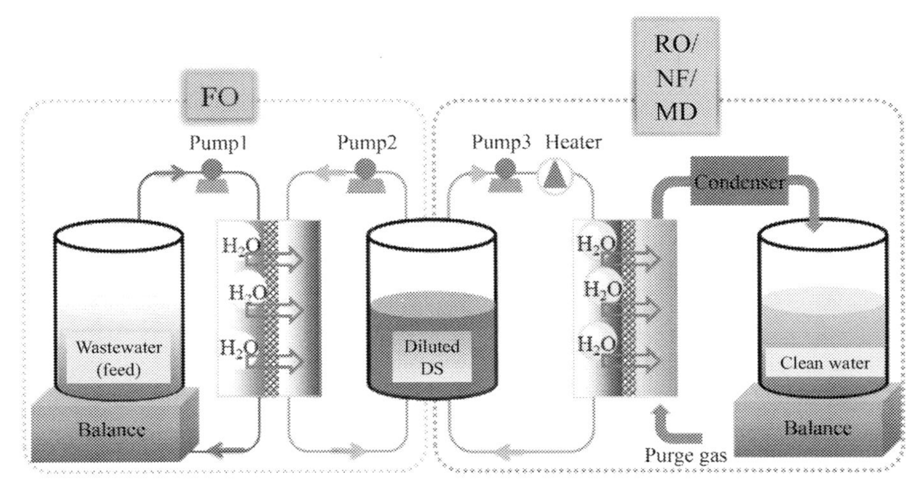

FIGURE 11–3 Schematic diagrams of the integrated FO with other systems for wastewater treatment [65]. *FO*, Forward osmosis.

11.4.2.1 Forward osmosis–reverse osmosis

The FO unit served as an effective pretreatment system prior to RO and the integrated FO–RO systems has a strong potential to successfully eliminate conventional pretreatment processes for RO. An integrated FO and RO system was applied to treat three different types of actual coal mine impacted waters (di-sodium hydrogen phosphate, sodium hexametaphosphate, and sodium lignosulfonate) [66]. Mine water was treated with the integrated system over an extended duration and the feed was concentrated to 20% of its starting volume. The integrated system was able to concentrate the brackish mine waters, recovering more than 80% of the volume of mine water and obtaining dischargeable quality treated water. Simple physical cleaning with clean water circulation was found to be effective in restoring the FO water flux. The osmotic gradient between two mine waters was also utilized to adopt mine water as a DS. The combination of FO with RO outperforms both FO- and RO-only systems. The combination of FO with RO provided a better performance than individual FO or RO in treating coal mine wastewater.

11.4.2.2 Forward osmosis–membrane distillation

A combination of osmotically and thermally driven membrane processes has been implemented for wastewater reclamation and reuse [57]. The integrated FO–MD systems are motivated by the following merits: (1) good solution to MD membrane fouling as substances in feedwater will be removed by FO and hence alleviate problems associated with fouling and wetting in the downstream MD, (2) convenient process, and (3) cut down capital and operating costs because DS is easily recycled and reduce the energy requirement of the wastewater concentration process [13,57,58,67]. The assessments of the available literature reveal that good stability in continuous operation of FO–MD applications are said to receive great attention until now and has been reliably proven to reduce the various contaminants [4,56,58]. They found that the hybrid FO–MD as promising solution that is capable of treating domestic wastewater [19], concentration of pharmaceuticals/protein solutions [13], oily wastewater separation [58,68], real dairy wastewater recycling [69], high salinity hazardous waste landfill leachate [70], etc.

The application of FO–MD hybrid process for the treatment of source-separated urine was recently investigated [56] to reuse the water and the nutrients in the urine in turn leading to the sustainability of the waste management system. FO was chosen as MD pretreatment to increase the overall nitrogen rejection and to prevent wetting of the MD membrane. The goal of this investigation was to tune the FO and MD operating parameters to reduce the nitrogen transport to the MD permeate. Feed temperature, nitrogen concentration, and membrane characteristics were investigated to optimize the MD process. The system presented a good separation performance, with 2.5 M NaCl (DS) commercial FO membranes, achieved a water flux between 31.5 and 28.7 LMH and a minimum nitrogen flux of 1.4 g/L. Applying minimal hydraulic pressure on the DS showed 33% reduction in the nitrogen transport and, at the same time, significantly reduced the net transmembrane water flux. To conclude, the hybrid FO–MD process is expected to be an effective solution for the production of clean water and concentrated fertilizer from human urine. This double barrier separation process could be suitable for both water reclamation in space application and resource recovery in urban application.

FO–MD system has been also found to successfully recover high amount of water from oily wastewater. For this purpose, Zhang et al. [58] employed cellulose triacetate thin-film composite hollow fiber membranes with relatively high water permeability and low salt permeability. The integrated FO–MD system had sustainably recovered at least 90% water for reuse and greatly retains NaCl and oil droplets (99.9%) while allowing only a small fraction of acetic acid to pass through. The work has demonstrated that not only water but also organic additives in the wastewater could be effectively retained by FO–MD systems for reuse or other utilizations. More recently, Al-Furaiji et al. [68] demonstrated the feasibility of FO–MD process in producing high-quality water from hyper-saline produced water containing hydrocarbons. Among four DSs [i.e., sodium chloride (NaCl), potassium chloride (KCl), lithium chloride (LiCl), and magnesium chloride ($MgCl_2$)], high osmotic pressure ($\sim$1600 atm, 10 M) of LiCl significantly leads to high FO flux. Contrary to FO, water flux of MD was relatively low due to the low vapor pressure. Best performance with stable fluxes for both FO and MD of 20 h operation was achieved by $MgCl_2$ (4.8 M) DSs. This study demonstrated that wastewater could be reclaimed by the FO–MD hybrid system for the reuse of urban recycled water or for higher grade utilization.

11.4.2.3 Forward osmosis–nanofiltration

Giagnorio et al. [71] carried out an experiment to assess the effectiveness of a hybrid FO–NF system for the extraction of high-quality water from wastewater. Real wastewater was used as the FS. The configuration of the process is shown in Fig. 11–4. With optimum configuration, overall FO–NF system can achieve up to 85% water recovery using Na_2SO_4 or $MgCl_2$ as the DS. Experimental tests with samples of the relevant wastewater showed that Cl^- and Mg^{2+} based draw solutes would be associated with larger membrane fouling, possibly due to their interaction with the other substances present in the FS. However, the results suggest that fouling would not significantly decrease the performance of the designed system. This study contributes to the further evaluation and potential implementation of FO in water reuse systems. Similarly, Pal et al. [72] found the modeled integrated FO–NF process to be versatile and effective process with low relative error (<0.1) for treatment and recirculation of leather-industry wastewater. Experimental analysis on a semipilot unit with real industrial wastewater summarizes the system has high performance capability owing to complete rejection of chromium and amount COD, chlorides and sulfates dropped by 98%. In addition, system recovered the NaCl-drawn solute by more than 98% for recycling in the FO loop.

11.4.3 Integration of pressure-retarded osmosis process

11.4.3.1 Pressure-retarded osmosis–reverse osmosis

He et al. [73] recently introduced a sustainable design and synthesis of hybrid PRO–RO systems for industrial wastewater treatment. The modules are connected in a series arrangement within a high-pressure vessel. Different industrial wastewaters discharged were used as feed streams of this hybrid system. Results from their study showed that optimal hybrid system can reduce 34.8% of CO_2 of its life cycle per kg permeate generated. Also, a simple case study on desalination was

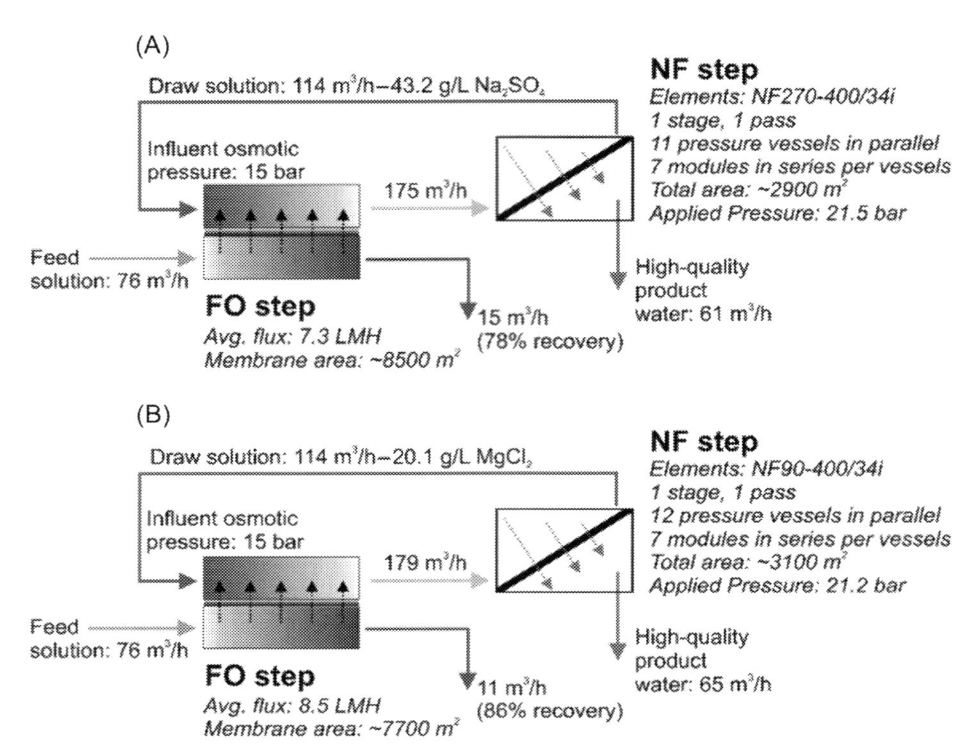

FIGURE 11–4 Configurations of the FO–NF hybrid system designed to treat real wastewater with (A) Na_2SO_4 and (B) $MgCl_2$ as the draw solute [71]. *FO*, Forward osmosis; *NF*, nanofiltration.

considered in this work, and the optimization results showed that the optimal RO–PRO hybrid system was highly useful for identifying energy-efficient designs for reducing the SEC.

11.4.3.2 Ultrafiltration–pressure-retarded osmosis, nanofiltration–pressure-retarded osmosis

The main challenge in PRO, membrane fouling, was tackled by applying UF and a sand filter to remove total organic matter (TOM), turbidity, and hardness. In trials, efficiency of the two methods was compared. Maximum TOM removal in UF and sand filter was 48.8% and 40.3% at 10.44 mg/L TOM initial concentration respectively. In all experiments, it was indicated that UF had better removal efficiency and consequently better performance in osmotic power generation process improvement.

Yang et al. [74] highlighted the system integration to mitigate fouling on the PRO and decrease the energy consumption. Wastewater was pretreated with three different modules, UF, NF, and LPRO before being used as FSs in the subsequent PRO tests. PRO process using NF and LPRO filtrates produced such a stable power density of 7.3 and 8.4 W/m^2, respectively. This study hence recommended NF as an optimal pretreatment method because NF removes most foulants and produce higher permeability than LPRO, while UF pretreatment was not helpful in fouling mitigation. Previous study by Chen et al. [75] used to pretreat the wastewater RO brine prior to feeding to the PRO process. Even though silica could not be effectively removed, low-pressure NF was able to mitigate the fouling potential from multivalent ions and organic matters. The effectiveness of NF (hollow fiber membrane module) as pretreatment was proven through the increased power density from 4.4 to 13.5 W/m^2 as well as high PRO water flux up to 30.5 LMH.

11.5 Future outlook and conclusion

Living ecosystems require water. The future resource availability is, however, highly contested, mostly due to global development that intersects dangerously with resource availability. Therefore, further research will always be needed to identify and evaluate the impacts of alternative paths toward this future, and the trade-offs that will be inevitable given our multiple, and not always compatible, dreams or goals. But clearly, the innovations on membrane-based technology play a significant role in determining water future resource supply and demand. Thanks to superior integration, membrane-based separation technology for desalination and wastewater has entered a booming phase, where the average number of publications per year keeps increasing. Such a large jump in publication number could be associated with the diversification of research, that is, integrated osmosis. Integration of osmosis has high potential to address various environmental problems as well as offer leapfrogging opportunities for solving water crisis around the world.

Based on the unique advantages of osmosis, that is, RO, FO, and PRO coupled with each other, it is important to exploit these to solve the challenges of wastewater remediation and even desalination. This integration is not smooth, however. Significant deficit spending is required to drive investment in energy storage and renewable sources in many countries. In regard to the industrial applications only the upscaling on pilot or full-scale will be the next step to optimize the operation and implement the optimal integration in industrial water and wastewater treatment. To further promote osmosis in industries, enhancements in different viewpoints and more research need to be done.

Acknowledgments

The authors would like to acknowledge the financial supports provided by Univerisiti Teknologi Malaysia under Research University Grant (18H35) and the Ministry of Education under FRGS (5F005).

References

[1] W.L. Ang, A. Wahab Mohammad, D. Johnson, N. Hilal, Forward osmosis research trends in desalination and wastewater treatment: a review of research trends over the past decade, J. Water Process. Eng. 31 (2019) 100886. <https://doi.org/10.1016/j.jwpe.2019.100886>.

[2] P.S. Goh, A.F. Ismail, B.C. Ng, M.S. Abdullah, Recent progresses of forward osmosis membranes formulation and design for wastewater treatment, Water (Switz.) 11 (2019). <https://doi.org/10.3390/w11102043>.

[3] G. Han, J. Zuo, C. Wan, T.S. Chung, Hybrid pressure retarded osmosis-membrane distillation (PRO-MD) process for osmotic power and clean water generation, Environ. Sci. Water Res. Technol. 1 (2015) 507–515. <https://doi.org/10.1039/c5ew00127g>.

[4] J. Kim, M. Park, H.K. Shon, J.H. Kim, Performance analysis of reverse osmosis, membrane distillation, and pressure-retarded osmosis hybrid processes, Desalination 380 (2016) 85–92. <https://doi.org/10.1016/j.desal.2015.11.019>.

[5] A. Altaee, A. Sharif, G. Zaragoza, A.F. Ismail, Evaluation of FO-RO and PRO-RO designs for power generation and seawater desalination using impaired water feeds, Desalination 368 (2015) 27–35. <https://doi.org/10.1016/j.desal.2014.06.022>.

[6] S.S. Shenvi, A.M. Isloor, A.F. Ismail, A review on RO membrane technology: developments and challenges, Desalination 368 (2015) 10–26. <https://doi.org/10.1016/J.DESAL.2014.12.042>.

[7] L. Chekli, S. Phuntsho, J.E. Kim, J. Kim, J.Y. Choi, J.S. Choi, et al., A comprehensive review of hybrid forward osmosis systems: performance, applications and future prospects, J. Membr. Sci. 497 (2016) 430–449. <https://doi.org/10.1016/j.memsci.2015.09.041>.

[8] L. Malaeb, G.M. Ayoub, Reverse osmosis technology for water treatment: state of the art review, Desalination 267 (2011) 1–8. <https://doi.org/10.1016/J.DESAL.2010.09.001>.

[9] P.S. Goh, A.F. Ismail, A review on inorganic membranes for desalination and wastewater treatment, Desalination 434 (2018) 60–80. <https://doi.org/10.1016/j.desal.2017.07.023>.

[10] X. Song, L. Wang, C.Y. Tang, Z. Wang, C. Gao, Fabrication of carbon nanotubes incorporated double-skinned thin film nanocomposite membranes for enhanced separation performance and antifouling capability in forward osmosis process, Desalination 369 (2015) 1–9. <https://doi.org/10.1016/j.desal.2015.04.020>.

[11] S. Zhao, L. Zou, C.Y. Tang, D. Mulcahy, Recent developments in forward osmosis: opportunities and challenges, J. Membr. Sci. 396 (2012) 1–21. <https://doi.org/10.1016/j.memsci.2011.12.023>.

[12] F. Garcia, D. Ciceron, A. Saboni, S. Alexandrova, Nitrate ions elimination from drinking water by nanofiltration: membrane choice, Sep. Purif. Technol. 52 (2006) 196–200. <https://doi.org/10.1016/j.seppur.2006.03.023>.

[13] K.Y. Wang, M.M. Teoh, A. Nugroho, T.S. Chung, Integrated forward osmosis-membrane distillation (FO-MD) hybrid system for the concentration of protein solutions, Chem. Eng. Sci. 66 (2011) 2421–2430. <https://doi.org/10.1016/j.ces.2011.03.001>.

[14] T. Matsuura, Progress in membrane scince and technology for seawater desalination—a review, Desalination 134 (2001) 47–54. <https://doi.org/10.1016/S0011-9164(01)00114-X>.

[15] C.A. Quist-Jensen, F. Macedonio, E. Drioli, Membrane technology for water production in agriculture: desalination and wastewater reuse, Desalination 364 (2015) 17–32. <https://doi.org/10.1016/j.desal.2015.03.001>.

[16] P.S. Goh, T. Matsuura, A.F. Ismail, N. Hilal, Recent trends in membranes and membrane processes for desalination, Desalination 391 (2016) 43–60. <https://doi.org/10.1016/j.desal.2015.12.016>.

[17] J.L. Prante, J.A. Ruskowitz, A.E. Childress, A. Achilli, RO-PRO desalination: an integrated low-energy approach to seawater desalination, Appl. Energy. 120 (2014) 104–114. <https://doi.org/10.1016/j.apenergy.2014.01.013>.

[18] S. Ho Chae, Y.M. Kim, H. Park, J. Seo, S.J. Lim, J.H. Kim, Modeling and simulation studies analyzing the pressure-retarded osmosis (PRO) and pro-hybridized processes, Energies 12 (2019). <https://doi.org/10.3390/en12020243>.

[19] J. Li, D. Hou, K. Li, Y. Zhang, J. Wang, X. Zhang, Domestic wastewater treatment by forward osmosis-membrane distillation (FO-MD) integrated system, Water Sci. Technol. 77 (2018) 1514–1523. <https://doi.org/10.2166/wst.2018.031>.

[20] Y. Choi, Y. Shin, H. Cho, Y. Jang, T.M. Hwang, S. Lee, Economic evaluation of the reverse osmosis and pressure retarded osmosis hybrid desalination process, Desalin. Water Treat. 57 (2016) 26680–26691. <https://doi.org/10.1080/19443994.2016.1190114>.

[21] G.P. Syed Ibrahim, A.M. Isloor, E. Yuliwati, A review: desalination by forward osmosis, Current Trends and Future Developments on (Bio-)Membranes, Elsevier, 2019, pp. 199–214. <https://doi.org/10.1016/B978-0-12-813551-8.00008-5>.

[22] N.J. Falizi, M.C. Hacıfazlıoğlu, İ. Parlar, N. Kabay, T.Ö. Pek, M. Yüksel, Evaluation of MBR treated industrial wastewater quality before and after desalination by NF and RO processes for agricultural reuse, J. Water Process. Eng. 22 (2018) 103–108. <https://doi.org/10.1016/J.JWPE.2018.01.015>.

[23] D.L. Shaffer, N.Y. Yip, J. Gilron, M. Elimelech, Seawater desalination for agriculture by integrated forward and reverse osmosis: improved product water quality for potentially less energy, J. Membr. Sci. 415–416 (2012) 1–8. <https://doi.org/10.1016/j.memsci.2012.05.016>.

[24] I. Petrinic, J. Korenak, D. Povodnik, C. Hélix-Nielsen, A feasibility study of ultrafiltration/reverse osmosis (UF/RO)-based wastewater treatment and reuse in the metal finishing industry, J. Cleaner Prod. 101 (2015) 292–300. <https://doi.org/10.1016/j.jclepro.2015.04.022>.

[25] M. Gündoğdu, Y.A. Jarma, N. Kabay, T. Pek, M. Yüksel, Integration of MBR with NF/RO processes for industrial wastewater reclamation and water reuse-effect of membrane type on product water quality, J. Water Process. Eng. 29 (2019). <https://doi.org/10.1016/j.jwpe.2018.02.009>.

[26] X. An, Y. Hu, N. Wang, Z. Zhou, Z. Liu, Continuous juice concentration by integrating forward osmosis with membrane distillation using potassium sorbate preservative as a draw solute, J. Membr. Sci. 573 (2019) 192–199. <https://doi.org/10.1016/j.memsci.2018.12.010>.

[27] B. Aftab, Y.S. Ok, J. Cho, J. Hur, Targeted removal of organic foulants in landfill leachate in forward osmosis system integrated with biochar/activated carbon treatment, Water Res. 160 (2019) 217–227. <https://doi.org/10.1016/j.watres.2019.05.076>.

[28] M. Gökçek, Integration of hybrid power (wind-photovoltaic-diesel-battery) and seawater reverse osmosis systems for small-scale desalination applications, Desalination 435 (2018) 210–220. <https://doi.org/10.1016/j.desal.2017.07.006>.

[29] J. Choi, S.J. Im, A. Jang, Application of volume retarded osmosis – low pressure membrane hybrid process for recovery of heavy metals in acid mine drainage, Chemosphere. 232 (2019) 264–272. <https://doi.org/10.1016/j.chemosphere.2019.05.209>.

[30] M. Sarai Atab, A.J. Smallbone, A.P. Roskilly, A hybrid reverse osmosis/adsorption desalination plant for irrigation and drinking water, Desalination 444 (2018) 44–52. <https://doi.org/10.1016/j.desal.2018.07.008>.

[31] V.K. Gupta, I. Tyagi, H. Sadegh, R. Shahryari-Ghoshekandi, A.S.H. Makhlouf, B. Maazinejad, Nanoparticles as adsorbent; a positive approach for removal of noxious metal ions: a review, Sci. Technol. Dev. 34 (2015) 195–214. <https://doi.org/10.3923/std.2015.195.214>.

[32] H. Sadegh, G.A.M. Ali, V.K. Gupta, A.S.H. Makhlouf, R. Shahryari-ghoshekandi, M.N. Nadagouda, et al., The role of nanomaterials as effective adsorbents and their applications in wastewater treatment, J. Nanostruct. Chem. 7 (2017) 1–14. <https://doi.org/10.1007/s40097-017-0219-4>.

[33] E.S. Ali, A.S. Alsaman, K. Harby, A.A. Askalany, M.R. Diab, S.M. Ebrahim, Yakoot, Recycling brine water of reverse osmosis desalination employing adsorption desalination: a theoretical simulation, Desalination 408 (2017) 13–24. <https://doi.org/10.1016/j.desal.2016.12.002>.

[34] E. Laskowska, M. Turek, K. Mitko, P. Dydo, Concentration of mine saline water in high-efficiency hybrid RO–NF system, Desalin. Water Treat. 128 (2018) 414–420. <https://doi.org/10.5004/dwt.2018.22877>.

[35] R. Valavala, J. Sohn, J. Han, N. Her, Y. Yoon, Pretreatment in reverse osmosis seawater desalination: a short review, Eng. Res. 16 (2011) 205–212. <https://doi.org/10.4491/eer.2011.16.4.205>.

[36] W.L. Ang, A.W. Mohammad, N. Hilal, C.P. Leo, A review on the applicability of integrated/hybrid membrane processes in water treatment and desalination plants, Desalination 363 (2015) 2–18. <https://doi.org/10.1016/j.desal.2014.03.008>.

[37] M. Clever, F. Jordt, R. Knauf, N. Räbiger, M. Ruedebusch, R. Hilker-Scheibel, Process water production from river water by ultrafiltration and reverse osmosis, Desalination 131 (2000) 325–336. <https://doi.org/10.1016/S0011-9164(00)90031-6>.

[38] A.W. Zularisam, A.F. Ismail, M. Sakinah, Application and challenges of membrane in surface water treatment, J. Appl. Sci. 10 (2010) 380–390. <https://doi.org/10.3923/jas.2010.380.390>.

[39] J.J. Qin, H.O. Maung, H. Lee, R. Kolkman, Dead-end ultrafiltration for pretreatment of RO in reclamation of municipal wastewater effluent, J. Membr. Sci. 243 (2004) 107–113. <https://doi.org/10.1016/j.memsci.2004.06.010>.

[40] O. Lorain, B. Hersant, F. Persin, A. Grasmick, N. Brunard, J.M. Espenan, Ultrafiltration membrane pretreatment benefits for reverse osmosis process in seawater desalting. Quantification in terms of capital investment cost and operating cost reduction, Desalination 203 (2007) 277–285. <https://doi.org/10.1016/j.desal.2006.02.022>.

[41] F. Knops, S. van Hoof, H. Futselaar, L. Broens, Economic evaluation of a new ultrafiltration membrane for pretreatment of seawater reverse osmosis, Desalination 203 (2007) 300–306. <https://doi.org/10.1016/j.desal.2006.04.013>.

[42] L.X. Gao, A. Rahardianto, H. Gu, P.D. Christofides, Y. Cohen, Novel design and operational control of integrated ultrafiltration — Reverse osmosis system with RO concentrate backwash, Desalination 382 (2016) 43–52.

[43] M. Talaeipour, J. Nouri, A.H. Hassani, A.H. Mahvi, An investigation of desalination by nanofiltration, reverse osmosis and integrated (hybrid NF/RO) membranes employed in brackish water treatment, J. Environ. Heal. Sci. Eng. 15 (2017) 1–9. <https://doi.org/10.1186/s40201-017-0279-x>.

[44] C. Kaya, Y.A. Jarma, A.M. Muhidin, E. Güler, N. Kabay, M. Arda, Seawater desalination by using nanofiltration (NF) and brackish water reverse osmosis (BWRO) membranes in sequential mode of operation graphical abstract keywords, J. Membr. Sci. Res. 6 (2020) 40–46. <https://doi.org/10.22079/JMSR.2019.107844.1264>.

[45] W. He, Y. Wang, V. Elyasigomari, M.H. Shaheed, Evaluation of the detrimental effects in osmotic power assisted reverse osmosis (RO) desalination, Renew. Energy 93 (2016) 608–619. <https://doi.org/10.1016/j.renene.2016.02.067>.

[46] M. Li, Optimization of multi-stage hybrid RO-PRO membrane processes at the water–energy nexus, Chem. Eng. Res. Des. 137 (2018) 1–9. <https://doi.org/10.1016/j.cherd.2018.06.042>.

[47] C.F. Wan, T.S. Chung, Energy recovery by pressure retarded osmosis (PRO) in SWRO-PRO integrated processes, Appl. Energy. 162 (2016) 687–698. <https://doi.org/10.1016/j.apenergy.2015.10.067>.

[48] C.F. Wan, T.S. Chung, Techno-economic evaluation of various RO + PRO and RO + FO integrated processes, Appl. Energy. 212 (2018) 1038–1050. <https://doi.org/10.1016/j.apenergy.2017.12.124>.

[49] F. Volpin, E. Fons, L. Chekli, J.E. Kim, A. Jang, H.K. Shon, Hybrid forward osmosis-reverse osmosis for wastewater reuse and seawater desalination: understanding the optimal feed solution to minimise fouling, Process. Saf. Environ. Prot. 117 (2018) 523–532. <https://doi.org/10.1016/J.PSEP.2018.05.006>.

[50] A.C. Mecha, Applications of reverse and forward osmosis processes in wastewater treatment: evaluation of membrane fouling, Osmotically Driven Membrane Processes - Approach, Development and Current Status, IntechOpen, 2018. <https://doi.org/10.5772/intechopen.72971>.

[51] J. Seo, Y.M. Kim, S.H. Chae, S.J. Lim, H. Park, J.H. Kim, An optimization strategy for a forward osmosis-reverse osmosis hybrid process for wastewater reuse and seawater desalination: a modeling study, Desalination 463 (2019) 40–49. <https://doi.org/10.1016/j.desal.2019.03.012>.

[52] Y. Choi, H. Cho, Y. Shin, Y. Jang, S. Lee, Economic evaluation of a hybrid desalination system combining forward and reverse osmosis, Membr. (Basel) 6 (2015). <https://doi.org/10.3390/membranes6010003>.

[53] J. Yaeli, Method and Apparatus for Processing Liquid Solutions of Suspensions Particularly Useful in the Desalination of Saline Water, Google Patents (1992).

[54] V. Yangali-Quintanilla, Z. Li, R. Valladares, Q. Li, G. Amy, Indirect desalination of Red Sea water with forward osmosis and low pressure reverse osmosis for water reuse, Desalination 280 (2011) 160–166. <https://doi.org/10.1016/j.desal.2011.06.066>.

[55] T.Y. Cath, J.E. Drewes, C.D. Lundin, N.T. Hancock, Forward osmosis—Reverse osmosis process offers a novel hybrid solution for water purification and reuse, IDA J. Desalination Water Reuse 2 (2013) 16–20.

[56] F. Volpin, L. Chekli, S. Phuntsho, N. Ghaffour, J.S. Vrouwenvelder, H.K. Shon, Optimisation of a forward osmosis and membrane distillation hybrid system for the treatment of source-separated urine, Sep. Purif. Technol. 212 (2019) 368–375. <https://doi.org/10.1016/J.SEPPUR.2018.11.003>.

[57] T. Husnain, Y. Liu, R. Riffat, B. Mi, Integration of forward osmosis and membrane distillation for sustainable wastewater reuse, Sep. Purif. Technol. 156 (2015) 424–431. <https://doi.org/10.1016/j.seppur.2015.10.031>.

[58] S. Zhang, P. Wang, X. Fu, T.S. Chung, Sustainable water recovery from oily wastewater via forward osmosis-membrane distillation (FO-MD), Water Res. 52 (2014) 112–121. <https://doi.org/10.1016/j.watres.2013.12.044>.

[59] M.M. Ling, T.-S. Chung, Surface-dissociated nanoparticle draw solutions in forward osmosis and the regeneration in an integrated electric field and nanofiltration system, Ind. Eng. Chem. Res. 51 (2012) 15463–15471.

[60] S. Dutta, K. Nath, Dewatering of brackish water and wastewater by an integrated forward osmosis and nanofiltration system for direct fertigation, Arab. J. Sci. Eng. 44 (2019) 9977–9986. <https://doi.org/10.1007/s13369-019-04102-3>.

[61] D.I. Kim, J. Kim, H.K. Shon, S. Hong, Pressure retarded osmosis (PRO) for integrating seawater desalination and wastewater reclamation: energy consumption and fouling, J. Membr. Sci. 483 (2015) 34–41. <https://doi.org/10.1016/j.memsci.2015.02.025>.

[62] M.R. Rahimpour, S. Mohsenpour, Integrating pressure-retarded osmosis and membrane distillation, Current Trends and Future Developments on (Bio-)Membranes, Elsevier, 2019, pp. 385–402. <https://doi.org/10.1016/b978-0-12-813545-7.00015-5>.

[63] M. Farrokh Shad, G.J.G. Juby, S. Delagah, M. Sharbatmaleki, M. Farrokh Shad, M. Sharbatmaleki, Evaluating occurrence of contaminants of emerging concerns in MF/RO treatment of primary effluent for water reuse—pilot study, J. Water Reuse Desalin. 9 (2019) 350–371. <https://doi.org/10.2166/wrd.2019.004>.

[64] X. Sun, C. Wang, Y. Li, W. Wang, J. Wei, Treatment of phenolic wastewater by combined UF and NF/RO processes, Desalination 355 (2015) 68–74. <https://doi.org/10.1016/j.desal.2014.10.018>.

[65] Q. Long, Y. Jia, J. Li, J. Yang, F. Liu, J. Zheng, et al., Recent advance on draw solutes development in forward osmosis, Processes 6 (2018) 165. <https://doi.org/10.3390/pr6090165>.

[66] R. Thiruvenkatachari, M. Francis, M. Cunnington, S. Su, Application of integrated forward and reverse osmosis for coal mine wastewater desalination, Sep. Purif. Technol. 163 (2016) 181–188. <https://doi.org/10.1016/j.seppur.2016.02.034>.

[67] G. Blandin, A.R.D. Verliefde, J. Comas, I. Rodriguez-Roda, P. Le-Clech, M. Stoller, et al., Efficiently combining water reuse and desalination through forward osmosis-reverse osmosis (FO-RO) hybrids: a critical review, Membranes 6 (2016) 37. <https://doi.org/10.3390/membranes6030037>.

[68] M. Al-Furaiji, N. Benes, A. Nijmeijer, J.R. McCutcheon, Use of a forward osmosis-membrane distillation integrated process in the treatment of high-salinity oily wastewater, Ind. Eng. Chem. Res. 58 (2019) 956–962. <https://doi.org/10.1021/acs.iecr.8b04875>.

[69] H. Song, F. Xie, W. Chen, J. Liu, FO/MD hybrid system for real dairy wastewater recycling, Environ. Technol. (UK) 39 (2018) 2411–2421. <https://doi.org/10.1080/09593330.2017.1377771>.

[70] Y. Zhou, M. Huang, Q. Deng, T. Cai, Combination and performance of forward osmosis and membrane distillation (FO-MD) for treatment of high salinity landfill leachate, Desalination 420 (2017) 99–105. <https://doi.org/10.1016/j.desal.2017.06.027>.

[71] M. Giagnorio, F. Ricceri, M. Tagliabue, L. Zaninetta, A. Tiraferri, Hybrid forward osmosis-nanofiltration for wastewater reuse: system design, Membranes 9 (2019) 61. <https://doi.org/10.3390/membranes9050061>.

[72] P. Pal, M. Sardar, M. Pal, S. Chakrabortty, J. Nayak, Modelling forward osmosis-nanofiltration integrated process for treatment and recirculation of leather industry wastewater, Comput. Chem. Eng. 127 (2019) 99–110. <https://doi.org/10.1016/j.compchemeng.2019.05.018>.

[73] C. He, Q. Zhu, B. Zhang, Q. Chen, M. Pan, Sustainable design and synthesis of hybrid RO-PRO systems for industrial wastewater treatment, Chem. Eng. Trans. 61 (2017) 817–822. <https://doi.org/10.3303/CET1761134>.

[74] T. Yang, C.F. Wan, J.Y. Xiong, T.S. Chung, Pre-treatment of wastewater retentate to mitigate fouling on the pressure retarded osmosis (PRO) process, Sep. Purif. Technol. 215 (2019) 390–397. <https://doi.org/10.1016/j.seppur.2019.01.032>.

[75] Y. Chen, C. Liu, L. Setiawan, Y.N. Wang, X. Hu, R. Wang, Enhancing pressure retarded osmosis performance with low-pressure nanofiltration pretreatment: membrane fouling analysis and mitigation, J. Membr. Sci. 543 (2017) 114–122. <https://doi.org/10.1016/j.memsci.2017.08.051>.

12 Development and implementations of integrated osmosis system

Saber Abdulhamid Alftessi, Twibi Mohamed, Jeganes Ravi, Siti Khadijah Hubadillah, Mohd Hafiz Dzarfan Othman, Ahmad Fauzi Ismail

ADVANCED MEMBRANE TECHNOLOGY RESEARCH CENTER, SCHOOL OF CHEMICAL AND ENERGY ENGINEERING, FACULTY OF ENGINEERING, UNIVERSITI TEKNOLOGI MALAYSIA, JOHOR, MALAYSIA

12.1 Introduction

The increasing demand for water and the limited availability of water sources have driven efforts toward desalination as an essential process for water production. Being that water scarcity is a global risk that is expected to increase in the next century, it has become necessary to develop techniques that are cost-effective and energy-efficient for pure water production from saline sources (such as brackish water or seawater). Attention has been given to the development and implementations of integrated osmosis systems (IOSs) that consist of several individual water treatment processes, such as reverse osmosis (RO), forward osmosis (FO), and membrane distillation (MD) in various combinations. This effort is geared toward reducing the cost and time of water treatment process. This will be detailed in this chapter.

RO is a separation process for the separation of a solvent from a feed solution. As an alternative membrane-based separation process, the RO process is based on the variation in osmotic pressure, where a concentrated solution that is under a high osmotic pressure draws water from a feed solution through a semipermeable membrane (SPM) [1]. Freshwater can be collected through the regeneration of the diluted drawn solution. The RO separation mechanism utilizes a SPM for clean water separation from a contaminated feed solution. The process involves the movement of the solvent from the less concentrated compartment to the higher concentrated compartment via a membrane with the aim of achieving a concentration equilibrium on both sides, as shown in Fig. 12–1. The introduction of an opposing hydrostatic pressure to the higher concentrated solution can hamper the transportation of the solvent. The magnitude of this hydrostatic pressure that is required to completely stop the flow of the solvent is referred to as the osmotic pressure [2]. RO is the separation of dissolvable solids from water by applying a pressure differential over a membrane layer that is penetrable to water but not to the dissolvable solids [3]. As it is so

Osmosis Engineering. DOI: https://doi.org/10.1016/B978-0-12-821016-1.00007-3

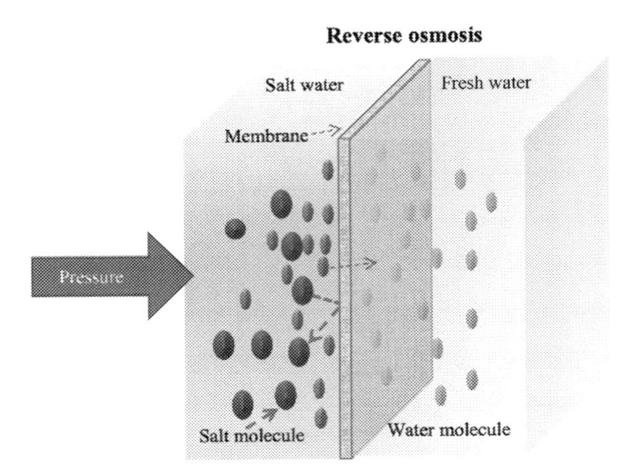

FIGURE 12–1 Illustration of the RO process. *RO*, Reverse osmosis.

appropriately named, this procedure is the precise inverse of the common osmosis phenomenon. In osmosis, the movement of solvents is from a region of low salt concentration to that of higher salt concentration via a SPM without any external impact. This stream flows until the internal pressure differentials are equivalent and thereby, making a zero-pressure differential and stopping the water flow. The water-powered hydraulic pressure is applied to the more concentrated solvent which causes water molecules to pass through a semiporous film to the less concentrated solution.

In FO, the driving force is generated by the difference in the pressure between the highly concentrated solution and the feed solution as shown in Fig. 12–2. As for RO, the separation process is propelled by the hydraulic pressure; hence, the energy required in the RO process is higher than that in FO [4]. Furthermore, the diffusion of the solute in FO is a two-dimensional movement that is dependent on the feedwater composition and that of the draw solution as well; this indicates the possibility of the draw solution diffusing into the feed solution and vice versa [5]. FO has been investigated several times for application in several areas, such as in power generation, food processing, desalination, water purification, and in the pharmaceutical sector [6]. One of the fundamental points of interest of feed-FO is the constrained measure of the required external energy to get water from the feed solution, using a low amount of energy to recycle the solution on one side of the film.

MD is another potential technology for saline water desalination. As a heat-driven process, MD is a microfiltration process that allows only the passage of vaporized molecules through a permeable hydrophobic membrane and the separation process is propelled by the variation in the vapor pressure between the surfaces of the permeable hydrophobic membrane. There are many attractive benefits of using MD as a separation process; these include its operation at low temperatures compared to the required temperature in the

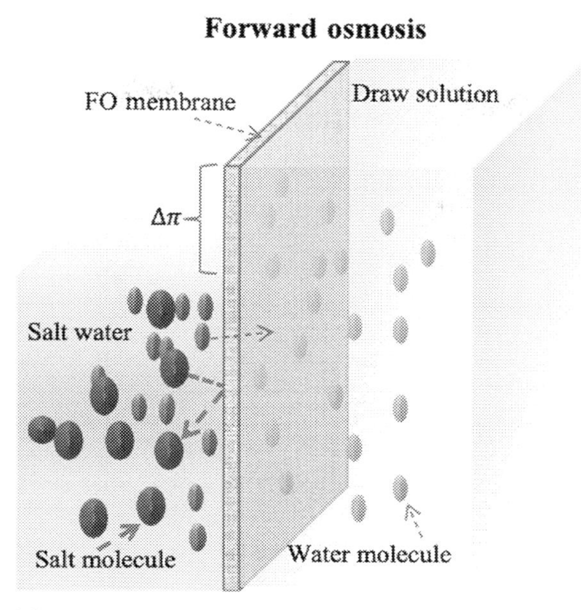

FIGURE 12–2 Illustration of the FO process. *FO*, Forward osmosis.

other traditional processes; MD processes do not require heating the solution (mainly water) to the boiling point. Furthermore, MD processes require lower hydrostatic pressure compared to the pressure needed in most of the pressure-driven membrane-based separation processes, such as RO. Hence, MD is presumably a cost-effective separation process that does not require most of the demanding membrane characteristics. As such, MD processes require less expensive resources such as plastic, thereby reducing the corrosion-related issues. Based on the vapor-liquid equilibrium principle, the rejection factor in MD processes is relatively high and as such, complete separation takes place theoretically. Additionally, MD processes require a relatively larger membrane pore size compared to that of RO. Therefore MD separation processes are less prone to fouling. An integrated separation system can be easily created by merging the MD system with other separation processes like ultrafiltration or RO [7,8]. Furthermore, MD processes can be performed using alternative power sources like solar energy [9]. The MD process is mainly applicable in seawater and brackish water desalination; it is equally appealing for heavy and organic metals removal from wastewater. It is also applicable to the treatment of radioactive materials before they can be safely discarded [10]. However, there are certain problems with MD processes; for instance, they are associated with low permeate flux when compared to RO. Another problem is the higher level of mass transfer resistance owing to air entrapment within the membrane, thereby limiting the permeate flux of MD processes. There is also a high level of conduction-related heat loss. Table 12–1 summarizes the comparison between RO, FO, and MD.

Table 12–1 Comparison among different desalination processes.

Process	Advantages	Drawbacks	Future improvement	References
RO	Good for seawater desalination Requires low energy compared to thermal processes; the operating cost and capital are lower compared to thermal processes; the water recovery rate is high The cost of the membrane is low, and the rate of thermal pollution is also low Mega-sized commercial plants can be operated with lower operating costs due to the economy of scale. Demand-based production is achievable	It requires external pressure to operate The maximum level of total dissolved solids (TDS) in the feed solution must not exceed 75,000 mg/L It is prone to membrane fouling and scaling. There is bound to membrane-related concentration polarization The discharged brine from the process is contaminated with process-related chemicals	Improvement of the process membrane; fabrication of new membrane materials That can be applicable in merged separation systems	[11–14]
FO	Separation is propelled by natural differences in osmotic pressure	Uses a concentrated draw solution	Emerging technology; pilot and commercial	[12,15,16]
MD	The feedwater has a high TDS of >200,000 mg/L Membrane fouling is less compared to RO Can be powered by renewable energy sources The rejection rate and nonvolatile compounds removal are high Portable desalination plants	Low water flux and recovery rate Potential membrane wetting Lack of specific MD membranes The module is more complex than RO	Emerging technology Membrane engineering Pilot and commercial plants Role in hybrid systems	[15,17–19]

Note: *RO*, Reverse osmosis; *MD*, membrane distillation.

12.2 Development of IOS

There are three common IOSs that are currently being investigated which are ROσFO, RO–MD, and FO–MD. In this section, the development of these integrated systems will be discussed.

12.2.1 Reverse osmosis–forward osmosis

RO is arguably the commonest technology used for clean water production from diluted draw solutions [20]. However, the integration of FO and RO is appealing, especially for open-loop desalination of seawater (once-through systems) in which seawater serves as the draw solution for subsequent clean water production from diluted seawater via RO [21]. It should also be noted that the FO–RO system is not favored thermodynamically in comparison to sewage direct RO because diluted seawater has a higher osmotic pressure compared to municipal sewage [22]. However, the advantages of the FO–RO system include its double barrier to pollutants, as well as the lower chances of fouling in the membranes. It is also beneficial because it requires less energy and lower osmotic potential for clean water production from diluted seawater compared to normal seawater [23].

12.2.2 Reverse osmosis–membrane distillation

RO is among the separation technologies that can be merged with MD; both processes can be integrated into any of the following two ways: (1) the RO brine can be used as feed to the MD, (2) the RO permeate can be used as feed to the MD. There is a great potential of utilizing RO brine as the feed for MD processes; it addresses the salt concentration-related effect on the MD process by regulating the maximum level of salt concentration of RO to about 70,000 mg/L. The RO–MD process is typically needed for increased water recovery in applications where brine disposal is a serious issue. The viability of applying RO–MD processes in groundwater concentrates has also been revealed, but the concept is prone to certain practical issues, such as membrane scaling [24].

12.2.3 Forward osmosis–membrane distillation

A separation technique that combines FO with MD can be appealing for the desalination of hypersaline-produced water as FO can efficiently handle extremely saline solutions with certain levels of hydrocarbons while MD is suitable for regenerating the draw solution. The efficiency of MD in treating nearly saturated solutions has been demonstrated [25]. A study by Zhang et al. [63] focused on the treatment of water using the FO–MD process. The feedwater in this study was synthetically produced with approximately 12,000 ppm salinity and oil content of 4000 ppm; the study also utilized a draw solution containing 5 M NaCl. From the results, FO and MD achieved $>$99.90% salt and oil rejection. Furthermore, the combined FO–MD technique has been studied for application in other areas, such as in the reclamation of wastewater [26], RO brine treatment [24], sewer mining [27], dye wastewater treatment [28], and the concentration of protein solutions [29].

12.3 Implementation of IOS

12.3.1 Integrated FO–RO system

FO is propelled by the variation in the osmotic pressure between a highly concentrated draw solution and the feed solution (such as seawater). This osmotic pressure difference is required for the transportation of the feed content into the draw solution through a salt-filtering barrier. In RO, the propelling force for the mass transport via the membrane is the pressure externally applied to the feed solution. However, FO relies on the difference in the osmotic pressure between the higher concentrated draw solution and the feed solution for mass transport [6]. For FO systems, the draw solution is normally made up of pure water and a draw solute that was selected based on its capability to establish high osmotic pressure, limit reverse flux of the solute from the membrane into the feed solution, as well as lower the chances of membrane scaling and fouling [30]. A continuous FO process generates a draw solution that can be diluted via the extraction of water from the feed, and this dilute draw solution can be regenerated into the expected product while the draw solution is recyclable in the FO unit.

Fig. 12–3 illustrates a merged FO–RO process for the desalination of seawater in which FO serves as the pretreatment phase for the RO that pervades the product water and recreates the draw solution for further recycling in the FO process. The FO process guarantees a lower chance of irreversible fouling compared to RO [31]. Therefore the FO unit can serve as the pretreatment stage for the subsequent RO unit as it requires only occasional osmotic backwashing or physical scouring for restoring the lost flux to fouling. An integrated FO–RO process can achieve low chloride and boron concentrations without needing any further posttreatment RO passes as the treatment scheme provides that the treated seawater must pass through two selective membranes.

Several studies have recently focused on the possibility of achieving an integrated FO–RO process, and the modeling of such system for seawater desalination has been performed based on the lab-scale data for an FO unit whereby seawater served as the feedwater while the draw solution is the concentrated sodium chloride solution. From these modeling and lab-scale experiments, the feasibility of an integrated FO–RO system as a desalination process has been proven, with possible benefits of lower energy input during the pretreatment step, easy pretreatment with less chemical use, as well as reduced chances of fouling, coupled with a prolonged service life of the subsequent RO process [33,34]. A pilot study targeted at an integrated FO–RO process for osmotic dilution, and wastewater treatment has also been reported.

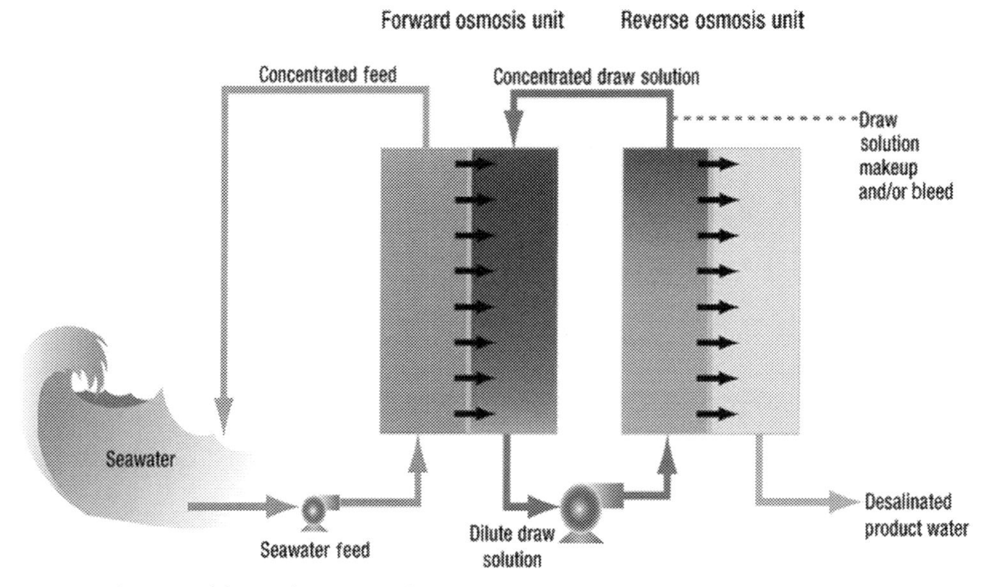

FIGURE 12–3 An integrated forward osmosis and reverse osmosis desalination process [32].

The sustained operation of an integrated FO–RO system has also been associated with certain challenges and the most significant of such challenges is the reverse permeation of the FO membrane by the concentrated draw solute into the less concentrated draw solution; this permeation decreases the concentration of the draw solution, as well as the osmotic pressure that propels the FO process, leading to the requirement for additional draw solute in order to sustain the systems' performance [33,35]. Fig. 12–1 illustrates a flow stream for the additional draw solution required for the sustenance of the osmotic pressure. Table 12–2 shows some of the draw solutions and recent applications of the FO–RO system.

An integrated FO–RO desalination process is more energy-efficient compared to the traditional two-pass seawater reverse osmosis (SWRO) process as depicted in Fig. 12–4. This is attributed to the pretreatment achieved from the RO process and the posttreatment achieved from the removal of chloride and boron in a single process. In the combined FO–RO system, the FO membranes replace the traditional pretreatment process of using coagulation, ultrafiltration, or granular media filtration, while the second-stage RO posttreatment is completely relegated.

Furthermore, the combined FO–RO process has other potential benefits when matched with the traditional SWRO; some of these benefits are as follows:

1. Reduced usage of feed systems and chemical storage, thereby saving capital, operation, and maintenance costs.
2. Improved water quality and reduced piping process costs.
3. Improved overall desalination process sustainability.

Table 12–2 Various applications with different draw solution of the FO–RO system.

Application	FO feed solution	FO draw solution	FO flux L/m/h^{-2}	RO rejection %	References
Wastewater treatment	Coal mine impacted water	Sodium hexametaphosphate (NaPO)	2.4	75.4	[36]
Wastewater treatment	Coal mine impacted water	Sodium lignosulfonate	4.0	96.3	[36]
Wastewater treatment	Coal mine impacted water	Disodium hydrogen phosphate Na_2HPO_4	3.3	85.39	[36]
Wastewater treatment	Sewage effluent	Engineered fertilizing solutions	13.2	99	[37]
Whey wastewater	Cheese whey	NaCl	10.6	96.4	[38]
Brackish water desalination	El-Salam Canal Water	NaCl	3.3	100	[39,40]
Wastewater treatment	Sewage effluent	NaCl	11.8	97	[37]
Wastewater treatment	Municipal wastewater effluent	Seawater	10	Not reported	[41]
Wastewater treatment	Wastewater	Seawater	22.5	80	[42]

Note: *FO*, Forward osmosis; *RO*, reverse osmosis.

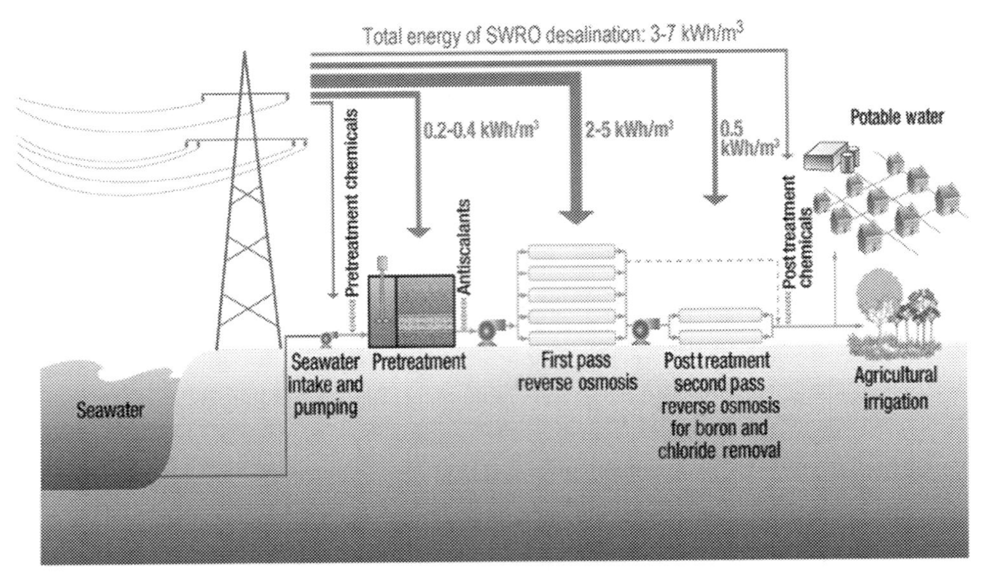

FIGURE 12–4 Specific energy consumption for a two-pass SWRO desalination [32]. *SWRO*, Seawater reverse osmosis.

12.3.2 Integrated RO–MD system

The application of RO technology as a final polishing method is becoming a common practice for wastewater recycling plants around the world. For example, a review by Tu et al. revealed the well-known application of RO in seawater desalination and boron removal for drinking water purposes [43]. In 1969 when RO technology was still in its infancy, the first scientific investigation into boron rejection by RO membranes took place [44]. Fig. 12–5 shows the study of the 'Scopus' search results revealing an increase in the number of articles published in the last decade (2009–19) for the RO study. From the figure, it is proved that RO is one of the important technologies that is widely used in purified water including drinking water production. While RO can sustain a high-quality water level, there is still a sizable amount of wastewater from the concentrates of RO that is generated, which typically comprises 20%–25% of the volume of the feed stream. Such processed discharge contains high levels of ions and salts in the initial drainage, including those that may be harmful or bioaccumulative [45]. As a consequence, in many countries, stricter concentrated pollution disposal laws are applied. According to the immediate dumping of accumulated wastewater containing micropollutants in water bodies, this would present long-term eco-toxicological danger and endanger the marine eco-system, in which the other posttreatment process would be needed to handle this issue before disposal.

Substantial focus has been given nowadays to various approaches for effective treatment of concentrated wastewater from RO discharge. For the treatment of this concentrated wastewater contaminant, conventional methods, such as coagulation and granular activated carbon (GAC) as well as advanced technologies such as ozonation, electrochemical oxidation,

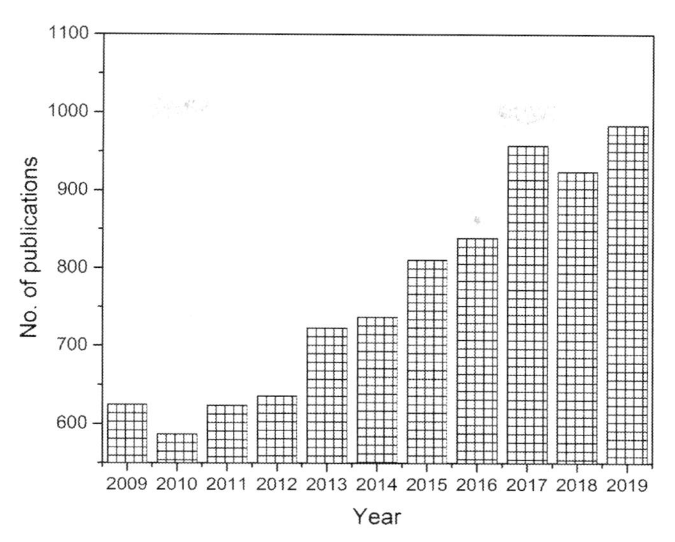

FIGURE 12–5 The number of publications on reverse osmosis studies (2009–19).

and photocatalysis, are successful, especially in reducing dissolved organic carbon (DOC) and selective micropollutants. A recent study by Jamil et al. investigated the DOCs removal from the RO's concentrated wastewater through a comparison of two approaches: (1) GAC and (2) ion-exchange resin adsorbents [46]. As a consequence, GAC had a higher capacity concentrated in extracting complete DOCs from RO than the two Purolite anion exchange resins. This is because GAC induced higher adsorption capacity than Purolites as based on the Freundlich constant. In addition, GAC had an ability to remove more hydrophobic DOCs at a high percentage, thereby, ion-exchange resins removed more humics. In the meantime, Yang et al. prepared Fe-based catalytics in conjunction with the ozonation and biodegradation cycle for the treatment of industrial wastewater concentrated by RO. Ozonation was commonly used here to eliminate trace toxins, humic compounds, and even nitrification-inhibiting contaminants by overt O_3 and indirect •OH oxidation [47]. Biodegradation is a main option for water treatment owing to its relatively cheap and more environmentally friendly properties, as opposed to other toxic methods of treatment. Accordingly, the study conducted three configurations of ozonation integrated process (Fig. 12–6). It can be inferred from the study that the combined procedure consisting of pretreatment ozonation to primarily improve biodegradability and biological posttreatment to fully eradicate the pollutants turned out to be a cost-effective option for effective ROC control. The only chemical or biological treatment option can have high operating costs or poor drawbacks in treatment performance.

In 2011, Bagastyo et al. applied electrochemical oxidation on RO concentrated through mixed metal oxide titanium coated electrodes [48]. In the study, five electrode materials were evaluated as anode in batch mode experiments. Electrochemical oxidation using Ti/SnO_2eSb and Ti/PteIrO_2 appeared to be sufficiently effective in removing the organics consisted in the

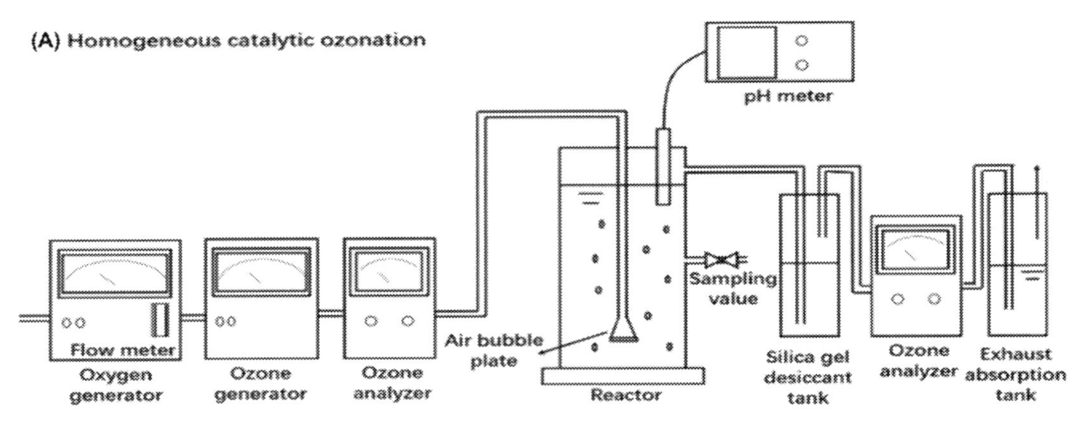

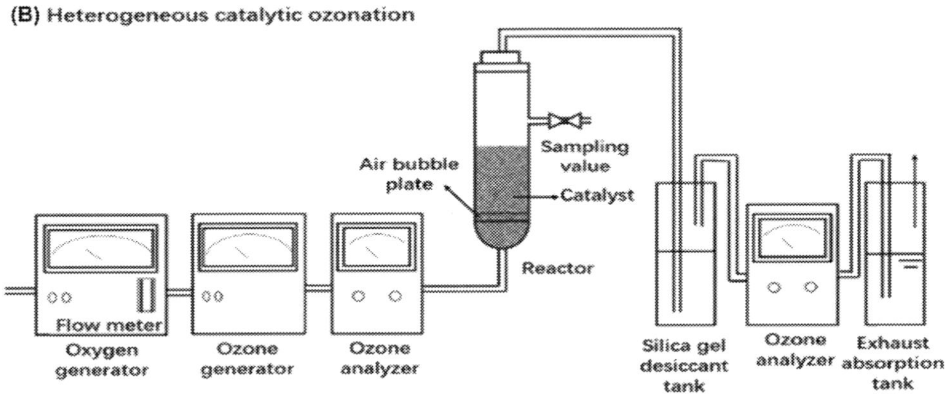

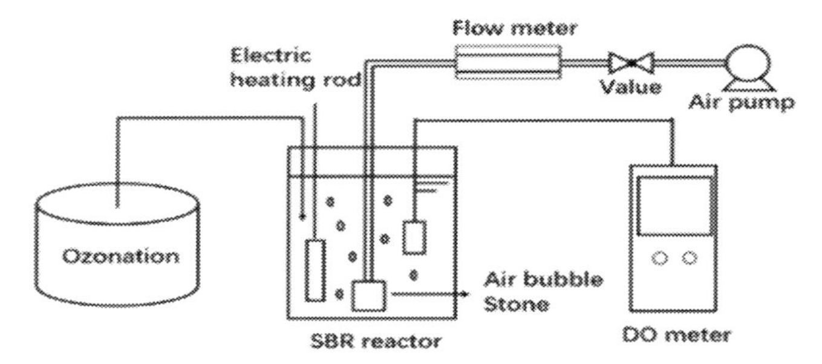

FIGURE 12–6 Three configurations of integrated ozonation process [47].

RO concentrates, nevertheless, the highest levels of dangerous y-products such as trihalomethanes and haloacetic acids were observed, particularly at Ti/SnO_2eSb and Ti/PteIrO_2 anodes. At the end of the experiment, it is proposed that accelerated RO electrochemical oxidation should be combined with a polishing solution to mitigate the leakage of halogenated repercussion. Later, in 2016 Weng and Pei studied the various electrode products by choosing Co as the doping agent to improve the different content of the Co-doped PbO_2 electrode for the treatment of RO derived from pharmaceutical wastewater [49]. The electrode was therefore prepared using the thermal decomposition and electrochemical deposition process and then used to degrade quinoline to explore quinolone and DOC elimination. It is worth mentioning here that quinoline is a common nitrogen that contains heterocyclic compounds and is commonly found in pharmaceutical wastewater. The test showed optimum quinoline, and DOC removals at the initial concentration of quinolone were 100% and 59.35% after 120 min. Nevertheless, the disadvantages of GAC and electrochemical oxidation are high operating costs and use in the operation of a massive scale [50].

MD is a heat combined membrane process that passes water vapor to the permeate side distillate via the pores of a hydrophobic membrane. According to Hubadillah et al. [51], MD is a technology that is able to produce pure and clean water efficiently at a lesser cost due to the lower operating pressure used. In view of the rejection percentage, theoretically speaking, the MD process is able to reject 100% of pollutants in water and produce pure water. Nowadays, a major global concern regarding scarcity of freshwater, at the same time, a conventional method like coagulation and flocculation, adsorption and membrane-based pressure driven show drawbacks, thus led MD into widespread attention. For example, Criscuoli and Bafaro compared the use of PP and PVDF as hydrophobic membranes in an MD system for arsenic removal [52]. Two membranes from PP and PVDF at different average pore size, membrane thickness, porosity, and LEP were applied. It was revealed that all membranes exhibited an excellent performance toward both As (III) and As (V). Almost 100% rejection of As (III) and As (V) was obtained with a high flux of 12.5 kg/hm^2 at 40 degrees. Compared to the conventional method such as coagulation/flocculation and ion exchange, no preoxidation step was required to convert As (III) into As (V) first in MD.

Solar or heat waste incorporation can fulfill the thermal criteria for the treatment of RO [53]. The vapor mass transfer system in MD probably provides complete ion rejection. Since concentrated RO wastewater typically produces low salinity and dissolved ions relative to concentrated RO seawater, MD may achieve higher concentrated water recovery and wastewater discharge levels from RO, creating a good quality permeate with likely fewer scaling problems. On the other hand, concentrated wastewater contains more DOCs and residual micropollutants [54]. When determining the maximum water recovery rate feasible under MD activity, due consideration must be given to the presence of these elements when RO-enriched wastewater is used. The removal of advantageous salts from wastewater will aid in counterbalancing the cost of treatment. Through adding combined electrodialysis and ion-exchange membranes, Zhang et al. tested the feasibility of retrieving useful compounds from RO-enriched wastewater [55]. This integrated system, however, contributed to the doubling of incidents that involved mitigating steps such as acidification. As MD is able to concentrate feed solutions, research has now gained interest in the precipitation of discrete salts and extracting useful ions from condensed MD. As a result, an advanced RO–MD system was developed and applied for the optimized treatment of wastewater from RO.

Qu et al. applied an integrated system of RO and MD to treat the RO-concentrated wastewater [56]. In the study, MD was also integrated with accelerated precipitation softening

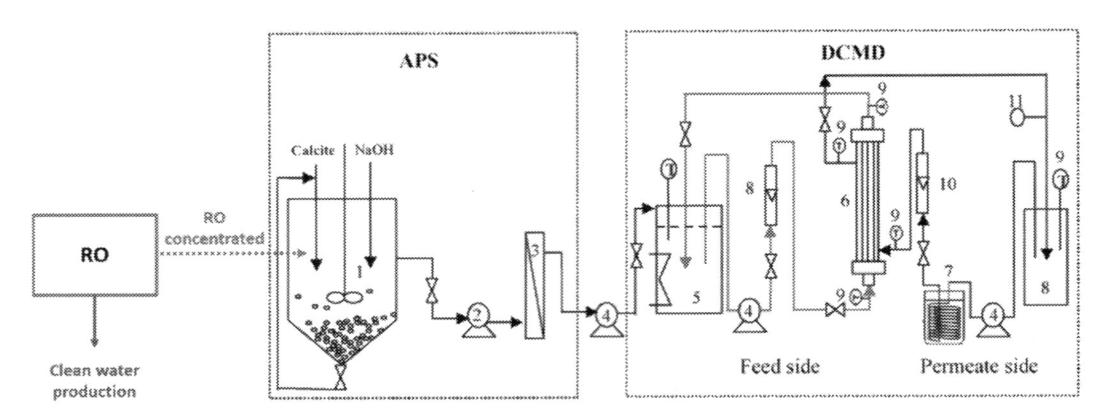

FIGURE 12–7 Schematic diagram for an integrated system of RO–APS–MD. *RO*, Reverse osmosis; *APS*, accelerated precipitation softening; *MD*, membrane distillation. *Modified from D. Qu, J. Wang, L. Wang, D. Hou, Z. Luan, B. Wang, Integration of accelerated precipitation softening with membrane distillation for high-recovery desalination of primary reverse osmosis concentrate, Sep. Purif. Technol. 67 (2009) 21–25.*

(APS) as APS could induce calcium carbonate crystallization through chemical dosing. Fig. 12–7 illustrates the schematic diagram for the integrated system of RO–APS–MD for the whole clean water production. This process can also be called a "zero discharge system." Correspondingly, the purpose of the study was to (1) assess the sodium dosage, seed dosage, and agitation rate through small-scale calcium removal tests; (2) demonstrate the laboratory-scale incorporation of APS with direct contact membrane distillation (DCMD) to ensure high water recovery; and (3) analyze the hydrophobic stability of the polyvinylidene fluoride (PVDF) membrane through permeate flux variation and conductivity over 300 h of continuous ruin. At the end of the experiment, there were 10.10, 5 g/L, and 200 r/min for optical solution pH, calcite dose, and agitation levels, respectively, for APS. Findings showed that APS treatment allowed calcium removal of up to 9% while reducing the scaling of $CaCO_3$ and $CaSO_4$ during the DCMD process. The permeate flux decreased by only 20% within 300 h, then the RO concentration was elevated 40 times and the total recovery improved to 98.8%.

12.3.3 Integrated FO–MD system

FO–MD is a new hybrid membrane technology that was patented in 2011 and has received increasing attention thereafter. The advent of this technology was motivated by the need to enhance the efficiency of both FO and MD systems. In FO, the net movement of water into the concentrated draw solution dilutes the solution gradually, resulting in the need to replace the draw solution. Meanwhile, MD is plagued by fouling of membranes by organic and inorganic substances in the feed solution which necessitates the replacement of the membrane from time to time. The ingenious FO–MD hybrid system works as shown in Fig. 12–8 where the draw solution of the FO system also acts as the feed solution of the MD system.

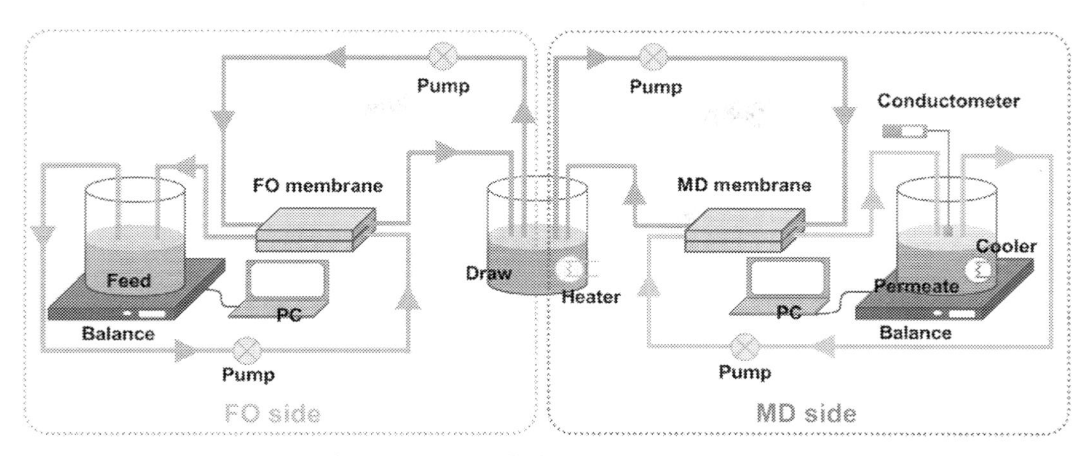

FIGURE 12–8 Schematic diagram of an FO–MD setup [57]. *FO*, Forward osmosis; *MD*, membrane distillation.

As water vapor evaporates through the MD membrane, the feed solution becomes more concentrated, thus regenerating the draw solution with high osmotic pressure. As organic and inorganic substances stay in the feed solution of the FO system, the MD membrane is saved from fouling. These features enable the FO–MD system to sustain itself for a long period of time, thus reducing the cost of replacing the membranes and draw solutions. The feasibility of this hybrid system is also made possible by the ability of MD to work with a very concentrated feed solution. DCMD is the most reported MD configuration in the FO–MD hybrid system due to the simplicity of the setup. The aforementioned benefits of the system have led to it being considered for the use in space stations.

Most of the FO–MD systems that have been reported in the literature involve two separate systems as shown in Fig. 12–8. The use of external compartments such as pumps to transport solutions translates to a high energy requirement and a large footprint. A simpler and more energy-friendly setup is a submerged FO–MD module as shown in Fig. 12–9. Nonetheless, the flux of the system can be lower than the usual setup. Also, the temperature of the draw solution for stable operation in the submerged module can be higher than the conventional setup due to polarization effects [58].

Over the years, the FO–MD system has been used in various applications such as desalination, wastewater treatment, and concentration of substances. Some of these applications are summarized in Table 12–3. Although NaCl is a popular choice when it comes to the draw solution, researchers have opted for other kinds of draw solutions that possess higher or equal osmotic pressure as well. Hydro-acid complexes such as $Na_5[Fe(C_6H_4O_7)_2]$ are claimed to have higher osmotic pressure than NaCl due to the hydrophilic and ionic moieties present in the complex [59]. Sodium propionate as a draw solution could produce higher flux for both the FO and MD systems compared to NaCl due to its lower reverse salt flux [60]. Trisodium phosphate is another choice of draw solution, which works best at pH 9 and concentration of 0.2 M [61].

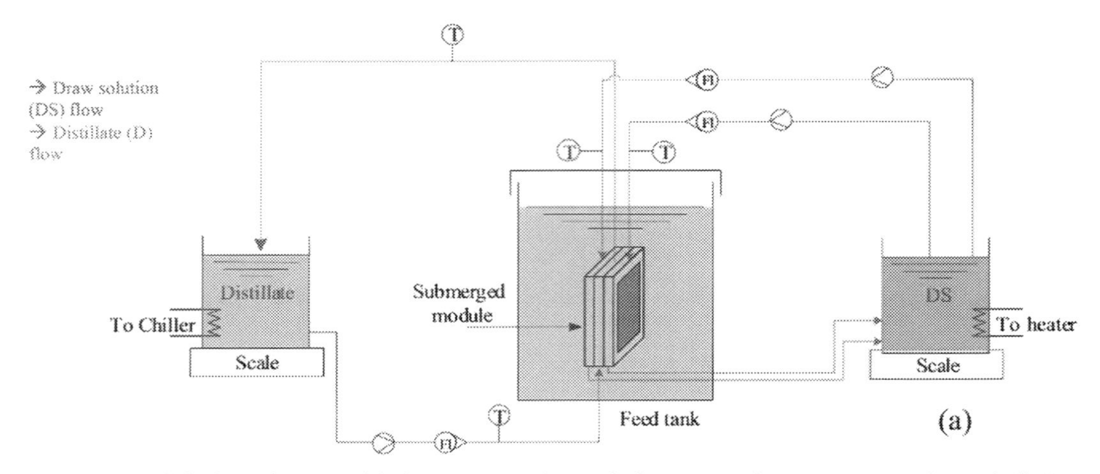

FIGURE 12–9 A hybrid membrane module for FO–MD application [58]. *FO*, Forward osmosis; *MD*, membrane distillation.

Table 12–3 Various applications of the FO–MD system.

Application	Feed solution of FO	Draw solution of FO	Flux of MD (LMH)	References
Brackish water desalination	Synthetic brackish water	Fertilizer solutions	5.7	[62]
Concentration of protein solutions	Bovine serum albumin	NaCl	~17	[29]
Water recovery from oily wastewater	Oil	NaCl or acetic acid	5.8	[63]
Treatment of fracking wastewater	Pretreated fracking wastewater	Sodium propionate (NaP)	13.82	[60]
Treatment of industrial wastewater produced by desulfurization process in a coal-fired power plant	Pretreated flue gas desulfurization wastewater	NaCl	18.5	[64]
Concentration of juice	Apple juice	Potassium sorbate preservative	5	[57]
Treatment of source-separated urine	Synthetic fresh urine and stored urine	NaCl	~16	[65]
Water recovery from hydraulic fracturing produced water	Produced water	NaCl	~70	[66]
Concentration of high-nutrient sludge	High-nutrient sludge solution	Trisodium phosphate	10.3	[61]
Seawater desalination	Simulated seawater	Poly(sodium styrene-4-sulfonate-co-n-isopropylacrylamide) (PSSS-PNIPAM)	2.5	[67]

Note: *FO*, Forward osmosis; *MD*, membrane distillation.

Interestingly, the increase in the temperature and flow rate of the draw solution can result in a more significant increase in the water flux of MD compared to that of FO. However, with the increase in the concentration of draw solution, the water flux of MD decreases while that of FO's increases due to increased concentration polarization and osmotic pressure, respectively [66].

While most applications of FO–MD focus on obtaining the product at the permeate side of the MD system, some applications, such as juice concentration, involve the recovery of the processed feed solution at the end of the operation. During the concentration of juice, water leaves the feed solution (juice) to the draw solution of higher osmotic pressure. As the draw solution can accumulate in the feed solution of FO by reverse flux, it should be a chemical that does not contaminate the feed solution as that will result in an extra step of separation or purification. Furthermore, the draw solution should be a substance that does not induce fouling in the MD membrane. For instance, potassium sorbate can be used as the draw solution in the concentration of apple juice as it is an essential additive for the juice and the MD system achieves a high rejection of the substance [57].

It is noteworthy that although the FO system is largely able to obviate the fouling of MD membranes, some small and hydrophilic contaminants, such as acetic acid, could still pass through the FO membrane and pose a fouling problem [63]. Also, despite the low propensity for fouling in the FO system, the use of feed solution with a very high content of suspended solids and ions, such as flue gas desulfurization wastewater, can induce significant fouling or scaling and result in flux decline. In order to overcome this setback, FO–MD systems can be integrated with other separation technologies such as microfiltration, ultrafiltration, or electrocoagulation to treat water before being used as a feed solution of the FO system. Antiscalant can also be blended with the draw solution to reduce the fouling of the FO membrane.

12.4 Conclusion and future research directions

FO–MD is a hybrid membrane technology that has a huge potential to solve the world's water crisis. The choice of draw solution and its concentration and temperature play a huge role in the performance of the system. Thus more research efforts on those topics could provide a better understanding of the system. The ways to reduce the size of the overall system should also be investigated to reduce the overall cost of the system and increase its portability. Meanwhile, RO–MD integrated system was developed for the treatment of RO-concentrated wastewater. In producing clean water through the RO membrane process, the treated water consisting of concentrated pollutants was discharged without any treatment. Therefore another treatment after RO process is needed. Prior to the development of RO–MD, several processes that are known as posttreatment were introduced including ozonation, electrochemical oxidation, and photocatalysis. However, these treatments showed drawbacks and are expensive. The RO–MD integrated system is not only a cost-effective system but could also lead to a zero discharge system as it effectively removed all the pollutants at the end of the process. The integrated FO–RO process is positioned to utilize less energy and achieve higher water quality requirements compared to the traditional RO process.

The possible advantage of the combined FO–RO process is beyond cost and energy savings as the use of less chemicals can prolong the service life of the treatment facilities and improve the sustainability of the process, thereby contributing significantly toward addressing the challenges of global water supply.

Although these integrated systems could provide zero discharge at the end of the process, there are problems that still remain unsolved. For example, wastewater containing heavy metals can be treated using these integrated systems and can produce clean water at high rejection. Unfortunately, heavy metals cannot be degraded and removed easily through these systems. What is more important, leaving these heavy metals untreated will harm living organisms and cause serious diseases and disorders. At the same time, microalgae have emerged as one of the potential approaches for completely removing heavy metals through a bioremediation process. This mechanism happens by biosorption of the heavy metals to the cell wall of microalgae. The efflux of metal ions will be outside the cell wall, thus reducing the heavy metals ions to a less toxic accumulation as complexation of heavy metal ions will be inside the cell wall. Therefore integrating any system with microalgae bioremediation is one of the promising integrated systems that not only promote zero discharge level but also introduce green technology into the world of separation technology.

References

[1] B. Ghosh, A.K. Ghosh, R.C. Bindal, P.K. Tewari, Studies on concentration of simulated ammonium-diuranate filtered effluent solution by forward osmosis using indigenously developed cellulosic osmosis membranes, Sep. Sci. Technol. 50 (2015) 324–331. Available from: https://doi.org/10.1080/01496395.2014.973517.

[2] J.W. Post, J. Veerman, H.V.M. Hamelers, G.J.W. Euverink, S.J. Metz, K. Nymeijer, et al., Salinity-gradient power: evaluation of pressure-retarded osmosis and reverse electrodialysis, J. Memb. Sci. 288 (2007) 218–230. Available from: https://doi.org/10.1016/j.memsci.2006.11.018.

[3] L.F. Greenlee, D.F. Lawler, B.D. Freeman, B. Marrot, P. Moulin, Reverse osmosis desalination: water sources, technology, and today's challenges, Water Res. 43 (9) (2009) 2317–2348.

[4] Z.L. Cheng, X. Li, Y. Da Liu, T.S. Chung, Robust outer-selective thin-film composite polyethersulfone hollow fiber membranes with low reverse salt flux for renewable salinity-gradient energy generation, J. Memb. Sci. 506 (2016) 119–129. Available from: https://doi.org/10.1016/j.memsci.2015.12.060.

[5] M.I. Dova, K.B. Petrotos, H.N. Lazarides, On the direct osmotic concentration of liquid foods. Part I: impact of process parameters on process performance, J. Food Eng. 78 (2007) 422–430. Available from: https://doi.org/10.1016/j.jfoodeng.2005.10.010.

[6] T.Y. Cath, A.E. Childress, M. Elimelech, Forward osmosis: principles, applications, and recent developments, J. Memb. Sci. 281 (2006) 70–87. Available from: https://doi.org/10.1016/j.memsci.2006.05.048.

[7] M. Gryta, K. Karakulski, A.W. Morawski, Purification of oily wastewater by hybrid UF/MD, Water Res. 35 (2001) 3665–3669. Available from: https://doi.org/10.1016/S0043-1354(01)00083-5.

[8] A. Criscuoli, E. Drioli, Energetic and exergetic analysis of an integrated membrane desalination system, Desalination 124 (1999) 243–249. Available from: https://doi.org/10.1016/S0011-9164(99)00109-5.

[9] J. Blanco Gálvez, L. García-Rodríguez, I. Martín-Mateos, Seawater desalination by an innovative solar-powered membrane distillation system: the MEDESOL project, Desalination 246 (2009) 567–576. Available from: https://doi.org/10.1016/J.DESAL.2008.12.005.

[10] S. Chou, L. Shi, R. Wang, C.Y. Tang, C. Qiu, A.G. Fane, Characteristics and potential applications of a novel forward osmosis hollow fiber membrane, Desalination 261 (2010) 365–372. Available from: https://doi.org/10.1016/j.desal.2010.06.027.

[11] M.H.I. Dore, Forecasting the economic costs of desalination technology, Desalination 172 (2005) 207–214. Available from: https://doi.org/10.1016/j.desal.2004.07.036.

[12] F. Macedonio, E. Drioli, A.A. Gusev, A. Bardow, R. Semiat, M. Kurihara, Efficient technologies for worldwide clean water supply, Chem. Eng. Process. Process Intensif. 51 (2012) 2–17. Available from: https://doi.org/10.1016/j.cep.2011.09.011.

[13] M.W. Shahzad, M. Burhan, L. Ang, K.C. Ng, Energy-water-environment nexus underpinning future desalination sustainability, Desalination 413 (2017) 52–64. Available from: https://doi.org/10.1016/j.desal.2017.03.009.

[14] R. Yasin, M. Dedmon, N. Dillon, N. Simaan, Investigating variability in cochlear implant electrode array alignment and the potential of visualization guidance, Int. J. Med. Rob. Comput. Assisted Surg. 15 (2019) e2009. Available from: https://doi.org/10.1002/rcs.2009.

[15] A. Subramani, J.G. Jacangelo, Emerging desalination technologies for water treatment: a critical review, Water Res. 75 (2015) 164–187. Available from: https://doi.org/10.1016/j.watres.2015.02.032.

[16] P.S. Goh, T. Matsuura, A.F. Ismail, N. Hilal, Recent trends in membranes and membrane processes for desalination, Desalination 391 (2016) 43–60. Available from: https://doi.org/10.1016/j.desal.2015.12.016.

[17] T. Tong, M. Elimelech, The global rise of zero liquid discharge for wastewater management: drivers, technologies, and future directions, Environ. Sci. Technol. 50 (2016) 6846–6855. Available from: https://doi.org/10.1021/acs.est.6b01000.

[18] G. Amy, N. Ghaffour, Z. Li, L. Francis, R.V. Linares, T. Missimer, et al., Membrane-based seawater desalination: present and future prospects, Desalination 401 (2017) 16–21. Available from: https://doi.org/10.1016/j.desal.2016.10.002.

[19] M. Baghbanzadeh, D. Rana, C.Q. Lan, T. Matsuura, Zero thermal input membrane distillation, a zero-waste and sustainable solution for freshwater shortage, Appl. Energy 187 (2017) 910–928. Available from: https://doi.org/10.1016/j.apenergy.2016.10.142.

[20] S. Vinardell, S. Astals, J. Mata-alvarez, J. Dosta, Techno-economic analysis of combining forward osmosis-reverse osmosis and anaerobic membrane bioreactor technologies for municipal wastewater treatment and water production, Bioresour. Technol. 2020 (2019) 122395. Available from: https://doi.org/10.1016/j.biortech.2019.122395.

[21] G. Blandin, A.R.D. Verliefde, J. Comas, I. Rodriguez-Roda, P. Le-Clech, Efficiently combining water reuse and desalination through forward osmosis-reverse osmosis (FO-RO) hybrids: a critical review, Membranes (Basel) 6 (2016). Available from: https://doi.org/10.3390/membranes6030037.

[22] D.L. Shaffer, J.R. Werber, H. Jaramillo, S. Lin, M. Elimelech, Forward osmosis: where are we now? Desalination 356 (2015) 271–284. Available from: https://doi.org/10.1016/j.desal.2014.10.031.

[23] C.F. Wan, T.S. Chung, Techno-economic evaluation of various RO + PRO and RO + FO integrated processes, Appl. Energy 212 (2018) 1038–1050. Available from: https://doi.org/10.1016/j.apenergy.2017.12.124.

[24] C.R. Martinetti, A.E. Childress, T.Y. Cath, High recovery of concentrated RO brines using forward osmosis and membrane distillation, J. Memb. Sci. 331 (2009) 31–39. Available from: https://doi.org/10.1016/j.memsci.2009.01.003.

[25] D.L. Shaffer, L.H. Arias Chavez, M. Ben-Sasson, S. Romero-Vargas Castrillón, N.Y. Yip, M. Elimelech, Desalination and reuse of high-salinity shale gas produced water: drivers, technologies, and future directions, Environ. Sci. Technol. 47 (2013) 9569–9583. Available from: https://doi.org/10.1021/es401966e.

[26] T.-S. Chung, J. Su, R.C. Ong, P. Wang, Advanced FO membranes from newly synthesized CAP polymer for wastewater reclamation through an integrated FO-MD hybrid system, AIChE J. 59 (4) (2012) 215–228.

[27] M. Xie, L.D. Nghiem, W.E. Price, M. Elimelech, A forward osmosis-membrane distillation hybrid process for direct sewer minings performance and limitations, Environ. Sci. Technol. 47 (2013) 13486–13493. Available from: https://doi.org/10.1021/es404056e.

[28] Q. Ge, P. Wang, C. Wan, T.S. Chung, Polyelectrolyte-promoted forward osmosis-membrane distillation (FO-MD) hybrid process for dye wastewater treatment, Environ. Sci. Technol. 46 (2012) 6236–6243. Available from: https://doi.org/10.1021/es300784h.

[29] K.Y. Wang, M.M. Teoh, A. Nugroho, T.S. Chung, Integrated forward osmosis-membrane distillation (FO-MD) hybrid system for the concentration of protein solutions, Chem. Eng. Sci. 66 (2011) 2421–2430. Available from: https://doi.org/10.1016/j.ces.2011.03.001.

[30] A. Achilli, T.Y. Cath, A.E. Childress, Selection of inorganic-based draw solutions for forward osmosis applications, J. Memb. Sci. 364 (2010) 233–241. Available from: https://doi.org/10.1016/j.memsci.2010.08.010.

[31] B. Mi, M. Elimelech, Organic fouling of forward osmosis membranes: fouling reversibility and cleaning without chemical reagents, J. Memb. Sci. 348 (2010) 337–345. Available from: https://doi.org/10.1016/j.memsci.2009.11.021.

[32] D.L. Shaffer, N. Yin, J. Gilron, M. Elimelech, Seawater desalination for agriculture by integrated forward and reverse osmosis: improved product water quality for potentially less energy, J. Memb. Sci. 415–416 (2012) 1–8. Available from: https://doi.org/10.1016/j.memsci.2012.05.016.

[33] O.A. Bamaga, A. Yokochi, E.G. Beaudry, Application of forward osmosis in pretreatment of seawater for small reverse osmosis desalination units, Desalin. Water Treat. 5 (2009) 183–191. Available from: https://doi.org/10.5004/dwt.2009.574.

[34] S. Zhao, L. Zou, D. Mulcahy, Brackish water desalination by a hybrid forward osmosis-nanofiltration system using divalent draw solute, Desalination 284 (2012) 175–181. Available from: https://doi.org/10.1016/j.desal.2011.08.053.

[35] T.Y. Cath, N.T. Hancock, C.D. Lundin, C. Hoppe-Jones, J.E. Drewes, A multi-barrier osmotic dilution process for simultaneous desalination and purification of impaired water, J. Memb. Sci. 362 (2010) 417–426. Available from: https://doi.org/10.1016/j.memsci.2010.06.056.

[36] R. Thiruvenkatachari, M. Francis, M. Cunnington, S. Su, Application of integrated forward and reverse osmosis for coal mine wastewater desalination, Sep. Purif. Technol. 163 (2016) 181–188. Available from: https://doi.org/10.1016/j.seppur.2016.02.034.

[37] M.A. Hafiz, A.H. Hawari, A. Altaee, A hybrid forward osmosis/reverse osmosis process for the supply of fertilizing solution from treated wastewater, J. Water Process. Eng. 32 (2019) 100975. Available from: https://doi.org/10.1016/j.jwpe.2019.100975.

[38] C. Aydiner, S. Topcu, C. Tortop, F. Kuvvet, D. Ekinci, N. Dizge, et al., A novel implementation of water recovery from whey: "forward-reverse osmosis" integrated membrane system, Desalin. Water Treat. 51 (2013) 786–799. Available from: https://doi.org/10.1080/19443994.2012.693713.

[39] H. Gadallah, H.M. Ali, S.S. Ali, R. Sabry, A. Gadallah, Application of forward/reverse osmosis hybrid system for brackish water desalination using El-Salam Canal Water, Sinai, Egypt, Part (1): FO performance Hanaa, Int. Conf. Environ. Sci. Eng. 32 (2014) 12–16. Available from: https://doi.org/10.7763/IPCBEE.

[40] R. Sabry, S. Ali, A.G. Gadallah, H.M. Ali, H. Gadallah, Application of forward/reverse osmosis hybrid system for brackish water desalination using El-Salam canal water, Sinai, Egypt, Part (2): pilot scale investigation, Int. J. ChemTech Res. 8 (2015) 102–112.

[41] R. Valladares Linares, Z. Li, V. Yangali-Quintanilla, N. Ghaffour, G. Amy, T. Leiknes, et al., Life cycle cost of a hybrid forward osmosis—low pressure reverse osmosis system for seawater desalination and wastewater recovery, Water Res. 88 (2016) 225–234. Available from: https://doi.org/10.1016/j.watres.2015.10.017.

[42] F. Volpin, E. Fons, L. Chekli, J.E. Kim, A. Jang, H.K. Shon, Hybrid forward osmosis-reverse osmosis for wastewater reuse and seawater desalination: Understanding the optimal feed solution to minimise fouling, Process. Saf. Environ. Prot. 117 (2018) 523–532. Available from: https://doi.org/10.1016/j.psep.2018.05.006.

[43] K.L. Tu, L.D. Nghiem, A.R. Chivas, Boron removal by reverse osmosis membranes in seawater desalination applications, Sep. Purif. Technol. 75 (2010) 87–101. Available from: https://doi.org/10.1016/J.SEPPUR.2010.07.021.

[44] F.M. Graber, H.K. Lonsdale, C.E. Milstead, B.P. Cross, Boron rejection by cellulose acetate reverse osmosis membranes, Desalination 7 (1970) 249–258. Available from: https://doi.org/10.1016/S0011-9164(00)80079-X.

[45] M. Umar, F.A. Roddick, L. Fan, O. Autin, B. Jefferson, Treatment of municipal wastewater reverse osmosis concentrate using UVC-LED/H_2O_2 with and without coagulation pre-treatment, Chem. Eng. J. 260 (2015) 649–656. Available from: https://doi.org/10.1016/J.CEJ.2014.09.028.

[46] S. Jamil, P. Loganathan, J. Kandasamy, A. Listowski, C. Khourshed, R. Naidu, et al., Removal of dissolved organic matter fractions from reverse osmosis concentrate: comparing granular activated carbon and ion exchange resin adsorbents, J. Environ. Chem. Eng. 7 (2019) 103126. Available from: https://doi.org/10.1016/J.JECE.2019.103126.

[47] L. Yang, M. Sheng, Y. Li, W. Xue, K. Li, G. Cao, A hybrid process of Fe-based catalytic ozonation and biodegradation for the treatment of industrial wastewater reverse osmosis concentrate, Chemosphere 238 (2020) 124639. Available from: https://doi.org/10.1016/J.CHEMOSPHERE.2019.124639.

[48] A.Y. Bagastyo, J. Radjenovic, Y. Mu, R.A. Rozendal, D.J. Batstone, K. Rabaey, Electrochemical oxidation of reverse osmosis concentrate on mixed metal oxide (MMO) titanium coated electrodes, Water Res. 45 (2011) 4951–4959. Available from: https://doi.org/10.1016/J.WATRES.2011.06.039.

[49] M. Weng, J. Pei, Electrochemical oxidation of reverse osmosis concentrate using a novel electrode: parameter optimization and kinetics study, Desalination 399 (2016) 21–28. Available from: https://doi.org/10.1016/J.DESAL.2016.08.002.

[50] E. Dialynas, D. Mantzavinos, E. Diamadopoulos, Advanced treatment of the reverse osmosis concentrate produced during reclamation of municipal wastewater, Water Res. 42 (2008) 4603–4608. Available from: https://doi.org/10.1016/J.WATRES.2008.08.008.

[51] S.K. Hubadillah, M.H.D. Othman, A.F. Ismail, M.A. Rahman, J. Jaafar, A low cost hydrophobic kaolin hollow fiber membrane (h-KHFM) for arsenic removal from aqueous solution via direct contact membrane distillation, Sep. Purif. Technol. 214 (2018) 31–39. Available from: https://doi.org/10.1016/j.seppur.2018.04.025.

[52] A. Criscuoli, P. Bafaro, Vacuum membrane distillation for purifying waters containing arsenic, Desalination 323 (2013) 17–21. Available from: https://doi.org/10.1016/J.DESAL.2012.08.004.

[53] S. Al-Obaidani, E. Curcio, F. Macedonio, G. Di Profio, H. Al-Hinai, E. Drioli, Potential of membrane distillation in seawater desalination: thermal efficiency, sensitivity study and cost estimation, J. Memb. Sci. 323 (2008) 85–98. Available from: https://doi.org/10.1016/J.MEMSCI.2008.06.006.

[54] A. Pérez-González, A.M. Urtiaga, R. Ibáñez, I. Ortiz, State of the art and review on the treatment technologies of water reverse osmosis concentrates, Water Res. 46 (2012) 267–283. Available from: https://doi.org/10.1016/J.WATRES.2011.10.046.

[55] Y. Zhang, B. Van der Bruggen, L. Pinoy, B. Meesschaert, Separation of nutrient ions and organic compounds from salts in RO concentrates by standard and monovalent selective ion-exchange membranes used in electrodialysis, J. Memb. Sci. 332 (2009) 104–112. Available from: https://doi.org/10.1016/J.MEMSCI.2009.01.030.

[56] D. Qu, J. Wang, L. Wang, D. Hou, Z. Luan, B. Wang, Integration of accelerated precipitation softening with membrane distillation for high-recovery desalination of primary reverse osmosis concentrate, Sep. Purif. Technol. 67 (2009) 21–25. Available from: https://doi.org/10.1016/J.SEPPUR.2009.02.021.

[57] X. An, Y. Hu, N. Wang, Z. Zhou, Z. Liu, Continuous juice concentration by integrating forward osmosis with membrane distillation using potassium sorbate preservative as a draw solute, J. Memb. Sci. 573 (2019) 192–199. Available from: https://doi.org/10.1016/j.memsci.2018.12.010.

[58] B.C. Ricci, B. Skibinski, K. Koch, C. Mancel, C.Q. Celestino, I.L.C. Cunha, et al., Critical performance assessment of a submerged hybrid forward osmosis—membrane distillation system, Desalination 468 (2019) 114082. Available from: https://doi.org/10.1016/j.desal.2019.114082.

[59] P. Wang, Y. Cui, Q. Ge, T. Fern Tew, T.S. Chung, Evaluation of hydroacid complex in the forward osmosis-membrane distillation (FO-MD) system for desalination, J. Memb. Sci. 494 (2015) 1–7. Available from: https://doi.org/10.1016/j.memsci.2015.07.022.

[60] M.S. Islam, K. Touati, M.S. Rahaman, Feasibility of a hybrid membrane-based process (MF-FO-MD) for fracking wastewater treatment, Sep. Purif. Technol. 229 (2019) 115802. Available from: https://doi.org/10.1016/j.seppur.2019.115802.

[61] N.C. Nguyen, H.T. Nguyen, S.T. Ho, S.S. Chen, H.H. Ngo, W. Guo, et al., Exploring high charge of phosphate as new draw solute in a forward osmosis-membrane distillation hybrid system for concentrating high-nutrient sludge, Sci. Total. Environ. 557–558 (2016) 44–50. Available from: https://doi.org/10.1016/j.scitotenv.2016.03.025.

[62] W. Suwaileh, D. Johnson, D. Jones, N. Hilal, An integrated fertilizer driven forward osmosis- renewables powered membrane distillation system for brackish water desalination: a combined experimental and theoretical approach, Desalination. 471 (2019) 114126. Available from: https://doi.org/10.1016/j.desal.2019.114126.

[63] S. Zhang, P. Wang, X. Fu, T.S. Chung, Sustainable water recovery from oily wastewater via forward osmosis-membrane distillation (FO-MD), Water Res. 52 (2014) 112–121. Available from: https://doi.org/10.1016/j.watres.2013.12.044.

[64] S. Lee, Y. Kim, S. Hong, Treatment of industrial wastewater produced by desulfurization process in a coal-fired power plant via FO-MD hybrid process, Chemosphere 210 (2018) 44–51. Available from: https://doi.org/10.1016/j.chemosphere.2018.06.180.

[65] F. Volpin, L. Chekli, S. Phuntsho, N. Ghaffour, J.S. Vrouwenvelder, H.K. Shon, Optimisation of a forward osmosis and membrane distillation hybrid system for the treatment of source-separated urine, Sep. Purif. Technol. 212 (2019) 368–375. Available from: https://doi.org/10.1016/j.seppur.2018.11.003.

[66] K. Sardari, P. Fyfe, S. Ranil, Wickramasinghe, Integrated electrocoagulation—Forward osmosis—membrane distillation for sustainable water recovery from hydraulic fracturing produced water, J. Memb. Sci. 574 (2019) 325–337. Available from: https://doi.org/10.1016/j.memsci.2018.12.075.

[67] D. Zhao, P. Wang, Q. Zhao, N. Chen, X. Lu, Thermoresponsive copolymer-based draw solution for seawater desalination in a combined process of forward osmosis and membrane distillation, Desalination 348 (2014) 26–32. Available from: https://doi.org/10.1016/j.desal.2014.06.009.

Conclusions and future prospects

Osmotic engineering (OE) processes have come a long way over the past half century or so and continue to advance in the treatment of natural and wastewaters streams and in the production of salinity gradient–based power. This advance has been the result of a substantial body of research, which is continually increasing. In the preceding chapters, written by some of the leading researchers in each field, we have intended to give a comprehensive overview of the basic principles, applications developments, and future direction of the major OE–based processes, including pressure-driven processes, such as reverse osmosis (RO) and nanofiltration (NF), osmotic gradient processes of forward osmosis (FO) and pressure retarded osmosis (PRO), and thermally driven techniques such as thermo-osmosis (TO) and osmotic membrane distillation. In addition, the constant striving for improved products and reduced costs has meant much recent research and development has been in integrated systems, which we have also addressed. The academic literature in these fields is immense and capable of filling several such books. As such, focus has been maintained on covering key advances as well as exemplar of different applications and processes.

Although the process technologies grouped under OE are varied, there are a number of future development strands that may be shared between several or all OE processes. In addition, some of the technologies presented are relatively mature and used at an industrial scale, such as RO and NF, while others are research intensive, but not yet used widely at an industrial level, such as FO and PRO. As such the future prospects look very different, with the well-established processes development being largely in process optimization, while for newer technologies there is much work still to be done on the development of the fundamentals of the process and in the theoretical understanding of those processes.

Common to all membrane separation-based technologies is the problem of fouling of membranes, both at the active layer surface and deep into the membrane materials. Fouling is grouped typically into the categories of inorganic scaling, organic fouling, biofouling, and colloidal or particulate fouling. In any single application, one or several of these types of fouling may be present depending upon the process waters being used, while the relative importance of each type and how they affect any engineering process can also depend upon the particular process and process parameters employed. As a result, much recent work, and presumably much future work because of the magnitude and complexity of the problem, is in the field of fouling mitigation, whether by treatment of feed waters or fabrication or modification of membranes to improve their fouling resistance as well as improved module design. Much fouling mitigation work has been in the production of membranes with surfaces that are smoother and more hydrophilic, with the aim of retaining a hydrated film in the immediate vicinity of the membrane to prevent fouling materials making a close approach. While this approach is promising or pressure and osmotic pressure-driven processes, it is not so useful for thermally driven processes, where the membrane typically needs to be hydrophobic. Other surface modification techniques, which have been reported extensively in the literature, include charge modification and addition of polymer brushes to the surfaces of polymer membranes. The former approach is generally aimed at creating a large negative charge at the membrane surface when in aqueous solution, as many common foulants, including bacterial cells and other colloids, are typically negatively charged. This is intended to lead to repulsive Coulombic interactions, reducing fouling. However, this approach has problems, namely that if the feed water contains a reasonable amount of salinity, such as brackish or seawater, then compression of the double layer due to dissolved ions will lead to the negation of this repulsive charge interaction. In addition, any simple chemical modification of the surface is likely to be altered from the intended due to further modification by components in any complex feed water, such as naturally sourced water or wastewater. Polymer brushes, employing steric interaction to prevent attachment of biofoulants do not have the problem of reduced efficacy due to electrolyte concentration, but all surface modifications when applied industrially are likely to not survive long-term filtration of real feed waters, due to being scoured from the membrane surface due to abrasion at the high flow rates employed. As such simple, and relatively delicate, surface modification of membranes is likely to be a dead end when it comes to long-term industrial application. As a result, it seems likely that the long-term development of fouling resistant materials may reside in the use of membranes fabricated from inherently resistant materials, including the use of nanofillers, to create membranes either containing biocidal or growth inhibitory substances in the case of biofouling resistance, or generally resistant to attachment of foulants in other ways.

While FO and PRO have similar fouling issues to pressure-driven processes, they have additional issues regarding the draw solution. This includes the selection of a suitable draw solution that gives a high osmotic pressure, while maintaining low reverse solute flux. For FO there is also the problem of regenerating the draw solution while keeping the combined energy expenditure from the FO plus regeneration processes below that of a more conventional rival process, such as RO. In terms of draw solutes with high osmotic pressure, there is much research into complex materials including polyelectrolytes, colloids, nanoparticles, and hydrogels, to name just a few. But as osmotic pressure is a colligative property dependent primarily upon the number concentration of dissolved species, large molecules and colloids are not often successful at achieving high osmotic pressures due to the requirement of high mass concentrations required to maintain high molar concentrations. As such, small molecules and ions are more effective at achieving high water fluxes for FO. However, the use of small ions comes at the cost of increased diffusion of solute across the membrane into the feed water. This reverse solute, or salt, flux can lead to increased scaling on the membrane-active surface, reduction of effective osmotic pressure difference across the active layer leading to reduced membrane flux, and the contamination of the feed water with the dissolved ions of the draw solution. This latter issue is particularly pertinent to where the application is for the concentration of a feedstock, such as fruit juices, where it will seriously impact the quality of the finished product. As such, much research in FO needs to be focused on producing tighter membranes that are resistant to the diffusion of solute molecules, albeit at the cost of reduced fluxes. The other major problem of FO, the cost-efficient regeneration of the draw solution, is difficult as the energy cost required to regenerate a draw solution is set by the osmotic pressure difference between draw solution in the concentrated and diluted states, which makes finding a regeneration technology sufficiently more low energy than, for example, RO to make the combined FO-regeneration technology viable in energy terms is not simple. As such much research in regeneration seems to be turning toward using either regeneration processes where free energy sources can be used, such as membrane distillation combined with solar heating, or where a regeneration step may not need to be used. This includes the use of fertilizers as draw solutes, with the dilute draw solution used in fertigation, or in the use of seawater or RO brine as draw solution. This latter brings in the potential for use of FO combined with seawater desalination as some sort of hybrid system combining treated effluent concentration with seawater or brine dilution. Much research is currently being carried out in this area, and it shows much future promise.

For PRO, obtaining an acceptable trade-off between power density and specific energy has proven to be challenging, which has hitherto restricted its commercialization. Much recent research has been in finding an optimal membrane configuration, which plays an important role in overall performance. However, there is still much room for improvement here. Hollow fiber membranes have so far been found to be superior to flat sheet membranes for PRO thus far, due to high packing density and excellent transport characteristics.

The salinity found in seawater is capable of sustaining a power production density of over 5.0 W/m^2. However, much research has been carried out using synthetic feedstocks, so more work needs to be done using real waters to develop this technology into a viable power generating technology to rival other renewable generation technologies.

With regard to power density, draw solutions high osmotic pressures are needed. Alternatively, the PRO process could be operated using a multistage setup where the diluted draw solution from each stage enters the following stage leading to an increase in extractable energy. Similarly, as for FO, a way forward may be to use concentrated RO brine as a draw solution. While this approach is promising for PRO, large membrane areas are needed to increase specific energy production, at the potential cost of lowered power density. The combined PRO-RO system could potentially create a large amount of energy, while eliminating the problem of brine discharge associated with RO desalination processes. However, more work is needed here, particularly in terms of estimating energy consumption and economics of the pretreatment and PRO processes, as well as for hybrid systems. Much research in this area has also been confined to the bench scale, rather than the pilot level using turbines. This is needed to optimize the efficiency of the turbines/generators and power recovery, as well as the membranes used, with the latter being the focus of lab research so far.

Osmotic distillation and osmotic membrane distillation are subjects still very much in their infancy. Very few experimental studies have been carried out into membrane development for these subjects. Instead, most research has been either theoretical in nature, concentrating on modeling, or for application studies concentrating on juice concentration, wastewater and removal of alcohol from drinks, as well as effects of various operating parameters on performance. The use of osmotic agents is an area that has yet to be explored in any great depth, with very few reports in the literature, compared to FO. In addition, very little industrial-scale use has been explored, due to lack of suitable membrane modules with sufficiently high surface areas to be of use beyond the pilot scale.

Thermo-osmosis is a non-isothermal separation process using a temperature gradient to drive flow. Research has been conducted over a number of years using a wide range of membrane materials, including ones of hydrophobic or hydrophilic materials with various types of surface charge characteristics, for a number of applications, including with water, electrolyte solutions, or other nonaqueous liquids.

Some of the theoretical basis for the transport mechanism of water across the membrane during TO is still missing. For instance, it is not clear why transport is from the cold to hot side of charged membranes, but it is in the other direction for uncharged membranes. In addition, flow can change depending upon temperature, pore size, or the nature of the solvent. This has led to confusing and contradictory research results. Other issues are the effects of membrane void fraction and membrane density and thickness. This has led to a scarcity of investigations of membranes specifically fabricated for TO.

In addition to separation by TO, much recent research work has been into the field of TO energy conversion, where TO can be used to generate electrical energy from low-quality waste heat sources. Here the permeate stream is pressurized, similarly to PRO, with the diffusion of vapor across the membrane leading to increased volume on the permeate side which can then be used to generate electricity via a turbine. This process shows great promise for the reduction of energy waste, with some research being carried out into the use of nanostructured membranes specifically for this process. With current interest in renewable energy and reduction of waste, this is an area likely to show much promise in the future.

Few industrial processes consist of single steps or unit processes. This is also reflected in OE applications, where optimizing a process solution has often involved the integration or hybridization of several systems. The integration of osmotic systems has great potential for solving a number of environmental problems, as well as increasing overall process efficiency and performance.

The integration of the different types of systems is not easy, as it necessarily needs to take place at the pilot or industrial scale to allow optimization or integration into water and wastewater treatment systems to be investigated. Development at such a scale requires significant investment.

Research into several types of integrated systems have been examined in this book, including FO-MD, RO-MD, and FO-RO. These systems have potential for not only overcoming process limitations of the individual systems but also for creating zero waste discharge processes (e.g., RO-MD), reduce energy consumption, and improve permeate water quality over RO combined with lowered chemical use (FO-RO).

One drawback of these systems is their inability to effectively deal with heavy metal containing waters. One possible route is through the combination of osmotic separation systems with microalgae remediation to remove heavy metals effectively within a zero discharge system.

Index

Note: Page numbers followed by "*f*" and "*t*" refer to figures and tables, respectively.

P